T0211880

Lecture Notes in Mathematics

Volume 2319

This series reports on new developments in all areas of mathematics and their applications - quickly, informally and at a high level. Mathematical texts analysing new developments in modelling and numerical simulation are welcome. The type of material considered for publication includes:

1. Research monographs
2. Lectures on a new field or presentations of a new angle in a classical field
3. Summer schools and intensive courses on topics of current research.

Texts which are out of print but still in demand may also be considered if they fall within these categories. The timeliness of a manuscript is sometimes more important than its form, which may be preliminary or tentative.

Titles from this series are indexed by Scopus, Web of Science, Mathematical Reviews, and zbMATH.

Rolf Schneider

Convex Cones

Geometry and Probability

 Springer

Rolf Schneider
Mathematisches Institut
University of Freiburg
Freiburg, Germany

ISSN 0075-8434   ISSN 1617-9692   (electronic)
Lecture Notes in Mathematics
ISBN 978-3-031-15126-2   ISBN 978-3-031-15127-9   (eBook)
https://doi.org/10.1007/978-3-031-15127-9

This Springer imprint is published by the registered company Springer Nature Switzerland AG
The registered company address is: Gewerbestrasse 11, 6330 Cham, Switzerland

Preface

In order to guide the reader into these notes, it seems appropriate to recall a word from the title: 'Geometry'. This emphasizes our viewpoint, and it indicates that we will not deal with the important role of convex cones in such fields as ordered vector spaces (e.g., [3]), measure theory (e.g., [25]), homogeneous or symmetric cones (e.g., [56]), or conic optimization. Rather, we concentrate on intuitive and elementary geometric appearances of convex cones, for example in the investigation of polyhedra and in stochastic geometry.

Besides introducing the reader to the fundamental facts about convex cones and their geometric functionals, this is a selection, illuminating different aspects of the geometry of convex cones. The principles guiding our choices are twofold. Some older or newer results about convex cones or their applications we found so remarkable that we think they should be pointed out, preserved also outside their original sources, and presented to a wider public. Other topics had the advantage that we were more familiar with them, after studying them in detail, and they were included at appropriate places. Accordingly, much of the material is close to various research articles, and the selection is rather subjective.

In the previous decade, some questions from applied mathematics, concerning, for instance, the average case analysis of conic optimization problems, or demixing by convex optimization under a probabilistic model, were treated in a way leading to increased interest in non-trivial intersections of convex cones. As an example, the following question is posed in [11, p. 227]: *"What is the probability that a randomly rotated convex cone shares a ray with a fixed cone?"* If 'randomly' is interpreted to imply uniform distribution, the probability in question can be calculated via the kinematic formula of conic (or, equivalently, spherical) integral geometry. While the spherical, and thus the conic, kinematic formula was already known, the new applications required additional information about the functionals appearing in it, the conic intrinsic volumes. These can be considered as the conic counterparts to the intrinsic volumes of convex bodies. The investigations of the new applications of the conic intrinsic volumes, of which we mention [5, 8, 9, 11, 127, 128, 129], required refined information, such as the explicit values of the conic intrinsic volumes for special cones or estimates of their asymptotic behavior in high dimensions. The new interest in conic intrinsic volumes and the conic kinematic formula was also a motivation for developing new and simplified approaches to known results, as in [6] and [10].

Another incentive for collecting observations about convex cones came from different publications on random cones, or on intersections of cones with random subspaces. The early paper [50] treated random cones generated by tessellations of $\mathbb{R}^d$ by independent random hyperplanes through the origin, or as the positive hull of independent random vectors, with quite general assumptions on the distributions. The article [52] investigated, among other things, random linear images of the nonnegative orthant in a higher-dimensional space. On the other hand, beautiful probabilistic applications were made of intersections of random linear subspaces with the cones of special conical tessellations; see [113, 114]. More recently, various different aspects of random cones were studied intensively, in [66, 67, 68, 69, 108, 109, 110, 111]. Reading such articles turned out to be a stimulus to have also a look at older publications about convex cones.

The use of convex cones is classical, of course, in the geometry of polyhedra. Normal cones and tangent cones are indispensable and familiar for anyone working with polyhedra. But also some rarer appearances of cones in the geometry of polyhedra deserve interest and should perhaps be more widely known. When reading new and older articles on convex cones, the idea arose of collecting various geometric facts on cones and of presenting them coherently. The selection of topics was, as mentioned, a matter of personal interests, though guided by the hope that greater diversity might help more readers to find something of interest for them.

Chapter 1 lays the foundations, mainly on convex cones and polyhedra. It collects some special results which are needed later. Particular emphasis is on valuations and, connected to this, on identities for characteristic functions. Polarity of convex cones is studied from various viewpoints.

Chapter 2 introduces the basic valuations that are used to measure polyhedral cones: the conic intrinsic volumes and the Grassmann angles (or conic quermassintegrals). It establishes relations for them and between them, making use of identities for characteristic functions and of a first integral-geometric formula. Valuations on polyhedral cones are then used to establish Gauss–Bonnet type theorems and tube formulas for compact general polyhedra.

A cone yields by intersection a subset of the unit sphere, and a subset of the unit sphere uniquely determines a cone. Therefore, the geometry of cones is equivalent to the geometry of subsets of the sphere. Treating cones in Euclidean space, where we can use the linear structure, has several advantages and makes the presentation easier. Sometimes, however, spherical geometry is needed, or is more appropriate. Chapter 3 treats, therefore, relations to spherical geometry. It provides some calculations for later applications, and also discusses some spherical inequalities, which can be re-interpreted in the geometry of cones.

Chapter 4 deals with substantial metric properties of and manipulations with convex cones. The 'Master Steiner formula' of McCoy and Tropp is proved, in a generalized local form, involving the local versions of the conic intrinsic volumes, the conic support measures. This is first done for polyhedral cones and then extended, by continuity, to general convex cones. As an outcome, the conic intrinsic volumes are thus defined as continuous functionals on general closed convex cones. Then the kinematic formula of integral geometry is proved for curvature measures of convex cones. Its global form provides the probability of the event that a uniform random cone has a common ray with a fixed cone. A concentration property of the conic intrinsic volumes around the statistical dimension then leads to a threshold phenomenon. The chapter deals briefly with inequalities for conic intrinsic volumes, and the weak continuity of the conic support measures is strengthened to Hölder continuity with respect to suitable metrics.

Finitely many hyperplanes through the origin generate a tessellation of the space into polyhedral convex cones. Chapter 5 treats these cones, in several different ways. First, a formula is proved giving the sum of the kth conic intrinsic volumes of the j-faces of such a tessellation. It reveals that this sum depends only on the combinatorics of the central arrangement provided by the hyperplanes. Special advantage is drawn from this fact in the determination of an absorption probability concerning a certain random walk. The rest of the chapter deals with the situation when the given hyperplanes are random, with a distribution satisfying some mild assumptions. They then give rise to different models of random cones, and for these, the expectations of various geometric functionals are determined. Similar results are obtained for lower-dimensional faces of the random tessellation. The final section is concerned with probabilities of non-trivial intersections for isotropic random cones.

Chapter 6 continues the investigation of random cones, under various different aspects. A simple way to generate a random cone is to take the image of a fixed cone under a random linear map. The first two sections deal with such random cones. The behavior of random cones in high dimensions is the topic of the next three sections. Random cones in halfspaces are briefly considered in the last section.

The role that convex cones play in Chapter 7 is quite different. Here a given convex cone serves as a cage for a convex hypersurface which is asymptotic to the cone. Examples appear in an old conjecture of Calabi, according to which every complete hyperbolic affine hypersphere is asymptotic to the boundary of a convex cone with apex at the center, and that every pointed convex cone with interior points determines a one-parameter family of affine hyperbolic hyperspheres which are asymptotic to the boundary of the cone. With a distinctly different motivation, Khovanskiĭ and Timorin [115] were led to consider convex sets K contained in a fixed cone C such that $C \setminus K$ is bounded. More generally, we shall study C-coconvex sets, by which we understand sets of the form $C \setminus K$, where C is a pointed closed convex cone with interior points, $K \subseteq C$ is a closed convex set, and $C \setminus K$ has finite volume. For such sets, we shall develop the first steps of a Brunn–Minkowski theory, relating volume and a kind of addition.

Freiburg im Breisgau
Spring 2022

Contents

1

Basic notions and facts

After fixing the notation and recalling some basic facts about incidence algebras, which will occasionally be needed, we use Section 1.3 to present introductory material about closed convex cones. Here we provide also some special lemmas, which will later be applied. Section 1.4 is devoted to polyhedra and deals with their normal cones and angle cones. In Section 1.5 we consider recession cones and show how they can be used in the description of unbounded polyhedra. Section 1.6, on valuations, begins with two classics, Groemer's extension theorem for valuations and Hadwiger's elementary approach to the Euler characteristic. The latter is extended in different directions. The Euler relation and several consequences are derived. The role of valuations on convex cones is pointed out. Identities between characteristic functions are the subject of Section 1.7. Such identities have appeared in different contexts in the literature. This topic is closely related to the theory of valuations. By integration of such identities, we later obtain classical angle sum relations and important relations between conic intrinsic volumes, one of which can be interpreted as a version of the Gauss–Bonnet theorem in spherical space. The chapter ends with some observations about the valuation property of polarity and with a characterization of polarity by its simplest properties.

1.1 Notation and prerequisites

We fix the notation and list the modest prerequisites for reading the following notes.

By $\mathbb{R}, \mathbb{R}_+, \mathbb{Z}, \mathbb{N}, \mathbb{N}_0$ we denote the set of, respectively, real numbers, nonnegative real numbers, integers, positive integers, nonnegative integers.

We work in a d-dimensional real vector space $(d \geq 2)$, for which we usually take $\mathbb{R}^d$, with standard scalar product $\langle \cdot, \cdot \rangle$ and induced norm $\|\cdot\|$. We use the scalar product to identify the dual space of $\mathbb{R}^d$ with $\mathbb{R}^d$ itself. The zero vector, or origin, of $\mathbb{R}^d$ is denoted by o. Whenever convenient, $\mathbb{R}^d$ is also considered as an affine space.

© The Author(s), under exclusive license to Springer Nature Switzerland AG 2022
R. Schneider, *Convex Cones*, Lecture Notes in Mathematics 2319,
https://doi.org/10.1007/978-3-031-15127-9_1

The set-theoretic Cartesian product $\mathbb{R}^{d_1} \times \mathbb{R}^{d_2}$ of $\mathbb{R}^{d_1}$ and $\mathbb{R}^{d_2}$ usually does not come along with a scalar product (unless it is tacitly identified with $\mathbb{R}^{d_1+d_2}$). Keeping the following convention in mind, we can dispense with introducing new notation.

Convention. For two Euclidean spaces $\mathbb{R}^{d_1}$, $\mathbb{R}^{d_2}$, their Cartesian product $\mathbb{R}^{d_1} \times \mathbb{R}^{d_2}$ is always equipped with the scalar product given by

$$\langle (x, y), (x', y') \rangle := \langle x, x' \rangle_1 + \langle y, y' \rangle_2, \quad (x, y), (x', y') \in \mathbb{R}^{d_1} \times \mathbb{R}^{d_2},$$

where $\langle \cdot, \cdot \rangle_i$ denotes the scalar product in $\mathbb{R}^{d_i}$, $i = 1, 2$.

The usual topological operators of interior, closure, boundary are denoted, respectively, by int, cl, bd. The relative interior of a set $A \subset \mathbb{R}^d$, denoted by relint A, is the interior of A with respect to the affine hull of A. Similarly, relbd A is the boundary of A relative to its affine hull.

Linear hull, affine hull, convex hull and positive hull are denoted, respectively, by lin, aff, conv, pos. The latter is defined by

$$\text{pos } A = \left\{ \sum_{i=1}^{k} \lambda_i a_i : k \in \mathbb{N}, \, a_i \in A, \, \lambda_i \geq 0 \right\}, \quad A \subset \mathbb{R}^d.$$

By $B^d = \{x \in \mathbb{R}^d : \|x\| \leq 1\}$ we denote the unit ball of $\mathbb{R}^d$. The unit sphere,

$$\mathbb{S}^{d-1} := \{u \in \mathbb{R}^d : \|u\| = 1\},$$

is endowed with the topology inherited from $\mathbb{R}^d$.

The group of all non-singular linear transformations of $\mathbb{R}^d$ is denoted by $\text{GL}(d)$. The *orthogonal group* $\text{O}(d)$ of $\mathbb{R}^d$ is the group of all linear mappings of $\mathbb{R}^d$ into itself preserving the scalar product, and $\text{SO}(d)$ is the subgroup of *rotations*, which preserve also the orientation. These groups are equipped with their usual topologies.

By $G(d, k)$ we denote the Grassmannian of k-dimensional linear subspaces of $\mathbb{R}^d$, endowed with its standard topology, $k = 0, \ldots, d$. We write

$$\mathcal{L}_\bullet := \bigcup_{k=0}^{d} G(d, k).$$

By a 'subspace', without adjective, of $\mathbb{R}^d$ we always mean a linear subspace. The orthogonal complement of a subspace L is denoted by $L^\perp$.

The closed segment with endpoints $x, y \in \mathbb{R}^d$ is denoted by $[x, y]$. We write hyperplanes of $\mathbb{R}^d$ often in the form

$$H(u, t) := \{x \in \mathbb{R}^d : \langle x, u \rangle = t\}$$

with $u \in \mathbb{S}^{d-1}$ and $t \in \mathbb{R}$. Such a hyperplane bounds the two closed halfspaces

$$H^+(u,t) := \{x \in \mathbb{R}^d : \langle x, u \rangle \geq t\},$$
$$H^-(u,t) := \{x \in \mathbb{R}^d : \langle x, u \rangle \leq t\}.$$

For $H(u,0)$ we usually write $u^\perp$.

Vector addition of subsets of $\mathbb{R}^d$ is denoted by $+$, thus

$$A + B := \{a + b : a \in A, b \in B\}$$

for $A, B \subseteq \mathbb{R}^d$. Also, $\lambda A := \{\lambda a : a \in A\}$ for $\lambda \in \mathbb{R}$; in particular, $-A = \{-a : a \in A\}$. Usually, $A + (-B)$ is abbreviated by $A - B$. We write $A + B$ in the form $A \oplus B$ if A and B lie in complementary affine subspaces of $\mathbb{R}^d$ and if we want to emphasize this fact.

A subset of $\mathbb{R}^d$ is *convex* if together with any two points it also contains the segment spanned by these points. The reader is assumed to be familiar with the basic notions and facts from the elementary geometry of convex sets (a general reference is [163]). In particular, support and separation theorems and properties of faces and support sets will be used without comment. For a closed convex set K, the support function is defined by

$$h(K, u) = \sup\{\langle x, u \rangle : x \in K\} \quad \text{for } u \in \mathbb{R}^d.$$

Here 'sup' can be replaced by 'max' if K is compact. For a vector $u \neq o$ for which $h(K, u) < \infty$, we write

$$H(K, u) = \{x \in \mathbb{R}^d : \langle x, u \rangle = h(K, u)\},$$
$$H^-(K, u) = \{x \in \mathbb{R}^d : \langle x, u \rangle \leq h(K, u)\},$$
$$H^+(K, u) = \{x \in \mathbb{R}^d : \langle x, u \rangle \geq h(K, u)\}.$$

Thus, $K \subseteq H^-(K, u)$, and $K \cap H(K, u) \neq \emptyset$ if K is compact. If $K \cap H(K, u) \neq \emptyset$, we call $H(K, u)$ the supporting hyperplane (or, briefly, support plane) of K with outer normal vector u.

We denote by $\mathcal{CC}^d$ the set of nonempty closed convex subsets of $\mathbb{R}^d$. By $\mathcal{K}^d \subset \mathcal{CC}^d$ we denote the subset of compact convex sets, also called *convex bodies* (though they need not have interior points).

We recall a useful tool, the nearest-point map (see, e.g., [163, Section 1.2], with different notation). For a set $K \in \mathcal{CC}^d$ and for $x \in \mathbb{R}^d$, there is a unique point $\Pi_K(x) \in K$ such that

$$\|x - \Pi_K(x)\| = \min\{\|x - y\| : y \in K\}.$$

This defines the *metric projection* $\Pi_K(\cdot)$, also called *nearest-point map*, of K.

The distance of a point $x \in \mathbb{R}^d$ from a set $K \in \mathcal{CC}^d$ is denoted by $\mathrm{dist}(x, K)$, thus

$$\mathrm{dist}(x, K) = \min\{\|x - y\| : y \in K\} = \|x - \Pi_K(x)\|.$$

The *Hausdorff metric* on $\mathcal{K}^d$ is defined by

$$d_H(K, L) = \max\{\sup_{x \in K} \inf_{y \in L} \|x - y\|, \sup_{x \in L} \inf_{y \in K} \|x - y\|\}$$

$$= \min\{\lambda \geq 0 : K \subseteq L + \lambda B^d, L \subseteq K + \lambda B^d\}$$

for $K, L \in \mathcal{K}^d$ (see, e.g., [163, Sect. 1.8]). The same definition can also be used to define the Hausdorff metric on the system of nonempty compact subsets of $\mathbb{R}^d$.

If a set U is given, then the *characteristic function* (on U) of a subset $A \subseteq U$ is defined by

$$\mathbb{1}_A(x) = \begin{cases} 1 \text{ if } x \in A, \\ 0 \text{ if } x \in U \setminus A. \end{cases}$$

If $A = \{x \in U : x \text{ has property } P\}$, then we also write

$$\mathbb{1}_A(x) = \mathbb{1}\{x \text{ has property } P\}.$$

We shall often use the notation

$$\mathbb{1}_A = \mathbb{1}[A],$$

for two reasons. When A will be replaced by more complicated expressions, the index notation might be hard to read. And several authors use $[A]$ instead of $\mathbb{1}_A$, so $\mathbb{1}[A]$ might be a good compromise and mnemonic device (whereas, in our context, $[A]$ is often used with a different meaning; see, e.g., McMullen [133, Chap. 3]).

If a set A is the disjoint union of the sets A_i, $i \in I$, we indicate this by

$$A = \biguplus_{i \in I} A_i.$$

1.2 Incidence algebras

Occasionally, we have to use the incidence algebra of a partially ordered set, and in particular its Möbius function (see, e.g., Rota [149], Stanley [184, Section 3.7]). Let $(M, \leq)$ be a finite partially ordered set. For such a poset, the *incidence algebra* $I(M)$ is defined as follows. Its elements are the functions $\xi : M \times M \to \mathbb{R}$ with $\xi(a, b) = 0$ if $a \not\leq b$. Addition of functions and multiplication by real numbers are defined pointwise, and multiplication of $\xi, \eta \in I(M)$ is defined by

$$(\xi * \eta)(a, b) = \sum_{a \leq x \leq b} \xi(a, x)\eta(x, b)$$

if $a \leq b$, and $(\xi * \eta)(a, b) = 0$ if $a \not\leq b$. This yields an associative algebra. The *delta function*, defined by

$$\delta(a, b) := \begin{cases} 1 \text{ if } a = b, \\ 0 \text{ if } a \neq b, \end{cases}$$

is the multiplicative identity element of the incidence algebra. The *zeta function* of $I(M)$ is defined by

$$\zeta(a, b) = 1 \quad \text{for } a \leq b.$$

The *Möbius function* μ of the poset M is defined recursively by $\mu(a, a) = 1$ for all $a \in M$ and

$$\mu(a, b) = -\sum_{a \leq x < b} \mu(a, x)$$

for $a, b \in M$ with $a < b$. One checks immediately that $\mu * \zeta = \delta$.

Defining $\mu'(a, a) = 1$ for all $a \in M$ and recursively

$$\mu'(a, b) = -\sum_{a < x \leq b} \mu'(x, b)$$

for $a, b \in M$ with $a < b$, one checks similarly that $\zeta * \mu' = \delta$. Since δ is the unit element of the incidence algebra, it follows that $\mu' = \mu$. Thus, μ is the inverse of ζ.

Since $\mu * \zeta = \delta = \zeta * \mu$, we have the equivalence

$$\beta = \alpha * \zeta \iff \alpha = \beta * \mu$$

for $\alpha, \beta \in I(M)$, or explicitly

$$\beta(z, y) = \sum_{z \leq x \leq y} \alpha(z, x) \ \forall z \leq y \iff \alpha(z, y) = \sum_{z \leq x \leq y} \beta(z, x)\mu(x, y) \ \forall z \leq y.$$

Let f, g be real functions on M. If $g(y) = \sum_{x \leq y} f(x)$ for all $y \in M$, then, for fixed y,

$$\sum_{x \leq y} g(x)\mu(x, y) = \sum_{x \leq y} \mu(x, y) \sum_{z \leq x} f(z) = \sum_{z \leq y} f(z) \sum_{z \leq x \leq y} \mu(x, y)$$

$$= \sum_{z \leq y} f(z)\delta(z, y) = f(y).$$

Conversely, if $f(y) = \sum_{x \leq y} g(x)\mu(x, y)$ for all $y \in M$, then, for fixed z,

$$\sum_{y \leq z} f(y) = \sum_{y \leq z} \sum_{x \leq y} g(x)\mu(x, y) = \sum_{x \leq z} g(x) \sum_{x \leq y \leq z} \mu(x, y)$$

$$= \sum_{x \leq z} g(x)\delta(x, z) = g(z).$$

Thus, we have

$$g(y) = \sum_{x \leq y} f(x) \ \forall y \in M \iff f(y) = \sum_{x \leq y} g(x)\mu(x, y) \ \forall y \in M. \qquad (1.1)$$

Similarly, one shows that

$$g(y) = \sum_{x \geq y} f(x) \ \forall \, y \in M \quad \Leftrightarrow \quad f(y) = \sum_{x \geq y} g(x)\mu(y, x) \ \forall \, y \in M. \qquad (1.2)$$

The equivalences (1.1) and (1.2) are known as the *Möbius inversion formula*.
For the simple proof of the following lemma, we refer to Stanley [184, 3.6.2]).

Lemma 1.2.1. *For an element $\xi \in I(M)$, the following conditions are equivalent: (a) ξ has a left inverse, (b) ξ has a right inverse, (c) ξ has a two-sided inverse (which is the unique left and right inverse), (d) $\xi(x, x) \neq 0$ for all $x \in M$.*

The next lemma is taken from Budach [40] (for extensions, we refer to Chen [49]).

Lemma 1.2.2. *Let $\xi \in I(M)$ satisfy $\xi(x, x) = 0$ for all $x \in M$. If $\xi^k(x, y) \neq 0$ for some $k \in \mathbb{N}$ and some $x, y \in M$ with $x \leq y$, then there are $x_0, \dots, x_k \in M$ such that*

$$x = x_0 < x_1 < \cdots < x_k = y.$$

Proof. If $\xi(x, y) \neq 0$, then $x < y$; thus the assertion holds for $k = 1$. Suppose the assertion has been proved for some $k \geq 1$. Assume that some $x, y \in M$ satisfy $x \leq y$ and $\xi^{k+1}(x, y) \neq 0$. Then

$$\sum_{x \leq a \leq y} \xi^k(x, a)\xi(a, y) = \xi^{k+1}(x, y) \neq 0.$$

Hence, there is at least one $a \in M$ with $\xi^k(x, a) \neq 0$ and $\xi(a, y) \neq 0$. The latter implies that $a < y$, and by the induction hypothesis, the former implies the existence of $x_0, \dots, x_k \in M$ with

$$x = x_0 < x_1 < \cdots < x_k = a < y =: x_{k+1}.$$

Thus, the assertion holds for $k + 1$, which completes the proof. □

For $\xi \in M$, we define $\xi^0 := \delta$.

Lemma 1.2.3. *If $\xi \in I(M)$ satisfies $\xi^{k+1} = 0$ for some $k \in \mathbb{N}_0$, then*

$$(\delta + \xi)^{-1} = \sum_{r=0}^{k} (-1)^r \xi^r.$$

Proof. From $\xi^{k+1} = 0$ we immediately get

$$(\delta + \xi) * \sum_{r=0}^{k} (-1)^r \xi^r = \delta.$$

This shows that $\delta + \xi$ is invertible, and it provides its inverse. □

1.3 Convex cones

The purpose of this section is to collect some basic properties of convex cones and to introduce the notions and some special results which will later be needed. Thus, this section is not very exciting; we have to lay the foundations, and we collect auxiliary assertions for later use.

Before treating cones, we have a short look at closed convex sets in general. Let $K \in \mathcal{CC}^d$. The *lineality space* $\mathrm{lineal}(K)$ of K is defined as the subspace of largest dimension of which a translate is contained in K (it is unique, since K is convex). Since K is convex and closed, it is clear that

$$x + \mathrm{lineal}(K) \subseteq K \quad \text{for } x \in K.$$

Further,

$$K = \mathrm{lineal}(K) \oplus K' \quad \text{with } K' = K \cap \mathrm{lineal}(K)^{\perp},$$

where K' does not contain a line. We call K' the *line-free kernel* of K. The ordered pair

$$\mathrm{type}(K) := (\dim \mathrm{lineal}(K), \epsilon) \quad \text{with } \epsilon = \begin{cases} 1 \text{ if } K' \text{ is bounded,} \\ 0 \text{ if } K' \text{ is unbounded,} \end{cases}$$

is called the *type* of K. It is easily seen that it is invariant under affine transformations of K.

Now we turn to cones. A subset $A \subseteq \mathbb{R}^d$ is a *cone* or a *conic set* if $a \in A$ implies $\lambda a \in A$ for $\lambda > 0$. With each $A \subseteq \mathbb{R}^d$ we associate its spanned cones

$$A^{\vee} := \{\lambda a : a \in A, \, \lambda \geq 0\}, \qquad A^{+} := \{\lambda a : a \in A, \, \lambda > 0\}.$$

A *convex cone* C is a cone satisfying $\lambda x + \mu y \in C$ for $x, y \in C$ and $\lambda, \mu > 0$. A subset $C \subseteq \mathbb{R}^d$ is a *closed convex cone* if it is closed, nonempty and is closed under vector addition and under multiplication by nonnegative real numbers. Thus, a closed convex cone always contains o. By $\mathcal{C}^d$ we denote the set of all closed convex cones in $\mathbb{R}^d$. A subspace of $\mathbb{R}^d$ belongs to $\mathcal{C}^d$; in particular, this holds for $\{o\}$ and $\mathbb{R}^d$. Occasionally, we have to exclude the cone $\{o\}$; for this, we denote by $\mathcal{C}^d_*$ the set of cones in $\mathcal{C}^d$ of positive dimension. Let $C \in \mathcal{C}^d$. We evidently have $\mathrm{lineal}(C) = C \cap (-C)$. If $\mathrm{lineal}(C) = \{o\}$, then C is called *pointed* (also called *salient* by some authors). We denote by $\mathcal{C}^d_p \subset \mathcal{C}^d$ the subset of all pointed cones.

With every closed convex set, we can associate a variety of convex cones.

Definition 1.3.1. *Let $A \in \mathcal{CC}^d$ be a closed convex set. For $x \in A$, the cone of feasible directions of A at x is given by*

$$\mathrm{cone}(A, x) := \{y \in \mathbb{R}^d : x + \lambda y \in A \text{ for some } \lambda > o\}$$

$$= \mathrm{pos}(A - x).$$

Clearly, $\mathrm{cone}(A, x)$ is a convex cone containing o, but it need not be closed.

We collect some observations about supporting hyperplanes of cones.

Lemma 1.3.1. *Let $C \in \mathcal{C}^d$, and let H be a supporting hyperplane of C. Then $\mathrm{lineal}(C) \subseteq H$.*

Proof. First we show that $o \in H$. By the definition of a supporting hyperplane, there is a point $x \in H \cap C$. If $x = o$, we are finished. Otherwise, the ray through x with endpoint o is contained in C. If $o \notin H$, then H would not support C, a contradiction.

Now suppose there would exist a point $y \in \mathrm{lineal}(C) \setminus H$. Then $y \neq o$, and the line through y and o belongs to $\mathrm{lineal}(C) \subseteq C$, thus H (which contains o) cannot support C, a contradiction. $\qquad \square$

Lemma 1.3.2. *Let $C \in \mathcal{C}^d$ be a cone which is not a line. Let H be a hyperplane with $H \cap C = \{o\}$. Then H supports C.*

Proof. Suppose that H would not support C. Then there are points $x, y \in C$ lying in different open halfspaces bounded by H. The line L through x and y intersects H in a point z, which belongs to C, hence $z = o$. Since $C \neq L$ by assumption, we can replace one of the points x, y by a point in the same open halfspace and in $C \setminus L$. This produces an intersection point in $(H \cap C) \setminus \{o\}$, a contradiction. $\qquad \square$

Lemma 1.3.3. *Let $C \in \mathcal{C}^d$, and let H be a hyperplane.*

(a) *If $H \cap C = \mathrm{lineal}(C)$, then H supports C.*

(b) *If $H \cap C \subseteq \mathrm{lineal}(C)$ and C is not a subspace, then H supports C.*

(c) *If H supports C and $H \cap C \subseteq \mathrm{lineal}(C)$, then $H \cap C = \mathrm{lineal}(C)$.*

Proof. (a) We use the orthogonal direct sum decomposition $C = \mathrm{lineal}(C) \oplus C'$, where C' is a pointed cone. We have $H \cap C' = \{o\}$ (if $x \in H \cap C'$, then $x \in \mathrm{lineal}(C)$, hence $x = o$). By Lemma 1.3.2 (and since C' is not a line), H supports C', and since $\mathrm{lineal}(C) \subseteq H$, the hyperplane H also supports $\mathrm{lineal}(C) \oplus C' = C$.

(b) Let $H \cap C \subseteq \mathrm{lineal}(C)$. Suppose that $H \cap C \neq \mathrm{lineal}(C)$. Since $\mathrm{lineal}(C) \subseteq C$, the intersection $H \cap \mathrm{lineal}(C)$ is a proper subspace of $\mathrm{lineal}(C)$. By assumption, $C \neq \mathrm{lineal}(C)$. Hence, there is a point $x \in C \setminus \mathrm{lineal}(C)$. Let R be the ray through x with endpoint o. Then $\mathrm{lineal}(C) + R \subseteq C$. There is a line L through o that is contained in H and not contained in $\mathrm{lineal}(C)$. It meets $\mathrm{lineal}(C) + R$ (and hence C) outside $\mathrm{lineal}(C)$, a contradiction. Thus, $H \cap C = \mathrm{lineal}(C)$, and the assertion follows from (a).

(c) If H supports C, then $o \in H$ and therefore $\mathrm{lineal}(C) \subseteq H$, hence $\mathrm{lineal}(C) \subseteq H \cap C$. $\qquad \square$

Lemma 1.3.4. *Let $C \in \mathcal{C}^d \setminus \mathcal{L}_\bullet$. Let L be a subspace with $L \cap C \subseteq \mathrm{lineal}(C)$. Then there is a supporting hyperplane H of C with $L \subseteq H$ and $H \cap C = \mathrm{lineal}(C)$.*

Proof. First we assume that C is a pointed cone, then the assumption says that $L \cap C = \{o\}$. We can easily construct a pointed closed convex cone D such that $C \setminus \{o\} \subset \operatorname{int} D$ and $L \cap D = \{o\}$. By [163, Thm. 1.3.8], D and L can be separated by a hyperplane H. Then $L \subseteq H$, and H supports D and hence C. Since $H \cap D$ is contained in the boundary of D, we have $H \cap C = \{o\}$.

Now we consider the general case, where $\operatorname{lineal}(C)$ can be any subspace. Let $E := [\operatorname{lineal}(C)]^{\perp}$. We use the direct orthogonal sum decomposition $C = \operatorname{lineal}(C) \oplus C'$ with $C' := C \cap E$. Then C' is a pointed cone with $(L \cap E) \cap C' = \{o\}$. By the first part of the proof, there is a hyperplane H' of the subspace E which supports C' and satisfies $L \cap E \subseteq H'$ and $H' \cap C' = \{o\}$. Then $H := H' \oplus \operatorname{lineal}(C)$ is a hyperplane in $\mathbb{R}^d$ which supports C and satisfies $L \subseteq H$ and $H \cap C = \operatorname{lineal}(C)$. $\qquad\square$

Next, we collect results on sums of cones. For $C_1, C_2 \in \mathcal{C}^d$ we have

$$C_1 + C_2 = \operatorname{conv}(C_1 \cup C_2). \tag{1.3}$$

In fact, if $x \in C_1 + C_2$, then $x = c_1 + c_2$ with $c_i \in C_i$ ($i = 1, 2$) and hence $x = 2(\frac{1}{2}c_1 + \frac{1}{2}c_2) \in 2\operatorname{conv}(C_1 \cup C_2) = \operatorname{conv}(C_1 \cup C_2)$. Conversely, suppose that $x \in \operatorname{conv}(C_1 \cup C_2)$, then $x = (1-\lambda)y + \lambda z$ with $\lambda \in [0,1]$ and $y, z \in C_1 \cup C_2$, say $y \in C_i$, $z \in C_j$, $i, j \in \{1, 2\}$. Then $(1-\lambda)y \in C_i$, $\lambda z \in C_j$, hence $x \in C_i + C_j \subseteq C_1 + C_2$ (since $2C_i = C_i$ and $o \in C_i, C_j$).

The sum of a cone (which is not a subspace) and a cone can be a subspace.

Lemma 1.3.5. *Let $C \in \mathcal{C}^d$ and $L \in \mathcal{L}_{\bullet}$. Then the following are equivalent:*

(a) $L + C$ *is a subspace,*

(b) $L \cap \operatorname{relint} C \neq \emptyset$.

Proof. If C is a subspace, then (a) and (b) hold. Hence, we assume that C is not a subspace.

Suppose that (a) does not hold. Then $L + C$ is a cone, hence it has a supporting hyperplane H that contains the lineality space of $L + C$ and hence L, but not $L + C$ (since $L + C$ is not a subspace). Then $L \cap \operatorname{relint} C = \emptyset$, thus (b) does not hold.

Suppose that (b) does not hold. Then there is a hyperplane H that properly separates L and C (see [163, Thm. 1.3.8]). Then $L \subset H$ (hence $o \in H$), and C is contained in one of the closed halfspaces bounded by H, but not in H. Then $L + C$ cannot be a subspace. $\qquad\square$

Note that the sum of two closed convex cones need not be closed. An example is given by a cone in $\mathbb{R}^3$ with a circular cross section and the linear hull of one of its boundary rays. Their sum is the union of an open halfspace and a line in its boundary. The following theorem allows us to exclude such cases.

Theorem 1.3.1. *Let $C, D \subset \mathbb{R}^d$ be closed convex cones with the property that their sum $C + D$ is not closed. Then there are a vector $u \in \mathbb{R}^d \setminus \{o\}$ and a hyperplane H through o such that C and D are contained in the same closed halfspace bounded by H and that $u \in C$ and $-u \in D$.*

Proof. Since $C + D$ is not closed, there are $c_i \in C$, $d_i \in D$ ($i \in \mathbb{N}$) such that $c_i + d_i \to x$ as $i \to \infty$ for some point $x \notin C + D$. If one of the sequences (c_i), (d_i) is bounded, then both are bounded and contain convergent subsequences, which yields that $x \in C + D$, a contradiction. Hence, both sequences are unbounded, and we can assume that $c_i \neq o \neq d_i$ for all i. Then $c_i = \lambda_i u_i$, $d_i = \mu_i v_i$ with unit vectors $u_i \in C$, $v_i \in D$ and $\lambda_i, \mu_i \to \infty$. A subsequence of (u_i) converges to some $u \in C$, and a subsequence of (v_i) converges to some $v \in D$. Then the left side of

$$\langle \lambda_i u_i + \mu_i v_i, u + v \rangle = \lambda_i \langle u_i, u + v \rangle + \mu_i \langle v_i, u + v \rangle$$

converges to $\langle x, u + v \rangle$, whereas on the right side we have $\langle u_i, u + v \rangle \to 1 + \langle u, v \rangle \geq 0$, $\langle v_i, u + v \rangle \to 1 + \langle u, v \rangle \geq 0$, $\lambda_i, \mu_i \to \infty$, which is only possible if $v = -u$. Hence, $u \in C$ and $-u \in D$. The point x satisfies $x \notin C + D$ and $x \in \mathrm{cl}(C + D)$, hence it is contained in a supporting hyperplane H of $\mathrm{cl}(C + D)$. Since the latter is a cone, we have $o \in H$. One halfspace H^- bounded by H contains $\mathrm{cl}(C + D)$, thus $C, D \subseteq H^-$. Since $u, -u \in H^-$, we have $u, -u \in H$. □

Further results on the closedness of $C + D$ are found in Waksman and Epelman [188, p. 95] and Pataki [144, Theorem 5.1].

For convex cones, polarity plays an important role.

Definition 1.3.2. *For an arbitrary subset $K \subset \mathbb{R}^d$, the* polar set *is defined by*

$$K^\circ := \{x \in \mathbb{R}^d : \langle x, y \rangle \leq 1 \text{ for all } y \in K\}.$$

If, in particular, C is a convex cone (not necessarily closed), then $y \in C$ implies $\lambda y \in C$ for all $\lambda \geq 0$ and hence the *dual* or *polar cone* of C is given by

$$C^\circ = \{x \in \mathbb{R}^d : \langle x, y \rangle \leq 0 \text{ for all } y \in C\}.$$

This can also be written as

$$C^\circ = \bigcap_{y \in C} H^-(y, 0) = \bigcap_{u \in C \cap \mathbb{S}^{d-1}} H^-(u, 0).$$

Trivially, $C_1 \subseteq C_2$ implies $C_2^\circ \subseteq C_1^\circ$. We have $C^\circ \in \mathcal{C}^d$, and if C is closed, then

$$C^{\circ\circ} := (C^\circ)^\circ = C$$

(see, e.g., [163, p. 35]). Clearly, if L is a subspace, then $L^\circ = L^\perp$. In particular, $\{o\}^\circ = \mathbb{R}^d$. For $C \in \mathcal{C}^d$ we have

$$\dim C + \dim \mathrm{lineal}(C^\circ) = d. \tag{1.4}$$

Proof. From lineal$(C^\circ) \subseteq C^\circ$ it follows that $C = C^{\circ\circ} \subseteq [\text{lineal}(C^\circ)]^\circ = [\text{lineal}(C^\circ)]^\perp$ and hence $\dim C \leq \dim \text{lineal}(C^\circ)^\perp = d - \dim \text{lineal}(C^\circ)$. Since $C \subseteq \text{lin}\, C$, we have $(\text{lin}\, C)^\perp = (\text{lin}\, C)^\circ \subseteq C^\circ$, thus $(\text{lin}\, C)^\perp$ is a subspace of C°, hence $\dim (\text{lin}\, C)^\perp \leq \dim \text{lineal}(C^\circ)$ and thus $d - \dim C \leq \dim \text{lineal}(C^\circ)$. This gives the assertion. □

The following is [163, Thm. 1.6.9]; see also Rockafellar [148, §14].

Theorem 1.3.2. *If $C_1, C_2 \in \mathcal{C}^d$, then*

$$(C_1 \cap C_2)^\circ = \text{cl}(C_1^\circ + C_2^\circ), \qquad (C_1 + C_2)^\circ = C_1^\circ \cap C_2^\circ.$$

If $C_1 \cup C_2$ is convex, then $C_1^\circ \cup C_2^\circ$ is convex, and

$$(C_1 \cap C_2)^\circ = C_1^\circ \cup C_2^\circ, \qquad (C_1 \cup C_2)^\circ = C_1^\circ \cap C_2^\circ.$$

For $C \in \mathcal{C}^d$ and $\vartheta \in O(d)$, we have

$$(\vartheta C)^\circ = \vartheta C^\circ.$$

This follows from $x \in (\vartheta C)^\circ \Leftrightarrow \langle x, y \rangle \leq 0 \ \forall y \in \vartheta C \Leftrightarrow \langle x, \vartheta y \rangle \leq 0 \ \forall y \in C \Leftrightarrow \langle \vartheta^{-1} x, y \rangle \leq 0 \ \forall y \in C \Leftrightarrow \vartheta^{-1} x \in C^\circ \Leftrightarrow x \in \vartheta C^\circ$.

We consider the nearest-point map for cones. For a closed convex cone C, the nearest-point map Π_C is homogeneous, that is

$$\Pi_C(\lambda x) = \lambda \Pi_C(x) \quad \text{for } \lambda \geq 0, \, x \in \mathbb{R}^d. \tag{1.5}$$

This follows from the homogeneity of the norm and the uniqueness of the nearest point.

The nearest-point maps of a closed convex cone and of its polar cone are related in an important way. The following theorem is due to Moreau [138], in a Hilbert space.

Theorem 1.3.3 (Moreau decomposition). *For a cone $C \in \mathcal{C}^d$ and for $x \in \mathbb{R}^d$,*

$$\Pi_C(x) + \Pi_{C^\circ}(x) = x \tag{1.6}$$

and

$$\langle \Pi_C(x), \Pi_{C^\circ}(x) \rangle = 0. \tag{1.7}$$

Proof. If $x \in C$, then $\Pi_C(x) = x$ and $\Pi_{C^\circ}(x) = o$, hence the assertion is trivial, similarly if $x \in C^\circ$. We assume, therefore, that $x \in \mathbb{R}^d \setminus (C \cup C^\circ)$.

By [163, Lemma 1.3.1], the vector $u := x - \Pi_C(x)$ is an outer normal vector of a supporting hyperplane H of C through $\Pi_C(x)$. Since C is a convex cone, the supporting hyperplane H contains o. Therefore, $u \in C^\circ$, and the vector $\Pi_C(x)$ is orthogonal to u. The hyperplane H' through o with normal vector $\Pi_C(x)$ is a supporting hyperplane of C°, and we have $u \in H'$. Therefore, the point $\Pi_{C^\circ}(x)$, which is the point in C° nearest to x, is also the point in H' nearest to x. This implies that $\Pi_{C^\circ}(x) = u$. Thus, the points $x, \Pi_C(x), o, \Pi_{C^\circ}(x)$ are the vertices of a rectangle. The assertions follow. □

Corollary 1.3.1. *Let $C \in \mathcal{C}_*^d$ and $x \in \mathbb{R}^d \setminus (C \cup C^\circ)$. Then the angles between x and $\Pi_C(x)$ and between x and $\Pi_{C^\circ}(x)$ add up to $\pi/2$.*

Corollary 1.3.2. *For $C \in \mathcal{C}^d$ and $x \in \mathbb{R}^d$,*

$$\operatorname{dist}(x, C) = \|\Pi_{C^\circ}(x)\|. \tag{1.8}$$

Proof. By the definition of Π_C, we have $\operatorname{dist}(x, C) = \|x - \Pi_C(x)\|$, and by (1.6) this is equal to $\|\Pi_{C^\circ}(x)\|$. $\qquad\square$

Theorem 1.3.3 has a converse.

Lemma 1.3.6. *Let $C \in \mathcal{C}^d$, $y \in C$, $z \in C^\circ$ be such that $\langle y, z \rangle = 0$. Then $x := y + z$ satisfies*

$$y = \Pi_C(x) \quad and \quad z = \Pi_{C^\circ}(x).$$

Proof. If $y = o$ or $z = o$, this is trivial; so let $y \neq o \neq z$. Write

$$H^- := \{w \in \mathbb{R}^d : \langle w, z \rangle \leq 0\}, \quad H := \{w \in \mathbb{R}^d : \langle w, z \rangle = 0\}.$$

For $w \in C$ we have $\langle w, z \rangle \leq 0$, since $z \in C^\circ$, thus $C \subseteq H^-$. By the assumptions, $y \in H \cap C$ and $\langle x, z \rangle = \|z\|^2$, thus $x \notin H^-$ and $\Pi_H(x) = y$. Therefore, $\Pi_C(x) = \Pi_H(x) = y$. A similar argument shows that $\Pi_{C^\circ}(x) = z$. $\qquad\square$

The following is a consequence.

Lemma 1.3.7. *Let $C, D \in \mathcal{C}^d$. Then*

$$\Pi_{C \times D}(x, y) = (\Pi_C(x), \Pi_D(y)) \quad for \ (x, y) \in \mathbb{R}^d \times \mathbb{R}^d.$$

Proof. Let $(x, y) \in \mathbb{R}^d \times \mathbb{R}^d$. Write $(\Pi_C(x), \Pi_D(y)) =: a$, $(\Pi_{C^\circ}(x), \Pi_{D^\circ}(y)) =: b$. Then $a + b = (x, y)$ by Theorem 1.3.3. By the definition of the scalar product in $\mathbb{R}^d \times \mathbb{R}^d$ (see the Convention in Sect. 1.1),

$$\langle a, b \rangle = \langle \Pi_C(x), \Pi_{C^\circ}(x) \rangle + \langle \Pi_D(y), \Pi_{D^\circ}(y) \rangle = 0.$$

We have $a \in C \times D$. Further, $b \in (C \times D)^\circ$, since any $(u, v) \in C \times D$ satisfies

$$\langle b, (u, v) \rangle = \langle \Pi_{C^\circ}(x), u \rangle + \langle \Pi_{D^\circ}(y), v \rangle \leq 0.$$

From $a \in C \times D$, $b \in (C \times D)^\circ$ and $\langle a, b \rangle = 0$ it follows, according to Lemma 1.3.6, that $\Pi_{C \times D}(x, y) = \Pi_{C \times D}(a + b) = a = (\Pi_C(x), \Pi_D(y))$. $\qquad\square$

The norm of the metric projection to $C \in \mathcal{C}^d$ can also be interpreted as the support function of the intersection of C with the unit ball. Namely, we have

$$\max_{y \in C \cap B^d} \langle y, x \rangle = \|\Pi_C(x)\| \quad \text{for } x \in \mathbb{R}^d. \tag{1.9}$$

This is clear if $x \in C$ or $x \in C^\circ$, so let us assume that $x \in \mathbb{R}^d \setminus (C \cap C^\circ)$. Let $u := \Pi_C(x)/\|\Pi_C(x)\|$. Let $H^-(C, x - \Pi_C(x))$ be the supporting halfspace of C with outer normal vector $x - \Pi_C(x)$, and let $H^-(B^d, u)$ be the supporting halfspace of the ball B^d with outer normal vector u. Then

$$C \cap B^d \subset H^-(C, x - \Pi_C(x)) \cap H^-(B^d, u) =: D$$

and hence

$$\max_{y \in C \cap B^d} \langle y, x \rangle = \max_{y \in D} \langle y, x \rangle = \langle u, x \rangle = \|\Pi_C(x)\|.$$

This proves (1.9).

We shall later need the following result.

Lemma 1.3.8. *Let $C \in \mathcal{C}^d$. The function $x \mapsto \|\Pi_C(x)\|^2$ is everywhere differentiable, and its gradient is given by*

$$\nabla\|\Pi_C(x)\|^2 = 2\Pi_C(x).$$

Proof. (After McCoy [127, p. 44]) By Corollary 1.3.2,

$$\|\Pi_C(x)\|^2 = \text{dist}^2(x, C^\circ). \tag{1.10}$$

For given $x \in \mathbb{R}^d$, we define

$$h(u) := \text{dist}^2(x + u, C^\circ) - \text{dist}^2(x, C^\circ) - 2\langle \Pi_C(x), u \rangle \quad \text{for } u \in \mathbb{R}^d.$$

By the definition of the distance and by (1.6), we have

$$\text{dist}^2(x + u, C^\circ) \leq \|x + u - \Pi_{C^\circ}(x)\|^2 = \|\Pi_C(x) + u)\|^2$$
$$= \|\Pi_C(x)\|^2 + 2\langle \Pi_C(x), u \rangle + \|u\|^2,$$

which gives

$$h(u) \leq \|\Pi_C(x)\|^2 - \text{dist}^2(x, C^\circ) + \|u\|^2 = \|u\|^2,$$

by (1.10). Since the function h is convex, we have

$$\frac{1}{2}h(u) + \frac{1}{2}h(-u) \geq h(o) = 0$$

and hence $h(u) \geq -h(-u) \geq -\|u\|^2$, thus $-\|u\|^2 \leq h(u) \leq \|u\|^2$, which shows that h is differentiable at $u = o$ and that $\nabla h(o) = o$. From this and (1.10) the assertion follows. $\qquad\square$

As we will be interested in intersections of cones, the following lemmas are relevant. The first one appears in Godland and Kabluchko [66, Lem. 2.14], with a different proof.

Lemma 1.3.9. *Let $C \in \mathcal{C}^d$ and $L \in \mathcal{L}_\bullet$. Then*

$$L \cap \operatorname{relint} C \neq \emptyset \;\Leftrightarrow\; L^\perp \cap C^\circ \subseteq \operatorname{lineal}(C^\circ).$$

Proof. We can assume that C is not a subspace, because otherwise the assertion is trivial. We use the direct orthogonal sum decomposition

$$C^\circ = \operatorname{lineal}(C^\circ) \oplus C',$$

where $C' = C^\circ \cap \operatorname{lineal}(C^\circ)^\perp$ is a pointed cone.

Suppose, first, that $L^\perp \cap C^\circ \subseteq \operatorname{lineal}(C^\circ)$. By Lemma 1.3.4, there is a supporting hyperplane H such that $L^\perp \subseteq H$ and $H \cap C^\circ = \operatorname{lineal}(C^\circ)$. Let u be the (with respect to C°) outer unit normal vector of H. Then $u \perp L^\perp$, hence $u \in L$. We have $H \cap C' = \{o\}$. Therefore, the vectors in a whole neighborhood, relative to $\operatorname{lin} C'$, of u are normal vectors of supporting hyperplanes of C°, hence they belong to C. Thus, $u \in \operatorname{relint} C$ and hence $L \cap \operatorname{relint} C \neq \emptyset$.

Second, suppose that $L \cap \operatorname{relint} C \neq \emptyset$. Choose a unit vector $u \in L \cap \operatorname{relint} C$. Since $v^\perp$ supports C' for all v in a neighborhood, relative to $\operatorname{lin} C'$, of u, we have $u^\perp \cap C' = \{o\}$ and hence $L^\perp \cap C^\circ \subseteq u^\perp \cap C^\circ \subseteq \operatorname{lineal}(C^\circ)$. $\qquad\square$

We say that a cone $C \in \mathcal{C}^d$ and a subspace $L \in \mathcal{L}_\bullet$ *touch* if $C \cap L \neq \{o\}$, but C and L can be separated by a hyperplane, that is, L is contained in some supporting hyperplane of C. Similarly, we say that two cones $C, D \in \mathcal{C}^d$ *touch* if $C \cap D \neq \{o\}$ but C and D can be separated (weakly) by a hyperplane.

Lemma 1.3.10. *Let $C \in \mathcal{C}^d$ and $L \in \mathcal{L}_\bullet$. If C and L touch, then also C° and $L^\perp$ touch. If C and L do not touch, then*

$$C \cap L \neq \{o\} \;\Leftrightarrow\; C^\circ \cap L^\perp = \{o\}.$$

Proof. Suppose that C and L touch. Then there are a vector $y \in C \cap L$, $y \neq o$, and a supporting hyperplane $H(C, v)$ of C such that $L \subseteq H(C, v)$ and $C \subseteq H^-(C, v)$. The latter implies that $v \in C^\circ$, and $L \subseteq H(C, v)$ implies that $\operatorname{lin}\{v\} \subseteq L^\perp$, thus $v \in C^\circ \cap L^\perp$ and hence $C^\circ \cap L^\perp \neq \{o\}$. Further, $\operatorname{lin}\{y\} \subseteq L$ implies that $L^\perp \subseteq \operatorname{lin}\{y\}^\perp$, and the latter is a supporting hyperplane of C°. Hence, C° and $L^\perp$ touch.

Assume now that C and L do not touch. Let $C \cap L \neq \{o\}$. Suppose that $C^\circ \cap L^\perp \neq \{o\}$. By duality, $C + L \neq \mathbb{R}^d$, hence there exists a closed halfspace H^-, bounded by some $H \in G(d, d-1)$, such that $C + L \subseteq H^-$. Then $L \subseteq H$. This means that C and L touch, a contradiction. Conversely, suppose that $C^\circ \cap L^\perp = \{o\}$. Then (as follows from [163, Thm. 1.3.8]) there is a supporting hyperplane $H(C^\circ, u)$ of C° such that $L^\perp \subseteq H(C^\circ, u)$ and $C^\circ \subseteq H^-(C^\circ, u)$. The latter implies that $u \in C$. From $L^\perp \subseteq H(C^\circ, u)$ we have $u \in L$, hence $u \in C \cap L$ with $u \neq o$. $\qquad\square$

The following lemmas concern the behavior of convex cones under linear maps. The first one, which will later be useful, is taken from [72, Lem. 5.1]). By $\ker \mathsf{A}$ we denote the kernel of a linear map A.

Lemma 1.3.11. *Let $C \in \mathcal{C}^d$, let $k \in \{1, \ldots, d-1\}$, and let $\mathsf{A} : \mathbb{R}^d \to \mathbb{R}^k$ be a linear map. The following are equivalent:*

(a) $(\text{relint}\, C) \cap \ker \mathsf{A} \neq \emptyset$,

(b) $\mathsf{A}C$ *is a subspace.*

Proof. Suppose that (a) holds. Then there are a neighborhood B of o relative to $\lim C$ and a vector $t \in \ker \mathsf{A}$ such that $B + t \subset C$. Applying A gives $\mathsf{A}B \subset \mathsf{A}C$. Since $\mathsf{A}C$ is a convex cone, this implies $\text{pos}(\mathsf{A}B) \subseteq \mathsf{A}C$. Therefore,

$$\lim(\mathsf{A}C) = \mathsf{A}\lim C = \mathsf{A}\,\text{pos}\,B = \text{pos}(\mathsf{A}B) \subseteq \mathsf{A}C,$$

which says that $\mathsf{A}C$ is a subspace. Thus (b) holds.

Suppose that (a) does not hold, thus $(\text{relint}\, C) \cap \ker \mathsf{A} = \emptyset$. By the separation theorem [163, Thm. 1.3.8], there exists a hyperplane H that properly separates C and $\ker \mathsf{A}$. Necessarily, $o \in H$ and $\ker \mathsf{A} \subseteq H$. Let u be the outer unit normal vector of H with respect to C, that is, $\langle x, u \rangle \leq 0$ for all $x \in C$. Since the separation is proper, we have $C \subsetneq H$, hence there exists a vector $x_0 \in C$ with $\langle x_0, u \rangle < 0$. Since the image of the transpose $\mathsf{A}^\top$ is the orthogonal complement of $\ker \mathsf{A}$ and thus contains u, there is a vector $v \in \mathbb{R}^k$ such that $\mathsf{A}^\top v = u$ (which implies $v \neq o$). For all $x \in C$ we then have

$$\langle \mathsf{A}x, v \rangle = \langle x, \mathsf{A}^\top v \rangle = \langle x, u \rangle \leq 0$$

and thus $\mathsf{A}C \subseteq \{y \in \mathbb{R}^k : \langle y, v \rangle \leq 0\}$. Since $\langle -\mathsf{A}x_0, v \rangle = -\langle x_0, u \rangle > 0$, we have $\lim \mathsf{A}C \subsetneq \{y \in \mathbb{R}^k : \langle y, v \rangle \leq 0\}$. This means that $\mathsf{A}C \neq \lim \mathsf{A}C$, that is, (b) does not hold. $\qquad\qquad\square$

The next lemma is a preparation for the subsequent one.

Lemma 1.3.12. *Let $\mathsf{A} : \mathbb{R}^d \to \mathbb{R}^k$ with $k < d$ be a surjective linear map. Let H be a hyperplane through o in $\mathbb{R}^d$ with*

$$\ker \mathsf{A} \subseteq H.$$

Then $\mathsf{A}H$ is a hyperplane through o in $\mathbb{R}^k$. If H^0 denotes one of the open halfspaces in $\mathbb{R}^d$ bounded by H, then $\mathsf{A}H^0$ is contained in one of the open halfspaces in $\mathbb{R}^k$ bounded by $\mathsf{A}H$.

Proof. Since A is surjective, we have $\dim \ker \mathsf{A} = d - k$. We choose a basis $(b_1, \ldots, b_d)$ of $\mathbb{R}^d$ such that

$$b_1, \ldots, b_{d-k} \in \ker \mathsf{A}, \quad b_{d-k+1}, \ldots, b_{d-1} \in H \setminus \ker \mathsf{A}, \quad b_d \in \mathbb{R}^n \setminus H.$$

The subspace $\mathsf{A}H$ is spanned by $\mathsf{A}b_{d-k+1}, \ldots, \mathsf{A}b_{d-1}$, hence $\dim \mathsf{A}H \leq k - 1$. Since $\mathsf{A}H$ and $\mathsf{A}b_d$ span $\mathbb{R}^k$, we must have $\dim \mathsf{A}H = k - 1$, thus $\mathsf{A}H$ is a hyperplane in $\mathbb{R}^k$.

Let H^0 be one of the open halfspaces in $\mathbb{R}^d$ bounded by H. Suppose that $\mathsf{A}H^0$ is not contained in one of the open halfspaces in $\mathbb{R}^k$ bounded by $\mathsf{A}H$.

Then there are points $x, y \in AH^0$ and a number $\lambda \in (0,1)$ such that $z := (1-\lambda)x + \lambda y \in AH$, say $z = Ah$ with $h \in H$. There are points $x', y' \in H^0$ with $Ax' = x$, $Ay' = y$, and the point $z' := (1-\lambda)x' + \lambda y'$ satisfies $Az' = z = Ah$. Thus, $z' - h \in \ker A \subseteq H$ and hence $z' \in H$. But $x', y' \in H^0$ implies $z' \in H^0$, a contradiction. $\qquad\square$

Lemma 1.3.13. *Let $C \in \mathcal{C}^d$, and let $A : \mathbb{R}^d \to \mathbb{R}^k$ with $k < d$ be a surjective linear map. If*

$$(\ker A) \cap C = \{o\},$$

then

$$\mathrm{lineal}(AC) = A\,\mathrm{lineal}(C) \tag{1.11}$$

and

$$\dim \mathrm{lineal}(AC) = \dim \mathrm{lineal}(C). \tag{1.12}$$

Proof. The image $A\,\mathrm{lineal}(C)$ is a subspace contained in AC, hence it is contained in $\mathrm{lineal}(AC)$. Suppose that

$$A\,\mathrm{lineal}(C) \subsetneqq \mathrm{lineal}(AC). \tag{1.13}$$

Since $(\ker A) \cap C = \{o\}$, by Lemma 1.3.4 there exists a supporting hyperplane H of C with $\ker A \subseteq H$ and $C \cap H = \mathrm{lineal}(C)$. By (1.13), there exists $y \in \mathrm{lineal}(AC)$ with $y \notin A\,\mathrm{lineal}(C)$. Then $y = Ax$ with $x \in C$. By Lemma 1.3.12, AH is a hyperplane in $\mathbb{R}^k$, and and if H^0 is one of the open halfspaces in $\mathbb{R}^d$ bounded by H, then AH^0 is contained in one of the open halfspaces bounded by AH. It follows that AH supports AC. Since $Ax \in \mathrm{lineal}(AC) \subseteq AH$, we conclude that $x \in H$ (since $x \in \pm H^0$ would imply that $Ax \notin AH$). Thus $x \in C \cap H = \mathrm{lineal}(C)$, but $Ax \notin A\,\mathrm{lineal}(C)$, a contradiction. This shows that (1.11) holds.

Let $\dim \mathrm{lineal}(C) = m$. If $m = 0$, then (1.11) shows that $\dim \mathrm{lineal}(AC) = 0$. Hence, we can assume that $m \geq 1$. Let $x_1, \ldots, x_m \in \mathrm{lineal}(C)$ be linearly independent. Suppose that $Ax_1, \ldots, Ax_m$ were linearly dependent. Then $\sum_{i=1}^m \lambda_i Ax_i = o$ with coefficients $\lambda_1, \ldots, \lambda_m$ not all zero. It follows that $\sum_{i=1}^m \lambda_i x_i \in (\ker A) \cap \mathrm{lineal}(C) = \{o\}$, a contradiction. This shows that $\dim [A\,\mathrm{lineal(C)}] \geq m = \dim \mathrm{lineal}(C)$. Since $\dim [A\,\mathrm{lineal(C)}] \leq \dim \mathrm{lineal}(C)$ is trivial, we get $\dim [A\,\mathrm{lineal(C)}] = \dim \mathrm{lineal}(C)$. Together with (1.11), this yields the assertion (1.12). $\qquad\square$

We shall need a metric on the set of closed convex cones. We define, in fact, two closely related metrics.

First, the metric δ_c is defined by

$$\delta_c(C, D) = d_H(C \cap B^d, D \cap B^d) \tag{1.14}$$

for $C, D \in \mathcal{C}^d$, where d_H is the usual Hausdorff metric on convex bodies of $\mathbb{R}^d$. Topological notions used in the following for the space $\mathcal{C}^d$ always refer to the topology induced by this metric. Observe that the topological space $\mathcal{C}^d$ has $d + 2$ connected components, namely $\mathcal{C}^d \setminus \mathcal{L}_\bullet$ and the Grassmannians $G(d, 0), \ldots, G(d, d)$.

Lemma 1.3.14. *Let $C, D \in \mathcal{C}^d$. Let $C_i, D_i \in \mathcal{C}^d$ ($i \in \mathbb{N}$) be cones such that $C_i \to C$ and $D_i \to D$ as $i \to \infty$. If $C \cap D = \{o\}$, then $C_i \cap D_i = \{o\}$ for all sufficiently large i. If $C \cap D \neq \{o\}$ and C, D do not touch, then $C_i \cap D_i \to C \cap D$ for $i \to \infty$.*

Proof. If $C \cap D = \{o\}$, then $C \cap D \cap \mathbb{S}^{d-1} = \emptyset$. It follows easily from the definition of the Hausdorff metric d_H that $C_i \cap D_i \cap \mathbb{S}^{d-1} = \emptyset$ for all sufficiently large i. For these i we have $C_i \cap D_i = \{o\}$.

If $C \cap D \neq \{o\}$, then C and D (which do not touch) cannot be separated by a hyperplane, hence $C \cap B^d$ and $D \cap B^d$ cannot be separated by a hyperplane (a separating hyperplane would have to contain o and hence would also separate C and D). It follows from [163, Thm. 1.8.10] that $C_i \cap D_i \cap B^d \to C \cap D \cap B^d$ with respect to the Hausdorff metric d_H and hence $C_i \cap D_i \to C \cap D$ with respect to the metric δ_c. $\qquad\square$

For a second metric, we denote by $d_a(x, y)$ the angle between two vectors $x, y \in \mathbb{R}^d \setminus \{o\}$, thus

$$d_a(x, y) = \arccos \left\langle \frac{x}{\|x\|}, \frac{y}{\|y\|} \right\rangle \quad \text{if } x, y \neq o. \tag{1.15}$$

This defines the usual geodesic metric on $\mathbb{S}^{d-1}$. We note that

$$\|x - y\| = 2 \sin \frac{1}{2} d_a(x, y) \quad \text{for } x, y \in \mathbb{S}^{d-1} \tag{1.16}$$

or, equivalently,

$$\|x - y\|^2 = 2(1 - \cos d_a(x, y)) \quad \text{for } x, y \in \mathbb{S}^{d-1}. \tag{1.17}$$

We supplement definition (1.15) by $d_a(o, o) = 0$ and $d_a(x, o) = d_a(o, x) = \pi/2$ if $x \neq o$.

Let $C \in \mathcal{C}^d_*$. The *angular distance* $d_a(x, C)$ of a point $x \in \mathbb{R}^d$ from C is defined by

$$d_a(x, C) := \arccos \frac{\|\Pi_C(x)\|}{\|x\|} \quad \text{if } x \neq o, \tag{1.18}$$

and $d_a(o, C) := \pi/2$. Thus,

$$d_a(x, C) = \begin{cases} \min\{d_a(x, y) : y \in C \setminus \{o\}\} & \text{if } x \notin C^\circ, \\ \pi/2 & \text{if } x \in C^\circ. \end{cases} \tag{1.19}$$

The connection with the Euclidean distance of x from C is given by

$$\operatorname{dist}(x, C) = \|x\| \sin d_a(x, C). \tag{1.20}$$

For $C \in \mathcal{C}_*^d$ we have

$$C^\circ = \{x \in \mathbb{R}^d : d_a(x, C) = \pi/2\}.$$

Corollary 1.3.1 implies that

$$d_a(x, C) + d_a(x, C^\circ) = \pi/2 \quad \text{for } x \in \mathbb{R}^d \setminus (C \cup C^\circ). \tag{1.21}$$

If $C, D \in \mathcal{C}^d$ are cones with $C \subset D$, then $C^\circ \supset D^\circ$, hence, for $x \in \mathbb{R}^d$,

$$\|\Pi_C(x)\| = \operatorname{dist}(x, C^\circ) \leq \operatorname{dist}(x, D^\circ) = \|\Pi_D(x)\|. \tag{1.22}$$

For $C \in \mathcal{C}_*^d$ and $\varepsilon \geq 0$, we define the *angular parallel set* of C at distance ε by

$$C_\varepsilon := \{x \in \mathbb{R}^d : d_a(x, C) \leq \varepsilon\}.$$

The *angular Hausdorff distance* of $C, D \in \mathcal{C}_*^d$ is then defined by

$$\delta_a(C, D) := \min\{\varepsilon \geq 0 : C \subseteq D_\varepsilon, \ D \subseteq C_\varepsilon\}.$$

Similarly as for the usual Hausdorff metric, one shows that this defines a metric on $\mathcal{C}_*^d$. From (1.17) one can deduce that δ_c and δ_a induce the same topology on $\mathcal{C}_*^d$.

The metric δ_a has some nice properties; for example, with respect to this metric, polarity is a local isometry. The following theorem and its proof are taken from Glasauer [63, Hilfssatz 2.2].

Theorem 1.3.4. *If $C, D \in \mathcal{C}_*^d$ are cones $\neq \mathbb{R}^d$ with $\delta_a(C, D) < \pi/2$, then*

$$\delta_a(C^\circ, D^\circ) = \delta_a(C, D).$$

Proof. Let $C, D \in \mathcal{C}_*^d$ be cones $\neq \mathbb{R}^d$ with $\alpha := \delta_a(C, D) < \pi/2$. Let $x \in \mathbb{R}^d \setminus \{o\}$ be a point with $d_a(x, C) < \pi/2 - \alpha$. There is a point $y \in C \setminus \{o\}$ with $d_a(x, y) < \pi/2 - \alpha$, and since $C \subseteq D_\alpha$, we have $y \in D_\alpha$. Further, there is a point $z \in D \setminus \{o\}$ with $d_a(z, y) \leq \alpha$. It follows that $d_a(x, D) < \pi/2$. We have shown that

$$\{x \in \mathbb{R}^d \setminus \{o\} : d_a(x, D) \geq \pi/2\} \subseteq \{x \in \mathbb{R}^d \setminus \{o\} : d_a(x, C) \geq \pi/2 - \alpha\}.$$

According to Corollary 1.3.1, the left side of the previous inclusion is equal to $D^\circ \setminus \{o\}$ and the right side is equal to $(C^\circ)_\alpha \setminus \{o\}$; thus $D^\circ \subseteq (C^\circ)_\alpha$. Similarly, $C^\circ \subseteq (D^\circ)_\alpha$. Therefore, $\delta_a(C^\circ, D^\circ) \leq \alpha = \delta_a(C, D) < \pi/2$. Applying the same reasoning to C°, D°, we get $\delta_a(C, D) \leq \delta_a(C^\circ, D^\circ)$ and hence the stated equality. $\qquad\square$

The following two lemmas (which are taken from [168]) estimate the distance between the projections $\Pi_C(x)$ and $\Pi_D(x)$, either Euclidean or angular, if C and D have small angular Hausdorff distance.

Lemma 1.3.15. *For $C, D \in \mathcal{C}_*^d$ and $x \in \mathbb{R}^d$, we have*

$$\|\Pi_C(x) - \Pi_D(x)\| \leq \|x\|\sqrt{10\delta_a(C, D)}.$$

Proof. By $B(o, r)$ we denote the ball with center o and radius r. Let $K :=$ $C \cap B(o, \|x\|)$ and $L := D \cap B(o, \|x\|)$. From $\Pi_C(x) \in B(o, \|x\|)$ it follows that $\Pi_C(x) = \Pi_K(x)$; similarly $\Pi_D(x) = \Pi_L(x)$. Elementary geometry together with a rough estimate shows that the Euclidean Hausdorff distance of K and L satisfies $d_H(K, L) \leq \delta_a(C, D)\|x\|$. By [163, Lem. 1.8.11], we obtain

$$\|\Pi_C(x) - \Pi_D(x)\| = \|\Pi_K(x) - \Pi_L(x)\|$$
$$\leq \sqrt{10\|x\|d_H(K, L)} \leq \|x\|\sqrt{10\delta_a(C, D)},$$

which was the assertion. □

Lemma 1.3.16. *Let $C, D \in \mathcal{C}_*^d$ and $x \in \mathbb{R}^d$. Suppose there exists a number $\varepsilon > 0$ with $d_a(x, C) \leq \pi/2 - \varepsilon$, $d_a(x, D) \leq \pi/2 - \varepsilon$, and $\delta_a(C, D) \leq \pi/2 - \varepsilon$. Then*

$$\cos d_a(\Pi_C(x), \Pi_D(x)) \geq 1 - c_\varepsilon \delta_a(C, D),$$

where $c_\varepsilon := 2(\pi^{-1} + \tan(\pi/2 - \varepsilon))$.

Proof. We set $\delta_a(C, D) = \delta_a$. If $x \in C \cup D$, say $x \in C$, then

$$d_a(\Pi_C(x), \Pi_D(x)) = d_a(x, \Pi_D(x)) \leq \delta_a,$$

hence the assertion follows from

$$\cos d_a(\Pi_C(x), \Pi_D(x)) \geq \cos \delta_a \geq 1 - \frac{2}{\pi}\delta_a \geq 1 - c_\varepsilon \delta_a.$$

Therefore, we assume now that $x \notin C \cup D$.

Since $d_a(x, C) < \pi/2$, we have $\Pi_C(x) \neq o$, and similarly $\Pi_D(x) \neq o$. We set $p := \Pi_C(x)/\|\Pi_C(x)\|$ and $q := \Pi_D(x)/\|\Pi_D(x)\|$. From $x \notin C$ it follows that $\Pi_{C^\circ}(x) \neq o$. Let $e := \Pi_{C^\circ}(x)/\|\Pi_{C^\circ}(x)\|$.

Without loss of generality, there is a two-dimensional unit sphere $\mathbb{S}^2$ that contains the three unit vectors p, q, e ($\mathbb{S}^2 \subseteq \mathbb{S}^{d-1}$ if $d \geq 3$; and if $d = 2$, we embed $\mathbb{R}^2$ in $\mathbb{R}^3$). Put $d_a(x, C) = d_a$ and $x_0 = x/\|x\|$. For $y \in \mathbb{S}^2$ and $\rho > 0$, let $S(y, \rho)$ denote the closed spherical cap in $\mathbb{S}^2$ with center y and (spherical $=$ angular) radius $\rho > 0$. Then

$$q \in S(x_0, d_a + \delta_a).$$

This holds because $d_a(x, \Pi_C(x)) = d_a$ and there is a point $z \in D$ with $d_a(z, \Pi_C(x)) \leq \delta_a$.

From $q \in D$ it follows that $d_a(q, C) \leq \delta_a$. By (1.21) we have $d_a(q, C) + d_a(q, C^\circ) = \pi/2$. Since $e \in C^\circ$, we get

$$d_a(q,e) \geq d_a(q,C^\circ) = \frac{\pi}{2} - d_a(q,C) \geq \frac{\pi}{2} - \delta_a,$$

which means that

$$q \notin \operatorname{int} S(e, \pi/2 - \delta_a)$$

(where int refers to $\mathbb{S}^2$).

From this, we see that an upper bound for the angular distance between p and q is attained if q is an intersection point of the boundaries of $S(x_0, d_a + \delta_a)$ and $S(e, \pi/2 - \delta_a)$, thus if

$$q = w \cos \delta_a + e \sin \delta_a \quad \text{with } w \in e^\perp \cap \mathbb{S}^2$$

and

$$\langle x_0, q \rangle = \cos(d_a + \delta_a).$$

For a point q satisfying these equations, we obtain

$$\cos(d_a + \delta_a) = \langle x_0, q \rangle = \langle p \cos d_a + e \sin d_a, w \cos \delta_a + e \sin \delta_a \rangle$$
$$= \langle p, w \rangle \cos d_a \cos \delta_a + \sin d_a \sin \delta_a$$

and

$$\cos d_a(p,q) = \langle p, q \rangle = \langle p, w \rangle \cos \delta_a$$
$$= \frac{1}{\cos d_a}[\cos(d_a + \delta_a) - \sin d_a \sin \delta_a]$$
$$= \cos \delta_a - 2(\tan d_a) \sin \delta_a$$
$$\geq 1 - 2\pi^{-1}\delta_a - 2(\tan d_a)\delta_a.$$

This yields the assertion. □

Notes for Section 1.3

1. The image of a closed convex cone under a projection is in general not closed. Consider, for example, a spherical cone and its projection along a boundary ray. The following was proved by Mirkil [137]. If a closed convex cone C has all its 2-dimensional projections closed, then C is polyhedral. For a more general version, see Klee [117, Thm. 4.11]. A very general investigation of the closedness of linear images of closed convex cones was undertaken by Pataki [144]. Stability of closedness of convex cones under linear mappings was investigated by Borwein and Moors [33, 34].

2. Let $C \in \mathcal{C}^d$ be a pointed cone of dimension d. It induces a partial order $\leq_C$ on $\mathbb{R}^d$ by means of $x \leq_C y \Leftrightarrow y - x \in C$ for $x, y \in \mathbb{R}^d$. This partial order can be related to properties of the nearest-point map Π_C. Let us say that Π_C is *C-isotone* if $x \leq_C y$ implies $\Pi_C(x) \leq_C \Pi_C(y)$ for all $x, y \in \mathbb{R}^d$, and that Π_C is *C-subadditive* if $\Pi_C(x + y) \leq_C \Pi_C(x) + \Pi_C(y)$ for $x, y \in \mathbb{R}^d$.

Theorem. *The following are equivalent:*

(1) Π_C *is C-isotone.*

(2) Π_{C° *is C°-subadditive.*

(3) $C^\circ = \text{pos}\{x_1, \ldots, x_d\}$ *with linearly independent vectors $x_1, \ldots, x_d$ satisfying $\langle x_i, x_j \rangle \leq 0$ for $i \neq j$.*

The equivalence of (1) and (2) was proved by Isac and Németh [101], and the equivalence of (2) and (3) by Németh and Németh [142]. The subject of characteristic properties of cones C for which Π_C is C-isotone was taken up in Isac and Németh [102].

3. The partial order $\leq_C$ appearing in the previous note is also the subject of the following theorem, which was proved by Artstein–Avidan and Slomka [13].

Theorem. *Let $d > 2$, and let $C \in \mathcal{C}^d$ be a pointed cone which has at least $d+1$ extremal vectors in general position. Let $T : C \to C$ be an order isomorphism, that is, a bijective map satisfying*

$$x \leq_C y \Leftrightarrow T(x) \leq_C T(y)$$

for all $x, y \in C$. Then there is a linear transformation $g \in \text{GL}(d)$ such that

$$T(x) = gx \text{ for } x \in C \quad \text{and} \quad gC = C.$$

Artstein–Avidan and Slomka have more general results and some applications.

4. The following theorem was proved by Sandgren [151].

Theorem. *Let $\{C_i : i \in I\}$ be a set of closed convex cones in $\mathbb{R}^d$ with the property that the convex hull of any d of the cones C_i is not the whole space $\mathbb{R}^d$. Then there is a hyperplane H through o such that each cone C_i is contained in one of the closed halfspaces bounded by H.*

5. Different metrics on $\mathcal{C}^d$ (or, more generally, on the set of nontrivial closed convex cones in a real normed vector space) were studied by Iusem and Seeger [105]. We mention some further investigations by Alberto Seeger and co-authors on convex cones, which, though interesting, will not be considered further in these notes. The papers [103, 104] of Iusem and Seeger deal with measuring the degree of pointedness of a closed convex cone. Gourion and Seeger [74] investigated solidity indices for convex cones. By definition, such an index is zero for a subspace and one for a halfspace, it is non-decreasing under set inclusion and invariant under orthogonal transformations. Henrion and Seegers [89] have investigated different notions of centers for convex cones. (See also [90, 91], where inradius and circumradius are defined, considered, and applied.) The angle between two of these centers, incenter and circumcenter, are used to define the *eccentricity* of a cone, which is treated in Seeger and Torki [173]. So-called volumetric centers are also studied in Seeger and

Torki [174, 175]. In analogy to measuring the degree of central symmetry of a convex body, Seeger and Torki [178] have investigated how to measure the axial symmetry of a convex cone. This work is continued and extended in [179]. In the papers [176, 177], Seeger and Torki treated conical analogues of the Loewner–John ellipsoid theorems and corresponding volume ratio estimates. Angles in cones are the subject of [73] and the work quoted there.

1.4 Polyhedra

This section is devoted to polyhedra. Later, polyhedral cones will play an essential role. On the other hand, various polyhedral cones are important to describe the local structure of a polyhedron.

Different authors may use the notion of 'polyhedron' with several different meanings (and also the same author at different occasions). Therefore, we emphasize that here by a *polyhedron* in $\mathbb{R}^d$ we understand the nonempty intersection of a finite family of closed halfspaces. (The family may be empty, so that by convention also $\mathbb{R}^d$ is considered as a polyhedron.) Thus, a polyhedron is always nonempty, closed and convex, but may be unbounded. A polyhedron that does not contain a line is called *line-free*. A bounded polyhedron is called a *polytope*. An *ro-polyhedron* is the relative interior of a polyhedron, and an *ro-polytope* is the relative interior of a polytope. We denote by $\mathcal{Q}^d, \mathcal{P}^d, \mathcal{Q}^d_{ro}, \mathcal{P}^d_{ro}$, respectively, the set of polyhedra, polytopes, ro-polyhedra, and ro-polytopes in $\mathbb{R}^d$. When it is necessary to replace $\mathbb{R}^d$ by some other affine space L, we write $\mathcal{P}(L)$ for the set of polytopes in L, and similarly for the other families of sets just considered.

We recall here the well-known fact that every polyhedron can be represented as the convex hull of finitely many points and rays, and that every such (nonempty) convex hull is a polyhedron. (See, e.g., the thorough investigation of polyhedra by Klee [117].)

The vector sum of two polyhedra is a polyhedron ([148, Cor. 19.3.2]). In particular, it is closed.

The intersection of a polyhedron P with a supporting hyperplane is again a polyhedron; it is a *support set* or a *face* of P. The polyhedron P is, by definition, also a face of itself. A polyhedron P has finitely many faces, of dimensions $0, \ldots, \dim P$. A face of P of dimension $\dim P - 1$ is called a *facet* of P. The empty set is here not considered as a face of a polyhedron. If P is a polyhedron, then the notation $F \leq P$ indicates that F is a face of P. In particular, a sum

$$\sum_{E \leq F \leq G} f(F),$$

where E and G are faces of a polyhedron P, will extend over all faces F of G (and thus of P) that contain E.

We denote by $\mathcal{F}_k(P)$ the set of k-dimensional faces of P, for $k = 0, \ldots, \dim P$, and we write

$$\mathcal{F}(P) = \bigcup_{k=0}^{\dim P} \mathcal{F}_k(P)$$

for the set of faces of P. The number of k-dimensional faces of P is denoted by $f_k(P)$. We use $\mathcal{F}_{bd}(P)$ for the set of bounded faces of a polyhedron P.

We emphasize again that $\mathcal{F}(P)$ is the set of nonempty faces of P. In the rare cases where the empty set has to be taken into consideration as a face, we mention this explicitly.

If F is a polyhedron, we denote by

$$\langle F \rangle = \lin (F - F)$$

the subspace parallel to the affine hull of F. We call $\langle F \rangle$ the *direction space* of F. Of course, if F is a cone, then $o \in F$ and hence $\langle F \rangle = \lin F$.

An important role is played in the following by polyhedral cones. A *polyhedral cone* is a cone $C \in \mathcal{C}^d$ that is a polyhedron. We denote the set of polyhedral cones in $\mathcal{C}^d$ by $\mathcal{PC}^d$, and $\mathcal{PC}_*^d$ is the subset of polyhedral cones different from $\{o\}$.

Each of the families $\mathscr{S} = \mathcal{Q}^d, \mathcal{P}^d, \mathcal{Q}_{ro}^d, \mathcal{P}_{ro}^d, \mathcal{C}^d, \mathcal{PC}^d$ is intersectional, that is, if $A, B \in \mathscr{S}$ and $A \cap B \neq \emptyset$, then $A \cap B \in \mathscr{S}$. We denote by $\mathsf{U}(\mathscr{S})$ the family of all finite unions of elements from $\mathscr{S}$, including $\emptyset$. Thus, in each case, $(\mathsf{U}(\mathscr{S}), \cup, \cap)$ is a lattice. The elements of $\mathsf{U}(\mathcal{Q}^d)$ will be called *general polyhedra*, those of $\mathsf{U}(\mathcal{Q}_{ro}^d)$ *general ro-polyhedra*, and those of $\mathsf{U}(\mathcal{P}_{ro}^d)$ *general ro-polytopes*. Thus, a general ro-polyhedron is a set that can be obtained from closed halfspaces by finitely many operations of intersection, union, and complementation.

It is an elementary fact that each polyhedron $P \in \mathcal{Q}^d$ is the disjoint union of the relative interiors of its faces. We write this as

$$P = \biguplus_{F \in \mathcal{F}(P)} \relint F,$$

and also as

$$\sum_{F \in \mathcal{F}(P)} \mathbb{1}[\relint F] = \mathbb{1}[P] \tag{1.23}$$

(recall that $\mathbb{1}[A] = \mathbb{1}_A$ denotes the characteristic function of A). This shows that $\mathcal{Q}^d \subset \mathsf{U}(\mathcal{Q}_{ro}^d)$, and similarly $\mathcal{P}^d \subset \mathsf{U}(\mathcal{P}_{ro}^d)$. It is also true that every general ro-polyhedron Q can be represented as the disjoint union of finitely many ro-polyhedra. To prove this, let $Q \in \mathsf{U}(\mathcal{Q}_{ro}^d)$; then Q is the union of finitely many ro-polyhedra $Q_1, \ldots, Q_r$. There are finitely many closed halfspaces $H_1^+, \ldots, H_s^+$ such that the closure of any Q_i is the intersection of some of these halfspaces, $i = 1, \ldots, r$. For $x, y \in \mathbb{R}^d$, we define $x \sim y$ if and only if for each $j \in \{1, \ldots, s\}$ we have $x \in H_j^+ \Leftrightarrow y \in H_j^+$ and $x \in \bd H_j^+ \Leftrightarrow y \in \bd H_j^+$. Then $\sim$ is an equivalence relation, each equivalence class is an ro-polyhedron, the finitely many equivalence classes are pairwise disjoint, and each point of Q belongs to some equivalence class. This proves the assertion.

As a polyhedron P is the disjoint union of the relative interiors of its faces, it will be convenient to define

$$\operatorname{skel}_k P := \bigcup_{F \in \mathcal{F}_k(P)} \operatorname{relint} F \qquad (1.24)$$

for $k = 0, \ldots, d$ (with $\operatorname{skel}_k P = \emptyset$ for $k > \dim P$). We call this set the k-*skeleton* of P, pointing out that (in contrast to other literature) it is not closed for $0 < k \leq \dim P$. This is irrelevant when the k-volume of the k-skeleton will be considered, but has the advantage that a polyhedron is the disjoint union of its skeletons of different dimensions, and it is important when the pre-image of the k-skeleton under the nearest-point map is considered.

Associated with a polyhedron are several polyhedral cones.

Definition 1.4.1. *Let $P \in \mathcal{Q}^d$ be a polyhedron. For $x \in P$, the* normal cone *of P at x is given by*

$$N(P, x) := \{u \in \mathbb{R}^d : \langle u, y - x \rangle \leq 0 \text{ for all } y \in P\}.$$

For a face F of P, the normal cone *of P at F is defined by*

$$N(P, F) := N(P, z)$$

for any point $z \in \operatorname{relint} F$.

This does not depend on the choice of the relatively interior point z. Note that $N(P, P) = \langle P \rangle^{\perp}$.

The normal cone of a polyhedron P at a face $F \neq P$ can be represented as follows. Let

$$P = \bigcap_{i=1}^{m} H_i^{-} \qquad (1.25)$$

with closed halfspaces $H_1^{-}, \ldots, H_m^{-}$. Without loss of generality, let $u_1, \ldots, u_k$ be the outer unit normal vectors of the halfspaces H_i^{-} that contain F in their boundary. Then

$$N(P, F) = \operatorname{pos}\{u_1, \ldots, u_k\}.$$

Clearly, this is a polyhedral cone, with lineality space $\langle P \rangle^{\perp}$. In particular,

$$\dim N(P, F) = d - \dim F.$$

Occasionally, it will be more appropriate to consider the d-dimensional cone

$$N(P, F) + \langle F \rangle,$$

which we call the *extended normal cone* of P at F.

The fact mentioned above, that every polyhedron is the disjoint union of the relative interiors of its faces, has several consequences. To introduce the first, let $P \in \mathcal{Q}^d$ be a polyhedron and recall that $\Pi_P(\cdot)$ denotes the metric projection to P, that is, $\Pi_P(x)$ is the unique point in P that is nearest to a given $x \in \mathbb{R}^d$. Then the following holds for each $x \in \mathbb{R}^d$.

(a) There is a unique face $F \in \mathcal{F}(P)$ with $\Pi_P(x) \in \operatorname{relint} F$, and we have $x - \Pi_P(x) \in N(P, F)$.

(b) There is a unique face $G \in \mathcal{F}(P)$ with $x - \Pi_P(x) \in \operatorname{relint} N(P, G)$, and we have $\Pi_P(x) \in G$.

Assertion (a) follows from the fact that P is the disjoint union of the relative interiors of its faces. Similarly, (b) follows from the fact that the normal fan (the polyhedral complex formed by the normal cones) of P is the disjoint union of the relative interiors of the normal cones of P and that $x \in P$ implies $x - \Pi_P(x) = o \in \operatorname{relint} N(P, P)$. Immediately from (a) we obtain that for each $x \in \mathbb{R}^d$ there is a unique face $F \in \mathcal{F}(P)$ such that $x \in (\operatorname{relint} F) + N(P, F)$, thus

$$\sum_{F \in \mathcal{F}(P)} \mathbb{1}[(\operatorname{relint} F) + N(P, F)] = 1. \tag{1.26}$$

Similarly, from (b) we get that

$$\sum_{F \in \mathcal{F}(P)} \mathbb{1}[F + \operatorname{relint} N(P, F)] = 1. \tag{1.27}$$

Less precisely (not caring about boundary points), this says that the polyhedra $F + N(P, F)$, $F \in \mathcal{F}(P)$, form a tessellation of $\mathbb{R}^d$.

Besides normal cones of a polyhedron at its faces, we consider angle cones.

Definition 1.4.2. *Let $P \in \mathcal{Q}^d$ be a polyhedron. For a face F of P, the* angle cone *of P at F is defined by*

$$A(F, P) = \operatorname{cone}(P, z) = \operatorname{pos}(P - z),$$

for any $z \in \operatorname{relint} F$.

In other words, the angle cone $A(F, P)$ is the cone of feasible directions (see Definition 1.3.1), $\operatorname{cone}(P, z)$, for $z \in \operatorname{relint} F$. It is clear that $A(F, P)$ is independent of the choice of the point $z \in \operatorname{relint} F$, and that $A(F, P)$ is a polyhedral cone, of the same dimension as P and with lineality space $\langle F \rangle$. It can be represented as follows. Let P be given by (1.25), and without loss of generality, let $H_1^-, \ldots, H_k^-$ be precisely those halfspaces that contain the face F in their boundary. Choose $z \in \operatorname{relint} F$. Then

$$A(F, P) = \bigcap_{i=1}^{k}(H_i^- - z).$$

It follows that

$$N(P, F) = A(F, P)^{\circ}. \tag{1.28}$$

Again, it will occasionally be appropriate to replace the angle cone by a d-dimensional cone, namely

$$A(F, P) + \langle P \rangle^{\perp},$$

which we call the *extended angle cone* of P at F. Note that

$$[N(P, F) + \langle F \rangle]^{\circ} = A(F, P) \cap \langle F \rangle^{\perp},$$
$$[A(F, P) + \langle P \rangle^{\perp}]^{\circ} = N(P, F) \cap \langle P \rangle,$$

by Theorem 1.3.2.

We turn to polyhedral cones. For the nearest-point map of a polyhedral cone, the following is a consequence of Theorem 1.3.3.

Lemma 1.4.1. *Let $C \in \mathcal{PC}^d$. If $F \in \mathcal{F}(C)$ and $x \in \mathbb{R}^d$, then*

$$\Pi_C(x) \in F \Leftrightarrow \Pi_{C^{\circ}}(x) \in N(C, F).$$

For polyhedral cones, the duality induces a correspondence between faces. Let $C \in \mathcal{PC}^d$ be a polyhedral cone, and let $F \in \mathcal{F}_k(C)$, $k \in \{0, \dots, \dim C\}$. For $u \in N(C, F)$, we have $C \subseteq H^-(u, 0)$, thus $N(C, F) \subseteq C^{\circ}$. For $x \in \langle F \rangle$ $(= \lim F)$, we have $\langle x, u \rangle = 0$ for all $u \in N(C, F)$, thus $N(C, F) \subseteq \langle F \rangle^{\perp}$. Therefore,

$$N(C, F) \subseteq C^{\circ} \cap \langle F \rangle^{\perp}.$$

Conversely, let $u \in C^{\circ} \cap \langle F \rangle^{\perp}$. Then $C \subseteq H^-(u, 0)$ and $F \subseteq H(u, 0)$, thus $u \in N(C, F)$. We conclude that

$$N(C, F) = C^{\circ} \cap \langle F \rangle^{\perp}.$$

This is a face of C°, since it is contained in all its supporting hyperplanes $H^-(z, 0)$, $z \in \operatorname{relint} F$, which intersect in $\langle F \rangle^{\perp}$.

We also note, using Theorem 1.3.2, that

$$N(C, F)^{\circ} = (C^{\circ} \cap \langle F \rangle^{\perp})^{\circ} = C + \langle F \rangle = C + (F - F),$$

thus (since $C + F = C$)

$$N(C, F)^{\circ} = C - F. \tag{1.29}$$

Since

$$\dim N(C, F) = d - \dim F,$$

we have $N(C, F) \in \mathcal{F}_{d-k}(C^{\circ})$, if $F \in \mathcal{F}_k(C)$.

Applying the preceding also to C° and $N(C, F)$, we conclude the following.

Lemma 1.4.2. *If $C \in \mathcal{PC}^d$ and $F \in \mathcal{F}_k(C)$ with $k \in \{0, \dots, d\}$, then $N(C, F) \in \mathcal{F}_{d-k}(C^{\circ})$ and*

$$N(C^{\circ}, N(C, F)) = F. \tag{1.30}$$

We also write

$$N(C, F) = \widehat{F}_C \tag{1.31}$$

and call this the face of $C°$ *conjugate* to F. Thus,

$$\widehat{F}_C = \{x \in C° : \langle x, y \rangle = 0 \text{ for all } y \in F\},$$

$$\dim F + \dim \widehat{F}_C = d,$$

and

$$\widehat{F}_C = G \iff \widehat{G}_{C°} = F. \tag{1.32}$$

Recall that (1.28) for a polyhedral cone C and a face F of C says that

$$N(C, F) = A(F, C)°. \tag{1.33}$$

Applying this to $C°$ and its face $N(C, F)$ and then using (1.30), we get

$$A(N(C, F), C°)° = N(C°, N(C, F)) = F \tag{1.34}$$

and hence

$$A(\widehat{F}_C, C°) = F°. \tag{1.35}$$

Lemma 1.4.3. *Let $F \subseteq G$ be faces of a polyhedral cone $C \in \mathcal{PC}^d$. Then*

$$N(G, F) = A(\widehat{G}_C, \widehat{F}_C)$$

and

$$A(F, G) = N(\widehat{F}_C, \widehat{G}_C).$$

Proof. First we assume that $G = C$. Then (1.33) and (1.35) (applied to $C°$ and $\widehat{F}_C$) give

$$N(C, F) = A(F, C)° = \widehat{F}_C = A(o, \widehat{F}_C) = A(\widehat{C}_C, \widehat{F}_C).$$

Now let G be a face of C different from C. We apply the preceding in $\lin G$, denoting normal cones and angle cones in $\lin G$ by N' and A', respectively. Then we get

$$N(G, F) = N'(G, F) + (\lin G)^{\perp} = A'(o, \widehat{F}_G) + (\lin G)^{\perp} = A(\widehat{G}_C, \widehat{F}_C),$$

as asserted. The second relation of the lemma follows by applying the first one to the polar cone.   □

Later, we consider intersections of polyhedral cones. We provide now some preparations.

Definition 1.4.3. *Let $C, D \in \mathcal{C}^d$. These cones intersect transversely, written $C \pitchfork D$, if*

$$\dim(C \cap D) = \dim C + \dim D - d \quad \text{and} \quad \relint C \cap \relint D \neq \emptyset.$$

Lemma 1.4.4. *Let $C, D \in \mathcal{PC}^d$ be polyhedral cones. Every face of $C \cap D$ is of the form $F \cap G$ with $F \in \mathcal{F}(C)$ and $G \in \mathcal{F}(D)$. The normal cones satisfy*

$$N(C \cap D, F \cap G) \supseteq N(C, F) + N(D, G). \tag{1.36}$$

If relint $F \cap$ relint $G \neq \emptyset$, *then equality holds in* (1.36). *If $F \pitchfork G$, then*

$$N(C \cap D, F \cap G) = N(C, F) \oplus N(D, G)$$

is a direct sum.

Proof. (We follow Amelunxen and Lotz [10, Prop. 2.5].) Let $E \in \mathcal{F}(C \cap D)$. Then $E = \{x \in \mathbb{R}^d : \langle x, u \rangle = 0\}$ for some $u \in (C \cap D)^\circ = C^\circ + D^\circ$, the latter by Theorem 1.3.2 (the sum of polyhedral cones is closed). Thus, $u = u_C + u_D$ with $u_C \in C^\circ$, $u_D \in D^\circ$. The faces $F := \{x \in C : \langle x, u_C \rangle = 0\} \in \mathcal{F}(C)$ and $G := \{x \in D : \langle x, u_D \rangle = 0\} \in \mathcal{F}(D)$ satisfy

$$F \cap G = \{x \in C \cap D : \langle x, u_C \rangle = \langle x, u_D \rangle = 0\}$$
$$= \{x \in C \cap D : \langle x, u \rangle = 0\} = E,$$

where it was used that $x \in C \cap D$ implies $\langle x, u_C \rangle \leq 0$ and $\langle x, u_D \rangle \leq 0$.

Further, with (1.29) we obtain

$$N(C \cap D, F \cap G)^\circ$$
$$= (C \cap D) - (F \cap G) \subseteq (C - F) \cap (D - G) \tag{1.37}$$
$$= N(C, F)^\circ \cap N(D, G)^\circ = (N(C, F) + N(D, G))^\circ,$$

and duality gives (1.36).

If $x \in$ relint F and $y \in C - F$, then it is easy to check that $y + \lambda x \in C$ for all sufficiently large $\lambda > 0$. Hence, if we can choose $x \in$ relint $F \cap$ relint G, then for each $y \in (C - F) \cap (D - G)$ we have $y + \lambda x \in C \cap D$ for sufficiently large $\lambda > 0$, hence $y = (y + \lambda x) - \lambda x \in (C \cap D) - (F \cap G)$. This shows that under the assumption relint $F \cap$ relint $G \neq \emptyset$, the inclusion (1.37) and hence (1.36) hold with equality.

If $F \pitchfork G$, then (1.36) holds with equality, and additionally $\dim F \cap G = \dim F + \dim G - d$. This gives

$$\dim (N(C, F) + N(D, G)) = \dim N(C \cap D, F \cap G) = d - \dim F \cap G$$
$$= d - \dim F - \dim G + d = \dim N(C, F) + \dim N(D, G),$$

thus the sum $N(C, F) + N(D, G)$ is direct. $\square$

Notes for Section 1.4

1. General references on polytopes and polyhedra are the books by Grünbaum [82], Brøndsted [39] and Ziegler [194].

2. Occasionally, we have profited from the presentation given in Amelunxen and Lotz [10].

3. A sporadic polyhedral cone. There exists a polyhedral cone, only in dimension three and unique up to rotations, which has a strange property. If $P \subset \mathbb{R}^d$ is a polyhedron and $K \subset \mathbb{R}^d$ is a convex body, we say that K is a *rotor* of P if K can be freely moved inside P, always touching all facets of P. More precisely, K is a rotor of P if to any rotation $\rho \in \mathrm{SO}(d)$ there exists a translation vector t such that $\rho K + t$ is contained in P and touches all its facets. The pair (P, K), where K is a rotor of the polyhedron P, is called *trivial* if the normal vectors of the facets of P are linearly independent or if K is a ball. All non-trivial pairs (P, K) of a polyhedron and a rotor of it were determined in [159]. The proof is reproduced in Groemer [79, Sect. 5.7]. If we now assume that (C, K) is a nontrivial pair consisting of a polyhedral cone C and a rotor K, then only the following situation is possible: $d = 3$ and, up to a rotation,

$$C = \bigcap_{i=1}^{4} H^-(u_i, 0)$$

is the polyhedral cone whose facets have the normal vectors $u_{1,2} = (\pm\sqrt{6}, 0, 1)$, $u_{3,4} = (0, \pm\sqrt{6}, 1)$ (with respect to an orthonormal basis). Opposite faces of this cone form an angle of $\arccos(5/7)$. The convex body K has the support function (restricted to the unit sphere) $Y_0 + Y_1 + Y_4$, where Y_i denotes a spherical harmonic of degree i. These spherical harmonics can be chosen such that K is convex and not a ball.

1.5 Recession cones

The study of unbounded polyhedra can in several respects be reduced to the study of polytopes and polyhedral cones. The key to this simplification is the notion of recession cone (see, e.g., Rockafellar [148, §8]).

For a nonempty subset $A \subseteq \mathbb{R}^d$, the *recession cone* is defined by

$$\mathrm{rec}\, A = \{y \in \mathbb{R}^d : x + \lambda y \in A \text{ for all } x \in A \text{ and all } \lambda \geq 0\}.$$

For a nonempty closed convex set $K \in \mathcal{CC}^d$, the recession cone can equivalently be defined, with an arbitrary $x \in K$, by

$$\mathrm{rec}\, K = \{y \in \mathbb{R}^d : x + \lambda y \in K \text{ for all } \lambda \geq 0\},$$

and by

$$\mathrm{rec}\, K = \bigcap_{\lambda > 0} \lambda(K - x). \tag{1.38}$$

In this case, it is a closed convex cone. Another description is given by the following lemma.

Lemma 1.5.1. *Let $K \in \mathcal{CC}^d$ and $o \in \operatorname{int} K$. Then*

$$\operatorname{rec} K = (\operatorname{pos} K^\circ)^\circ. \tag{1.39}$$

Proof. If K is bounded, then both sides of (1.39) are equal to $\{o\}$; hence we can assume that K is unbounded. Then K contains a ray with endpoint o, hence K° is contained in a closed halfspace with o in the boundary. In that case,

$$\operatorname{pos} K^\circ = \bigcap_{K^\circ \subseteq H^-(y,0)} H^-(y,0) = \bigcap_{\operatorname{pos}\{y\} \subseteq K} H^-(y,0).$$

On the other hand, for $x \in \mathbb{R}^d$ we have

$$x \in (\operatorname{rec} K)^\circ \Leftrightarrow \langle x,y \rangle \le 0 \text{ for all } y \in \operatorname{rec} K$$

$$\Leftrightarrow \langle x,y \rangle \le 0 \text{ for all } y \text{ with } \operatorname{pos}\{y\} \subseteq K$$

$$\Leftrightarrow x \in \bigcap_{\operatorname{pos}\{y\} \subseteq K} H^-(y,0).$$

It follows that $(\operatorname{rec} K)^\circ = \operatorname{pos} K^\circ$, which gives the assertion. □

Lemma 1.5.2. *Let $K_1, K_2 \in \mathcal{CC}^d$ be closed convex sets. If $K_1 \cap K_2 \ne \emptyset$, then*

$$\operatorname{rec}(K_1 \cap K_2) = \operatorname{rec} K_1 \cap \operatorname{rec} K_2.$$

If $K_1 \cup K_2$ is convex, then

$$\operatorname{rec}(K_1 \cup K_2) = \operatorname{rec} K_1 \cup \operatorname{rec} K_2;$$

in particular, $\operatorname{rec} K_1 \cup \operatorname{rec} K_2$ is convex.

Proof. To prove the first assertion, let $x \in K_1 \cap K_2$. Let $y \in \operatorname{rec}(K_1 \cap K_2)$. Then $x + y \in K_1 \cap K_2$. Thus, $x \in K_1$, $x + y \in K_1$, hence $y \in \operatorname{rec} K_1$. Similarly, $y \in \operatorname{rec} K_2$, thus $y \in \operatorname{rec} K_1 \cap \operatorname{rec} K_2$. Conversely, let $y \in \operatorname{rec} K_1 \cap \operatorname{rec} K_2$. Then $x + y \in K_1$ and $x + y \in K_2$, hence $x + y \in K_1 \cap K_2$, thus $y \in \operatorname{rec}(K_1 \cap K_2)$.

To prove the second assertion, let $y \in \operatorname{rec}(K_1 \cup K_2)$. Since $K_1 \cup K_2$ is assumed to be convex, we have $K_1 \cap K_2 \ne \emptyset$ and hence we can choose $x \in K_1 \cap K_2$. Then $x + y \in K_1 \cup K_2$. If $x + y \in K_1$, then $y \in \operatorname{rec} K_1$, otherwise $y \in \operatorname{rec} K_2$; in any case, $y \in \operatorname{rec} K_1 \cup \operatorname{rec} K_2$. Conversely, let $y \in \operatorname{rec} K_1 \cup \operatorname{rec} K_2$. If $y \in \operatorname{rec} K_1$, then $x + y \in K_1$, otherwise $x + y \in K_2$; in any case, $x + y \in K_1 \cup K_2$. Since $K_1 \cup K_2$ is closed and convex, this is sufficient to conclude that $y \in \operatorname{rec}(K_1 \cup K_2)$. □

Lemma 1.5.3. *Let $K \in \mathcal{CC}^d$, and let $u \in \mathbb{S}^{d-1}$ be such that $H(K,u)$ is a support plane of K. Then*

$$\operatorname{rec}(K \cap H(K,u)) = (\operatorname{rec} K) \cap H(u,0). \tag{1.40}$$

In particular,

$$K \cap H(K, u) \text{ is bounded } \Leftrightarrow (\operatorname{rec} K) \cap H(u, 0) = \{o\}. \tag{1.41}$$

To each support set F of K there exists a unique support set F' of $\operatorname{rec} K$ such that

$$F = K \cap H(K, u) \Rightarrow F' = (\operatorname{rec} K) \cap H(u, 0). \tag{1.42}$$

We have $\dim F' \leq \dim F$.

Proof. Relation (1.40), and thus (1.41), follows immediately from the definition of recession cones.

Let F be a support set of K, say $F = K \cap H(K, u)$ with $u \in \mathbb{S}^{d-1}$.

If F is bounded, then $F' := (\operatorname{rec} K) \cap H(u, 0)$ is bounded, since a ray R contained in F' would satisfy $F + R \subseteq K \cap H(K, u) = F$, a contradiction. Hence, in this case, $F' = \{o\}$.

Suppose that F is unbounded. Let $x \in F$, and define F' as the union of all rays R with endpoint o such that $x + R \subseteq F$. Then F' is a closed convex cone, $\dim F' \leq \dim F$, $F' \subseteq \operatorname{rec} K$, and $F' \subseteq H(u, 0)$, thus $F' \subseteq (\operatorname{rec} K) \cap H(u, 0)$. The equality sign holds here, since any ray $R \subseteq (\operatorname{rec} K) \cap H(u, 0)$ satisfies $x + R \subseteq K \cap H(K, u)$.

The latter argument shows also that F' does not depend on the choice of the vector u (for which $F = K \cap H(K, u)$). $\qquad\square$

Now we consider polyhedra. If $P \in \mathcal{Q}^d$ is a polyhedron, its recession cone is a polyhedral cone and can be represented as follows. If P is given by (1.25) and if $H_i^- + t_i$ is the translate of H_i^- with o in the boundary, then

$$\operatorname{rec} P = \bigcap_{i=1}^{m} (H_i^- + t_i).$$

As said initially, the study of unbounded polyhedra can sometimes be reduced to the study of polytopes and polyhedral cones. A first step in this direction is done in the following two lemmas.

Lemma 1.5.4. *Let $P \in \mathcal{Q}^d$ be a line-free polyhedron, and let B be the union of its bounded faces. Then*

$$P = B + \operatorname{rec} P.$$

Proof. We can assume that $\dim P = d$. The inclusion $P \subseteq B + \operatorname{rec} P$ is clear from the definitions. For the converse, let $x \in P$. If P is bounded, we are finished. If P is unbounded, there is a vector $y_1 \in \operatorname{rec} P \setminus \{o\}$ with $x + \mathbb{R}_+ y_1 \subseteq P$. Since P does not contain a line, there is a point $a_1 = x - \lambda_1 y_1$, with $\lambda_1 \geq 0$, in the boundary of P and hence in some face F_1 of P with $\dim F_1 < d$. If F_1 is unbounded, there is a vector $y_2 \in \operatorname{rec} P \setminus \{o\}$ with $a_1 + \mathbb{R}_+ y_2 \subseteq F_1$. Since F_1 does not contain a line, there is a point $a_2 = a_1 - \lambda_2 y_2$, with $\lambda_2 \geq 0$, in the relative boundary of F_1 and hence in some face F_2 of F_1 with $\dim F_2 < \dim F_1$. We continue this procedure, which ends after finitely many steps with a vector

$y_k \in \operatorname{rec} P \setminus \{o\}$ and a point $a_k = a_{k-1} - \lambda_k y_k$, where $\lambda_k \geq 0$ and $a_k \in F_k$ for some bounded face F_k of P (possibly one-pointed). Then

$$x = a_k + \lambda_k y_k + \lambda_{k-1} y_{k-1} + \cdots + \lambda_1 y_1 \in B + \operatorname{rec} P,$$

which completes the proof. $\square$

In particular, a polyhedron is the sum of its recession cone and a polytope.

We shall now show that a line-free polyhedron is the union of direct sums of polytopes and polyhedral subcones of the recession cone.

Let $P \in \mathcal{Q}^d$ be a line-free polyhedron, without loss of generality of dimension d. Its recession cone is pointed and hence has a supporting plane, $H(u, 0)$, such that $\operatorname{rec} P \subset H^-(u, 0)$ and $\operatorname{rec} P \cap H(u, 0) = \{o\}$. Then each intersection $H(u, t) \cap P$ is bounded. Since P has only finitely many bounded faces, we can choose $t \in \mathbb{R}$ such that $P \cap H^-(u, t)$ has only one bounded facet, namely $P \cap H(u, t)$. The intersection $P \cap H^+(u, t)$ is bounded, and each bounded face of $P \cap H^-(u, t)$ is a face of $P \cap H(u, t)$. Thus, by cutting off a suitable polytope from a line-free polyhedron, we obtain a polyhedron with only one bounded facet, and we may restrict ourselves to such polyhedra in the following. (Note that $P \cap H^+(u, t)$ is the direct sum of itself and the subcone $\{o\}$ of the recession cone.)

Lemma 1.5.5. *Let* $P \in \mathcal{Q}^d$ *be a line-free polyhedron of dimension d with exactly one bounded facet, denoted by B. Then to each face $F \in \mathcal{F}(B)$ there is a polyhedral cone $C_F \subseteq \operatorname{rec} P$ such that*

$$\bigcup_{F \in \mathcal{F}(B)} F \oplus C_F = P \tag{1.43}$$

and

$$\bigcup_{F \in \mathcal{F}(B)} C_F = \operatorname{rec} P. \tag{1.44}$$

Proof. Since P has dimension d and only one bounded facet, it is unbounded. For each unbounded face G (including P) of P we choose $v_G \in \operatorname{relint} \operatorname{rec} G$. For $F \in \mathcal{F}(B)$ we define $C_F = \operatorname{pos}\{v_G : F < G\}$ (with $F < G$ meaning that F is a face of G different from G).

Now let $x \in P$. Since $v_P \in \operatorname{rec} P$ and P is line-free, there is a point $a_1 = x - \lambda_1 v_P \in \operatorname{bd} P$ with $\lambda_1 \geq 0$. We have $a_1 \in G_1$ for some face $G_1 \in \mathcal{F}(P)$. Since $v_P \in \operatorname{int} \operatorname{rec} P$, the face G_1 and $\operatorname{lin}\{v_P\}$ lie in complementary subspaces. If G_1 is unbounded, there is similarly a point $a_2 = a_1 - \lambda_2 v_{G_1} \in \operatorname{bd} P$ with $\lambda_2 \geq 0$, which lies in a face $G_2 \in \mathcal{F}(P)$. Again, the face G_2 and $\operatorname{lin}\{v_P, v_{G_1}\}$ lie in complementary subspaces. After finitely many steps, the procedure ends with a point $a_k = a_{k-1} - \lambda_k v_{G_{k-1}} \in \operatorname{bd} P$ with $\lambda_k \geq 0$, where a_k lies in a bounded face, say $G_k =: F$, of P and hence of B. We have

$$x = a_k + \lambda_k v_{G_{k-1}} + \cdots + \lambda_1 v_P \in F + C_F.$$

The latter sum is direct, since the face F and $\lin\{v_P, v_{G_1}, \ldots, v_{G_{k-1}}\}$ lie in complementary subspaces. Thus, (1.43) holds.

To prove (1.44), we note that trivially each cone C_F satisfies $C_F \subseteq \rec P$. Conversely, let $w \in \rec P$. Choose x in P. Since $\mathcal{F}(B)$ is finite, there is a sequence $(\lambda_n)_{n \in \mathbb{N}}$ tending to infinity such that $x + \lambda_n w \in F \oplus C_F$ for a fixed $F \in \mathcal{F}(B)$. It follows that $\lambda_n^{-1} x + w \in (\lambda_n^{-1} F) + C_F$, which for $n \to \infty$ gives $w \in C_F$. This completes the proof.   □

1.6 Valuations

Many of the interesting functionals appearing in convex geometry are valuations. In the present section we introduce and study this notion, first abstractly and then mainly for polyhedra. The consideration of valuations will be continued in later sections.

We recall that a family $\mathcal{S}$ of sets is called *intersectional* if $A, B \in \mathcal{S}$ and $A \cap B \neq \emptyset$ implies $A \cap B \in \mathcal{S}$. If $\mathcal{S}$ is an intersectional family, we denote by $\mathsf{U}(\mathcal{S})$ the set of all finite unions of elements from $\mathcal{S}$, together with the empty set. Then $(\mathsf{U}(\mathcal{S}), \cup, \cap)$ is a lattice (which means that $\cup$ and $\cap$ are binary operations on $\mathcal{S}$ that are commutative, associative, and satisfy the absorption laws). A function φ from this lattice into an abelian group (with addition $+$ and neutral element 0) is called a *valuation* if $\varphi(\emptyset) = 0$ and

$$\varphi(A \cup B) + \varphi(A \cap B) = \varphi(A) + \varphi(B)$$

for all $A, B \in \mathsf{U}(\mathcal{S})$ (see, e.g., Birkhoff [27, p. 230]).

This notion of a valuation on a lattice is too restrictive for the use in convexity, and is therefore extended as follows.

Definition 1.6.1. *Let $\mathcal{S}$ be an intersectional family of sets. A mapping φ from $\mathcal{S}$ into an abelian group is called* additive *or a* valuation *if*

$$\varphi(A \cup B) + \varphi(A \cap B) = \varphi(A) + \varphi(B)$$

whenever $A, B, A \cup B \in \mathcal{S}$, and $\varphi(\emptyset) = 0$ if $\emptyset \in \mathcal{S}$.

A function φ on an intersectional family $\mathcal{S}$ with values in an abelian group is said to be *fully additive* (or to satisfy the *inclusion-exclusion principle*) if

$$\varphi(A_1 \cup \cdots \cup A_m) = \sum_{r=1}^{m} (-1)^{r-1} \sum_{1 \leq i_1 < \cdots < i_r \leq m} \varphi(A_{i_1} \cap \cdots \cap A_{i_r}) \qquad (1.45)$$

for all $m \in \mathbb{N}$ and all $A_1, \ldots, A_m \in \mathcal{S}$ with $A_1 \cup \cdots \cup A_m \in \mathcal{S}$. For $m = 2$, equation (1.45) reduces to the definition of the valuation property.

The characteristic function mapping, $A \mapsto \mathbb{1}_A$, is a trivial, though important, example of a fully additive function.

Theorem 1.6.1. *Let $\mathscr{S}$ be an intersectional family of sets, let S be the union of the sets in $\mathscr{S}$, and let $\mathbb{1}_A = \mathbb{1}[A]$ be the characteristic function (defined on S) of $A \in \mathscr{S}$. Then the mapping $A \mapsto \mathbb{1}[A]$ from $\mathscr{S}$ into the abelian group of real functions on S is fully additive, that is, it satisfies*

$$\mathbb{1}[A_1 \cup \cdots \cup A_m] = \sum_{r=1}^{m} (-1)^{r-1} \sum_{1 \le i_1 < \cdots < i_r \le m} \mathbb{1}[A_{i_1} \cap \cdots \cap A_{i_r}] \qquad (1.46)$$

for $A_1, \ldots, A_m \in \mathscr{S}$ with $A_1 \cup \cdots \cup A_m \in \mathscr{S}$.

Proof. Let $A_1, \ldots, A_m \in \mathscr{S}$ and $A_1 \cup \cdots \cup A_m \in \mathscr{S}$. If $x \notin A_1 \cup \cdots \cup A_m$, both sides of (1.46) are zero at x. Let $x \in A_1 \cup \cdots \cup A_m$. Then the left side of (1.46) at x is equal to 1. Let k be the number of indices $i \in \{1, \ldots, m\}$ for which $x \in A_i$. Then the right side of (1.46) at x is equal to

$$\sum_{r=1}^{k} (-1)^{r-1} \binom{k}{r} = 1 - (1-1)^k = 1,$$

as stated. □

If $\mathscr{S}$ is a lattice and φ is a valuation on $\mathscr{S}$, then (1.45) holds for arbitrary $A_1, \ldots, A_m \in \mathscr{S}$, by induction; thus φ is fully additive. Since a lattice is a more useful domain for a valuation than an intersectional family, it is of interest to know whether a given valuation φ on an intersectional family $\mathscr{S}$ has an extension, as a valuation, to the lattice $\mathsf{U}(\mathscr{S})$. A necessary condition, as just seen, is that φ be fully additive. Groemer's extension theorem says that this condition is also sufficient. By $\mathsf{U}^{\bullet}(\mathscr{S})$ we denote the $\mathbb{Z}$-module spanned by the characteristic functions of the elements of $\mathscr{S}$, and by $\overline{\mathsf{U}}(\mathscr{S})$ the family of subsets of S with characteristic function in $\mathsf{U}^{\bullet}(\mathscr{S})$. Then $\mathsf{U}(\mathscr{S}) \subseteq \overline{\mathsf{U}}(\mathscr{S})$, by (1.46). The advantage of the family $\overline{\mathsf{U}}(\mathscr{S})$ is that it is a *ring* in S, that is, it satisfies $\emptyset \in \overline{\mathsf{U}}(\mathscr{S})$ and $A \cup B \in \overline{\mathsf{U}}(\mathscr{S})$, $A \setminus B \in \overline{\mathsf{U}}(\mathscr{S})$ for $A, B \in \overline{\mathsf{U}}(\mathscr{S})$.

By $V(\mathscr{S})$ we denote the real vector space spanned by the characteristic functions of the elements of $\mathscr{S}$. The following result is due to Groemer [78].

Theorem 1.6.2 (Groemer's extension theorem, first version). *Let $\mathscr{S}$ be an intersectional family of sets (including $\emptyset$), and let φ be a mapping from $\mathscr{S}$ into a real vector space X, such that $\varphi(\emptyset) = 0$. Then the following statements are equivalent.*

(a) *φ is fully additive on $\mathscr{S}$;*

(b) *If*

$$c_1 \mathbb{1}[A_1] + \cdots + c_k \mathbb{1}[A_k] = 0$$

with $A_i \in \mathscr{S}$ and $c_i \in \mathbb{R}$ for $i = 1, \ldots, k$, then

$$c_1 \varphi(A_1) + \cdots + c_k \varphi(A_k) = 0;$$

(c) *There exists a linear mapping $\phi : V(\mathscr{S}) \to X$ with*

$$\phi(\mathbb{1}[A]) = \varphi(A) \quad for \ A \in \mathscr{S};$$

(d) φ has an additive extension to the ring $\overline{U}(\mathscr{S})$;

(e) φ has an additive extension to the lattice $U(\mathscr{S})$.

In connection with Property (c) of Theorem 1.6.2 we point out that some authors <u>define</u> a valuation on $\mathscr{S}$ as a linear mapping $V(\mathscr{S}) \to X$, where X is a real vector space (see, e.g., Barvinok [18, p. 29], [19, p. 11]).

We need not reproduce the proof of Theorem 1.6.2 here, since we can refer to [163, Thm. 6.2.1 and Remark 6.2.2]. There, the following variant is shown, with essentially the same arguments. Here the range need not be a vector space.

Theorem 1.6.3 (Groemer's extension theorem, second version). *Let* $\mathscr{S}$ *be an intersectional family of sets (including $\emptyset$), and let φ be a mapping from $\mathscr{S}$ into an abelian group (considered as a $\mathbb{Z}$-module), such that $\varphi(\emptyset) = 0$. Then the following statements are equivalent.*

(a) φ *is fully additive on* $\mathscr{S}$;

(b) *If*

$$n_1 \mathbb{1}[A_1] + \cdots + n_k \mathbb{1}[A_k] = 0$$

with $A_i \in \mathscr{S}$ *and* $n_i \in \mathbb{Z}$ *for* $i = 1, \ldots, k$, *then*

$$n_1 \varphi(A_1) + \cdots + n_k \varphi(A_k) = 0;$$

(c) *The function* ϕ *defined by* $\phi(\mathbb{1}[A]) = \varphi(A)$ *for* $A \in \mathscr{S}$ *has a $\mathbb{Z}$-linear extension to* $U^\bullet(\mathscr{S})$;

(d) φ *has an additive extension to the ring* $\overline{U}(\mathscr{S})$;

(e) φ *has an additive extension to the lattice* $U(\mathscr{S})$.

The preceding general theorems will here mainly be applied to valuations on sets of polyhedra. Clearly, each of the families $\mathscr{S} = \mathcal{Q}^d, \mathcal{P}^d, \mathcal{PC}^d, \mathcal{Q}^d_{ro}, \mathcal{P}^d_{ro}, \mathcal{PC}^d_{ro}$ is intersectional; here $\mathcal{PC}^d_{ro}$ denotes the set of relatively open polyhedral cones in $\mathbb{R}^d$. The valuation property on general ro-polyhedra is particularly easy to handle, due to the following simple fact. If φ is a valuation on $U(\mathcal{Q}^d_{ro})$ and if a general ro-polyhedron Q is represented as the disjoint union of ro-polyhedra $Q_1, \ldots, Q_r$ (which is always possible), then

$$\varphi(Q) = \varphi(Q_1) + \cdots + \varphi(Q_r).$$

This follows from the inclusion-exclusion principle, together with $\varphi(\emptyset) = 0$.

The valuation property on general ro-polyhedra can often be derived from a seemingly weaker property.

Definition 1.6.2. *A function* φ *on the set* $\mathcal{Q}^d$ *of polyhedra with values in an abelian group is weakly additive if, after setting $\varphi(\emptyset) = 0$, it satisfies*

$$\varphi(P) = \varphi(P \cap H^+) + \varphi(P \cap H^-) - \varphi(P \cap H)$$

for all $P \in \mathcal{Q}^d$ and all hyperplanes $H \subset \mathbb{R}^d$, where H^+, H^- are the closed halfspaces bounded by H.

Clearly, every valuation on $\mathcal{Q}^d$ is weakly additive. The converse is also true, in a stronger sense.

Theorem 1.6.4. *Every weakly additive function on $\mathcal{Q}^d$ with values in an abelian group is fully additive on $\mathcal{Q}^d$. Hence, it can be extended to a valuation on $\mathsf{U}(\mathcal{Q}^d_{ro})$.*

With $\mathcal{Q}^d$ replaced by the set $\mathcal{P}^d$ of polytopes in Definition 1.6.2 and Theorem 1.6.4, this is well known; see, e.g., [163], Theorem 6.2.3 and Corollary 6.2.4. The proof given there can verbally be carried over from the set $\mathcal{P}^d$ of polytopes to the set $\mathcal{Q}^d$ of polyhedra.

An analogous situation is presented by polyhedral cones.

Definition 1.6.3. *A function φ on the set $\mathcal{PC}^d$ of polyhedral cones with values in an abelian group is* weakly additive *if it satisfies*

$$\varphi(C) = \varphi(C \cap H^+) + \varphi(C \cap H^-) - \varphi(C \cap H)$$

for all $C \in \mathcal{PC}^d$ and all hyperplanes $H \subset \mathbb{R}^d$ through o.

We recall that the relative interior of a cone in $\mathcal{PC}^d$ is called a *relatively open polyhedral cone*, and that the set of all such cones is denoted by $\mathcal{PC}^d_{ro}$.

Theorem 1.6.5. *Every weakly additive function on $\mathcal{PC}^d$ with values in an abelian group is fully additive on $\mathcal{PC}^d$. Hence, it can be extended to a valuation on $\mathsf{U}(\mathcal{PC}^d_{ro})$.*

Proof. Again, the proof given for the case of polytopes (that is, with $\mathcal{P}^d$ instead of $\mathcal{PC}^d$) in [163, Thm. 6.2.3, Cor. 6.2.4] can verbally be carried over to the present case, replacing hyperplanes by hyperplanes through o.   □

Now we introduce a basic valuation, the Euler characteristic. The subsequent proof is an extension of a well-known proof by Hadwiger [84] (reproduced in [85, p. 239] and, e.g., in [163, Thm. 4.3.1]), who considered the lattice $\mathsf{U}(\mathcal{K}^d)$. Here we consider finite unions of (not necessarily bounded) closed convex sets.

Recall that type(K) of a closed convex set $K \in \mathcal{CC}^d$ is defined as the pair $(\dim \operatorname{lineal}(K), \epsilon)$, where $\epsilon = 1$ or 0 according to whether the line-free kernel of K is bounded or not.

Theorem 1.6.6 (and Definition). *There is a unique real valuation χ on $\mathsf{U}(\mathcal{CC}^d)$, the* Euler characteristic, *with*

$$\chi(K) = (-1)^k \epsilon \quad \text{if type}(K) = (k, \epsilon), \quad K \in \mathcal{CC}^d. \tag{1.47}$$

It satisfies

$$\chi(K \oplus M) = \chi(K)\chi(M) \tag{1.48}$$

for $K, M \in \mathcal{CC}^d$ lying in complementary subspaces.

Proof. The existence is proved by induction with respect to the dimension. The zero-dimensional case being trivial, we assume that $d \geq 1$ and that the existence of χ has been proved in affine spaces of dimension less than d. We use the hyperplanes $H_\lambda = H(u, \lambda) = \{x \in \mathbb{R}^d : \langle u, x \rangle = \lambda\}$ with some fixed $u \in \mathbb{S}^{d-1}$ and varying $\lambda \in \mathbb{R}$. For a set $K \in \mathsf{U}(\mathcal{CC}^d)$ we define

$$\chi(K) := -\lim_{\mu \to -\infty} \chi(K \cap H_\mu) + \sum_{\lambda \in \mathbb{R}} \left[\chi(K \cap H_\lambda) - \lim_{\mu \downarrow \lambda} \chi(K \cap H_\mu) \right] \tag{1.49}$$

This definition makes sense, for the following reasons. The set K is the union of finitely many closed convex sets $K_1, \ldots, K_m$. Since χ on the right-hand side of (1.49) is a valuation on the lattice of finite unions of closed convex sets in H_λ, we have

$$\chi(K \cap H_\lambda) = \sum_{r=1}^{m} (-1)^{r-1} \sum_{1 \leq i_1 < \cdots < i_r \leq m} \chi(K_{i_1} \cap \cdots \cap K_{i_r} \cap H_\lambda).$$

There are finitely many numbers $\lambda_1, \ldots, \lambda_s$ such that for λ in any of the components of $\mathbb{R} \setminus \{\lambda_1, \ldots, \lambda_s\}$, either $K_{i_1} \cap \cdots \cap K_{i_r} \cap H_\lambda = \emptyset$ or the type of $K_{i_1} \cap \cdots \cap K_{i_r} \cap H_\lambda$ is constant, for $1 \leq i_1 < \cdots < i_r \leq m$. This shows that all limits in (1.49) exist and that the sum is finite. The induction hypothesis implies that the function χ thus defined on $\mathsf{U}(\mathcal{CC}^d)$ is a valuation.

Now let $K \in \mathcal{CC}^d$. We have $\dim(H_0 \cap \text{lineal}(K)) = k - 1$ or k. Let

$$\lambda_{\sup} := \sup\{\lambda \in \mathbb{R} : K \cap H_\lambda \neq \emptyset\},$$

$$\lambda_{\inf} := \inf\{\lambda \in \mathbb{R} : K \cap H_\lambda \neq \emptyset\}.$$

For $\lambda \in (\lambda_{\inf}, \lambda_{\sup})$, the type of $K \cap H_\lambda$ is constant.

Case 1: $\dim(H_0 \cap \text{lineal}(K)) = k - 1$.

In this case, $\lambda_{\sup} = \infty$ and $\lambda_{\inf} = -\infty$. If $\text{type}(K) = (k, 1)$, then $\text{type}(K \cap H_\lambda) = (k-1, 1)$ for all λ, and (1.49) gives $\chi(K) = -\chi(K \cap H_\lambda) = -(-1)^{k-1} = (-1)^k \cdot 1$. If $\text{type}(K) = (k, 0)$, then $\text{type}(K \cap H_\lambda) = (k-1, 0)$, and (1.49) gives $\chi(K) = 0$.

Case 2: $\dim(H_0 \cap \text{lineal}(K)) = k$.

Subcase a: $\lambda_{\sup} = \infty$ and $\lambda_{\inf} = -\infty$.

Then $\text{type}(K) = (k, 0)$. For each $\lambda \in \mathbb{R}$, we have $\text{type}(K \cap H_\lambda) = (k, 0)$ (if all the sections $K \cap H_\lambda$ were bounded, then K would contain a line not parallel to H_0). From (1.49) it follows that $\chi(K) = 0$.

Subcase b: $\lambda_{\sup} < \infty$ and $\lambda_{\inf} = -\infty$.

If $\lambda_{\sup}$ is attained, then K contains a halfline which is not part of a line in K, hence $\text{type}(K) = (k, 0)$. Since $\text{type}(K \cap H_\lambda)$ is constant for $\lambda \leq \lambda_{\sup}$, (1.49) gives $\chi(K) = 0$. If $\lambda_{\sup}$ is not attained, then from the fact that $\text{type}(K \cap H_\lambda)$ is constant for $\lambda < \lambda_{\sup}$ it follows by (1.49) that $\chi(K) = 0$.

Subcase c: $\lambda_{\sup} = \infty$ and $\lambda_{\inf} > -\infty$.

If $\lambda_{\inf}$ is attained, it follows as in Subcase b that $\text{type}(K) = (k, 0)$, and $\text{type}(K \cap H_\lambda)$ is constant for $\lambda \geq \lambda_{\inf}$. Hence (1.49) gives $\chi(K) = 0$. If $\lambda_{\inf}$ is not attained, the fact that $\text{type}(K \cap H_\lambda)$ is constant for $\lambda > \lambda_{\inf}$ implies that $\chi(K) = 0$.

Subcase d: $\lambda_{\sup} < \infty$ and $\lambda_{\inf} > -\infty$.

If $\text{type}(K) = (k, 1)$, then $\text{type}(K \cap H_\lambda) = (k, 1)$ and thus $\chi(K \cap H_\lambda) = (-1)^k$, for $\lambda_{\inf} \leq \lambda \leq \lambda_{\sup}$. This gives $\chi(K) = (-1)^k$. If $\text{type}(K) = (k, 0)$, then $\text{type}(K \cap H_\lambda) = (k, 0)$ and hence $\chi(K \cap H_\lambda) = 0$ for all λ, which gives $\chi(K) = 0$.

We have proved that the valuation χ on $\mathsf{U}(\mathcal{CC}^d)$ satisfies (1.47).

The uniqueness of χ is clear by the inclusion-exclusion principle.

If $K, M \in \mathcal{CC}^d$ lie in complementary subspaces, then (1.48) can be proved by induction with respect to the dimension, using (1.49) and choosing the vector u in one of the complementary subspaces. □

The following special values of the Euler characteristic follow immediately from Theorem 1.6.6. We have

$$\chi(K) = 1 \quad \text{for } K \in \mathcal{K}^d \tag{1.50}$$

and

$$\chi(K) = 0 \quad \text{if } K \in \mathcal{CC}^d \text{ is unbounded and line-free.} \tag{1.51}$$

For an arbitrary cone $C \in \mathcal{C}^d$,

$$\chi(C) = \begin{cases} 0 & \text{if } C \notin \mathcal{L}_\bullet, \\ (-1)^{\dim C} & \text{if } C \in \mathcal{L}_\bullet. \end{cases} \tag{1.52}$$

From this, we can deduce the following.

Theorem 1.6.7. *Let $P \in \mathcal{P}^d$ be a d-dimensional polytope. Then*

$$\chi(\text{bd } P) = 1 - (-1)^d.$$

Proof. Without loss of generality, we assume that $o \in \text{int } P$. Let $F_1, \ldots, F_m$ be the facets of P, and define $C_i := \text{pos } F_i$ for $i = 1, \ldots, m$. Then $C_i \in \mathcal{PC}^d$ and

$$\chi(C_{i_1} \cap \cdots \cap C_{i_m}) = 0 \Leftrightarrow F_{i_1} \cap \cdots \cap F_{i_m} \neq \emptyset \Leftrightarrow \chi(F_{i_1} \cap \cdots \cap F_{i_m}) = 1,$$
$$\chi(C_{i_1} \cap \cdots \cap C_{i_m}) = 1 \Leftrightarrow F_{i_1} \cap \cdots \cap F_{i_m} = \emptyset \Leftrightarrow \chi(F_{i_1} \cap \cdots \cap F_{i_m}) = 0.$$

Hence, by the inclusion-exclusion principle,

$$\chi(\mathrm{bd}\,P) = \sum_{r=1}^{m}(-1)^{r-1} \sum_{1\leq i_1 < \cdots < i_r \leq m} \chi(F_{i_1} \cap \cdots \cap F_{i_r})$$

$$= \sum_{r=1}^{m}(-1)^{r-1} \sum_{1\leq i_1 < \cdots < i_r \leq m} (1 - \chi(C_{i_1} \cap \cdots \cap C_{i_r}))$$

$$= 1 - \chi(C_1 \cup \cdots \cup C_m),$$

which gives the assertion since $\chi(\mathbb{R}^d) = (-1)^d$. $\square$

For a closed convex set $K \in \mathcal{CC}^d$, we have

$$\chi(K) = \chi(\mathrm{rec}\,K). \tag{1.53}$$

This is clear if K is compact. Suppose that $\mathrm{rec}\,K$ is a pointed cone $\neq \{o\}$. Then there is a supporting hyperplane $H(u,0)$ of $\mathrm{rec}\,K$ with $\mathrm{rec}\,K \subset H^+(u,0)$ and $\mathrm{rec}\,K \cap H(u,0) = \{o\}$. The closed convex set K has a supporting hyperplane $H(u,t)$ with outer normal vector $-u$, since K does not contain a ray $x + \mathbb{R}^+ v$ with $x \in K$ and $\langle u,v \rangle > 0$. Each intersection $K \cap H(u,\lambda)$ with $\lambda \geq t$ is compact, convex and nonempty, and each intersection $K \cap H(u,\lambda)$ with $\lambda < t$ is empty. It follows from (1.49) that $\chi(K) = 0 = \chi(\mathrm{rec}\,K)$. Now suppose that $\mathrm{rec}\,K$ has a lineality space $F := \mathrm{lineal}(\mathrm{rec}\,K) \neq \{o\}$. Then $\mathrm{rec}\,K = F \oplus C$ with a pointed cone C. For each $x \in K$ we have $x + F \subseteq K$, and $K' := \langle F \rangle^{\perp} \cap K$ has recession cone C, and $K = K' \oplus F$. Now the assertion (1.53) follows from (1.48).

For $K \in \mathcal{CC}^d$, the Euler characteristic satisfies $\chi(K) = 1$ if K is bounded, and $\chi(K) = 0$ if the line-free kernel of K is unbounded. In contrast, there is also a valuation on $\mathsf{U}(\mathcal{CC}^d)$ which is equal to 1 on all $K \in \mathcal{CC}^d$.

Theorem 1.6.8 (and Definition). *There is a unique real valuation $\overline{\chi}$ on $\mathsf{U}(\mathcal{CC}^d)$ with*

$$\overline{\chi}(K) = 1 \quad \text{for } K \in \mathcal{CC}^d. \tag{1.54}$$

Proof. We need only define

$$\overline{\chi}(K) := \lim_{r\to\infty} \chi(K \cap B(o,r)) \quad \text{for } K \in \mathsf{U}(\mathcal{CC}^d),$$

where $B(o,r)$ denotes the ball with center o and radius r. It is clear that the limit exists, that this defines a valuation $\overline{\chi}$ on $\mathsf{U}(\mathcal{CC}^d)$ (since χ is a valuation on $\mathsf{U}(\mathcal{CC}^d)$), and that (1.54) holds. $\square$

We prove now a counterpart to Theorem 1.6.6 for the lattice $\mathsf{U}(\mathcal{Q}_{ro}^d)$ of general ro-polyhedra. The valuations defined in Theorems 1.6.6 and 1.6.9 coincide where both are defined (see the explanations after the proof); hence either of them may be called Euler characteristic and denoted by χ. The proof, as carried out in [164], closely follows that of Hadwiger.

Theorem 1.6.9 (and Definition). *There is a unique real valuation χ on* $\mathsf{U}(\mathcal{Q}_{ro}^d)$, *the* Euler characteristic, *with*

$$\chi(Q) = (-1)^{\dim Q} \quad \text{for } Q \in \mathcal{Q}_{ro}^d. \tag{1.55}$$

It satisfies $\chi(P) = 1$ for $P \in \mathcal{P}^d$, and

$$\chi(A \oplus B) = \chi(A)\,\chi(B) \tag{1.56}$$

for $A, B \in \mathcal{Q}^d$ lying in complementary subspaces.

Proof. The first part of the proof proceeds as before. We assume that $d \geq 1$ and that the existence of χ has been proved in affine spaces of dimension less than d (the case $d = 0$ being trivial). Again we use the hyperplanes H_λ, and for a general ro-polyhedron $Q \in \mathsf{U}(\mathcal{Q}_{ro}^d)$ we define

$$\chi(Q) := -\lim_{\mu \to -\infty} \chi(Q \cap H_\mu) + \sum_{\lambda \in \mathbb{R}} \left[\chi(Q \cap H_\lambda) - \lim_{\mu \downarrow \lambda} \chi(Q \cap H_\mu) \right]. \tag{1.57}$$

This definition makes sense, for the following reasons. First, each $Q \cap H_\lambda$, $\lambda \in \mathbb{R}$, is a general ro-polyhedron in an affine space of dimension $d - 1$, so that $\chi(Q \cap H_\lambda)$ is defined. Second, since Q is the disjoint union of finitely many ro-polyhedra $Q_1, \ldots, Q_r$, there are finitely many numbers $\lambda_1, \ldots, \lambda_s$ such that for λ in any of the components of $\mathbb{R} \setminus \{\lambda_1, \ldots, \lambda_s\}$, the dimension of $Q_i \cap H_\lambda$ is independent of λ, for $i = 1, \ldots, r$ (where $\dim \emptyset = -1$, by definition). Thus, $\lambda \mapsto \chi(Q \cap H_\lambda)$ is constant on each such component. This shows, third, that all limits in (1.6.3) exist and that the sum is finite. The induction hypothesis implies that the function χ thus defined on $\mathsf{U}(\mathcal{Q}_{ro}^d)$ is a valuation. Now let $Q \in \mathcal{Q}_{ro}^d$. If Q is contained in some H_λ, then $\chi(Q) = (-1)^{\dim Q}$ by the induction hypothesis. If Q is not contained in some H_λ, then the right-hand side of (1.57) gives $-(-1)^{\dim Q - 1} + 0 = (-1)^{\dim Q}$ if $Q \cap H_\lambda \neq \emptyset$ for all large $-\lambda$, and otherwise it gives $0 + (0 - (-1)^{\dim Q - 1}) = (-1)^{\dim Q}$. Similarly, one obtains that $\chi(P) = 1$ for $P \in \mathcal{P}^d$.

The uniqueness of χ is clear, since each $Q \in \mathsf{U}(\mathcal{Q}_{ro}^d)$ is a disjoint union of ro-polyhedra. Relation (1.56) can be proved as in Theorem 1.6.6. $\qquad\square$

To show that the valuation just defined coincides with that of Theorem 1.6.6 where both are defined, let us denote, for the moment, the second one by χ'. By Theorems 1.6.6 and 1.6.9, we have $\chi(\mathcal{P}) = \chi'(P)$ if P is either an affine subspace or a compact polytope. Let P be an unbounded, closed polyhedron. Since both valuations are multiplicative under direct sums, we may assume that P is line-free. Then it has a vertex v, and hence a supporting hyperplane H with $H \cap P = \{v\}$. Each hyperplane parallel to H meets P either in a compact polytope or in the empty set. Now it follows from (1.57) that $\chi'(P) = 0 = \chi(P)$. Since χ and χ' coincide on polyhedra, they also coincide on finite unions of polyhedra. (The closure of the union of finitely many ro-polyhedra is the union of finitely many closed polyhedra.)

Theorem 1.6.9 immediately gives another proof of Theorem 1.6.7. If P is a d-dimensional polytope, then bd P and int P are disjoint, and both sets belong to $\mathsf{U}(\mathcal{Q}_{ro}^d)$. It follows that

$$\chi(\operatorname{bd} P) = \chi(P) - \chi(\operatorname{int} P) = 1 - (-1)^d,$$

by (1.55).

The following is easy, but useful.

Lemma 1.6.1. *If a general ro-polyhedron $Q \in \mathsf{U}(\mathcal{Q}_{ro}^d)$ is the disjoint union of the ro-polyhedra $Q_1, \ldots, Q_m \in \mathcal{Q}_{ro}^d$, then*

$$\sum_{i=1}^m (-1)^{\dim Q_i} = \chi(Q).$$

In fact, since $Q_i \cap Q_j = \emptyset$ for $i \neq j$, the additivity of χ yields

$$\sum_{i=1}^m (-1)^{\dim Q_i} = \sum_{i=1}^m \chi(Q_i) = \chi\left(\bigcup_{i=1}^m Q_i\right) = \chi(Q).$$

Since a polyhedron is the disjoint union of the relative interiors of its faces, the preceding immediately yields the following.

Theorem 1.6.10 (Euler relation). *For $P \in \mathcal{Q}^d$,*

$$\sum_{F \in \mathcal{F}(P)} (-1)^{\dim F} = \chi(P). \tag{1.58}$$

We use angle cones to 'localize' the Euler relation, that is, generalize it to the faces containing a given face. (Also in the following, we talk of local relations for a polytope if they involve only faces containing a given nonempty face.) Let $P \in \mathcal{Q}^d$ be a polyhedron, and let E be a k-face of P, for some $k \in \{0, \ldots, \dim P - 1\}$. The i-faces F of P containing E are in one-to-one correspondence with the i-faces of the angle cone $A(E, P)$, so that a translate of F is contained in the corresponding face of $A(E, P)$. Therefore,

$$\sum_{E \leq F \in \mathcal{F}(P)} (-1)^{\dim F} = \sum_{F \in \mathcal{F}(A(E,P))} (-1)^{\dim F}.$$

(Recall that $E \leq F$ denotes that E is a face of the polyhedron F.) Since $A(E, P)$ is a polyhedral cone and not a subspace (since $E \neq P$), (1.58) and (1.52) give the following.

Theorem 1.6.11 (Local Euler relation). *If $P \in \mathcal{Q}^d$ is a polyhedron and $E \in \mathcal{F}(P)$ is a face with $E \neq P$, then*

$$\sum_{E \leq F \in \mathcal{F}(P)} (-1)^{\dim F} = 0. \tag{1.59}$$

This can be used to derive a counterpart to the Euler relation. For $P \in \mathcal{Q}^d$ we have

$$\sum_{F \in \mathcal{F}(P)} (-1)^{\dim F} \chi(F) = \sum_{F \in \mathcal{F}(P)} (-1)^{\dim F} \sum_{E \in \mathcal{F}(F)} (-1)^{\dim E}$$

$$= \sum_{E \in \mathcal{F}(P)} (-1)^{\dim E} \sum_{E \leq F \in \mathcal{F}(P)} (-1)^{\dim F}$$

$$= (-1)^{\dim P} \cdot (-1)^{\dim P} = 1,$$

where (1.59) for $E \neq P$ was used. Hence

$$\sum_{F \in \mathcal{F}(P)} (-1)^{\dim F} \chi(F) = 1 \quad \text{for } P \in \mathcal{Q}^d. \tag{1.60}$$

For compact P, this is the same as (1.58), but not for unbounded P. Using the valuation $\overline{\chi}$ defined in Theorem 1.6.8, we can write (1.60) as

$$\sum_{F \in \mathcal{F}(P)} (-1)^{\dim F} \chi(F) = \overline{\chi}(P) \quad \text{for } P \in \mathcal{Q}^d. \tag{1.61}$$

The local Euler relation yields the Möbius function of the poset of faces of a polyhedron. Let $P \in \mathcal{Q}^d$, and let $\mathcal{F}(P)$ be partially ordered by inclusion. By definition (see Section 1.1), the Möbius function of the incidence algebra $I(\mathcal{F}(P))$ is given by $\mu(F, F) = 1$ for all $F \in \mathcal{F}(P)$ and

$$\mu(E, G) = - \sum_{E \leq F < G} \mu(E, F)$$

for $E, G \in \mathcal{F}(P)$ with $E < G$. From Theorem 1.6.11 we immediately get the following.

Theorem 1.6.12. *Let $P \in \mathcal{Q}^d$ be a polyhedron. The Möbius function of the face lattice $\mathcal{F}(P)$ is given by*

$$\mu(F, G) = (-1)^{\dim G - \dim F} \quad \text{for } F \leq G \tag{1.62}$$

(and $\mu(F, G) = 0$ if $F \not\leq G$).

The following is, therefore, a special case of the Möbius inversion formula.

Lemma 1.6.2. *If ψ is a function on $\mathcal{Q}^d$ with values in an abelian group and if φ is defined by*

$$\varphi(P) = \sum_{F \in \mathcal{F}(P)} \psi(F) \quad \text{for } P \in \mathcal{Q}^d,$$

then

$$\psi(P) = (-1)^{\dim P} \sum_{F \in \mathcal{F}(P)} (-1)^{\dim F} \varphi(F) \quad \text{for } P \in \mathcal{Q}^d.$$

Proof. If the first relation holds, then

$$\sum_{F\in\mathcal{F}(P)} (-1)^{\dim F}\varphi(F) = \sum_{F\in\mathcal{F}(P)} (-1)^{\dim F} \sum_{E\in\mathcal{F}(F)} \psi(E)$$

$$= \sum_{E\in\mathcal{F}(P)} \psi(E) \sum_{E\leq F\in\mathcal{F}(P)} (-1)^{\dim F}$$

$$= \psi(P)(-1)^{\dim P}$$

by (1.59). □

The preceding suggests the following definition. Derived valuations were first introduced and studied by Sallee [150].

Theorem 1.6.13 (and Definition). *For a valuation φ on $\mathcal{Q}^d$, its* derived valuation *φ^* is defined by*

$$\varphi^*(P) := \sum_{F\in\mathcal{F}(P)} (-1)^{\dim F}\varphi(F), \quad P\in\mathcal{Q}^d. \tag{1.63}$$

Then φ^* is a valuation on $\mathcal{Q}^d$, and $\varphi^{**} = \varphi$.
Further, the extensions of φ and φ^* to $\mathsf{U}(\mathcal{Q}^d_{ro})$ satisfy

$$\varphi(\operatorname{relint} P) = (-1)^{\dim P}\varphi^*(P), \tag{1.64}$$

$$\varphi^*(\operatorname{relint} P) = (-1)^{\dim P}\varphi(P) \tag{1.65}$$

for $P\in\mathcal{Q}^d$.

Proof. We can assume that φ is a valuation on $\mathsf{U}(\mathcal{Q}^d_{ro})$. Each $P\in\mathcal{Q}^d$ is the disjoint union of the relative interiors of its faces, hence

$$\varphi(P) = \sum_{F\in\mathcal{F}(P)} \varphi(\operatorname{relint} F) = \sum_{F\in\mathcal{F}(P)} \psi(F)$$

with $\psi(F) := \varphi(\operatorname{relint} F)$. By Lemma 1.6.2,

$$\psi(P) = (-1)^{\dim P} \sum_{F\in\mathcal{F}(P)} (-1)^{\dim F}\varphi(F),$$

which is (1.64).
From $\varphi^*(P) = (-1)^{\dim P}\varphi(\operatorname{relint} P)$ it follows immediately that φ^* is a weak valuation on $\mathcal{Q}^d$ and hence a valuation.
For $P\in\mathcal{Q}^d$ we have

$$\varphi^*(P) = \sum_{F\in\mathcal{F}(P)} \psi(F) \quad \text{with } \psi(F) = (-1)^{\dim F}\varphi(F),$$

hence Lemma 1.6.2 gives

$$\psi(P) = (-1)^{\dim P} \sum_{F \in \mathcal{F}(P)} (-1)^{\dim F} \varphi^*(P) = (-1)^{\dim P} \varphi^{**}(P)$$

and therefore $\varphi^{**} = \varphi$.

Relation (1.65) follows from (1.64).   □

Equation (1.61) now shows that

$$\overline{\chi} = \chi^* \quad \text{on } \mathcal{Q}^d.$$

The function $\overline{\chi}$ (defined in Theorem 1.6.8) with $\overline{\chi}(P) = 1$ for all $P \in \mathcal{Q}^d$ is a valuation on $\mathcal{Q}^d$ and hence has an extension to a valuation on $\mathsf{U}(\mathcal{Q}^d_{ro})$.

Definition 1.6.4. $\overline{\chi}$ *denotes also the valuation on* $\mathsf{U}(\mathcal{Q}^d_{ro})$ *with* $\overline{\chi}(P) = 1$ *for all* $P \in \mathcal{Q}^d$.

Of course, this definition is consistent with the previous one. We have

$$\overline{\chi}(\text{relint } P) = (-1)^{\dim P} \chi(P) \quad \text{for } P \in \mathcal{Q}^d \tag{1.66}$$

by Theorem 1.6.13.

The Euler relation can be extended to polyhedral complexes, and a similar relation can be proved for the valuation $\overline{\chi}$. By a *polyhedral complex* in $\mathbb{R}^d$ we understand a finite set $\mathcal{Z}$ of polyhedra in $\mathcal{Q}^d$ such that each face of a polyhedron of $\mathcal{Z}$ belongs to $\mathcal{Z}$ and that the intersection of any two polyhedra from $\mathcal{Z}$ is a face of each or empty. The union of all polyhedra in $\mathcal{Z}$, which is an element of $\mathsf{U}(\mathcal{Q}^d)$, is denoted by $|\mathcal{Z}|$ and called the *underlying space* (or *carrier*, or *support*) of $\mathcal{Z}$.

Theorem 1.6.14. *If* $\mathcal{Z}$ *is a polyhedral complex in* $\mathbb{R}^d$, *then*

$$\chi(|\mathcal{Z}|) = \sum_{P \in \mathcal{Z}} (-1)^{\dim P} = \sum_{r=0}^{\infty} (-1)^r f_r(\mathcal{Z}), \tag{1.67}$$

where $f_r(\mathcal{Z})$ *denotes the number of r-dimensional polyhedra in* $\mathcal{Z}$ *(the* **Euler–Poincaré formula***), and*

$$\overline{\chi}(|\mathcal{Z}|) = \sum_{P \in \mathcal{Z}} (-1)^{\dim P} \chi(P). \tag{1.68}$$

Proof. It follows from the properties of a complex that $|\mathcal{Z}|$ is the disjoint union of the relative interiors of the elements of $\mathcal{Z}$, hence

$$\chi(|\mathcal{Z}|) = \sum_{P \in \mathcal{Z}} \chi(\text{relint } P) = \sum_{P \in \mathcal{Z}} (-1)^{\dim P}$$

by (1.55), and

$$\overline{\chi}(|\mathcal{Z}|) = \sum_{P \in \mathcal{Z}} \overline{\chi}(\text{relint } P) = \sum_{P \in \mathcal{Z}} (-1)^{\dim P} \chi(P)$$

by (1.66).   □

We reformulate now some of the previous observations in terms of characteristic functions.

The assertions about the correspondence $\varphi \mapsto \varphi^*$ can be given a more general form. Recall that for an intersectional family $\mathscr{S}$ we have denoted by $V(\mathscr{S})$ the real vector space spanned by the characteristic functions of the sets of $\mathscr{S}$. This vector space appears in Theorem 1.6.2. Exploiting the fact that the mapping $A \mapsto \mathbb{1}[A]$ is a valuation, we can establish the existence of some linear mappings of $V(\mathscr{S})$ into itself, where $\mathscr{S}$ is a set of polyhedra.

Theorem 1.6.15. *There is a linear mapping $\phi^* : V(\mathcal{Q}^d) \to V(\mathcal{Q}^d)$ with*

$$\phi^*(\mathbb{1}[P]) = \sum_{F \in \mathcal{F}(P)} (-1)^{\dim F} \mathbb{1}[F] \quad for\ P \in \mathcal{Q}^d.$$

It satisfies $\phi^* \circ \phi^* = \mathrm{id}$ and

$$\phi^*(\mathbb{1}[P]) = (-1)^{\dim P} \mathbb{1}[\mathrm{relint}\ P],$$
$$\phi^*(\mathbb{1}[\mathrm{relint}\ P]) = (-1)^{\dim P} \mathbb{1}[P]$$

for $P \in \mathcal{Q}^d$.

Proof. Since $\varphi[P] := \mathbb{1}[P]$ defines a valuation on $\mathcal{Q}^d$, by Theorem 1.6.13 also $\varphi^*(P) := \sum_{F \in \mathcal{F}(P)} (-1)^{\dim F} \mathbb{1}[F]$ yields a valuation. The existence of the linear mapping ϕ^* now follows from Theorem 1.6.2. For $P \in \mathcal{Q}^d$,

$$\phi^*(\phi^*(\mathbb{1}[P]))$$
$$= \sum_{F \in \mathcal{F}(P)} (-1)^{\dim F} \phi^*(\mathbb{1}[P]) = \sum_{F \in \mathcal{F}(P)} (-1)^{\dim F} \sum_{E \in \mathcal{F}(F)} (-1)^{\dim E} \mathbb{1}[E]$$
$$= \sum_{E \in \mathcal{F}(P)} (-1)^{\dim E} \mathbb{1}[E] \sum_{E \leq F \in \mathcal{F}(P)} (-1)^{\dim F} = \mathbb{1}[P]$$

by the local Euler relation. By Theorem 1.6.13, for $P \in \mathcal{P}^d$,

$$\phi^*(\mathbb{1}[P]) = \varphi^*(P) = (-1)^{\dim P} \varphi(\mathrm{relint}\ P) = (-1)^{\dim P} \mathbb{1}[\mathrm{relint}\ P].$$

Application of ϕ^* to this relation yields the last assertion. □

Theorem 1.6.16. *There is a linear mapping $\phi_{\mathrm{rec}} : V(\mathcal{Q}^d) \to V(\mathcal{Q}^d)$ with*

$$\phi_{\mathrm{rec}}(\mathbb{1}[P]) = \mathbb{1}[\mathrm{rec}\ P] \quad for\ P \in \mathcal{Q}^d.$$

Proof. The mapping defined by $\varphi(P) = \mathbb{1}[\mathrm{rec}\ P]$, $P \in \mathcal{Q}^d$, is a valuation, by Lemma 1.5.2 and the valuation property of the characteristic function. The existence of the linear mapping ϕ_{rec} now follows from Theorem 1.6.2. □

We conclude this section with some more observations on the recession cone in connection to valuations. We can take a valuation ψ on the set $\mathcal{PC}^d$ of polyhedral cones and define

$$\varphi(P) := \psi(\operatorname{rec} P)$$

for $P \in \mathcal{Q}^d$. It follows from Lemma 1.5.2 that this defines a valuation on $\mathcal{Q}^d$. Theorem 1.6.17 below shows that all translation invariant valuations on $\mathcal{Q}^d$ are generated in this way. Without the assumption of translation invariance, this would not be true; an example is given by $\varphi(P) = \mathbb{1}[P](o)$.

Lemma 1.6.3. *Let φ be a translation invariant valuation on $\mathcal{Q}^d$. Then the following holds.*

(a) *The valuation φ is constant on $\mathcal{P}^d$.*

(b) *If $P = C \oplus A$ with a polyhedral cone C of dimension $\dim C \geq 1$ and a polytope $A \in \mathcal{P}^d$, then $\varphi(P) = \varphi(C)$.*

Proof. By Theorem 1.6.4, the valuation φ can be extended, as a translation invariant valuation, to $\mathsf{U}(\mathcal{Q}_{ro}^d)$. The extension is also denoted by φ.

On one-pointed sets, φ attains a constant value c.

For given $a > 0$, a closed half-line R can be written as the disjoint union of a half-open segment S_a of length a and a translate of R. It follows that $\varphi(R) = \varphi(S_a) + \varphi(R)$, hence $\varphi(S_a) = 0$ and, therefore, $\varphi(S) = c$ for every closed segment S.

In particular, this proves assertion (a) for $d \leq 1$, and (b) is trivial for $d \leq 1$.

We assume now that $d \geq 2$ and that (a) has been proved in dimension $d - 1$.

Let $C \in \mathcal{PC}^d$ be a cone with $\dim C \geq 1$. Let $L = (\operatorname{lin} C)^{\perp}$. We define

$$\psi(A) := \varphi(C \oplus A) \quad \text{for } A \in \mathsf{U}(\mathcal{Q}_{ro}(L)).$$

Then ψ is a translation invariant valuation on $\mathsf{U}(\mathcal{Q}_{ro}(L))$. By the induction hypothesis, it is constant on $\mathcal{P}(L)$. In particular, ψ attains the same value at $A \in \mathcal{P}(L)$ and at $\{o\}$. Therefore, $\varphi(C \oplus A) = \psi(A) = \psi(\{o\}) = \varphi(C + \{o\}) = \varphi(C)$, which is assertion (b) in dimension d.

With linearly independent vectors $a_1, \ldots, a_d \in \mathbb{R}^d$, we define, for $k = 1, \ldots, d$,

$$A_k = \operatorname{conv}\{a_1, \ldots, a_k\}, \quad A_0 = \emptyset, \quad A_{d+1} = \operatorname{conv}\{o, a_1, \ldots, a_d\},$$
$$C_k = \operatorname{pos}\{a_k, \ldots, a_d\}, \quad C_{d+1} = \{o\}.$$

Then $A_k \in \mathcal{P}^d$ is a $(k-1)$-dimensional simplex for $k = 1, \ldots, d+1$, and $C_k \in \mathcal{C}^d$ is a $(d+1-k)$-dimensional simple cone. The sets A_k and C_k lie in complementary affine subspaces, and

$$C_1 = \biguplus_{k=1}^{d+1} [(A_k \setminus A_{k-1}) \oplus C_k] \tag{1.69}$$

is a disjoint union. For $k = 2, \ldots, d$ we have

$$(A_k \setminus A_{k-1}) \oplus C_k = (A_k \oplus C_k) \setminus (A_{k-1} \oplus C_k)$$

and hence, by (b),

$$\varphi((A_k \setminus A_{k-1}) \oplus C_k) = \varphi(A_k \oplus C_k) - \varphi(A_{k-1} \oplus C_k) = \varphi(C_k) - \varphi(C_k) = 0.$$

Therefore, (1.69) gives $\varphi(C_1) = \varphi(A_1 \oplus C_1) + \varphi(A_{d+1} \setminus A_d)$ and hence $\varphi(A_{d+1} \setminus A_d) = 0$. Since $\varphi(A_d) = c$ by (a) in dimension $d - 1$, we conclude that $\varphi(A_{d+1}) = c$. Here A_{d+1} can be any d-dimensional simplex. Since the valuation φ coincides with χc on simplices, it coincides with χc on polytopes. This proves (a) in dimension d. $\qquad\square$

The following theorem is due to McMullen [131, Thm. 7], but we give a different proof. In our context, this theorem is of interest as it stresses the role of valuations on cones.

Theorem 1.6.17. *Let φ be a translation invariant valuation (with values in some abelian group) on $\mathcal{Q}^d$. Then $\varphi(P) = \varphi(\mathrm{rec}\, P)$ for all $P \in \mathcal{Q}^d$.*

Proof. Let $P \in \mathcal{Q}^d$. We can assume that $\dim P = d$. First we assume that P is line-free. Then we can choose a hyperplane $H(u, t)$ such that $P \cap \mathrm{int}\, H^-(u, t)$ contains all bounded faces of P and thus $P \cap H^+(u, t)$ is a polyhedron with exactly one bounded facet, and with recession cone $\mathrm{rec}\, P$. By Lemma 1.6.3, the valuation φ is constant on polytopes, hence $\varphi(P \cap \mathrm{int}\, H^-(u, t)) = \varphi(P) - \varphi(P \cap H(u, t)) = 0$ and thus $\varphi(P) = \varphi(P \cap H^+(u, t))$. Therefore, we may assume in the following that P is a polyhedron with exactly one bounded facet, which we denote by B.

By Lemma 1.5.5, to each face $F \in \mathcal{F}(B)$ there is a polyhedral cone $C_F \le \mathrm{rec}\, P$ such that

$$\bigcup_{F \in \mathcal{F}(B)} F \oplus C_F = P \tag{1.70}$$

and

$$\bigcup_{F \in \mathcal{F}(B)} C_F = \mathrm{rec}\, P. \tag{1.71}$$

We may assume that $o \in B$. Let $F_1, F_2 \in \mathcal{F}(B)$, and suppose that $x \in (F_1 \oplus C_{F_1}) \cap (F_2 \oplus C_{F_2})$. Then $x = y_1 + c_1 = y_2 + c_2$ with $y_i \in F_i$ and $c_i \in C_{F_i}$ for $i = 1, 2$. The vector $y_1 - y_2 = c_2 - c_1$ is contained in the hyperplane H through o containing B and also in $\mathrm{rec}\, P$. Since $H \cap \mathrm{rec}\, P = \{o\}$, we conclude that $y_1 = y_2 \in F_1 \cap F_2$ and and $c_1 = c_2 \in C_{F_1} \cap C_{F_2}$. Thus, $(F_1 \oplus C_{F_1}) \cap (F_2 \oplus C_{F_2}) \subseteq (F_1 \cap F_2) \oplus (C_{F_1} \cap C_{F_2})$. The reverse inclusion is trivial. By induction,

$$(F_1 \oplus C_{F_1}) \cap \cdots \cap (F_k \oplus C_{F_k}) = (F_1 \cap \cdots \cap F_k) \oplus (C_{F_1} \cap \cdots \cap C_{F_k})$$

for $F_1, \ldots, F_k \in \mathcal{F}(B)$. From Lemma 1.6.3 we conclude that

$$\varphi((F_1 \oplus C_{F_1}) \cap \cdots \cap (F_k \oplus C_{F_k})) = \varphi(C_{F_1} \cap \cdots \cap C_{F_k}).$$

Let $\mathcal{F}(B) = \{F_1, \ldots, F_m\}$ and set $F_i \oplus C_{F_i} =: A_i$ for $i = 1, \ldots, m$. By the inclusion-exclusion principle (1.45), and using (1.70) and (1.71),

$$\varphi(P) = \varphi \left(\bigcup_{i=1}^m A_i \right) = \sum_{r=1}^m (-1)^{r-1} \sum_{1 \leq i_1 < \cdots < i_r \leq m} \varphi(A_{i_1} \cap \cdots \cap A_{i_r})$$

$$= \sum_{r=1}^m (-1)^{r-1} \sum_{1 \leq i_1 < \cdots < i_r \leq m} \varphi(C_{F_{i_1}} \cap \cdots \cap C_{F_{i_r}})$$

$$= \varphi \left(\bigcup_{i=1}^m C_{F_i} \right) = \varphi(\operatorname{rec} P),$$

as stated.

If P is not line-free, we have $P = L \oplus P'$ with a subspace L and a line-free polyhedron P' in an affine subspace L' complementary to L. We have $\operatorname{rec} P = L \oplus \operatorname{rec} P'$. Defining $\psi(Q) := \varphi(L \oplus Q)$ for polyhedra $Q \in \mathcal{Q}(L')$, we get a valuation ψ on $\mathcal{Q}(L')$, which is invariant under translations of L' into itself. By what we have just proved, $\psi(Q) = \psi(\operatorname{rec} Q)$. Therefore,

$$\varphi(P) = \varphi(L \oplus P') = \psi(P') = \psi(\operatorname{rec} P') = \varphi(L + \operatorname{rec} P' = \varphi(\operatorname{rec} P),$$

which completes the proof. $\square$

Notes for Section 1.6

1. The inductive proof given above for Theorems 1.6.6 and 1.6.8 is essentially due to Hadwiger [84], except that he proved the result for finite unions of convex bodies. That his proof can be extended to relatively open, not necessarily bounded convex sets, was pointed out (and generalized) by Lenz [122]. For an extension of Hadwiger's approach to not necessarily bounded closed convex sets, see Chen [48]. A proof by Nef [141] for polytopes can be extended to unbounded polyhedra. Another short proof was given by Lawrence [121].

2. The valuation $\overline{\chi}$ (Theorem 1.6.8) is called the *second Euler characteristic* by Chen [48] and the *extended Euler characteristic* by McMullen [131, p. 182]. However, the reader should be aware that some authors understand by the Euler characteristic χ on $\overline{\mathsf{U}}(\mathcal{CC}^d)$ the valuation that is equal to 1 on each nonempty closed convex set (e.g., Barvinok [18]).

For a comparison of the two Euler characteristics, we also refer to Section 3.5 of the book by Beck and Sanyal [24].

3. A systematic study of 'linearizations' in the style of Theorems 1.6.15 and 1.6.16 was made by Przesławski [146]. There one also finds more general versions (not restricted to polyhedra) of these two theorems.

1.7 Identities for characteristic functions

In order to motivate this section, we reformulate part of Theorem 1.6.3.

Lemma 1.7.1. *Let $\mathscr{S}$ be an intersectional family of sets (including $\emptyset$) and let φ be a fully additive valuation on $\mathscr{S}$ with values in an abelian group, such that $\varphi(\emptyset) = 0$. If*

$$n_1 \mathbb{1}[A_1] + \cdots + n_k \mathbb{1}[A_k] = 0$$

holds for some $A_1, \ldots, A_k \in \mathscr{S}$ and $n_1, \ldots, n_k \in \mathbb{Z}$, then

$$n_1 \varphi(A_1) + \cdots + n_k \varphi(A_k) = 0$$

holds.

Thus, a linear relation for the values of a general (fully additive) valuation holds if only the corresponding relation for the characteristic function is valid. This has been useful in several different contexts; therefore it is worthwhile to study identities for characteristic functions. For example, they are related to decompositions of $\mathbb{R}^d$, possibly with integer multiplicities, that are induced by a given polyhedron. Certain identities for characteristic functions yield, by integration, some well-known angle sum relations; others yield relations between conic intrinsic volumes.

We begin with a relation for so-called tangent cones, whose proof requires little more than the Euler–Poincaré formula.

Definition 1.7.1. *Let $P \in \mathcal{Q}^d$ be a polyhedron and F a face of P. Then*

$$T(F, P) := \operatorname{pos}(P - z) + z = \{z + \lambda v : [z, z + v] \subseteq P, \lambda \geq 0\}$$

with $z \in \operatorname{relint} F$ is the tangent cone *of P at F.*

Clearly, we just have $T(F, P) = A(F, P) + z$ with $z \in \operatorname{relint} F$, where $A(F, P)$ is the angle cone according to Definition 1.4.2. Strictly speaking, a tangent cone is not a cone in our sense, but the translate of a cone. However, we stick to this terminology, which is quite common.

The following theorem expresses the characteristic function of a polyhedron as a linear combination of the characteristic functions of its tangent cones.

Theorem 1.7.1. *If $P \in \mathcal{Q}^d$ is a polyhedron, then*

$$\sum_{F \in \mathcal{F}(P)} (-1)^{\dim F} \chi(F) \mathbb{1}[T(F, P)] = \mathbb{1}[P]. \tag{1.72}$$

Proof. If $x \in P$, then $x \in T(F, P)$ for each $F \in \mathcal{F}(P)$, hence

$$\sum_{F \in \mathcal{F}(P)} (-1)^{\dim F} \chi(F) \mathbb{1}[T(F, P)](x) = \sum_{F \in \mathcal{F}(P)} (-1)^{\dim F} \chi(F) = \overline{\chi}(P) = 1$$

by (1.68), applied to the complex of faces of P. Let $x \in \mathbb{R}^d \setminus P$. Let $\mathcal{F}_x$ denote the subcomplex of $\mathcal{F}(P)$ that is visible from x. Then for $F \in \mathcal{F}(P)$ we have $F \notin \mathcal{F}_x \Leftrightarrow x \in T(F, P)$. Let H be a hyperplane that strictly separates x and P. The central projection from x maps the complex $\mathcal{F}_x$ to a polyhedral complex in H, whose support is a polyhedron. The cells of this complex are in one-to-one correspondence with the faces in $\mathcal{F}_x$. It follows that

$$\sum_{F \in \mathcal{F}_x} (-1)^{\dim F} \chi(F) = 1,$$

by (1.68). Therefore,

$$\sum_{F \in \mathcal{F}(P)} (-1)^{\dim F} \chi(F) \mathbb{1}[T(F, P)](x)$$

$$= \sum_{F \in \mathcal{F}(P)} (-1)^{\dim F} \chi(F) - \sum_{F \in \mathcal{F}_x} (-1)^{\dim F} \chi(F) = 0,$$

which completes the proof. $\qquad\square$

If $P \in \mathcal{P}^d$ is a polytope, then (1.72) reads

$$\sum_{F \in \mathcal{F}(P)} (-1)^{\dim F} \mathbb{1}[T(F, P)] = \mathbb{1}[P]. \tag{1.73}$$

The following specialization of Theorem 1.7.1 is useful for some applications.

Corollary 1.7.1. *Let φ be a valuation on $\mathcal{Q}^d$ which vanishes on polyhedra containing a line. Then, for each polytope $P \in \mathcal{P}^d$,*

$$\varphi(P) = \sum_{v \in \text{vert } P} \varphi(T(\{v\}, P)).$$

Proof. By Theorem 1.6.4, the valuation φ is fully additive on $\mathcal{Q}^d$. By Groemer's extension theorem (Thm. 1.6.3), any linear relation with integer coefficients between characteristic functions of elements of $\mathcal{Q}^d$ implies the corresponding relation for a valuation, hence Theorem 1.7.1 yields

$$\sum_{F \in \mathcal{F}(P)} (-1)^{\dim F} \varphi(T(F, P)) = \varphi(P),$$

since $\chi(F) = 1$ for each face F of the polytope P (recall that, by our definition, the elements of $\mathcal{F}(P)$ are nonempty). Here $\varphi(T(F, P)) = 0$ if $\dim F \geq 1$, since in that case, the polyhedron $T(F, P)$ contains a line. This gives the assertion. $\qquad\square$

Corollary 1.7.1 has been used to derive different versions of Brion's theorem. That theorem deals with valuations on polyhedra with values in the abelian group of rational functions in d indeterminates over $\mathbb{R}$, and which vanish on polyhedra containing lines. Brion's theorem for rational polytopes is of importance for lattice point counting, but there are also versions for general polyhedra. We mention here only the original paper by Brion [38], the survey article by Barvinok and Pommersheim [20], and point out that the subject has already been treated in several books; see Barvinok [18, 19], Beck and Robins [23], Beck and Sanyal [24].

Our next goal is to prove a variant of Theorem 1.7.1 for simple polytopes, where the summation is only over the vertices of the polytope. This requires a modification of the tangent cones at the vertices.

Let $P \in \mathcal{P}^d$ be a d-dimensional polytope which is simple, that is, each vertex of P is contained in precisely d facets, and hence in precisely d edges, of P. The tangent cones at the vertices of P are modified with respect to a given vector $\xi \in \mathbb{R}^d \setminus \{o\}$. We assume that the linear function $\ell_\xi := \langle \xi, \cdot \rangle$ is not constant on any edge of P. Let $u_1, \ldots, u_k$ be the outer unit normal vectors of the facets of P. Let v be a vertex of P, and let $u_{i_1}, \ldots, u_{i_d}$ be the normal vectors of the facets of P containing v. There is a unique representation

$$\xi = \sum_{j=1}^{d} \gamma_j u_{i_j} \tag{1.74}$$

with $\gamma_j \in \mathbb{R}$. We have $\gamma_j \neq 0$ for $j = 1, \ldots, d$, since otherwise ξ would be orthogonal to an edge of P, which we have excluded. Now we define the *forward cone* (with respect to ξ) of P at v by

$$T_\xi(v, P) := \bigcap_{\gamma_j < 0} H^-(P, u_{i_j}) \cap \bigcap_{\gamma_j > 0} H^-(P, u_{i_j})^c. \tag{1.75}$$

Here $H^-(P, u)$ denotes the supporting halfspace of P with outer normal vector u, and $H^-(P, u)^c$ denotes its complement. Thus, $H^-(P, u)^c$ is an open halfspace, and we note that in general the cone $T_\xi(v, P)$ is neither closed nor open. Its closure is a polyhedral cone. On this closed cone, the linear function ℓ_ξ attains its minimum at v.

We define the *index* $\operatorname{ind}_\xi(v)$ of v (with respect to ξ) as the number of indices j for which $\gamma_j > 0$.

The following theorem and its proof are due to Lawrence [120].

Theorem 1.7.2. *Let $P \in \mathcal{P}^d$ be a simple polytope. Let $\xi \in \mathbb{R}^d \setminus \{o\}$ be a vector that is not orthogonal to any edge of P. Then*

$$\sum_{v \in \operatorname{vert} P} (-1)^{\operatorname{ind}_\xi(v)} \mathbb{1}[T_\xi(v, P)] = \mathbb{1}[P].$$

Proof. For each face F of P, the linear function ℓ_ξ attains its maximum on F at a unique vertex of F, since ℓ_ξ is not constant on any edge of P. We denote this vertex by $max(F)$.

Let $v \in \text{vert } P$ be given. First we state the identity

$$\sum_{\substack{F \in \mathcal{F}(P) \\ max(F)=v}} (-1)^{\dim F} \mathbb{1}[T(F,P)] = (-1)^{\text{ind}_\xi(v)} \mathbb{1}[T_\xi(v,P)]. \qquad (1.76)$$

As above, we denote by $u_{i_1}, \ldots, u_{i_d}$ the normal vectors of the facets of P containing v, and we recall that the numbers γ_j are defined by (1.74). For the proof of (1.76), we denote by F_i the facet of P with outer normal vector u_i. With each nonempty subset $S \subseteq \{1, \ldots, d\}$ we associate the face

$$F(S) := \bigcap_{j \in S} F_{i_j},$$

and we define $F(\emptyset) := P$. Then $F(S)$ is a face of P with $v \in F(S)$, and $\dim F(S) = d - |S|$, where $|S| := \text{card } S$. Each face of P containing v is a face $F(S)$ with unique S.

For $x \in \mathbb{R}^d$ we define

$$S_x := \{j \in \{1, \ldots, d\} : x \in H^-(P, u_{i_j})\}.$$

Since

$$T(F(S), P) = \bigcap_{j \in S} H^-(P, u_{i_j}),$$

we have

$$x \in T(F(S), P) \Leftrightarrow S \subseteq S_x.$$

Define

$$W := \{j \in \{1, \ldots, d\} : \gamma_{i_j} < 0\}.$$

From

$$\ell_\xi(x) = \ell_\xi(v) + \sum_{j=1}^{d} \gamma_j \langle x - v, u_{i_j} \rangle$$

it follows that the function ℓ_ξ attains its maximum on $F(S)$ at v if and only if $W \subseteq S$. Therefore, we obtain

$$\sum_{\substack{F \in \mathcal{F}(P) \\ max(F)=v}} (-1)^{\dim F} \mathbb{1}[T(F,P)](x)$$

$$= \sum_{W \subseteq S \subseteq \{1,\ldots,d\}} (-1)^{\dim F(S)} \mathbb{1}[T(F(S),P)](x)$$

$$= \sum_{W \subseteq S \subseteq S_x} (-1)^{d-|S|} = \begin{cases} (-1)^{d-|W|} & \text{if } S_x = W, \\ 0 & \text{otherwise} \end{cases}$$

$$= (-1)^{\text{ind}_\xi(v)} \mathbb{1}[T_\xi(v,P)](x).$$

For the last equality, observe that (1.75) implies that $x \in T_\xi(v, P)$ if and only if $x \in H^-(P, u_{i_j})$ for $j \in W$ and $x \notin H^-(P, u_{i_j})$ for $j \in S_x \setminus W$. This proves relation (1.76).

Summing now relation (1.76) over all vertices of P, we obtain

$$\sum_{v \in \text{vert } P} (-1)^{\text{ind}_\xi(v)} \mathbb{1}[T_\xi(v, P)] = \sum_{F \in \mathcal{F}(P)} (-1)^{\dim F} \mathbb{1}[T(F, P)] = \mathbb{1}[P],$$

by Theorem 1.7.1.   □

The preceding proof, as mentioned, follows Lawrence [120]. In later literature, the modified tangent cone $T_\xi(v, P)$ is often called *polarized tangent cone* and is defined in a different, though equivalent way. As above, let $P \in \mathcal{P}^d$ be a d-dimensional simple polytope. Let $\xi \in \mathbb{R}^d$ be a vector that is not orthogonal to any edge of P. To a given vertex v of P, let $e_1, \ldots, e_d$ be nonzero vectors parallel to the edges emanating from v, pointing from v to the neighboring vertices. Then the polarized tangent cone of P at v can be represented by

$$T_\xi(v, P) = v + \sum_{\langle \xi, e_j \rangle > 0} \mathbb{R}_{\geq 0} e_j + \sum_{\langle \xi, e_j \rangle < 0} \mathbb{R}_{<0} e_j.$$

Coming back to relation (1.72), we note that for a pointed polyhedral cone it reduces to a tautology, since $\chi(F) = 0$ for every face $F \neq \{o\}$ of C, and $T(\{o\}, C) = C$. For polyhedral cones, however, tangent cones are angle cones, and their indicator functions satisfy linear relations for general polyhedra. So, we turn now to angle cones.

According to our definition, the angle cone of a polyhedron P at a face F (also called the cone of feasible directions of P at F) is defined by

$$A(F, P) = \text{pos}(P - z), \quad z \in \text{relint } F.$$

Let $P \in \mathcal{P}^d$ be a polytope. If we apply to (1.72) the mapping ϕ_{rec} from Theorem 1.6.16 and observe that $\text{rec } T(F, P) = A(F, P)$, we obtain

$$\sum_{F \in \mathcal{F}(P)} (-1)^{\dim F} \mathbb{1}[A(F, P)] = \mathbb{1}[\{o\}]. \tag{1.77}$$

The following theorem generalizes this to polyhedra.

Theorem 1.7.3 (General Brianchon–Gram–Sommerville relation). *If $P \in \mathcal{Q}^d$ is a polyhedron, then*

$$\sum_{F \in \mathcal{F}(P)} (-1)^{\dim F} \mathbb{1}[\text{relint } A(F, P)] = (-1)^{\dim P} \mathbb{1}[-\text{rec } P], \tag{1.78}$$

or equivalently,

$$\sum_{F \in \mathcal{F}(P)} (-1)^{\dim F} \mathbb{1}[A(F, P)] = (-1)^{\dim \text{rec } P} \mathbb{1}[-\text{relint rec } P]. \tag{1.79}$$

Proof. Let $P \in \mathcal{Q}^d$. If $x \notin \text{aff } P$, then at x both sides of (1.78) are zero. Therefore, we need only consider points in aff P. This means that without loss of generality we can (and will) assume that dim $P = d$. At $x = o$, (1.78) holds, since $o \in \text{relint } A(F, P)$ if and only if $F = P$, and $o \in \text{rec } P$. For $x \in \mathbb{R}^d \setminus \{o\}$, let H_x be a hyperplane orthogonal to x, let Π_x be the orthogonal projection to H_x, and let $P_x = \Pi_x(P)$. Let

$$\mathcal{F}(P, x) := \{F \in \mathcal{F}(P) : \dim F \leq d - 1,\, x \in \text{int } A(F, P)\}.$$

Suppose, first, that $x \notin -\text{rec } P$. For each $F \in \mathcal{F}(P, x)$, the projection $\Pi_x(F)$ is a polyhedron in H_x whose relative interior is contained in the relative interior of P_x. The ro-polyhedra

$$\Pi_x(\text{relint } F) \quad \text{with } F \in \mathcal{F}(P, x)$$

form a disjoint decomposition of relint P_x. Therefore, Lemma 1.6.1 gives

$$\sum_{j=0}^{d-1} (-1)^j \sum_{F \in \mathcal{F}_j(P)} \mathbb{1}[\text{int } A(F, P)](x) = \chi(\text{relint } P_x) = (-1)^{d-1}.$$

This holds if $x \notin -\text{rec } P$. If $x \in -\text{rec } P \setminus \{o\}$, then $\mathcal{F}(P, x) = \emptyset$, hence $\mathbb{1}[\text{int } A(F, P)](x) = 0$ for all $F \in \mathcal{F}_j(P)$, $j \in \{0, \ldots, d-1\}$. Thus, for arbitrary $x \in \mathbb{R}^d \setminus \{o\}$ we have

$$\sum_{j=0}^{d-1} (-1)^j \sum_{F \in \mathcal{F}_j(P)} \mathbb{1}[\text{int } A(F, P)](x) = (-1)^{d-1} (1 - \mathbb{1}[-\text{rec } P](x)),$$

which, because of int $A(P, P) = \mathbb{R}^d$, can be written as

$$\sum_{j=0}^{d} (-1)^j \sum_{F \in \mathcal{F}_j(P)} \mathbb{1}[\text{int } A(F, P)](x) = (-1)^d \mathbb{1}[-\text{rec } P](x).$$

This proves (1.78). Relation (1.79) is obtained by applying the mapping ϕ^* of Theorem 1.6.15 to (1.78). $\qquad\square$

The proof given above for (1.78) follows McMullen [131] (who in turn followed Shephard [180]), except that we have carried it out for characteristic functions.

Equation (1.78) is a general version of the *Brianchon–Gram–Sommerville relation* for polyhedra. It is 'general' because it is at the level of characteristic functions. From it, the classical versions for angle functions follow immediately by integration (see Theorem 2.2.1).

Often in the literature, relation (1.72) is called the 'Brianchon–Gram relation'. This seems unjustified, in so far as neither the paper of Brianchon [36] nor that of Gram [75] contains anything in this direction. On the other hand, the formulas are related; for example, (1.72) implies (1.77), which yields the Brianchon–Gram relation for polytopes by integration.

We remark that for a polyhedral cone $C \in \mathcal{PC}^d$, relation (1.79) reads

$$\sum_{F \in \mathcal{F}(C)} (-1)^{\dim F} \mathbb{1}[A(F, C)] = (-1)^{\dim C}[-\text{relint}\, C].$$

Applied to the dual cone C°, this can be written as

$$\sum_{F \in C} (-1)^{d-\dim F} \mathbb{1}[A(\widehat{F}_C, C^\circ)] = (-1)^{d-\dim \text{lineal}(C)} \mathbb{1}[-\text{relint}\, C^\circ],$$

where (1.4) was applied to C°. By (1.35), this is equivalent to

$$\sum_{F \in \mathcal{F}(C)} (-1)^{\dim F} \mathbb{1}[F^\circ] = (-1)^{\dim \text{lineal}(C)} \mathbb{1}[-\text{relint}\, C^\circ]. \tag{1.80}$$

Theorem 1.7.3 has also a localized version. Roughly speaking, relations for polyhedral cones yield local relations for polyhedra. Let $P \in \mathcal{Q}^d$ and let E be a face of P. Let $z \in \text{relint}\, E$. The faces F' of $A(E, P)$ are in one-to-one correspondence with the faces F of P satisfying $E \leq F$, such that $F' = \text{pos}\,(F - z)$, thus $A(E, F) = F'$ and hence

$$A(F', A(E, P)) = A(F, P). \tag{1.81}$$

Similarly,

$$N(A(E, P), F') = N(P, F). \tag{1.82}$$

Applying (1.78) to the cone $A(E, P)$ and observing (1.81) and rec $A(E, P) = A(E, P)$, we obtain the following localized version.

Theorem 1.7.4. *If $P \in \mathcal{Q}^d$ and $E \in \mathcal{F}(P)$, then*

$$\sum_{E \leq F \in \mathcal{F}(P)} (-1)^{\dim F} \mathbb{1}[\text{relint}\, A(F, P)] = (-1)^{\dim P} \mathbb{1}[-A(E, P)]. \tag{1.83}$$

For an application, we reformulate this for extended angle cones. Let $E \leq G$ be faces of the polyhedron P. Relation (1.83) for $P = G$ reads

$$\sum_{E \leq F \leq G} (-1)^{\dim F} \mathbb{1}[\text{relint}\, A(F, G)](x) = (-1)^{\dim G} \mathbb{1}[-A(E, G)](x) \tag{1.84}$$

for $x \in \mathbb{R}^d$. For $x \notin \langle G \rangle$, both sides of (1.84) are zero. For $x \in \langle G \rangle$, we have

$$x \in (\text{relint}\, A(F, G)) + \langle G \rangle^\perp \Leftrightarrow x \in \text{relint}\, A(F, G),$$
$$x \in -A(F, G) + \langle G \rangle^\perp \Leftrightarrow x \in -A(F, G).$$

Therefore, (1.84) can equivalently be written as

$$\sum_{E \leq F \leq G} (-1)^{\dim F} \mathbb{1}[(\operatorname{relint} A(F,G)) + \langle G \rangle^{\perp}]$$

$$= (-1)^{\dim G} \mathbb{1}[-A(E,G) + \langle G \rangle^{\perp}]. \tag{1.85}$$

We need also a relation involving normal cones. Recall that (1.26) states for a polyhedron $P \in \mathcal{Q}^d$ that

$$\sum_{F \in \mathcal{F}(P)} \mathbb{1}[(\operatorname{relint} F) + N(P,F)] = 1. \tag{1.86}$$

Also from this, we can derive a local version. Let $P \in \mathcal{Q}^d$ and let E be a face of P. Applying (1.86) to the angle cone $A(E,P)$, we obtain

$$\sum_{F' \in \mathcal{F}(A(E,P))} \mathbb{1}[(\operatorname{relint} F') + N(A(E,P), F')] = 1. \tag{1.87}$$

Using (1.82), relation (1.87) can be written as

$$\sum_{E \leq F \in \mathcal{F}(P)} \mathbb{1}[(\operatorname{relint} A(E,F)) + N(P,F)] = 1. \tag{1.88}$$

We have $A(E,F) \subseteq \langle F \rangle$ and $N(P,F) \subseteq \langle F \rangle^{\perp}$. Hence, if $x = x_1 + x_2$ with $x_1 \in \langle F \rangle$ and $x_2 \in \langle F \rangle^{\perp}$, then

$$x \in (\operatorname{relint} A(E,F)) + N(P,F)$$

$$\Leftrightarrow x_1 \in \operatorname{relint} A(E,F) \wedge x_2 \in N(P,F)$$

$$\Leftrightarrow x \in (\operatorname{relint} A(E,F)) + \langle F \rangle^{\perp} \wedge x \in N(P,F) + \langle F \rangle.$$

It follows that (1.88) is equivalent to

$$\sum_{E \leq F \in \mathcal{F}(P)} \mathbb{1}[(\operatorname{relint} A(E,F)) + \langle F \rangle^{\perp}] \mathbb{1}[N(P,F) + \langle F \rangle] = 1. \tag{1.89}$$

We can now derive the following.

Theorem 1.7.5. *Let $P \in \mathcal{Q}^d$ be a polyhedron, and let $E \neq P$ be a face of P. Then*

$$\sum_{E \leq F \in \mathcal{F}(P)} (-1)^{\dim F} \mathbb{1}[A(E,F) - N(P,F)] = 0. \tag{1.90}$$

Proof. The subsequent argument becomes more perspicuous if, following Mc-Mullen [130], we use the incidence algebra $I(\mathcal{F}(P))$ of the poset of faces of P (partially ordered by inclusion). We recall that by (1.59) its Möbius function is given by

$$\mu(E, F) = (-1)^{\dim F - \dim E} \quad \text{for } E, F \in \mathcal{F}(P) \text{ with } E \le F,$$

and that the delta, zeta and Möbius function of $I(\mathcal{F}(P))$ (see Section 1.1) satisfy $\mu \circ \zeta = \delta = \zeta \circ \mu$.

Now we fix a vector $x \in \mathbb{R}^d$ and define elements $B, \overline{B}, \Gamma$ of the incidence algebra $I(\mathcal{F}(P))$ by

$$B(F, G) = \mathbb{1}[(\operatorname{relint} A(F, G)) + \langle G \rangle^{\perp}](x),$$
$$\overline{B}(F, G) = (-1)^{\dim G - \dim F} \mathbb{1}[-A(F, G) + \langle G \rangle^{\perp}](x),$$
$$\Gamma(F, G) = \mathbb{1}[N(G, F) + \langle F \rangle](x)$$

for faces $F \le G$ of P. Then relations (1.85) and (1.89) (for $P = G$) say that

$$\mu * B = \overline{B}, \qquad B * \Gamma = \zeta.$$

Therefore,

$$\overline{B} * \Gamma = (\mu * B) * \Gamma = \mu * (B * \Gamma) = \mu * \zeta = \delta,$$

hence

$$(\overline{B} * \Gamma)(E, G) = 0 \quad \text{for } E < G.$$

Explicitly, this reads

$$\sum_{E \le F \le G} (-1)^{\dim F - \dim E} \mathbb{1}[-A(E, F) + \langle F \rangle^{\perp}](x) \mathbb{1}[N(G, F) + \langle F \rangle](x) = 0.$$

We apply this with $G = P$ and x replaced by $-x$ and note that $\mathbb{R}^d = \langle F \rangle \oplus \langle F \rangle^{\perp}$, where $A(E, F) \subseteq \langle F \rangle$ and $N(P, F) \subseteq \langle F \rangle^{\perp}$. Therefore,

$$\mathbb{1}[-A(E, F) + \langle F \rangle^{\perp}](-x) \mathbb{1}[N(P, F) + \langle F \rangle](-x) = 1$$
$$\Leftrightarrow -x \in -A(E, F) + \langle F \rangle^{\perp} \wedge -x \in N(P, F) + \langle F \rangle$$
$$\Leftrightarrow -x = -a + b \text{ with } a \in A(E, F) \text{ and } b \in N(P, F)$$
$$\Leftrightarrow \mathbb{1}[A(E, F) - N(P, F)](x) = 1.$$

This yields the assertion. $\qquad\qquad\qquad\qquad\qquad\qquad\qquad\qquad\qquad\qquad\quad \square$

Corollary 1.7.2. *If $C \in \mathcal{C}^d \setminus \mathcal{L}_{\bullet}$ is a polyhedral cone, then*

$$\sum_{F \in \mathcal{F}(C)} (-1)^{\dim F} \mathbb{1}[F - N(C, F)] = 0. \qquad (1.91)$$

Proof. If $P = C$ is a cone and $E = \operatorname{lineal}(C)$, then (1.91) is obtained from (1.90), since $A(\operatorname{lineal}(C), F) = F$ for the cones $F \in \mathcal{F}(C)$. $\qquad\qquad \square$

It is relation (1.91) which later (in Section 2.3) will give, by integration, an important relation between conic intrinsic volumes.

Notes for Section 1.7

1. Relation (1.72) appears in McMullen [134], in the language of the polyhedron group.

2. Theorem 1.7.2 was proved by Lawrence [120], who used it for an algorithm to compute the volume of a simple polytope. A similar decomposition, for different purposes, was presented earlier by Varchenko [186]. An extension to non-simple polytopes was proved by Haase [83].

Relations between the formulas of Brion and of Varchenko–Lawrence were discussed by Haase [83] and by Beck, Haase and Sottile [22].

3. Although the angle sum relations named after Brianchon and Gram will not be considered before Section 2.2, we give already here some hints to the literature.

Let P be a three-dimensional convex polytope, and let σ_i denote the sum of the internal angles of P at its i-dimensional faces (as defined in Section 2.2). Brianchon [36] proved (with different notation) the relation

$$\sigma_0 - \sigma_1 + \sigma_2 - \sigma_3 = 0. \tag{1.92}$$

For the proof, he used the formula for the area of a spherical convex n-gon in terms of its internal angles, and the Euler relation for the face numbers of P. Gram [75] used the first tool to derive a formula which, together with the second tool, yields (1.92). For a tetrahedron, (1.92) was proved before by de Gua. For this, and for further historical remarks, we refer to Grünbaum [82, 14.4], who also gave a proof of the Brianchon–Gram angle sum relation for polytopes in d dimensions. An extension, for not necessarily bounded polyhedra, was shown by McMullen [131].

4. An abstract version of the Brianchon–Gram–Sommerville relation in terms of scissors congruence appears in McMullen [131, Theorem 1′]. A detailed study of identities related to characteristic functions of polyhedra, including combinatorial versions of the Gram–Sommerville and the Gauss–Bonnet relations (for decompositions of Euclidean polyhedra) was made by Chen [47]. The proof of Theorem 1.7.3 given above is taken from [164].

5. For a full-dimensional polyhedral cone, formula (1.79) appears in Adiprasito and Sanyal [1, Lem. 4.1].

6. The relations (1.79) and (1.80) motivated Zheng and Zydor [193] to consider, for $x, y \in \mathbb{R}^d$, the mapping

$$C \mapsto \sum_{F \in \mathcal{F}(C)} (-1)^{\dim F} \mathbb{1}[A(F, C)](x) \mathbb{1}[-F^\circ](y), \quad C \in \mathcal{P}^d,$$

and to derive a family, parameterized by $y \in \mathbb{R}^d$, of valuations on $\mathcal{PC}^d$ with values in the vector space $V(\mathcal{P}^d)$ spanned by the characteriatic functions of polytopes in $\mathbb{R}^d$.

7. Corollary 1.7.2 on polyhedral cones is a special case of the following general theorem on polyhedra.

Theorem. *Let $P \in \mathcal{Q}^d$, and write $P = \text{lineal}(P) \oplus P'$, so that P' is a line-free polyhedron. Then*

$$\sum_{F \in \mathcal{F}(P)} (-1)^{\dim F} \mathbb{1}[F - N(P,F)] = \begin{cases} (-1)^{\dim \text{lineal}(P)} & \text{if } P' \text{ is bounded,} \\ 0 & \text{if } P' \text{ is unbounded.} \end{cases}$$

This theorem was proved by Hug and Kabluchko [94].

Concerning the history of this result, we mention that a weaker form of Corollary 1.7.2 was first stated by McMullen [130], namely (1.91) outside some exceptional set of measure zero (but not empty), determined by C. The proof given by McMullen [130, §3] was only sketched; it was carried out in [170, Thm. 6.5.5]. The proof of Corollary 1.7.2 given above appeared in [164].

For a polytope $P \in \mathcal{P}^d$, the theorem above reads

$$\sum_{F \in \mathcal{F}(P)} (-1)^{\dim F} \mathbb{1}[F - N(P,F)] = 1 \tag{1.93}$$

That this holds almost everywhere on $\mathbb{R}^d$, was proved by Glasauer [65].

We can integrate relation (1.93) over P with respect to the Lebesgue measure (for this, it is sufficient that the relation holds almost everywhere). Writing

$$W(P,F) := (F - N(P,F)) \cap P \quad \text{for } F \in \mathcal{F}(P)$$

and denoting by V_d the volume in $\mathbb{R}^d$, we obtain

$$\sum_{i=0}^{d-1} (-1)^i \sum_{F \in \mathcal{F}_i(P)} V_d(W(P,F)) = (1 - (-1)^d) V_d(P). \tag{1.94}$$

(Note that $W(P,P) = P$.) For an elementary approach to this equation in three dimensions we refer to Callahan and Hann [43].

Equation (1.94) has a version that holds for general convex bodies (using support measures). This was first obtained for sufficiently smooth bodies by Hann [88], for general convex bodies by Hug [93], and was proved in a simpler way by Glasauer [65].

1.8 Polarity as a valuation

In this section, we have a closer look at the valuation property of polarity on closed convex sets. We begin with some more general remarks on polarity.

For $\emptyset \neq A \subseteq \mathbb{R}^d$, the *polar set* is defined by

$$A^\circ := \{x \in \mathbb{R}^d : \langle x, y \rangle \leq 1 \text{ for all } y \in A\}. \tag{1.95}$$

If A is a cone, then (1.95) is equivalent to

$$A^\circ = \{x \in \mathbb{R}^d : \langle x, y \rangle \leq 0 \text{ for all } y \in A\}.$$

Let $A^{\circ\circ} := (A^\circ)^\circ$ and

$$\mathrm{conv}(o, A) := \mathrm{conv}(A \cup \{o\}).$$

Lemma 1.8.1. *Let $\emptyset \neq A \subseteq \mathbb{R}^d$. Then A° is closed and convex, and $o \in A^\circ$. Further, $A^{\circ\circ} = \mathrm{cl}\,\mathrm{conv}(o, A)$.*

Proof. If $x_1, x_2 \in A^\circ$ and $\lambda \in [0, 1]$, then $\langle (1-\lambda)x_1 + \lambda x_2, y \rangle \leq 1$ for all $y \in A$, hence $(1 - \lambda)x_1 + \lambda x_2 \in A^\circ$, thus A° is convex. From the continuity of the scalar product it follows that A° is closed. Clearly, $o \in A^\circ$.

Let $y \in A$. For all $x \in A^\circ$ we have $\langle x, y \rangle \leq 1$, hence $y \in A^{\circ\circ}$. Thus, $A \subseteq A^{\circ\circ}$. Since $o \in A^{\circ\circ}$ and $A^{\circ\circ}$ is closed and convex, also $\mathrm{cl}\,\mathrm{conv}(o, A) \subseteq A^{\circ\circ}$.

Let $z \in \mathbb{R}^d \setminus \mathrm{cl}\,\mathrm{conv}(o, A)$. Since $\mathrm{cl}\,\mathrm{conv}(o, A)$ is closed and convex, $\mathrm{cl}\,\mathrm{conv}(o, A)$ and z can be strongly separated by a hyperplane. Hence, there are $u \in \mathbb{S}^{d-1}$ and $\alpha \in \mathbb{R}$ with $\langle z, u \rangle > \alpha$ and $\langle y, u \rangle < \alpha$ for all $y \in \mathrm{cl}\,\mathrm{conv}(o, A)$. Since $o \in \mathrm{cl}\,\mathrm{conv}(o, A)$, this implies $\alpha > 0$. Since $\langle y, u/\alpha \rangle < 1$ for all $y \in A$, we have $u/\alpha \in A^\circ$. Since $\langle z, u/\alpha \rangle > 1$, we have $z \notin A^{\circ\circ}$. This shows that $A^{\circ\circ} = \mathrm{cl}\,\mathrm{conv}(o, A)$. $\qquad\square$

As a consequence, if $K \in \mathcal{CC}^d$ and $o \in K$, then $K^{\circ\circ} = K$.

Theorem 1.8.1. *Let $K_1, K_2 \in \mathcal{CC}^d$ be closed convex sets such that $K_1 \cup K_2$ is convex. Then*

$$(K_1 \cup K_2)^\circ = K_1^\circ \cap K_2^\circ, \qquad (K_1 \cap K_2)^\circ = K_1^\circ \cup K_2^\circ.$$

Proof. For $i = 1, 2$ we have $K_1 \cap K_2 \subseteq K_i$, hence $K_i^\circ \subseteq (K_1 \cap K_2)^\circ$, thus $K_1^\circ \cup K_2^\circ \subseteq (K_1 \cap K_2)^\circ$.

Let $x \in \mathbb{R}^d \setminus (K_1^\circ \cup K_2^\circ)$. For $i = 1, 2$, the point x and the convex body K_i° can be strongly separated by a hyperplane, hence there are $u_i \in \mathbb{S}^{d-1}$ and $\alpha_i \in \mathbb{R}$ such that $\langle x, u_i \rangle > \alpha_i$ and $\langle y, u_i \rangle < \alpha_i$ for all $y \in K_i^\circ$. Since $o \in K_i^\circ$, this implies $\alpha_i > 0$. Thus, $\langle y, u_i/\alpha_i \rangle < 1$ for all $y \in K_i^\circ$, hence $u_i/\alpha_i \in K_i^{\circ\circ} = \mathrm{cl}\,\mathrm{conv}(o, K_i)$. There exists $\lambda_i \geq 1$ with $\lambda_i u_i/\alpha_i \in K_i$. Since $K_1 \cup K_2$ is convex, there exists $z \in K_1 \cap K_2$ with $z \in [\lambda_1 u_1/\alpha_1, \lambda_2 u_2/\alpha_2]$. Then $\langle x, z \rangle > 1$ and hence $x \notin (K_1 \cap K_2)^\circ$. We have proved that $(K_1 \cap K_2)^\circ \subseteq K_1^\circ \cup K_2^\circ$ and thus $(K_1 \cap K_2)^\circ = K_1^\circ \cup K_2^\circ$.

From $K_i \subseteq K_1 \cup K_2$ we get $(K_1 \cup K_2)^\circ \subseteq K_1^\circ \cap K_2^\circ$. Let $x \in K_1^\circ \cap K_2^\circ$. Then $\langle x, y \rangle \leq 1$ for all $y \in K_1 \cup K_2$. Hence, $K_1^\circ \cap K_2^\circ \subseteq (K_1 \cup K_2)^\circ$. Here equality holds, which completes the proof. $\qquad\square$

For $L, M \in \mathcal{CC}^d$ with $L \cup M \in \mathcal{CC}^d$ we have $(L \cup M) + (L \cap M) = L + M$; see [163, (3.4)] (where the boundedness is not needed). Hence, for $K_1, K_2 \in \mathcal{CC}^d$ with convex union $K_1 \cup K_2$, Theorem 1.8.1 gives

$$(K_1 \cup K_2)^\circ + (K_1 \cap K_2)^\circ = K_1^\circ + K_2^\circ.$$

In this sense, polarity can be viewed as a valuation.

Now recall that $V(\mathcal{CC}^d)$ denotes the real vector space spanned by the characteristic functions of closed convex sets in $\mathbb{R}^d$. If we define a mapping $\psi : \mathcal{CC}^d \to V(\mathcal{CC}^d)$ by

$$\psi(K) = \mathbb{1}[K^\circ] \quad \text{for } K \in \mathcal{CC}^d,$$

then for $K_1, K_2 \in \mathcal{CC}^d$ with convex union $K_1 \cup K_2$ we have

$$\psi(K_1 \cup K_2) + \psi(K_1 \cap K_2) = \mathbb{1}[(K_1 \cup K_2)^\circ] + \mathbb{1}[(K_1 \cap K_2)^\circ]$$
$$= \mathbb{1}[K_1^\circ \cap K_2^\circ] + \mathbb{1}[K_1^\circ \cup K_2^\circ] = \mathbb{1}[K_1^\circ] + \mathbb{1}[K_2^\circ] = \psi(K_1) + \psi(K_2).$$

It is not a priori clear that ψ is fully additive, but this is implied by the following theorem, in view of Theorem 1.6.2.

Theorem 1.8.2. *There is a linear mapping $\phi_{\mathrm{pol}} : V(\mathcal{CC}^d) \to V(\mathcal{CC}^d)$ such that*

$$\phi_{\mathrm{pol}}(\mathbb{1}[K]) = \mathbb{1}[K^\circ] \quad \text{for } K \in \mathcal{CC}^d.$$

Proof. (After Barvinok [18, Sect. IV.1], [19, Chap. 5]) By Theorems 1.6.8 and 1.6.2 (and using the same notation), there is a linear mapping $\overline{\chi} : V(\mathcal{CC}^d) \to \mathbb{R}$ such that

$$\overline{\chi}(\mathbb{1}[K]) = 1 \quad \text{for } K \in \mathcal{CC}^d.$$

For $\varepsilon > 0$ and $y \in \mathbb{R}^d$, we set $H_{y,\varepsilon} := \{x \in \mathbb{R}^d : \langle x, y \rangle \geq 1 + \varepsilon\}$. We define a mapping ϕ_ε from $V(\mathcal{CC}^d)$ into the vector space of real functions on $\mathbb{R}^d$ by

$$\phi_\varepsilon(g)(y) := \overline{\chi}(g) - \overline{\chi}(g\mathbb{1}[H_{y,\varepsilon}]) \quad \text{for } g \in V(\mathcal{CC}^d).$$

This is possible, since $V(\mathcal{CC}^d)$ is closed under multiplication. Clearly, ϕ_ε is linear. For $K \in \mathcal{CC}^d$ we have

$$\phi_\varepsilon(\mathbb{1}[K])(y) = \overline{\chi}(\mathbb{1}[K]) - \overline{\chi}(\mathbb{1}[K \cap H_{y,\varepsilon}])$$
$$= \begin{cases} 1 \text{ if } K \cap H_{y,\varepsilon} = \emptyset, \\ 0 \text{ if } K \cap H_{y,\varepsilon} \neq \emptyset, \end{cases} = \begin{cases} 1 \text{ if } \langle x, y \rangle < 1 + \varepsilon \,\forall\, x \in K, \\ 0 \text{ otherwise.} \end{cases}$$

It follows that

$$\lim_{\varepsilon \downarrow 0} \phi_\varepsilon(\mathbb{1}[K])(y) = \begin{cases} 1 \text{ if } \langle x, y \rangle \leq 1 \,\forall\, x \in K, \\ 0 \text{ otherwise,} \end{cases} = \mathbb{1}[K^\circ](y).$$

Hence, $\phi_{\mathrm{pol}} := \lim_{\varepsilon \downarrow 0} \phi_\varepsilon$ exists and defines a linear mapping $\phi_{\mathrm{pol}} : V(\mathcal{CC}^d) \to V(\mathcal{CC}^d)$ with $\phi_{\mathrm{pol}}(\mathbb{1}[K]) = \mathbb{1}[K^\circ]$. $\quad\square$

By Theorem 1.8.2, any linear relation

$$\sum_{i=1}^{m} \alpha_i \mathbb{1}[K_i] = 0$$

with $m \in \mathbb{N}$, $\alpha_i \in \mathbb{R}$, $K_i \in \mathcal{CC}^d$ $(i = 1, \ldots, m)$ implies a corresponding linear relation

$$\sum_{i=1}^{m} \alpha_i \mathbb{1}[K_i^{\circ}] = 0.$$

To give an example how this can be applied, we first state, for a polytope $P \in \mathcal{P}^d$ with $o \in \operatorname{int} P$, the identity

$$\sum_{F \in \mathcal{F}(P) \cup \{\emptyset\}} (-1)^{\dim F} \mathbb{1}[\operatorname{conv}(o, F)] = 0. \tag{1.96}$$

(where $\dim \emptyset := -1$). For $x \in \mathbb{R}^d \setminus P$, this holds trivially. At $x = o$, the left side of (1.96) gives

$$-1 + \sum_{F \in \mathcal{F}(P)} (-1)^{\dim F} = 0,$$

by the Euler relation. Let $x \in P \setminus \{o\}$. Let $x' \in \operatorname{bd} P$ be the unique point with $x \in [o, x']$. There is a unique face $E \in \mathcal{F}(P)$ with $x' \in \operatorname{relint} E$, and $E \neq P$. We have

$$\mathbb{1}[\operatorname{conv}(o, F)](x) = 1 \Leftrightarrow E \leq F,$$

hence

$$\sum_{F \in \mathcal{F}(P) \cup \{\emptyset\}} (-1)^{\dim F} \mathbb{1}[\operatorname{conv}(o, F)](x) = \sum_{E \leq F \in \mathcal{F}(P)} (-1)^{\dim F} = 0,$$

by the local Euler relation. Thus, (1.96) is established.

To (1.96) we now apply the mapping $\phi_{\operatorname{pol}}$ of Theorem 1.8.2, to obtain

$$\sum_{F \in \mathcal{F}(P) \cup \{\emptyset\}} (-1)^{\dim F} \mathbb{1}[\operatorname{conv}(o, F)^{\circ}] = 0.$$

We note that for a face $F \in \mathcal{F}(P)$, $F \neq P$, we have

$$\operatorname{conv}(o, F)^{\circ} = T(\widehat{F}, P^{\circ}),$$

where $\widehat{F}$ denotes the face of the polar polytope P° that is conjugate to F (see, e.g., [163, Sect. 2.4]). We have $\dim \widehat{F} + \dim F = d - 1$. Further,

$$\operatorname{conv}(o, \emptyset)^{\circ} = \mathbb{R}^d = T(P^{\circ}, P^{\circ}), \qquad \operatorname{conv}(o, P)^{\circ} = P^{\circ}.$$

Hence, we obtain

$$-\mathbb{1}[T(P^\circ, P^\circ)] + \sum_{F \in \mathcal{F}(P),\, F \neq P} (-1)^{\dim F} \mathbb{1}[T(\widehat{F}, P^\circ)] + (-1)^d \mathbb{1}[P^\circ] = 0,$$

which can be written as

$$\sum_{G \in \mathcal{F}(P^\circ)} (-1)^{\dim G} \mathbb{1}[T(G, P^\circ)] = \mathbb{1}[P^\circ]. \tag{1.97}$$

This is relation (1.72) for the polytope P°.

Also other relations found in Section 1.7 can be derived via polarity. In the next example, we replace the convex hulls of the faces and the origin in (1.96) by the positive hulls of the faces. Let $P \in \mathcal{P}^d$ and $o \in \operatorname{int} P$. Then

$$\sum_{F \in \mathcal{F}(P)} (-1)^{\dim F} \mathbb{1}[\operatorname{pos} F] = \mathbb{1}[\{o\}]. \tag{1.98}$$

At $x = o$, (1.98) follows from the Euler relation. Let $x \neq o$. The ray through x with endpoint o intersects the boundary of P in $\operatorname{relint} E$ for a unique face E, and $E \neq P$. For $F \in \mathcal{F}(P)$ we have $\mathbb{1}[\operatorname{pos} F](x) = 1 \Leftrightarrow E \leq F$. Therefore,

$$\sum_{F \in \mathcal{F}(P)} (-1)^{\dim F} \mathbb{1}[\operatorname{pos} F](x) = \sum_{E \leq F \in \mathcal{F}(P)} (-1)^{\dim F} = 0 = \mathbb{1}[\{o\}](x),$$

where the local Euler relation was used.

To (1.98), we can apply the mapping ϕ_{pol} of Theorem 1.8.2. Noting that

$$(\operatorname{pos} F)^\circ = A(\widehat{F}, P^\circ), \qquad A(P^\circ, P^\circ) = \mathbb{R}^d,$$

we get

$$\sum_{F \in \mathcal{F}(P),\, F \neq P} (-1)^{\dim F} \mathbb{1}[A(\widehat{F}, P)] + (-1)^d \mathbb{1}[\{o\}] = \mathbb{1}[\mathbb{R}^d].$$

This can be written as

$$\sum_{G \in \mathcal{F}(P^\circ)} (-1)^{\dim G} \mathbb{1}[A(G, P^\circ)] = \mathbb{1}[\{o\}]. \tag{1.99}$$

This is relation (1.77) for the polytope P°.

Notes for Section 1.8

1. A version of Theorem 1.8.2 for cones was first proved by Lawrence [119, Sect. 7], accompanied by the remark that analogous results hold for other polarities. That equations (1.97) and (1.99) can be obtained via polarity, was pointed out to the author by A. Barvinok. The derivation of formula (1.97) via polarity is also explained in the article by Beck, Haase, and Sottile [22].

1.9 A characterization of polarity

The polarity of convex cones can essentially be characterized by its simplest properties. In dimensions at least three, polarity is, up to a linear mapping, the only idempotent mapping of the set of closed convex cones into itself that inverts inclusions. This is made precise by the following theorem.

Theorem 1.9.1. *Let $d \geq 3$. Let $\psi : \mathcal{C}^d \to \mathcal{C}^d$ be a mapping satisfying*

$$\psi(\psi(C)) = C \quad \text{for } C \in \mathcal{C}^d \tag{1.100}$$

and

$$C \subset D \Rightarrow \psi(C) \supset \psi(D) \quad \text{for } C, D \in \mathcal{C}^d. \tag{1.101}$$

Then there is a selfadjoint linear transformation $g \in \mathrm{GL}(d)$ such that

$$\psi(C) = gC^\circ \quad \text{for all } C \in \mathcal{C}^d.$$

We reproduce here, with modifications, the proof given in [162]. It borrows some arguments from Böröczky and Schneider [32].

Theorem 1.9.1 will be deduced from a classification of the endomorphisms of the lattice of closed convex cones in dimensions at least three. The set $\mathcal{C}^d$ of closed convex cones in $\mathbb{R}^d$ is a lattice with the operations $\cap$ of intersection and $\vee$ defined by

$$C \vee D := \mathrm{cl}\,\mathrm{conv}(C \cup D) = \mathrm{cl}(C + D), \quad C, D \in \mathcal{C}^d.$$

Theorem 1.9.2. *Let $d \geq 3$. Let $\varphi : \mathcal{C}^d \to \mathcal{C}^d$ be a mapping satisfying*

$$\varphi(C \cap D) = \varphi(C) \cap \varphi(D), \tag{1.102}$$

$$\varphi(C \vee D) = \varphi(C) \vee \varphi(D), \tag{1.103}$$

for all $C, D \in \mathcal{C}^d$. Then either φ is constant, or there is a linear transformation $g \in \mathrm{GL}(d)$ such that $\varphi(C) = gC$ for all $C \in \mathcal{C}^d$.

Let us first show why there is no corresponding characterization for $d = 2$. Let $h : \mathbb{R} \to \mathbb{R}$ be a strictly increasing, continuous function with $h(\alpha + \pi) = h(\alpha) + \pi$ for $\alpha \in \mathbb{R}$. Using standard coordinates in $\mathbb{R}^2$, define the ray $R_\alpha := \{\lambda(\cos\alpha, \sin\alpha) : \lambda \geq 0\}$. Every cone $\in \mathcal{C}^2$ different from $\mathbb{R}^2$ and $\{o\}$ is of the form $C = \bigcup_{\beta \leq \alpha \leq \gamma} R_\alpha$ with $\beta \leq \gamma \leq \beta + \pi$, where the angle β is uniquely determined $\mod 2\pi$. We define $\varphi(C) := \bigcup_{h(\beta) \leq \alpha \leq h(\gamma)} R_\alpha$ and supplement this by $\varphi(\mathbb{R}^2) := \mathbb{R}^2$ and $\varphi(\{o\}) := \{o\}$. Then φ is an endomorphism of the lattice $(\mathcal{C}^2, \cap, \vee)$, which is not of the form stated in the theorem above.

Before proving Theorem 1.9.2, we show that it implies Theorem 1.9.1. For this, we first deduce from Theorem 1.9.2 the following corollary.

Corollary 1.9.1. *Let $d \geq 3$. Let $\sigma : \mathcal{C}^d \to \mathcal{C}^d$ be a surjective mapping satisfying*

$$C_1 \subset C_2 \Leftrightarrow \sigma(C_1) \subset \sigma(C_2) \quad \text{for all } C_1, C_2 \in \mathcal{C}^d. \tag{1.104}$$

Then there exists a linear transformation $g \in \mathrm{GL}(d)$ such that $\sigma(C) = gC$ for $C \in \mathcal{C}^d$.

Proof. Let $C_1, C_2 \in \mathcal{C}^d$. Then (1.104) gives $\sigma(C_1 \cap C_2) \subseteq \sigma(C_1) \cap \sigma(C_2)$. Suppose this inclusion is strict. Then by the surjectivity of σ there exists a cone $D \in \mathcal{C}^d$ such that $\sigma(C_1 \cap C_2) \subsetneqq \sigma(D) \subseteq \sigma(C_1) \cap \sigma(C_2)$, and (1.104) gives $C_1 \cap C_2 \subsetneqq D \subseteq C_1 \cap C_2$, a contradiction.

Similarly, (1.104) gives $\sigma(C_1) \vee \sigma(C_2) \subseteq \sigma(C_1 \vee C_2)$. Suppose this inclusion is strict. Then by the surjectivity of σ there exists a cone $D \in \mathcal{C}^d$ such that $\sigma(C_1) \vee \sigma(C_2) \subseteq \sigma(D) \subsetneqq \sigma(C_1 \vee C_2)$. Now (1.104) gives $C_1 \vee C_2 \subseteq D \subsetneqq C_1 \vee C_2$, a contradiction.

Thus, σ is a lattice endomorphism, and the assertion follows from Theorem 1.9.2. □

Proof of Theorem 1.9.1. The mapping $C \mapsto C^\circ$ reverses inclusions. Hence, if ψ satisfies the assumptions of Theorem 1.9.1, then the map σ defined by $\sigma(C) := \psi(C^\circ)$ satisfies the assumptions of Corollary 1.9.1 (the surjectivity follows from (1.100)). Therefore, there exists a linear transformation $g \in \mathrm{GL}(d)$ such that $\sigma(C) = gC$ for $C \in \mathcal{C}^d$. Since $\psi(C) = gC^\circ$ and $(gC)^\circ = g^{-\top}C^\circ$, it follows from (1.100) that g is self-adjoint. □

Before the proof of Theorem 1.9.2, we state a simple lemma.

Lemma 1.9.1. *Let $M \subset \mathbb{R}^d$ be a nonempty closed convex set. Let $\mathcal{F}$ be a family of d-dimensional closed convex sets such that $K \neq M$ for $K \in \mathcal{F}$ and $K_1 \cap K_2 = M$ for $K_1, K_2 \in \mathcal{F}$ with $K_1 \neq K_2$. Then $\mathcal{F}$ is at most countable.*

Proof. Let S be a dense sequence in $\mathbb{R}^d$. Let $K \in \mathcal{F}$. Then $K \setminus M$ contains points of S, and we give K the smallest number of a point of S in $K \setminus M$. This yields an enumeration of $\mathcal{F}$. □

Proof of Theorem 1.9.2. Suppose that φ satisfies the assumptions (1.102) and (1.103). Then $C \subset D$ with $C, D \in \mathcal{C}^d$ implies $\varphi(C) = \varphi(C \cap D) = \varphi(C) \cap \varphi(D) \subset \varphi(D)$, thus φ is inclusion preserving.

In the following, we assume that $d \geq 3$. We wite $\mathcal{R}$ for the set of all rays with endpoint o in $\mathbb{R}^d$. We distinguish several cases.

Case 1. $\varphi(R) = \varphi(\{o\})$ for all $R \in \mathcal{R}$.

We choose rays $R_1, \ldots, R_{d+1}$ with $R_1 \vee \cdots \vee R_{d+1} = \mathbb{R}^d$. For $C \in \mathcal{C}^d$ we then have $\{o\} \subseteq C \subseteq R_1 \vee \cdots \vee R_{d+1}$ and hence $\varphi(\{o\}) \subseteq \varphi(C) \subseteq \varphi(R_1) \vee \cdots \vee \varphi(R_{d+1}) = \varphi(\{o\}$, thus

$$\varphi(C) = \varphi(\{o\}) \quad \text{for } C \in \mathcal{C}^d.$$

We see that in this case the mapping φ is constant.

Case 2. $\varphi(R) = \varphi(\{o\})$ for some $R \in \mathcal{R}$, but not for all $R \in \mathcal{R}$.

Let $\mathcal{A} := \{R \in \mathcal{R} : \varphi(R) = \varphi(\{o\})\}$ and $A := \bigcup_{R \in \mathcal{A}} R$. If $S, T \in \mathcal{A}$ and $U \in \mathcal{R}$ satisfies $U \subseteq S \vee T$, then $\varphi(\{o\}) \subseteq \varphi(U) \subseteq \varphi(S) \vee \varphi(T) = \varphi(\{o\})$, hence $U \in \mathcal{A}$; thus A is a convex cone. By the separation theorem, it lies in a halfspace, since there are rays not belonging to it. Its closure is different from $\mathbb{R}^d$ and $\{o\}$, hence there are rays $R \in \mathcal{A}$ and $T, U \in \mathcal{R} \setminus \mathcal{A}$ such that $T \cap U = \{o\}$ and $U \subseteq R \vee T$. Then $\varphi(\{o\}) \subseteq \varphi(U) \subseteq [\varphi(R) \vee \varphi(T)] \cap \varphi(U) = [\varphi(\{o\}) \vee \varphi(T)] \cap \varphi(U) = \varphi(\{o\} \vee T) \cap \varphi(U) = \varphi(T) \cap \varphi(U) = \varphi(\{o\})$, hence $\varphi(U) = \varphi(\{o\})$, a contradiction. Thus, Case 2 cannot occur.

It remains to consider the situation where $\varphi(R) \neq \varphi(\{o\})$ for all $R \in \mathcal{R}$. If this holds, we have

$$\dim \varphi(R) \geq 1 \quad \text{for all } R \in \mathcal{R}.$$

In fact, the equality $\varphi(R) = \{o\}$ for some $R \in \mathcal{R}$ would imply $\varphi(\{o\}) \subseteq \varphi(R) = \{o\}$ and hence $\varphi(R) = \varphi(\{o\})$, a contradiction.

Case 3. $\varphi(R) \neq \varphi(\{o\})$ for all $R \in \mathcal{R}$, $\varphi(\{o\}) \neq \{o\}$.

Case 4. $\varphi(R) \neq \varphi(\{o\})$ for all $R \in \mathcal{R}$, $\varphi(\{o\}) = \{o\}$; there exists a ray $P \in \mathcal{R}$ or a line P through o with $\dim \varphi(P) \geq 2$.

We shall lead both cases 3 and 4 simultaneously to a contradiction, but this requires some preparations. We assume in the following that $\varphi(R) \neq \varphi(\{o\})$ for all $R \in \mathcal{R}$. In Case 3 we set $P := \{o\}$. Let $B := \varphi(P)$.

A ray $R \in \mathcal{R}$ is called *free* if $R \not\subseteq \lim P$. Let R be a free ray. Then $\varphi(R) \cap \varphi(P) = \varphi(R \cap P) = \varphi(\{o\})$. Suppose that $\varphi(R \vee P) = B$. Then $\varphi(R) \vee \varphi(P) = \varphi(P)$, thus $\varphi(R) \subseteq \varphi(P)$ and hence $\varphi(R) \cap \varphi(P) = \varphi(R) \neq \varphi(\{o\})$, the latter by assumption. This contradiction shows that $\varphi(R \vee P) \neq B$ for any free ray R.

By a *sheet* we understand a set $R \vee \lim P$ with a free ray R. (Thus, in Case 3, a sheet is just an arbitrary ray, and in Case 4 it is any halfplane bounded by $\lim P$.) A sheet is called *bad* if it contains a free ray R with $\varphi(P \vee R) \subseteq \lim B$; otherwise it is called *good*. Let $R_i \vee \lim P$, $i = 1, 2$, be different bad sheets with free rays R_i satisfying $\varphi(P \vee R_i) \subseteq \lim B$. Then $(R_1 \vee P) \cap (R_2 \vee P) = P$, hence $\varphi(R_1 \vee P) \cap \varphi(R_2 \vee P) = B$. As shown above, $\varphi(R_i \vee P) \neq B$. Now it follows from Lemma 1.9.1 (applied in $\lim B$) that there are at most countably many bad sheets.

Let $b := \dim B$, thus $b \geq 1$ in Case 3 and $b \geq 2$ in Case 4.

Suppose that $b \geq d - 1$. Let the set $\mathcal{S}$ contain precisely one free ray from every good sheet, and no other elements, and let $\mathcal{F} := \{\varphi(R \vee P) : R \in \mathcal{S}\}$. If $R \in \mathcal{S}$, then $\varphi(R \vee P) \not\subseteq \lim B$, hence $\dim \varphi(R \vee P) = d$. For $R_1, R_2 \in \mathcal{S}$ we have $\varphi(R_1 \vee P) \cap \varphi(R_2 \vee P) = B$ and $\varphi(R_i \vee P) \neq B$, as shown above, hence $\varphi(R_1 \vee P) \neq \varphi(R_2 \vee P)$. Thus, $\mathcal{F}$ is uncountable, since there are uncountably many good sheets. But Lemma 1.9.1 shows that the family $\mathcal{F}$ is countable. This contradiction proves that $b \leq d - 2$.

For $x \in \mathbb{R}^d \setminus \{o\}$, we denote by $R^x := \{\lambda x : \lambda \geq 0\}$ the ray generated by x. Let $k \in \{1, \ldots, d - b\}$. A set $\{x_1, \ldots, x_k\}$ of points in $\mathbb{R}^d \setminus \{o\}$, briefly a k-set, is called *full* if

$$\dim \varphi(R^{x_1} \vee \cdots \vee R^{x_k} \vee P) \geq b + k.$$

A k-flat $E \subset \mathbb{R}^d$ is called *general* if $o \notin E$ in Case 3 and $\dim \mathrm{aff}(E \cup \mathrm{lin}\, P) = k + 2$ in Case 4.

Now we need the following lemma.

Lemma 1.9.2. *Let $k \in \{2, \ldots, d - b\}$. In every general $(k - 1)$-flat $E \subset \mathbb{R}^d$ there is a full k-set.*

Proof. We use induction on k. First let $k = 2$. Let $E \subset \mathbb{R}^d$ be a general 1-flat. Since E is general, each of its points, with at most countably many exceptions, is contained in a good sheet, and different points of E are in different sheets. Choose a point $x_1 \in E$ in a good sheet. Then $B \subseteq \varphi(R^{x_1} \vee P) \not\subseteq \mathrm{lin}\, B$, hence $\{x_1\}$ is a full 1-set. If there exists $y \in E \setminus \{x_1\}$ such that $\{x_1, y\}$ is a full 2-set, we are done. Otherwise, for each $y \in E \setminus \{x_1\}$ the 2-set $\{x_1, y\}$ is not full. This implies that the linear subspace $A := \mathrm{lin}\, \varphi(R^{x_1} \vee P)$ is of dimension $b + 1$ and that $\varphi(R^y \vee P) \subseteq A$. Since $\varphi(R^{y_1} \vee P) \cap \varphi(R^{y_2} \vee P) = B$ for different $y_1, y_2 \in E \setminus \{x_1\}$, this is a contradiction to Lemma 1.9.1 (applied in A).

Now let $k \in \{2, \ldots, d - b - 1\}$ and suppose that the assertion has been proved for this number. Let $E \subset \mathbb{R}^d$ be a general k-flat. Let $F \subset E$ be a general $(k - 1)$-flat, and choose a full k-set $\{x_1, \ldots, x_k\}$ in F. If there exists $y \in E \setminus F$ such that $\{x_1, \ldots, x_k, y\}$ is a full $(k + 1)$-set, then the assertion is proved for $k + 1$. Otherwise, for each $y \in E \setminus F$, the $(k + 1)$-set $\{x_1, \ldots, x_k, y\}$ is not full. This implies that the linear subspace $A := \mathrm{lin}\, \varphi(R^{x_1} \vee \cdots \vee R^{x_k} \vee P)$ is of dimension $b + k$ and that $\varphi(R^y) \subseteq A$. Let $F' \subset E$ be a $(k - 1)$-flat which is parallel to F but different from it; it is also general. Choose a full k-set $\{y_1, \ldots, y_k\}$ in F' and set $C_{F'} := R^{y_1} \vee \cdots \vee R^{y_k} \vee P$. Then $\varphi(C_{F'}) \subset A$ and $\dim \varphi(C_{F'}) = b + k$. W can choose uncountably many such F' such that any two of them, say F_1 and F_2, satisfy $C_{F_1} \cap C_{F_2} = P$ (observe that $d - b - 1 \leq d - 2$ in Case 3 and $d - b - 1 \leq d - 3$ in Case 4), thus $\varphi(C_{F_1}) \cap \varphi(C_{F_2}) = B$. Since $\dim A = b + k$, Lemma 1.9.1 yields a contradiction. This completes the induction and thus the proof of the lemma. $\qquad\square$

Now we can deal with Cases 3 and 4. Let $k = d - b$ and choose a general k-flat $F \subset \mathbb{R}^d$. Let $\mathcal{E}$ be the family of $(k - 1)$-flats contained in F and parallel to a fixed one. Since $E \in \mathcal{E}$ is general, by Lemma 1.9.2 there exists a full k-set $\{x_1, \ldots, x_k\}$ in E. We set $C_E := R^{x_1} \vee \cdots \vee R^{x_k} \vee P$. Then $\dim C_E = d$ and $C_{E_1} \cap C_{E_2} = P$ for $E_1, E_2 \in \mathcal{E}$ with $E_1 \neq E_2$, hence $\varphi(C_{E_1}) \cap \varphi(C_{E_2}) = B$. By Lemma 1.9.1, this is a contradiction, since $\mathcal{E}$ is uncountable. This shows that Cases 3 and 4 cannot occur.

We are thus, finally, left with the following case.

Case 5. $\varphi(\{o\}) = \{o\}$ and $\dim \varphi(R) = 1$ if $R \in \mathcal{R}$ or R is a line through o.

In this case, the φ-image of a ray is either a ray or a line through the origin. We show that it is always a ray. Let $R \in \mathcal{R}$. Write $L = R \cup (-R)$. The images $\varphi(R)$, $\varphi(-R)$, $\varphi(L)$ are all one-dimensional. Since $\varphi(R) \subseteq \varphi(L)$, $\varphi(-R) \subseteq \varphi(L)$, $\varphi(R) \cap \varphi(-R) = \{o\}$, this is only possible if $\varphi(R)$ is a ray.

Now let $E \subset \mathbb{R}^d$ be a two-dimensional linear subspace. We show that the rays in E are mapped under φ into a two-dimensional subspace. Let R, S be different rays in E. If $\varphi(S) = -\varphi(R)$, then a ray $T \subset E$ different from R and S satisfies $\varphi(T) \neq -\varphi(R)$, since $\varphi(T) \cap \varphi(S) = \{o\}$. Hence, we can assume that $\varphi(S) \neq -\varphi(R)$. Let E' be the two-dimensional subspace spanned by $\varphi(R)$ and $\varphi(S)$. Let $Z \in \mathcal{R}$. If $Z \subset R \vee S$, then $\varphi(Z) \subset \varphi(R) \vee \varphi(S) \subset E'$. Let $Z \subset R \vee -S$, but $Z \neq -S$. Then $\varphi(R) \subset \varphi(Z) \vee \varphi(S)$, which implies that $\varphi(Z) \subset E'$. Similarly, $Z \subset S \vee -R$, but $Z \neq -R$ implies that $\varphi(Z) \subset E'$. Finally, if $Z \subset -R \vee -S$, we can choose rays $U \subset R \vee -S$ and $V \subset S \vee -R$ with $Z \subset U \vee V$, which gives $\varphi(Z) \subset E'$. We have proved that $\varphi(Z) \subset E'$ for each ray $Z \subset E$.

Let $R \in \mathcal{R}$. We assert that $\varphi(-R) = -\varphi(R)$. For the proof, choose $S \in \mathcal{R}$ with $S \neq \pm R$. Let A, B, C be any three of the rays $R, -R, S, -S$. Then $(A \vee B) \cap C = \{o\}$, hence $(\varphi(A) \vee \varphi(B)) \cap \varphi(C) = \{o\}$ and thus $\varphi(C) \not\subseteq \varphi(A) \vee \varphi(B)$. Since this holds for all choices of A, B, C from $\{R, -R, S, -S\}$ and since $\varphi(R)$, $\varphi(-R)$, $\varphi(S)$, $\varphi(-S)$ lie in a two-dimensional subspace, the set $\{\varphi(R), \varphi(-R), \varphi(S), \varphi(-S)\}$ must be of the form $\{\pm U, \pm V\}$ with two rays U, V satisfying $U \neq \pm V$. Suppose that $\varphi(-R) = -\varphi(S)$. Choose a ray $T \subset S \vee -R$ with $T \neq S, -R$. Then $\varphi(T) \subset \varphi(S) \vee \varphi(-R)$ and $\varphi(T) \cap \varphi(-R) = \varphi(T) \cap \varphi(S) = \{o\}$, a contradiction. Thus, $\varphi(-R) = -\varphi(R)$ is the only possibility.

Our next aim is to construct the required linear map. First let $\mathcal{L}$ be the set of one-dimensional linear subspaces of $\mathbb{R}^d$. For $L \in \mathcal{L}$, there are two rays $R, -R$ with $L = R \cup -R$; we call them the *generating rays* of L. Since $\varphi(-R) = -\varphi(R)$, the line $f(L) := \varphi(R) \cup -\varphi(R)$ does not depend on the choice of R. Thus, we have defined a map $f : \mathcal{L} \to \mathcal{L}$. Let $L_1, L_2 \in \mathcal{L}$, $L_1 \neq L_2$. Let R_i be a generating ray of L_i, for $i = 1, 2$. If $f(L_1) = f(L_2)$, then either $\varphi(R_1) = \varphi(R_2)$ or $\varphi(R_1) = -\varphi(R_2)$. But $R_1 \cap R_2 = R_1 \cap -R_2 = \{o\}$, hence $\varphi(R_1) \cap \varphi(R_2) = \varphi(R_1) \cap -\varphi(R_2) = \{o\}$, a contradiction. Thus, the map f is injective.

Suppose that $L_1, L_2, L_3 \in \mathcal{L}$ are pairwise different and are coplanar, that is, contained in a two-dimensional linear subspace of $\mathbb{R}^d$. Then we can choose generating rays R_i of L_i ($i = 1, 2, 3$) such that $R_2 \subset R_1 \vee R_3$. Then $\varphi(R_2) \subset \varphi(R_1) \vee \varphi(R_2)$. Therefore, the lines $f(L_1), f(L_2), f(L_3)$ are coplanar.

Since $d \geq 3$, it now follows from the Fundamental Theorem of Projective Geometry (see, e.g., Faure [57] for a proof of a general version), together with the injectivity of f, that there exists a linear transformation $g \in \mathrm{GL}(d)$ such that $f(L) = gL$ for all $L \in \mathcal{L}$.

Let $R \in \mathcal{R}$. Then $\varphi(R)$ and $\varphi(-R) = -\varphi(R)$ are rays with $gR \vee -gR = g(R \vee -R) = f(R \vee -R) = \varphi(R) \vee -\varphi(R)$. It follows that either $\varphi(R) = gR$ or $\varphi(R) = -gR$. Let $\mathcal{A}_+ := \{R \in \mathcal{R} : \varphi(R) = gR\}$, and let A_+ be the union of the rays in $\mathcal{A}_+$. Similarly $\mathcal{A}_-$ and A_- are defined. Then A_+ and A_- are convex cones with $A_+ \cup A_- = \mathbb{R}^d$. If both are non-empty, they have a common boundary ray S and we can choose rays $R_+ \in \mathcal{A}_+$ and $R_- \in \mathcal{A}_-$ such that $S \subset R_+ \vee R_-$. Then $\varphi(S) \subset \varphi(R_+) \vee \varphi(R_-)$, hence either $gS \subset gR_+ \vee -gR_-$ or $-gS \subset gR_+ \vee -gR_-$, which is a contradiction, since neither gS nor $-gS$ is contained in the cone $gR_+ \vee -gR_-$. It follows that A_+ or A_- is empty, and we can assume (replacing g by $-g$ if necessary) that $A_- = \emptyset$. Then $\varphi(R) = gR$ for all $R \in \mathcal{R}$.

Let $C \in \mathcal{C}^d$ and $R \in \mathcal{R}$. If $R \subseteq C$, then $gR = \varphi(R) \subseteq \varphi(C)$, hence $gC \subseteq \varphi(C)$. Let S be a ray with $S \not\subseteq gC$. Then $g^{-1}S \not\subseteq C$, hence $g^{-1}S \cap C = \{o\}$. This gives $S \cap \varphi(C) = g(g^{-1}S) \cap \varphi(C) = \varphi(g^{-1}S) \cap \varphi(C) = \{o\}$ and thus $S \not\subseteq \varphi(C)$. We have obtained $gC = \varphi(C)$, which completes the proof of the theorem. $\qquad\square$

2

Angle functions

Whereas the considerations of the first chapter were essentially combinatorial in character, we begin now with measuring convex polytopes and polyhedral cones. In Section 2.1 we deal briefly with invariant measures, as needed later. In Section 2.2 we introduce angle functions, prove the classical Brianchon–Gram–Sommerville relations and some consequences, and present McMullen's [130] relations between inner and outer angles. Then we introduce, in Section 2.3, one of our main topics, the conic intrinsic volumes and the closely related Grassmann angles, which are fundamental quantities to measure a polyhedral cone. In the fourth chapter, they will be extended to general closed convex cones and investigated in greater detail. We prove various relations for these functionals.

The last two sections of the present chapter are included since they make essential use of the angle functions of polyhedral cones. Section 2.4 proves the Gauss–Bonnet theorem for finite unions of polytopes (and for polytopal cell complexes) and discusses a large class of extensions of this theorem, using valuations on polyhedral cones. Section 2.5 deals with a polyhedral counterpart to another classical differential-geometric result, namely Weyl's tube formula. Here the Lipschitz–Killing curvatures of general polyhedra, as introduced by Cheeger, Müller and Schrader [45, (3.7)], are treated with the more elementary approach suggested by Budach [40].

2.1 Invariant measures

In this section, we collect basic facts about the standard invariant measures of Euclidean and spherical geometry, which we frequently use in the following. By a measure, without further specification, on a topological space X we always understand a measure on the σ-algebra $\mathcal{B}(X)$ of Borel sets of X. Measurability of a function on a measure space will be understood, where necessary, with respect to the completion of the measure. Invariance will refer to one of the following groups: the translation group T_d of $\mathbb{R}^d$, the orthogonal group $O(d)$ or

© The Author(s), under exclusive license to Springer Nature Switzerland AG 2022
R. Schneider, *Convex Cones*, Lecture Notes in Mathematics 2319,
https://doi.org/10.1007/978-3-031-15127-9_2

its subgroup $SO(d)$ of proper (orientation preserving) rotations, or the group G_d of rigid motions (orientation preserving isometries of $\mathbb{R}^d$). All these groups carry their standard topologies.

Generally, $\mathscr{H}^k$ denotes k-dimensional Hausdorff measure, but in special cases we use different notation. So we denote Lebesgue measure on $\mathcal{B}(\mathbb{R}^d)$ by λ_d, and the Lebesgue measure on a k-dimensional affine subspace L of $\mathbb{R}^d$ ($k \in \{0, \ldots, d\}$) is denoted by λ_L, or also by λ_k if the domain is clear from the context. The Lebesgue measure λ_d is invariant under isometries of $\mathbb{R}^d$, and the following characterization is well known.

Theorem 2.1.1. *Every translation invariant, locally finite measure on $\mathcal{B}(\mathbb{R}^d)$ is a constant multiple of the Lebesgue measure λ_d.*

Here a measure on $\mathbb{R}^d$ is called *locally finite* if it is finite on compact sets. A translation invariant measure which is finite on the unit cube $[0,1]^d$ is locally finite, hence we can also state the following corollary.

Corollary 2.1.1. *The Lebesgue measure λ_d is the only translation invariant measure on $\mathbb{R}^d$ that attains the value one at $[0,1]^d$.*

The spherical Lebesgue measure on $\mathbb{S}^{d-1}$ is denoted by σ_{d-1}. It can be defined by

$$\sigma_{d-1}(B) = \frac{1}{d}\lambda_d\left(\bigcup_{u \in B}[o, u]\right)$$

for $B \in \mathcal{B}(\mathbb{S}^{d-1})$. The k-dimensional spherical Lebesgue measure on a k-dimensional great subsphere S of $\mathbb{S}^{d-1}$ is denoted by σ_S, or also by σ_k. The isometry group $O(d)$ acts transitively on the unit sphere $\mathbb{S}^{d-1}$, and the spherical Lebesgue measure is invariant under this group. The following is well known.

Theorem 2.1.2. *Every $SO(d)$ invariant finite signed measure on $\mathcal{B}(\mathbb{S}^{d-1})$ is a constant multiple of the spherical Lebesgue measure σ_{d-1}.*

The Lebesgue measure of the unit ball is denoted by κ_d and the spherical Lebesgue measure of the unit sphere by ω_d, thus

$$\kappa_d = \lambda_d(B^d) = \frac{\pi^{\frac{d}{2}}}{\Gamma\left(1 + \frac{d}{2}\right)}, \qquad \omega_d = \sigma_{d-1}(\mathbb{S}^{d-1}) = d\kappa_d = \frac{2\pi^{\frac{d}{2}}}{\Gamma\left(\frac{d}{2}\right)}.$$

Euclidean and spherical Lebesgue measure are connected by the formula for transformation to spherical coordinates, namely

$$\lambda_d(A) = \int_0^\infty \int_{\mathbb{S}^{d-1}} \mathbb{1}_A(ru) r^{d-1} \sigma_{d-1}(\mathrm{d}u)\,\mathrm{d}r \qquad (2.1)$$

for $A \in \mathcal{B}(\mathbb{R}^d)$.

When working with a cone C, it is often convenient to express the spherical measure of $C \cap \mathbb{S}^{d-1}$ in terms of the Gaussian measure. The standard *Gaussian measure* on a k-dimensional subspace $L \in \mathcal{L}_\bullet$ is the probability measure defined by

$$\gamma_L(A) = \frac{1}{\sqrt{2\pi}^k} \int_A e^{-\frac{1}{2}\|x\|^2} \lambda_k(\mathrm{d}x)$$

for $A \in \mathcal{B}(L)$. We write $\gamma_{\mathbb{R}^d} = \gamma_d$.

If now $A \in \mathcal{B}(\mathbb{S}^{d-1})$ and

$$A^\vee := \{\lambda u : u \in A, \ \lambda \geq 0\}$$

is the cone spanned by A, then (2.1) gives

$$\gamma_d(A^\vee) = \frac{1}{\sqrt{2\pi}^d} \int_{A^\vee} e^{-\frac{1}{2}\|x\|^2} \lambda_d(\mathrm{d}x)$$

$$= \frac{1}{\sqrt{2\pi}^d} \int_0^\infty \int_A e^{-\frac{1}{2}r^2} r^{d-1} \sigma_{d-1}(\mathrm{d}u)\,\mathrm{d}r$$

$$= \frac{1}{\sqrt{2\pi}^d} \sigma_{d-1}(A) \int_0^\infty e^{-\frac{1}{2}r^2} r^{d-1}\,\mathrm{d}r,$$

and since $\int_0^\infty e^{-\frac{1}{2}r^2} r^{d-1}\,\mathrm{d}r = 2^{\frac{d}{2}-1}\Gamma\left(\frac{d}{2}\right)$, we have

$$\gamma_d(A^\vee) = \frac{\sigma_{d-1}(A)}{\omega_d}. \tag{2.2}$$

A different way to prove this is to use Theorem 2.1.2.

We can use the connection (2.2) to state a corollary of Theorem 2.1.2. For this, we need the following notations.

If $L \in \mathcal{L}_\bullet$ is a subspace, we denote by SO_L the subgroup of $\mathrm{SO}(d)$ that fixes $L^\perp$ pointwise and hence maps L into itself. It is isomorphic to the group of proper rotations of L.

Further, for $L \in \mathcal{L}_\bullet$ we write

$$\widehat{\mathcal{B}}(L) := \{A \in \mathcal{B}(L) : A^+ = A\},$$

which is known as the conic σ-algebra in L.

Corollary 2.1.2. *Let $L \in \mathcal{L}_\bullet$, and let ψ be a finite signed measure on $\widehat{\mathcal{B}}(L)$ which is invariant under SO_L. Then there is a real constant c such that*

$$\psi(A) = c\gamma_L(A) \quad \text{for } A \in \widehat{\mathcal{B}}(L).$$

We continue with some properties of the Gaussian measure on cones.

Lemma 2.1.1. *Let $k \in \{1, \ldots, d\}$. On the set of k-dimensional cones in $\mathcal{C}^d$, the function $C \mapsto \gamma_{\langle C \rangle}(C)$ is continuous (with respect to the metric δ_c defined by (1.14)). On $\mathcal{C}^d$, the function $C \mapsto \gamma_d(C)$ is continuous.*

Proof. For a convex cone C of dimension $k \geq 1$ it follows from (2.2) that $\gamma_{\langle C \rangle}(C) = k\omega_k^{-1}\lambda_k(C \cap B^d)$. This is a constant multiple of the kth Euclidean intrinsic volume of $C \cap B^d$, and each Euclidean intrinsic volume is continuous with respect to the Hausdorff metric. $\qquad\square$

Suppose that C_1, C_2 are cones which are Borel sets and lie in totally orthogonal subspaces E_1, E_2 of $\mathbb{R}^d$, say $C_i \subset E_i$, $i = 1, 2$, $\dim E_1 = k$, $\dim E_2 = d-k$. Then

$$\gamma_d(C_1 \oplus C_2) = \gamma_{\langle C_1 \rangle}(C_1)\gamma_{\langle C_2 \rangle}(C_2). \tag{2.3}$$

In fact, writing $x \in \mathbb{R}^d$ in the form $x = x_1 + x_2$ with $x_i \in E_i$, $i = 1, 2$, we have $\|x\|^2 = \|x_1\|^2 + \|x_2\|^2$ and hence

$$\gamma_d(C_1 \oplus C_2) = \frac{1}{\sqrt{2\pi}^d} \int_{C_1 \oplus C_2} e^{-\frac{1}{2}\|x\|^2} \lambda_d(\mathrm{d}x)$$

$$= \frac{1}{\sqrt{2\pi}^d} \int_{C_1} \int_{C_2} e^{-\frac{1}{2}\|x_1\|^2} e^{-\frac{1}{2}\|x_2\|^2} \lambda_{d-k}(\mathrm{d}x_2)\, \lambda_k(\mathrm{d}x_1),$$

which gives the assertion.

Some of the relations to be studied later can conveniently be formulated in terms of probabilities involving a Gaussian random vector. For this purpose, we denote, for a given subspace $L \in \mathcal{L}_\bullet$, by $\mathbf{g}_L$ a random vector (on some probability space $(\Omega, \mathcal{A}, \mathbb{P})$) in L with distribution γ_L. We write $\mathbf{g}_{\mathbb{R}^d} = \mathbf{g}$.

For example, if $C \in \mathcal{PC}^d$ and $F \in \mathcal{F}(C)$, then the nearest point map Π_C satisfies, for $x \in \mathbb{R}^d$,

$$\Pi_C(x) \in \mathrm{relint}\, F \Leftrightarrow x \in (\mathrm{relint}\, F) + N(C, F).$$

Therefore,

$$\mathbb{P}\left(\Pi_C(\mathbf{g}) \in \mathrm{relint}\, F\right) = \gamma_d(F + N(C, F)) \tag{2.4}$$

(since $\gamma_d(\mathrm{bd}(F + N(C, F))) = 0$). This quantity will later be of importance.

Further, we shall have to use the Haar measures on rotation groups and Grassmannians. The theory of Haar measures can be found in many books, of which we mention here only [51], [92], [140], [147]. For our purposes, the elementary introduction given in [170, Ch. 13] should be enough.

By ν we denote the unique normalized Haar measure on the rotation group $\mathrm{SO}(d)$. Thus, ν is a Borel probability measure on the topological group $\mathrm{SO}(d)$ which is invariant under left and right multiplication and under inversion. An elementary existence proof is given in [170, Thm. 13.2.9], and the uniqueness follows from [170, Thm. 13.1.3]. The Haar probability measure on SO_L, where $L \in \mathcal{L}_\bullet$, is denoted by ν_L.

The following remark will be useful. If $A \subseteq \mathbb{S}^{d-1}$ is a Borel set, then, for arbitrary $u \in \mathbb{S}^{d-1}$,

$$\int_{\mathrm{SO}(d)} \mathbb{1}_A(\vartheta u)\, \nu(\mathrm{d}\vartheta) = \frac{\sigma_{d-1}(A)}{\omega_d}. \tag{2.5}$$

In fact, defining $\mu(A)$ by the left side of (2.5) for arbitrary $A \in \mathcal{B}(\mathbb{S}^{d-1})$, we obtain a finite measure μ which is $\mathrm{SO}(d)$ invariant, by the invariance properties of ν. Therefore, (2.5) follows from Theorem 2.1.2 and the fact that $\mu(\mathbb{S}^{d-1}) = 1$.

If $A \in \mathcal{B}(\mathbb{R}^d)$ is a conic Borel set, that is, $\lambda A = A$ for $\lambda \geq 0$, then, for arbitrary $x \in \mathbb{R}^d \setminus \{o\}$,

$$\int_{\mathrm{SO}(d)} \mathbb{1}_A(\vartheta x)\, \nu(\mathrm{d}\vartheta) = \gamma_d(A). \tag{2.6}$$

This follows from (2.5) and (2.2), since $\mathbb{1}_A(\vartheta(x/\|x\|)) = \mathbb{1}_A(\vartheta x)$.

Let $q \in \{0, \ldots, d\}$. We recall that the *Grassmannian* $G(d, q)$ is the set of all q-dimensional subspaces of $\mathbb{R}^d$, equipped with the standard topology, for which we refer, e.g., to [170, Section 13.2]. The rotation group $\mathrm{SO}(d)$ operates continuously and transitively on $G(d, q)$, and there is a unique probability measure on $G(d, q)$ which is invariant under this operation (see, e.g., [170, Thms. 13.2.11, 13.1.5]). We denote this normalized invariant measure on the homogeneous space $G(d, q)$ by ν_q. If $L \in \mathcal{L}_\bullet$ is a subspace, we denote the Grassmannian of q-dimensional subspaces of L by $G(L, q)$ and its invariant probability measure by ν_q^L, for $q \in \{0, \ldots, \dim L\}$.

The uniqueness of the rotation invariant measures ν_q has some immediate consequences. For example, ν_{d-q} is the image measure of ν_q under the mapping $L \mapsto L^\perp$ from $G(d, d-q)$ to $G(d, q)$. Further, for any nonnegative measurable function f on $G(d, q)$ and any fixed $E \in G(d, q)$ we have

$$\int_{\mathrm{SO}(d)} f(\vartheta E)\, \nu(\mathrm{d}\vartheta) = \int_{G(d,q)} f(L)\, \nu_q(\mathrm{d}L). \tag{2.7}$$

We shall need repeatedly that in integrations with respect to the invariant measure ν on the rotation group, certain sets are negligible. Let $L, L' \in \mathcal{L}_\bullet$ be subspaces. They are said to be *in general position* if

$$\dim(L \cap L') = \max\{0, \dim L + \dim L' - d\}.$$

In other words, $L \in G(d, k)$ and $L' \in G(d, m)$ are in general position if and only if

$$\dim(L \cap L') = \begin{cases} k + m - d & \text{if } k + m \geq d, \\ 0 & \text{if } k + m < d. \end{cases}$$

The subspaces L, L' are *in special position* if and only if they are not in general position, and this is equivalent to

$$\mathrm{lin}(L \cup L') \neq \mathbb{R}^d \quad \text{and} \quad L \cap L' \neq \{o\}.$$

Lemma 2.1.2. *Let $L, L' \in \mathcal{L}_\bullet$ be subspaces. The set of all rotations $\vartheta \in \mathrm{SO}(d)$ for which L and $\vartheta L'$ are in special position has ν-measure zero.*

Two different elementary proofs can be found in [163, Lemma 4.4.1] and [170, Lemma 13.2.1].

The following two consequences will be needed. We recall that two cones $C, D \in \mathcal{C}^d$ *touch* if $C \cap D \neq \{o\}$, but C and D can be separated by a hyperplane.

Lemma 2.1.3. *Let* $C, D \in \mathcal{P}\mathcal{C}^d$ *be polyhedral cones. The set of all rotations* $\vartheta \in \mathrm{SO}(d)$ *for which* C *and* ϑD *touch has* ν-*measure zero.*

Proof. Suppose that C and ϑD touch. Then they can be separated by a hyperplane H, necessarily through o. The faces $F = H \cap C$ of C and $\vartheta G = H \cap \vartheta D$ of ϑD satisfy $F \cap \vartheta G \neq \{o\}$ and $\mathrm{lin}(F \cup \vartheta G) \neq \mathbb{R}^d$, hence the subspaces $\mathrm{lin}\, F$ and $\vartheta \mathrm{lin}\, G$ are in special position. Since F and G have only finitely many faces, the assertion follows from Lemma 2.1.2. □

The following lemma and its proof are essentially taken from Amelunxen and Lotz [10]. Recall that two cones $C, D \in \mathcal{C}^d$ *intersect transversely*, written $C \pitchfork D$, if

$$\dim(C \cap D) = \dim C + \dim D - d \quad \text{and} \quad \mathrm{relint}\, C \cap \mathrm{relint}\, D \neq \emptyset.$$

Lemma 2.1.4. *Let* $C, D \in \mathcal{P}\mathcal{C}^d$ *be polyhedral cones. The set*

$$A := \{\vartheta \in \mathrm{SO}(d) : C \cap \vartheta D = \{o\} \text{ or } C \pitchfork \vartheta D\}$$

has ν-*measure one. If* C *and* D *are not both subspaces, then the set*

$$B := \{\vartheta \in \mathrm{SO}(d) : C \cap \vartheta D = \{o\} \text{ or } C \cap \vartheta D \text{ is not a subspace}\}$$

has ν-*measure one.*

Proof. Let S be the set of all $\vartheta \in \mathrm{SO}(d)$ such that, for some faces $F \in \mathcal{F}(C)$, $G \in \mathcal{F}(D)$, the subspaces $\langle F \rangle$ and $\langle G \rangle$ are in special position. It follows from Lemma 2.1.2 that $\nu(S) = 0$.

We claim that $\mathrm{SO}(d) \setminus S \subseteq A$. Let $\vartheta \in \mathrm{SO}(d) \setminus S$. If $C \cap \vartheta D = \{o\}$, then $\vartheta \in A$. Let $C \cap \vartheta D \neq \{o\}$. If $\mathrm{relint}\, C \cap \mathrm{relint}\, \vartheta D \neq \emptyset$, then $\vartheta \notin S$ implies $\dim(C \cap \vartheta D) = \dim C + \dim D - d$ and hence $C \pitchfork \vartheta D$, thus $\vartheta \in A$. Hence, assume that $\mathrm{relint}\, C \cap \mathrm{relint}\, \vartheta D = \emptyset$. There exists a separating hyperplane H such that $C \subseteq H^+$ and $\vartheta D \subseteq H^-$. Let $F := C \cap H$ and $G := D \cap \vartheta^{-1}H$. Since $\langle F \rangle \cup \vartheta \langle G \rangle \subseteq H$ and $\langle F \rangle \cap \vartheta \langle G \rangle \supseteq C \cap \vartheta D \cap H \neq \{o\}$, the subspaces $\langle F \rangle$ and $\vartheta \langle G \rangle$ are in special position, which means that $\vartheta \in S$, a contradiction. Thus, $\mathrm{SO}(d) \setminus S \subseteq A$, hence $\nu(A) = 1$.

To prove the second assertion, suppose that C is not a subspace. We claim that $A \subseteq B$, which then gives $\nu(B) = 1$. Let $\vartheta \in A$. If $C \cap \vartheta D = \{o\}$, then $\vartheta \in B$. Assume that $C \cap \vartheta D \neq \{o\}$ and hence $C \pitchfork \vartheta D$, thus $\mathrm{relint}\, C \cap \mathrm{relint}\, \vartheta D \neq \emptyset$. There exists $o \neq x \in (\mathrm{relint}\, C) \cap \vartheta D$. Since C is not a subspace, we have $x \notin \mathrm{lineal}(C)$ (the lineality space of C) and hence $x \notin \mathrm{lineal}(C \cap \vartheta D) = \mathrm{lineal}(C) \cap \mathrm{lineal}(\vartheta D)$. Since $x \in C \cap \vartheta D$, it follows that $C \cap \vartheta D$ is not a subspace, thus $\vartheta \in B$. □

2.2 Angles

To measure polyhedra locally, several angle functions have to be introduced. For a cone $C \in \mathcal{C}^d$, we define the *solid angle* of C by

$$\alpha(C) := \gamma_{\langle C \rangle}(C) \tag{2.8}$$

if $\dim C \geq 1$, and $\alpha(\{o\}) := 1$. Thus, if $\dim C = k > 0$, then

$$\alpha(C) = \frac{\sigma_{k-1}(C \cap \mathbb{S}^{d-1})}{\omega_k}. \tag{2.9}$$

By its definition, the function α is intrinsic, that is, $\alpha(C)$ depends only on C and not on the dimension of the space in which it is embedded. We can also write

$$\alpha(C) = \mathbb{P}\left(\mathbf{g}_{\langle C \rangle} \in C\right).$$

On the cones of a fixed dimension, the function α is continuous. This follows from Lemma 2.1.1.

For a cone $C \in \mathcal{C}^d$, we define the *outer angle* of C by

$$v_0(C) := \gamma_d(C^\circ). \tag{2.10}$$

Also the function v_0 is intrinsic. In fact, if C lies in a subspace L of $\mathbb{R}^d$, and $C^{\circ L}$ denotes the polar cone of C with respect to L, then $C^\circ = C^{\circ L} \oplus L^\perp$ (an orthogonal direct sum) and hence

$$\gamma_d(C^\circ) = \gamma_d(C^{\circ L} \oplus L^\perp) = \gamma_{\langle C \rangle}(C^{\circ L}),$$

by (2.3).

Definition 2.2.1. *Let $P \in \mathcal{Q}^d$ be a polyhedron and F a face of P. The* internal angle *of P at F is defined by*

$$\beta(F, P) = \gamma_{\langle P \rangle}(A(F, P)) \tag{2.11}$$

and the external angle *of P at F by*

$$\gamma(F, P) = \gamma_{\langle F \rangle^\perp}(N(P, F)). \tag{2.12}$$

If $F, G \in \mathcal{F}(P)$, but $F \not\leq G$, one defines $\beta(F, G) = 0 = \gamma(F, G)$.

We note that also $\beta(F, P)$ and $\gamma(F, P)$ depend only on P and F and not on the dimension of the ambient space. Although the normal cone $N(P, F)$ depends on the space in which P is embedded, this does not hold for $\gamma(F, P)$. In fact, if P lies in a subspace L of $\mathbb{R}^d$ and if $N_L(P, F)$ denotes the normal cone of P at F with respect to L, then $N(P, F) = N_L(P, F) \oplus L^\perp$ (an orthogonal direct sum), and the assertion follows from (2.3).

In the particular case of a cone $C \in \mathcal{PC}^d$, we have $\{o\} \in \mathcal{F}(C)$, and we write $\beta(o, C) := \beta(\{o\}, C)$, thus

$$\beta(o, C) = \gamma_{\langle C \rangle}(C) = \alpha(C) = \beta(\mathrm{lineal}(C), C), \tag{2.13}$$

where $\mathrm{lineal}(C)$ is the lineality space of C.

Under polarity of cones, the angles behave as follows (recall that, by (1.31), $N(C, F) = \widehat{F}_C$ is the face of C° conjugate to F). For faces $F \leq G$ of a cone $C \in \mathcal{PC}^d$ it follows from Definition 2.2.1 and Lemma 1.4.3 that

$$\gamma(\widehat{G}_C, \widehat{F}_C) = \gamma_{\langle \widehat{G}_C \rangle^{\perp}}(N(\widehat{F}_C, \widehat{G}_C)) = \gamma_{\langle G \rangle}(A(F, G)) = \beta(F, G),$$

$$\beta(\widehat{G}_C, \widehat{F}_C) = \gamma_{\langle \widehat{F}_C \rangle}(A(\widehat{G}_C, \widehat{F}_C)) = \gamma_{\langle F \rangle^{\perp}}(N(G, F)) = \gamma(F, G),$$

thus

$$\beta(F, G) = \gamma(\widehat{G}_C, \widehat{F}_C), \qquad \gamma(F, G) = \beta(\widehat{G}_C, \widehat{F}_C). \tag{2.14}$$

Now let $P \in \mathcal{Q}^d$ be a polyhedron. Integrating the relation (1.78) with respect to the Gaussian measure $\gamma_{\langle P \rangle}$ (and observing that $\dim A(F, P) = \dim P$ for $F \in \mathcal{F}(P)$), we obtain the following.

Theorem 2.2.1 (Brianchon–Gram–Sommerville relation). *If $P \in \mathcal{Q}^d$ is a polyhedron, then*

$$\sum_{F \in \mathcal{F}(P)} (-1)^{\dim F} \beta(F, P) = (-1)^{\dim P} \alpha(\mathrm{rec}\, P). \tag{2.15}$$

In particular, if $P \in \mathcal{P}^d$ is a polytope, then the Brianchon–Gram relation

$$\sum_{F \in \mathcal{F}(P)} (-1)^{\dim F} \beta(F, P) = 0 \tag{2.16}$$

holds. If $C \in \mathcal{C}^d$ is a polyhedral cone, then the Sommerville relation

$$\sum_{F \in \mathcal{F}(C)} (-1)^{\dim F} \beta(F, C) = (-1)^{\dim C} \beta(o, C) \tag{2.17}$$

holds.

The special cases in Theorem 2.2.1 are clear, since $\mathrm{rec}\, P = \{o\}$ if P is a polytope, and $\mathrm{rec}\, C = C$ if C is a cone. (To compare the formulae with other formulations in the literature, note that the internal angles as used here are intrinsic and that, by our convention, $P \in \mathcal{F}(P)$ and $\beta(P, P) = 1$.)

Since the Brianchon–Gram relation is obtained from an identity for characteristic functions, it is not restricted to interior angles obtained via integration. More generally, we may adopt the following definition of Backman, Manecke and Sanyal [15].

Definition 2.2.2. *A* cone angle *is a simple valuation* $\widehat{\alpha}$ *on* $\mathcal{PC}^d$ *that satisfies* $\widehat{\alpha}(\mathbb{R}^d) = 1$.

Given such a cone angle $\widehat{\alpha}$, we define for each polytope $P \in \mathcal{P}^d$ and for $i = 0, \ldots, d$,

$$\widehat{\alpha}_i(P) := \sum_{F \in \mathcal{F}_i(P)} \widehat{\alpha}(A(F, P)).$$

Then it follows from (1.79) and Lemma 1.7.1 that

$$\sum_{i=0}^{d} (-1)^i \widehat{\alpha}_i(P) = 0. \tag{2.18}$$

This can be considered as a general form of the Brianchon–Gram relation. In [15] it is shown that for any cone angle $\widehat{\alpha}$ this is, up to a factor, the only linear relation between $\widehat{\alpha}_0(P), \ldots, \widehat{\alpha}_d(P)$ that holds for all polytopes P.

We add some remarks on the Brianchon–Gram relation.

Remark 2.2.1. For odd d and a d-dimensional pointed cone C, relation (2.17) reads

$$\beta(o, C) = \frac{1}{2} \sum_{\{o\} \neq F \in \mathcal{F}(C)} (-1)^{\dim F - 1} \beta(F, C).$$

Since

$$\beta(o, C) = \frac{\sigma_{d-1}(C \cap \mathbb{S}^{d-1})}{\omega_d},$$

this shows that in even-dimensional spheres the spherical volume of a convex spherical polytope (contained in an open halfsphere) can be expressed in terms of internal angles of the spherical polytope.

From (1.83), we obtain the following local version.

Theorem 2.2.2. *If* $P \in \mathcal{Q}^d$ *and* $E \in \mathcal{F}(P)$, *then*

$$\sum_{E \leq F \leq P} (-1)^{\dim F - \dim E} \beta(F, P) = \beta(E, P). \tag{2.19}$$

If we apply the preceding to the polar cone of C and use (2.14), we obtain

Theorem 2.2.3. *If* $P \in \mathcal{Q}^d$ *and* $E \in \mathcal{F}(P)$, *then*

$$\sum_{E \leq F \leq P} (-1)^{\dim F - \dim E} \gamma(E, F) = \gamma(E, P). \tag{2.20}$$

Now we turn to some important non-linear angle sum relations.

Theorem 2.2.4. *If* $P \in \mathcal{Q}^d$ *and* $E \in \mathcal{F}(P)$, *then*

$$\sum_{E \leq F \leq P} \beta(E, F) \gamma(F, P) = 1. \tag{2.21}$$

Proof. First we assume that P is a polyhedral cone C. We integrate relation (1.86) with respect to the Gaussian measure γ_d and observe (2.3), to obtain

$$\sum_{F \in \mathcal{F}(C)} \beta(o, F)\gamma(F, C) = 1. \tag{2.22}$$

Let E be a face of C and denote by $\cdot|\langle E \rangle^{\perp}$ the orthogonal projection to the subspace $\langle E \rangle^{\perp}$. Then

$$\beta(E|\langle E \rangle^{\perp}, F|\langle E \rangle^{\perp}) = \beta(E, F), \qquad \gamma(F|\langle E \rangle^{\perp}, C|\langle E \rangle^{\perp}) = \gamma(F, C).$$

Applying (2.22) in $\langle E \rangle^{\perp}$, we get

$$\sum_{E \leq F \leq P} \beta(o, F|\langle E \rangle^{\perp})\gamma(F|\langle E \rangle^{\perp}, C|\langle E \rangle^{\perp}) = 1,$$

which gives the assertion for a polyhedral cone C.

If now $P \in \mathcal{Q}^d$ and $E \in \mathcal{F}(P)$, we apply the preceding to the angle cone $A(E, P)$. $\qquad \square$

The obtained angle sum relations can conveniently be summarized in terms of the incidence algebra $I(\mathcal{F}(P))$. Recall that this is the incidence algebra of the poset of faces of the polyhedron P, partially ordered by inclusion. By Theorem 1.6.12, its Möbius function is given by

$$\mu(F, G) = (-1)^{\dim G - \dim F}$$

for faces $F \leq G$ of P. For given P, the angle functions β and γ, restricted to faces of P, are elements of $I(\mathcal{F}(P))$, and so are the functions defined by

$$\overline{\beta}(F, G) := (-1)^{\dim G - \dim F}\beta(F, G), \tag{2.23}$$

$$\overline{\gamma}(F, G) := (-1)^{\dim G - \dim F}\gamma(F, G) \tag{2.24}$$

for faces $F \leq G$.

Theorem 2.2.5. *The angle functions of a polyhedron $P \in \mathcal{Q}^d$ satisfy the following relations in the incidence algebra $I(\mathcal{F}(P))$:*

$$\overline{\beta} = \mu * \beta, \qquad \overline{\gamma} = \gamma * \mu, \tag{2.25}$$

$$\beta * \gamma = \zeta, \tag{2.26}$$

$$\overline{\beta} * \gamma = \delta = \beta * \overline{\gamma}. \tag{2.27}$$

Proof. Relations (2.25) are just short forms of Theorems 2.2.2 and 2.2.3. Relation (2.26) follows from Theorem 2.2.4. With (2.25) and (2.26), one gets $\overline{\beta} * \gamma = (\mu * \beta) * \gamma = \mu * (\beta * \gamma) = \mu * \zeta = \delta$ and $\beta * \overline{\gamma} = \beta * (\gamma * \mu) = (\beta * \gamma) * \mu = \zeta * \mu = \delta$. $\qquad \square$

Our next goal will be to express external angles in terms of internal angles, and conversely. With Theorems 2.2.5–2.2.7, we follow McMullen [130].

Let $P \in \mathcal{Q}^d$. For the angle function $\beta \in I(\mathcal{F}(P))$ and for $\xi = \beta - \delta$ we have $\xi(F, F) = 0$ for $F \in \mathcal{F}(P)$, hence Lemma 1.2.2 gives $\xi^{\dim P + 1} = 0$. From Lemma 1.2.3 we see that

$$\beta^{-1} = \sum_{r=0}^{\dim P} (-1)^r (\beta - \delta)^r.$$

Since $(\beta - \delta)(E, E) = 0$ and $(\beta - \delta)(E, G) = \beta(E, G)$ for $E < G$, it follows from the definition of the product in $I(\mathcal{F}(P))$ that

$$(\beta - \delta)^r (E, G) = \sum{}^* \beta(E, F_1) \beta(F_1, F_2) \cdots \beta(F_{r-1}, G) =: \sigma_r(E, G)$$

for $r \geq 2$, where the sum $\sum^*$ is taken over all ordered $(r - 1)$-tuples $(F_1, \ldots, F_{r-1})$ with

$$E < F_1 < F_2 < \cdots < F_{r-1} < G.$$

We put $\sigma_0 := \delta$. By (2.27) we have $\overline{\gamma} = \beta^{-1}$, hence

$$\overline{\gamma} = \sum_{r=0}^{\dim P} (-1)^r \sigma_r.$$

Since $\sigma_r(E, G) = 0$ for $r > \dim G - \dim E$ and since

$$\overline{\gamma}(E, G) = (-1)^{\dim G - \dim E} \gamma(E, G),$$

we have obtained the following result.

Theorem 2.2.6. *If $E \leq G$ are faces of $P \in \mathcal{Q}^d$ with $\dim G - \dim E = k$, then*

$$\gamma(E, G) = \sum_{r=0}^{k} (-1)^{k-r} \sigma_r(E, G)$$

with

$$\sigma_r(E, G) = \sum{}^* \beta(E, F_1) \beta(F_1, F_2) \cdots \beta(F_{r-1}, G),$$

where the sum $\sum^*$ extends over all ordered $(r - 1)$-tuples $(F_1, \ldots, F_{r-1})$ with

$$E < F_1 < F_2 < \cdots < F_{r-1} < G.$$

In a similar way, exchanging the roles of β and γ, or by applying the previous theorem to the polar cone of C, one obtains the following.

Theorem 2.2.7. *If $E \leq G$ are faces of $P \in \mathcal{Q}^d$ with $\dim G - \dim E = k$, then*

$$\beta(E, G) = \sum_{r=0}^{k} (-1)^{k-r} \tau_r(E, G)$$

with

$$\tau_r(E, G) = \sideset{}{^*}\sum \gamma(E, F_1)\gamma(F_1, F_2) \cdots \gamma(F_{r-1}, G),$$

where the sum $\sum^*$ extends over all ordered $(r-1)$-tuples $(F_1, \ldots, F_{r-1})$ with

$$E < F_1 < F_2 < \cdots < F_{r-1} < G.$$

Notes for Section 2.2

1. Concerning the Brianchon–Gram–Sommerville relation, see Notes 3, 4, 5 for Section 1.7. Additionally, we mention Sommerville [182], Perles and Shephard [145], Welzl [189].

2.3 Conic intrinsic volumes and Grassmann angles

In this section, we introduce a fundamental sequence of functionals to measure polyhedral cones. Later, these functionals will be extended to general convex cones.

Let $C \in \mathcal{PC}^d$ be a polyhedral cone. First we define, for each face $F \in \mathcal{F}(C)$, the quantity

$$v_F(C) = \mathbb{P}\left(\Pi_C(\mathbf{g}) \in \operatorname{relint} F\right), \tag{2.28}$$

recalling that $\mathbf{g}$ is a standard Gaussian random vector in $\mathbb{R}^d$. More explicitly,

$$v_F(C) = \alpha(F + N(C, F)) = \alpha(F)\,\alpha(N(C, F)) = \beta(o, F)\,\gamma(F, C), \tag{2.29}$$

where (2.3) was used. By the same argument as used after Definition 2.2.1, it follows that v_F is intrinsic. Since $\dim(F + N(C, F)) = d$, we also have

$$v_F(C) = \gamma_d(F + N(C, F))$$
$$= \sigma_{d-1}((F + N(C, F)) \cap \mathbb{S}^{d-1})\,\omega_d^{-1}$$

by (2.4). If C is a subspace, then C is its only face, and $v_C(C) = 1$.

For later application, we note the following, which immediately follows from Lemma 1.4.4.

Lemma 2.3.1. *Let $C, D \in \mathcal{PC}^d$, $F \in \mathcal{F}(C)$, $G \in \mathcal{F}(D)$, and suppose that F and G intersect transversely. Then*

$$v_{F \cap G}(C \cap D) = \alpha(F \cap G)\alpha(N(C, F) + N(D, G)).$$

Now we define:

Definition 2.3.1. *The conic intrinsic volumes* $v_0(C), \ldots, v_d(C)$ *of a polyhedral cone* $C \in \mathcal{PC}^d$ *are defined by*

$$v_k(C) = \sum_{F \in \mathcal{F}_k(C)} v_F(C) = \sum_{F \in \mathcal{F}_k(C)} \beta(o, F) \gamma(F, P) \qquad (2.30)$$

for $k = 0, \ldots, d$.

As usual, an empty sum is defined as 0. Thus, $v_k(C) = 0$ if $\dim C < k$. For formal reasons, we define $v_k(C) = 0$ for all integers $k > \dim C$. We note that the definition of v_0 included in Definition 2.3.1 is consistent with the one given by (2.10).

Notation. We have to warn the reader that different notation is in use. In particular, what was denoted by $v_k(P)$ for a spherical polytope P in [170, Section 6.5], is now denoted by $v_{k+1}(P^\vee)$ (and later, in Section 3.3, by $v_k^s(P)$).

In particular, if C is a subspace, then

$$v_k(C) = \left\{ \begin{array}{l} 1 \text{ if } k = \dim C, \\ 0 \text{ otherwise,} \end{array} \right\} = \delta_{k, \dim C}, \qquad (2.31)$$

where $\delta_{k,j}$ is the Kronecker symbol.

Corollary 2.3.1. *The functions* $v_0, \ldots, v_d$ *on* $\mathcal{PC}^d$ *are linearly independent.*

We also note that

$$\dim C = k \quad \Rightarrow \quad v_k(C) = \alpha(C) = \gamma_{\langle C \rangle}(C). \qquad (2.32)$$

As the name of the conic intrinsic volumes already indicates, $v_k(C)$ does not depend on the dimension of the space in which C is embedded. This is clear, since the function v_F used for the definition is intrinsic. Also clear from the definition is that

$$v_k(\vartheta C) = v_k(C) \quad \text{for } \vartheta \in O(d).$$

We have now several equivalent representations of the conic intrinsic volumes, and it may be good to summarize them here:

$$v_k(C) = \sum_{F \in \mathcal{F}_k(C)} v_F(C)$$

$$= \sum_{F \in \mathcal{F}_k(C)} \gamma_d(F + N(C, F))$$

$$= \sum_{F \in \mathcal{F}_k(C)} \gamma_{\langle F \rangle}(F) \gamma_{\langle F \rangle^\perp}(N(C, F))$$

$$= \sum_{F \in \mathcal{F}_k(C)} \alpha(F) \alpha(N(C, F))$$

$$= \sum_{F \in \mathcal{F}_k(C)} \beta(o, F)\, \gamma(F, C)$$

$$= \mathbb{P}\left(\Pi_C(\mathbf{g}) \in \mathrm{skel}_k C\right),$$

where $\mathrm{skel}_k C$ was defined in (1.24).

We note that the use of the Gaussian distribution above is convenient (for example, since (2.3) can be used), but not necessary; it could be replaced by other rotation invariant probability distributions.

If $F \in \mathcal{F}_k(C)$, then $\dim N(C, F) = d - k$ and hence

$$v_k(F) = \alpha(F), \quad v_{d-k}(N(C, F)) = \alpha(N(C, F)),$$

thus

$$v_k(C) = \sum_{F \in \mathcal{F}_k(C)} v_k(F) v_{d-k}(N(C, F)). \tag{2.33}$$

Example ([5, Example 4.4.7]) Let $(e_1, \ldots, e_d)$ be an orthonormal basis of $\mathbb{R}^d$, and let $C = \mathrm{pos}\{e_1, \ldots, e_d\}$, the nonnegative orthant. There are $\binom{d}{k}$ faces $F \in \mathcal{F}_k(C)$, and for each of them, the set $F + N(C, F)$ is congruent to C. It follows that $\gamma_d(F + N(C, F)) = 2^{-d}$ and thus

$$v_k(C) = 2^{-d}\binom{d}{k}.$$

If $C, D \in \mathcal{PC}^d$, then $C \times D$ is a polyhedral cone in $\mathbb{R}^d \times \mathbb{R}^d$. We recall the Convention made in Section 1.1, about the standard scalar product of $\mathbb{R}^d \times \mathbb{R}^d$. With respect to this scalar product, $\mathbb{R}^d \times \mathbb{R}^d$ is the orthogonal direct sum of the subspaces $\mathbb{R}^d \times \{o\}$ and $\{o\} \times \mathbb{R}^d$. We can assume, therefore, that $C \subseteq \mathbb{R}^d \times \{o\}$ and $D \subset \{o\} \times \mathbb{R}^d$, and replace $C \times D$ by $C \oplus D$, a direct orthogonal sum. Then each face $F \in \mathcal{F}_k(C \oplus D)$ is of the form $F = F_i \oplus G_j$ with $F_i \in \mathcal{F}_i(C)$, $G_j \in \mathcal{F}_j(D)$, $i + j = k$; further, $N(C \oplus D, F_i \oplus G_j) = N(C, F_i) \oplus N(D, G_j)$. This gives

$$\alpha(F) = \alpha(F_i)\alpha(G_j), \quad \alpha(N(C \oplus D, F)) = \alpha(N(C, F_i))\alpha(N(D, G_j)),$$

hence $v_F(C \oplus D) = v_{F_i}(C) v_{G_j}(D)$ and, therefore,

$$v_k(C \oplus D) = \sum_{F \in \mathcal{F}_k(C \oplus D)} v_F(C \oplus D)$$

$$= \sum_{i+j=k} \sum_{F_i \in \mathcal{F}_i(C)} \sum_{G_j \in \mathcal{F}_j(D)} v_{F_i}(C) v_{G_j}(D).$$

We write this as

$$v_k(C \times D) = \sum_{i+j=k} v_i(C) v_j(D). \tag{2.34}$$

The conic intrinsic volumes are valuations. This follows later easily from more general representations. However, already at this stage we can give an elementary proof. For this, we use that v_k can evidently be represented by

$$v_k(C) = \sum_{F \in \mathcal{F}_k(C)} \gamma_d((\operatorname{relint} F) + N(C, F)) \tag{2.35}$$

for $C \in \mathcal{PC}^d$.

Lemma 2.3.2. *Let* $k \in \{0, \ldots, d\}$, *and let* $C_1, C_2 \in \mathcal{PC}^d$ *be cones such that* $C_1 \cup C_2$ *is convex. Then*

$$v_k(C_1 \cup C_2) + v_k(C_1 \cap C_2) = v_k(C_1) + v_k(C_2).$$

Proof. Let $C \in \mathcal{PC}^d$ and $x \in \mathbb{R}^d$. Then there is a unique face $F \in \mathcal{F}(C)$ with $\Pi_C(x) \in \operatorname{relint} F$ (since C is the disjoint union of the relative interiors of its faces). We denote this face by $F(C, x)$. For $k \in \{0, \ldots, d\}$ we define

$$\psi_k(C, x) := \begin{cases} 1 & \text{if } \dim F(C, x) = k, \\ 0 & \text{otherwise.} \end{cases}$$

Now let $C_1, C_2 \in \mathcal{PC}^d$ be cones with convex union $C_1 \cup C_2$. We choose a finite set $\mathcal{U}$ of hyperplanes through o with the following properties:
(a) The hyperplanes of $\mathcal{U}$ cover the boundaries of C_1, C_2.
(b) If $F \in \mathcal{F}_k(C_i)$ ($i \in \{1, 2\}$) and $G \in \mathcal{F}_{k-1}(F)$, then $(\operatorname{lin} F)^\perp + \operatorname{lin} G \in \mathcal{U}$.
Let U be the union of the hyperplanes in $\mathcal{U}$.
Now let $x \in \mathbb{R}^d \setminus U$. We state that

$$\psi_k(C_1 \cup C_2, x) + \psi_k(C_1 \cap C_2, x) = \psi_k(C_1, x) + \psi_k(C_2, x). \tag{2.36}$$

For the proof, suppose, first, that $x \in C_1 \cup C_2$, say $x \in C_1$. Then $x \in \operatorname{int} C_1$, since $x \notin U$. If also $x \in C_2$, then $x \in \operatorname{int} C_2$. In this case, $\dim F(C, x) = d$ for $C = C_1, C_2, C_1 \cup C_2, C_1 \cap C_2$, hence both sides of (2.36) are equal to 2 for $k = d$ and equal to 0 for $k \neq d$. Suppose that $x \notin C_2$. Then $\dim F(C_2, x) =: m < d$. We have $\psi_m(C_2, x) = \psi_m(C_1 \cap C_2, x) = 1$ and $\psi_m(C_1, x) = \psi_m(C_1 \cup C_2, x) = 0$, thus (2.36) holds for $k = m$. Further, $\psi_d(C_1, x) = \psi_d(C_1 \cup C_2, x) = 1$ and $\psi_d(C_2, x) = \psi_d(C_1 \cap C_2, x) = 0$, thus (2.36) holds for $k = d$. For $k \neq m, d$, both sides of (2.36) are zero.

Suppose, second, that $x \notin C_1 \cup C_2$. Let $y := \Pi_{C_1 \cup C_2}(x)$. Suppose that y is contained in precisely one of the cones, say $y \in C_1$ and $y \notin C_2$. Then $\Pi_{C_1 \cup C_2}(x) = \Pi_{C_1}(x)$, $\Pi_{C_1 \cap C_2}(x) = \Pi_{C_2}(x)$ and hence

$$\psi_k(C_1 \cup C_2, x) = \psi_k(C_2, x), \qquad \psi_k(C_1 \cap C_2, x) = \psi_k(C_1, x)$$

for any $k \in \{0, \ldots, d\}$. Thus (2.36) holds. Suppose now that y is contained in both cones, $y \in C_1 \cap C_2$. Let $\dim F(C_1 \cup C_2, x) =: m$. Thus, there is a face $F \in \mathcal{F}_m(C_1 \cup C_2)$ with $y \in \operatorname{relint} F$. We have $F = (C_1 \cup C_2) \cap H(C_1 \cup C_2, x - y)$, where $H(C_1 \cup C_2, x - y)$ is the supporting hyperplane of $C_1 \cup C_2$ with outer normal vector $x - y$. Let

$$F_1 := C_1 \cap H(C_1, x - y), \qquad F_2 := C_2 \cap H(C_2, x - y).$$

Then $F = F_1 \cup F_2$. At least one of these faces has dimension m, say $\dim F_1 = m$. We distinguish two cases.

Case (1): $\dim F_2 = m$. If the point y were in the relative boundary of F_1 or F_2, then the point x would lie in one of the hyperplanes of $\mathcal{U}$, which is excluded. Hence, $y \in \mathrm{relint}\, F_i$ for $i = 1, 2$. Thus $\psi_m(C_1, x), \psi_m(C_2, x), \psi_m(C_1 \cup C_2, x), \psi_m(C_1 \cap C_2, x)$ are all equal to 1. Therefore, (2.36) holds for $k = m$, and for $k \neq m$ both sides of (2.36) are equal to 0.

Case (2): $\dim F_2 < m$. In this case, for each $k \in \{0, \ldots, d\}$ we have $\psi_k(C_1 \cup C_2, x) = \psi_k(C_1, x)$ ($= 1$ if $k = m$ and $= 0$ if $k \neq m$) and $\psi_k(C_1 \cap C_2, x) = \psi_k(C_2, x)$ ($= 1$ if $k = \dim F_2$ and $= 0$ if $k \neq \dim F_2$). Thus, (2.36) holds also in this case.

Integrating (2.36) over $\mathbb{R}^d \setminus U$ and observing (2.35), we obtain the assertion of the lemma. □

The conic intrinsic volumes satisfy some linear relations and a duality relation.

Theorem 2.3.1. *The conic intrinsic volumes of a polyhedral cone $C \in \mathcal{PC}^d$ satisfy*

$$\sum_{k=0}^{d} v_k(C) = 1 \tag{2.37}$$

and

$$\sum_{k=0}^{d} (-1)^k v_k(C) = \chi(C). \tag{2.38}$$

Further,

$$v_k(C) = v_{d-k}(C^\circ) \tag{2.39}$$

for $k = 0, \ldots, d$.

Proof. Relation (1.86), integrated with respect to the Gaussian measure γ_d, immediately yields

$$\sum_{F \in \mathcal{F}(C)} \alpha(F + N(C, F)) = 1,$$

which is (2.37).

Similarly, if C is not a subspace, we obtain from (1.91) by integration (and observing that $F - N(C, F)$ is the image of $F + N(C, F)$ under an isometry) that

$$\sum_{k=0}^{d} (-1)^k v_k(C) = 0.$$

If C is a subspace, then (2.31) gives

$$\sum_{k=0}^{d} (-1)^k v_k(C) = (-1)^{\dim C}.$$

In view of (1.52), both relations together can be written as (2.38), valid for arbitrary polyhedral cones $C \in \mathcal{C}^d$.

Finally, if $F \in \mathcal{F}_k(C)$, then $G := N(C, F) \in \mathcal{F}_{d-k}(C^\circ)$ and $N(C^\circ, G) = F$ by Lemma 1.4.2 , hence $F + N(C, F) = G + N(C^\circ, G)$. It follows that

$$v_k(C) = v_{d-k}(C^\circ) \tag{2.40}$$

for $k = 0, \ldots, d$. $\qquad\square$

On arbitrary polyhedral cones in $\mathcal{PC}^d$, relations (2.37) and (2.38) together are equivalent to the relations

$$v_0 + v_2 + v_4 + \cdots = \frac{1}{2}(1 + \chi) \tag{2.41}$$

and

$$v_1 + v_3 + v_5 + \cdots = \frac{1}{2}(1 - \chi). \tag{2.42}$$

We recall from (1.52) that $\chi(C) = 0$ for $C \in \mathcal{C}^d$ if C is not a subspace, and $\chi(C) = (-1)^k$ if $C \in G(d, k)$. Equations (2.41) and (2.42) are related to the spherical *Gauss–Bonnet theorem* (see Section 3.3).

Now we relate the conic intrinsic volumes to another series of functionals, the Grassmann angles. For $C \in \mathcal{C}^d \setminus \mathcal{L}_\bullet$ they can be defined by

$$U_j(C) := \frac{1}{2} \mathbb{P}\left(C \cap \mathbf{L}_{d-j} \neq \{o\}\right),$$

where $\mathbf{L}_{d-j}$ is a random $(d - j)$-subspace with distribution ν_{d-j}, the rotation invariant probability measure on the Grassmannian $G(d, d-j)$. In other words, we define:

Definition 2.3.2. *For a closed convex cone $C \in \mathcal{C}^d \setminus \mathcal{L}_\bullet$ and for $j = 0, \ldots, d$, the j-th Grassmann angle of C is defined by*

$$U_j(C) = \frac{1}{2} \int_{G(d, d-j)} \mathbb{1}\{C \cap L \neq \{o\}\}\, \nu_{d-j}(\mathrm{d}L). \tag{2.43}$$

The measurability of the integrand is clear, since the set where it is zero is open in $G(d, d - j)$. Trivially, $U_0(C) = 1/2$ (since $C \neq \{o\}$). It follows from Lemma 2.1.2 that

$$U_j(C) = 0 \quad \text{if } \dim C \leq j,\ C \in \mathcal{C}^d \setminus \mathcal{L}_\bullet. \tag{2.44}$$

In particular, $U_d(C) = 0$. For formal reasons, we define also $U_j(C) = 0$ for integers $j > d$.

Grassmann angles were introduced by Grünbaum [81], though with a different notation. For polyhedral cones $C \in \mathcal{PC}^d \setminus \mathcal{L}_\bullet$, he defined

$$\gamma^{m,d}(C) = \int_{G(d,m)} (1 - \mathbb{1}\{C \cap L \neq \{o\}\})\, \nu_m(\mathrm{d}L).$$

Thus, the connection to our notation is

$$\gamma^{m,d}(C) = 1 - 2U_{d-m}(C).$$

We take the liberty to call U_j the jth Grassmann angle, to avoid the clumsy (though in principle more appropriate) name of 'conic quermassintegral'.

Let $0 \leq j \leq m \leq d-1$, let $E \in G(d,m)$ and $C \in \mathcal{C}^d$ be a cone with $C \subset E$. The image measure of ν_{d-j} under the map $L \mapsto L \cap E$ from $G(d, d-j)$ to the Grassmannian of $(m-j)$-subspaces in E is the normalized Haar measure on the latter space. Here we use the fact that

$$\nu_{d-j}(\{L \in G(d, d-j) : L \cap E \notin G(d, m-j)\}) = 0,$$

by Lemma 2.1.2. Therefore, it follows that $U_j(C)$ does not depend on whether it is computed in $\mathbb{R}^d$ or in E.

We extend the definition of the Grassmann angle to arbitrary $C \in \mathcal{C}^d$ (including subspaces) by

$$U_j(C) = \frac{1}{2} \int_{G(d,d-j)} (1 - \chi(C \cap L))\, \nu_{d-j}(\mathrm{d}L). \qquad (2.45)$$

This is consistent with (2.43), because if $C \in \mathcal{C}^d$ is not a subspace, then

$$1 - \chi(C \cap L) = \mathbb{1}\{C \cap L \neq \{o\}\}$$

for ν_{d-j}-almost all $L \in G(d, d-j)$. In fact, the set of all $L \in G(d, d-j)$ for which $C \cap L$ is a subspace $\neq \{o\}$ has ν_{d-j}-measure zero, by Lemma 2.1.2 and (2.7) (if $C \cap L$ is a subspace $\neq \{o\}$, then L and the lineality space of C are in special position). If $C \cap L$ is not a subspace, then $C \cap L \neq \{o\}$ and $\chi(C \cap L) = 0$, by (1.52).

From (2.45) it is clear that U_j is a valuation on $\mathcal{C}^d$.

For a subspace $L_k \in G(d,k)$, we obtain from (2.45) and (1.52), together with Lemma 2.1.2, that

$$U_j(L_k) = \begin{cases} \frac{1}{2}(1 - (-1)^{k-j}) & \text{if } j \leq k, \\ 0 & \text{if } j > k. \end{cases} \qquad (2.46)$$

If $C \in \mathcal{PC}^d \setminus \mathcal{L}_\bullet$, then the duality relation

$$U_k(C) + U_{d-k}(C^\circ) = \frac{1}{2} \tag{2.47}$$

holds for $k = 0, \ldots, d$. In fact, it follows from Lemma 1.3.10 and Lemma 2.1.2 that for ν_{d-k}-almost all $L \in G(d, d-k)$ we have

$$\chi(C \cap L) + \chi(C^\circ \cap L^\perp) = 1.$$

Together with the fact that ν_{d-k} is the image measure of ν_k under the mapping $L \mapsto L^\perp$, this yields (2.47).

To relate the Grassmann angles to the conic intrinsic volumes, we need an integral-geometric result. More precisely, we shall show that the conic intrinsic volumes satisfy a Crofton type formula. A more general integral-geometric formula will be proved in Section 4.6.

Theorem 2.3.2. *Let $C \in \mathcal{PC}^d$ and $j \in \{1, \ldots, d-1\}$. Then*

$$\int_{G(d,j)} v_k(C \cap L) \, \nu_j(\mathrm{d}L) = v_{d+k-j}(C) \tag{2.48}$$

for $k = 1, \ldots, j$, and

$$\int_{G(d,j)} v_0(C \cap L) \, \nu_j(\mathrm{d}L) = \sum_{i=0}^{d-j} v_i(C). \tag{2.49}$$

Relation (2.48) can equivalently be written as

$$\int_{SO(d)} v_k(C \cap \vartheta L) \, \nu(\mathrm{d}\vartheta) = v_{d+k-j}(C), \tag{2.50}$$

with a fixed subspace $L \in G(d, j)$; this follows from (2.7).

First we have to show that the integrand in (2.50) is measurable. By Definition 2.3.1,

$$v_k(C \cap \vartheta L) = \sum_{J \in \mathcal{F}_k(C \cap \vartheta L)} v_J(C \cap \vartheta L).$$

By Lemma 2.1.4 (applied to all faces of C and to L), it holds for ν-almost all $\vartheta \in SO(d)$ that each k-face J of $C \cap \vartheta L$ is of the form $J = F \cap \vartheta L$ with $F \in \mathcal{F}_i(C)$, $i+j = d+k$, and $F \pitchfork \vartheta L$, that is, F and ϑL intersect transversely. Therefore,

$$v_k(C \cap \vartheta L) = \sum_{F \in \mathcal{F}_{d+k-j}(C)} v_{F \cap \vartheta L}(C \cap \vartheta L) \mathbb{1}\{F \pitchfork \vartheta L\} \tag{2.51}$$

for ν-almost all $\vartheta \in SO(d)$. By Lemma 2.1.4,

$$v_{F \cap \vartheta L}(C \cap \vartheta L) \mathbb{1}\{F \pitchfork \vartheta L\} = v_k(F \cap \vartheta L) v_{d-k}(N(C, F) + \vartheta L^\perp)$$

(note that $v_{d-k}(N(C,F) + \vartheta L^\perp) = 0$ if $\dim F \cap \vartheta L > k$). On the set $S_{F,L} := \{\vartheta \in SO(d) : F \pitchfork \vartheta L\}$, the mapping $\vartheta \mapsto F \cap \vartheta L$ is continuous. On the set of k-dimensional cones in $\mathcal{C}^d$, the mapping $C \mapsto v_k(C)$ is continuous, by Lemma 2.1.1. Hence $\vartheta \mapsto v_k(F \cap \vartheta L)$ is continuous on $S_{F,L}$. Similarly, on $S_{F,L}$ the mapping $\vartheta \mapsto N(C,F) + \vartheta L^\perp$ is continuous, and $N(C,F) + \vartheta L^\perp$ is of dimension $d - k$. As above, it follows that $\vartheta \mapsto v_{d-k}(N(C,F) + \vartheta L^\perp)$ is continuous on $S_{F,L}$. Thus, $SO(d)$ is the union of an open set on which the function $\vartheta \mapsto v_{F \cap \vartheta L}(C \cap \vartheta L)$ vanishes, a closed set of measure zero, and an open set on which the function is continuous. We deduce that the function $\vartheta \mapsto v_k(C \cap \vartheta L)$ is measurable.

Our proof of Theorem 2.3.2 follows Amelunxen and Lotz [10]. First we show two lemmas. In the following special case of (2.50), the index k of v_k coincides with $\dim C + \dim L - d$.

Lemma 2.3.3. *Let $C \in \mathcal{PC}^d$ be a polyhedral cone with $\dim C = i$ and let $L \in G(d,j)$ be a subspace, where $i + j = d + k > d$. Then*

$$\int_{SO(d)} v_k(C \cap \vartheta L)\, \nu(d\vartheta) = v_i(C). \tag{2.52}$$

Proof. By Lemma 2.1.2, for ν-almost all $\vartheta \in SO(d)$ we have $\dim(\langle C \rangle \cap \vartheta L) = k$. Therefore, the left side of (2.52) is equal to

$$\int_{SO(d)} \gamma_{\langle C \rangle \cap \vartheta L}(C \cap \vartheta L)\, \nu(d\vartheta) = \int_{SO(d)} \int_{\langle C \rangle \cap \vartheta L} \mathbb{1}[C](x)\, \gamma_{\langle C \rangle \cap \vartheta L}(dx)\, \nu(d\vartheta)$$

(recall that we write $\mathbb{1}[C] = \mathbb{1}_C$). Here, by the invariance of ν, the outer integral does not change if we replace ϑ by $\rho\vartheta$ with $\rho \in SO_{\langle C \rangle}$. Therefore (and since $\rho\langle C \rangle = \langle C \rangle$ and $\nu_{\langle C \rangle}(SO_{\langle C \rangle}) = 1$), we get

$$\int_{SO(d)} v_k(C \cap \vartheta L)\, \nu(d\vartheta)$$

$$= \int_{SO_{\langle C \rangle}} \int_{SO(d)} \int_{\rho(\langle C \rangle \cap \vartheta L)} \mathbb{1}[C](x) \gamma_{\rho(\langle C \rangle \cap \vartheta L)}(dx)\, \nu(d\vartheta)\, \nu_{\langle C \rangle}(d\rho)$$

$$= \int_{SO_{\langle C \rangle}} \int_{SO(d)} \int_{\langle C \rangle \cap \vartheta L} \mathbb{1}[C](\rho x) \gamma_{\langle C \rangle \cap \vartheta L}(dx)\, \nu(d\vartheta)\, \nu_{\langle C \rangle}(d\rho)$$

$$= \int_{SO(d)} \int_{\langle C \rangle \cap \vartheta L} \left[\int_{SO_{\langle C \rangle}} \mathbb{1}[C](\rho x)\, \nu_{\langle C \rangle}(d\rho) \right] \gamma_{\langle C \rangle \cap \vartheta L}(dx)\, \nu(d\vartheta)$$

$$= \gamma_{\langle C \rangle}(C) = v_i(C),$$

where we have applied Fubini's theorem and then, to evaluate the integral in brackets, used (2.6) in $\langle C \rangle$. For the latter, we need $x \neq o$, therefore we point out that for ν-almost all $\vartheta \in SO(d)$ we have $\dim(\langle C \rangle \cap \vartheta L) = k > 0$, and if this holds, then $\gamma_{\langle C \rangle \cap \vartheta L}$-almost all $x \in \langle C \rangle \cap \vartheta L$ satisfy $x \neq o$. This completes the proof of (2.52). $\square$

Lemma 2.3.4. *Let $C \in \mathcal{PC}^d$ be a polyhedral cone with $\dim C = i$ and let $L \in G(d, j)$ be a subspace, where $i + j = d - k < d$. Then*

$$\int_{SO(d)} v_{d-k}(C + \vartheta L) \, \nu(d\vartheta) = v_i(C). \tag{2.53}$$

Proof. The measurability of the integrand follows from $v_{d-k}(C + \vartheta L) = v_k(C^\circ \cap \vartheta L^\perp)$.

By Lemma 2.1.2, for ν-almost all ϑ we have $\langle C \rangle \cap \vartheta L = \{o\}$ and $\dim(\langle C \rangle + \vartheta L) = i + j = d - k$. Using (2.3), we get

$$v_{d-k}(C + \vartheta L) = \int_{\langle C \rangle + \vartheta L} \mathbb{1}[C + \vartheta L](x) \, \gamma_{\langle C \rangle + \vartheta L}(dx)$$

$$= \int_{\mathbb{R}^d} \mathbb{1}[C + \vartheta L + (\langle C \rangle + \vartheta L)^\perp](x) \, \gamma_d(dx).$$

For ν-almost all ϑ, we have a unique decomposition

$$x = x_{C,\vartheta} + x_{L,\vartheta} + x_\vartheta \tag{2.54}$$

with $x_{C,\vartheta} \in \langle C \rangle$, $x_{L,\vartheta} \in \vartheta L$, $x_\vartheta \in (\langle C \rangle + \vartheta L)^\perp$, which we call the ϑ-decomposition of x. By the uniqueness of the decomposition,

$$x \in C + \vartheta L + (\langle C \rangle + \vartheta L)^\perp \quad \Leftrightarrow \quad x_{C,\vartheta} \in C,$$

hence

$$\int_{SO(d)} v_{d-k}(C + \vartheta L) \, \nu(d\vartheta)$$

$$= \int_{SO(d)} \int_{\mathbb{R}^d} \mathbb{1}[C + \vartheta L + (\langle C \rangle + \vartheta L)^\perp](x) \, \gamma_d(dx) \, \nu(d\vartheta)$$

$$= \int_{SO(d)} \int_{\mathbb{R}^d} \mathbb{1}[C](x_{C,\vartheta}) \, \gamma_d(dx) \, \nu(d\vartheta). \tag{2.55}$$

Now let $\rho \in SO_{\langle C \rangle}$. Applying ρ to both sides of (2.54), we get

$$\rho x = \rho x_{C,\vartheta} + \rho x_{L,\vartheta} + \rho x_\vartheta$$

with $\rho x_{C,\vartheta} \in \langle C \rangle$, $\rho x_{L,\vartheta} \in \rho \vartheta L$, $\rho x_\vartheta \in (\langle C \rangle + \rho \vartheta L)^\perp$, because $\rho \langle C \rangle = \langle C \rangle$. On the other hand, the $\rho\vartheta$-decomposition of ρx reads

$$\rho x = (\rho x)_{C,\rho\vartheta} + (\rho x)_{L,\rho\vartheta} + (\rho x)_{\rho\vartheta}$$

with $(\rho x)_{C,\rho\vartheta} \in \langle C \rangle$, $(\rho x)_{L,\rho\vartheta} \in \rho \vartheta L$, $(\rho x)_{\rho\vartheta} \in (\langle C \rangle + \rho \vartheta L)^\perp$. Thus, the two decompositions are identical.

The integral (2.55) does not change if we replace ϑ by $\rho\vartheta$ and x by ρx, by the invariance properties of ν and γ_d. Therefore,

$$\int_{SO(d)} \int_{\mathbb{R}^d} \mathbb{1}[C](x_{C,\vartheta}) \, \gamma_d(dx) \, \nu(d\vartheta)$$

$$= \int_{SO(d)} \int_{\mathbb{R}^d} \mathbb{1}[C]((\rho x)_{C,\rho\vartheta}) \, \gamma_d(dx) \, \nu(d\vartheta)$$

$$= \int_{SO(d)} \int_{\mathbb{R}^d} \mathbb{1}[C](\rho x_{C,\vartheta}) \, \gamma_d(dx) \, \nu(d\vartheta)$$

$$= \int_{SO_{\langle C \rangle}} \int_{SO(d)} \int_{\mathbb{R}^d} \mathbb{1}[C](\rho x_{C,\vartheta}) \, \gamma_d(dx) \, \nu(d\vartheta) \, \nu_{\langle C \rangle}(d\rho)$$

$$= \int_{SO(d)} \int_{\mathbb{R}^d} \left[\int_{SO_{\langle C \rangle}} \mathbb{1}[C](\rho x_{C,\vartheta}) \, \nu_{\langle C \rangle}(d\rho) \right] \gamma_d(dx) \, \nu(d\vartheta)$$

$$= v_i(C).$$

We have applied (2.6) in $\langle C \rangle$, which is possible since $x_{C,\vartheta} \in \langle C \rangle$ and $x_{C,\vartheta} \neq o$ for ν-almost all ϑ and γ_d-almost all x. $\qquad\square$

Proof of Theorem 2.3.2. By (2.51),

$$\int_{SO(d)} v_k(C \cap \vartheta L) \, \nu(d\vartheta) = \sum_{F \in \mathcal{F}_{d+k-j}(C)} \int_{SO(d)} v_{F \cap \vartheta L}(C \cap \vartheta L) \mathbb{1}\{F \pitchfork \vartheta L\} \, \nu(d\vartheta).$$

Hence, if we show that

$$\int_{SO(d)} v_{F \cap \vartheta L}(C \cap \vartheta D) \mathbb{1}\{F \pitchfork \vartheta L\} \, \nu(d\vartheta) = v_F(C), \qquad (2.56)$$

then the proof of (2.48) is complete. By Lemmas 2.1.4 and 2.3.1 and by (2.32), for ν-almost all $\vartheta \in SO(d)$ we have

$$v_{F \cap \vartheta L}(C \cap \vartheta L) \mathbb{1}\{F \pitchfork \vartheta L\} = v_k(F \cap \vartheta L) v_{d-k}(N(C,F) + \vartheta L^\perp)$$

(note that $v_{d-k}(N(C,F) + \vartheta L^\perp) = 0$ if $\dim(F \cap \vartheta L) > k$). Thus, we have to prove that

$$I := \int_{SO(d)} v_k(F \cap \vartheta L) v_{d-k}(N(C,F) + \vartheta L^\perp) \, \nu(d\vartheta) = v_F(C). \qquad (2.57)$$

For the proof, let $F \in \mathcal{F}_i(C)$ with $i + j = d + k$ be given. In the following, we replace ϑ by $\rho\vartheta$ with $\rho \in SO_{\langle F \rangle}$ (noting that $N(C,F) = \rho N(C,F)$), which does not change the integral, then integrate over all ρ with respect to $\nu_{\langle F \rangle}$, and use the $SO(d)$ invariance of the v_i and Fubini's theorem. We obtain

$$I = \int_{SO_{\langle F \rangle}} \int_{SO(d)} v_k(F \cap \rho\vartheta L) v_{d-k}(N(C,F) + \rho\vartheta L^\perp) \, \nu(d\vartheta) \, \nu_{\langle F \rangle}(d\rho)$$

$$= \int_{\mathrm{SO}_{\langle F \rangle}} \int_{\mathrm{SO}(d)} v_k(F \cap \rho \vartheta L) v_{d-k}(N(C,F) + \vartheta L^{\perp}) \, \nu(\mathrm{d}\vartheta) \, \nu_{\langle F \rangle}(\mathrm{d}\rho)$$

$$= \int_{\mathrm{SO}(d)} \left[\int_{\mathrm{SO}_{\langle F \rangle}} v_k(F \cap \rho \vartheta L) \, \nu_{\langle F \rangle}(\mathrm{d}\rho) \right] v_{d-k}(N(C,F) + \vartheta L^{\perp}) \, \nu(\mathrm{d}\vartheta).$$

For the integral in brackets, we state that

$$\int_{\mathrm{SO}_{\langle F \rangle}} v_k(F \cap \rho \vartheta L) \, \nu_{\langle F \rangle}(\mathrm{d}\rho) = v_i(F). \tag{2.58}$$

For the proof, note that $F \cap \rho \vartheta L = F \cap \rho(\vartheta L \cap \langle F \rangle)$ and that for ν-almost all ϑ we have $\dim(\vartheta L \cap \langle F \rangle) = k$. Therefore, (2.58) follows from (2.52), applied in $\langle F \rangle$.

With (2.58) we get

$$I = v_i(F) \int_{\mathrm{SO}(d)} v_{d-k}(N(C,F) + \vartheta L^{\perp}) \, \nu(\mathrm{d}\vartheta)$$

$$= v_i(F) v_{d-i}(N(C,F)) = v_F(C),$$

where we have finally applied (2.53). We have shown (2.57), which finishes the proof of (2.48).

Finally, we remark that (2.49) follows from (2.48), by using (2.39), that is, $\sum_{i=0}^{d} v_i(C \cap \vartheta L) = 1$. $\qquad\square$

Theorem 2.3.2 yields a relation between the Grassmann angles and the conic intrinsic volumes. For $C \in \mathcal{PC}^d$ and $j \in \{0, \ldots, d\}$, we obtain from (2.45), (2.42) and (2.48) that

$$U_j(C) = \frac{1}{2} \int_{G(d,d-j)} (1 - \chi(C \cap L)) \, \nu_{d-j}(\mathrm{d}L)$$

$$= \sum_{k=0}^{\lfloor \frac{d-1}{2} \rfloor} \int_{G(d,d-j)} v_{2k+1}(C \cap L) \, \nu_{d-j}(\mathrm{d}L)$$

$$= \sum_{k=0}^{\lfloor \frac{d-1-j}{2} \rfloor} v_{j+2k+1}(C),$$

or

$$U_j(C) = v_{j+1}(C) + v_{j+3}(C) + \ldots \tag{2.59}$$

If $C \in \mathcal{PC}^d \setminus \mathcal{L}_{\bullet}$, then it follows from (2.59) together with (2.41) and (2.42) that also

$$1 - 2U_j(C) = 2[v_{j-1}(C) + v_{j-3}(C) + \ldots]. \tag{2.60}$$

From (2.59) we obtain, conversely, that

$$v_j \quad = U_{j-1} - U_{j+1} \quad \text{for } j = 1, \ldots, d-2,$$
$$v_{d-1} = U_{d-2},$$
$$v_d \quad = U_{d-1}.$$

Since the Grassmann angles U_j are valuations on $\mathcal{PC}^d$, the same holds for the conic intrinsic volumes on $\mathcal{PC}^d$.

The following theorem repeats as a summary the most important properties of the conic intrinsic volumes and the Grassmann angles of polyhedral cones.

Theorem 2.3.3. *On the space $\mathcal{PC}^d$ of polyhedral convex cones, the conic intrinsic volumes $v_0, \ldots, v_d$, defined by*

$$v_k(C) = \sum_{F \in \mathcal{F}_k(C)} \beta(o, F)\gamma(F, C),$$

and the Grassmann angles $U_0, \ldots, U_d$, defined by

$$U_k(C) = \frac{1}{2} \int_{G(d,d-k)} (1 - \chi(C \cap L)) \, \nu_{d-k}(\mathrm{d}L),$$

are $O(d)$-invariant valuations. They satisfy the duality relations

$$v_j(C) = v_{d-j}(C^\circ), \tag{2.61}$$

$$U_j(C) + U_{d-j}(C^\circ) = \frac{1}{2}, \quad C \notin \mathcal{L}_\bullet, \tag{2.62}$$

for $j = 0, \ldots, d$ and the linear relations

$$\sum_{k=0}^{d} v_k = 1, \tag{2.63}$$

$$\sum_{k=0}^{d} (-1)^k v_k(C) = \chi, \tag{2.64}$$

equivalently

$$\sum_{k=0}^{\lfloor \frac{d}{2} \rfloor} v_{2k} = \frac{1}{2}(1 + \chi) \tag{2.65}$$

and

$$\sum_{k=0}^{\lfloor \frac{d-1}{2} \rfloor} v_{2k+1} = \frac{1}{2}(1 - \chi). \tag{2.66}$$

If $C \oplus D$ is a direct orthogonal sum of cones $C, D \in \mathcal{PC}^d$, then

$$v_k(C \oplus D) = \sum_{i+j=k} v_i(C)v_j(D). \tag{2.67}$$

Further,

$$U_j(C) = \sum_{k=0}^{\lfloor \frac{d-1-j}{2} \rfloor} v_{j+2k+1}(C) \tag{2.68}$$

and, for $C \notin \mathcal{L}_\bullet$,

$$U_j(C) = \frac{1}{2} - \sum_{k=0}^{\lfloor \frac{j-1}{2} \rfloor} v_{j-2k-1}(C). \tag{2.69}$$

Finally,

$$v_j = U_{j-1} - U_{j+1}. \tag{2.70}$$

On $\mathcal{PC}^d \setminus \mathcal{L}_\bullet$, the functionals U_j are increasing under set inclusion.

The last statement is clear from (2.43). However, we point out that 'increasing' is meant in the weak sense: it may happen that $C \subsetneq D$ but $U_j(C) = U_j(D)$. For example, every three-dimensional cone C with a one-dimensional lineality space satisfies $U_1(C) = 1/2$. On the set of pointed cones, each U_j with $j \in \{1, \ldots, d-1\}$ is strictly increasing.

We could use the relations (2.70) to define the conic intrinsic volumes on the set $\mathcal{C}^d$ of all closed convex cones in $\mathbb{R}^d$. However, a different approach in Chapter 4 will yield more, namely localizations of the conic intrinsic volumes, continuity properties of these, and considerably more general integral-geometric formulas.

We deduce some more integral-geometric results. Using (2.68), we get, for $k = 1, \ldots, d-1$ and $j = 0, \ldots, k-1$,

$$\int_{G(d,k)} U_j(C \cap L)\, \nu_k(\mathrm{d}L) = \int_{G(d,k)} \sum_{r=0}^{\lfloor \frac{k-1-j}{2} \rfloor} v_{j+2r+1}(C \cap L)\, \nu_k(\mathrm{d}L)$$

$$= \sum_{r=0}^{\lfloor \frac{k-1-j}{2} \rfloor} v_{n-k+j+2r+1}(C) = U_{n-k+j}(C).$$

The result,

$$\int_{G(d,k)} U_j(C \cap L)\, \nu_k(\mathrm{d}L) = U_{n-k+j}(C) \tag{2.71}$$

can also be obtained by a direct integral-geometric argument, as follows. We have (say, for nonnegative, measurable functions f on $G(d, k-j)$), where $j < k$,

$$\int_{G(d,k)} \int_{G(L,k-j)} f(E)\, \nu_{k-j}^L(\mathrm{d}E)\, \nu_k(\mathrm{d}L) = \int_{G(d,k-j)} f(E)\, \nu_{k-j}(\mathrm{d}E),$$

where $G(L, k-j)$ denotes the Grassmannian of $(k-j)$-dimensional subspaces of L and ν_{k-j}^L is its rotation invariant probability measure. Up to a constant, this follows from the uniqueness of invariant measures, and the constant must be equal to one, because all involved measures are probability measures. It follows that

$$\int_{G(d,k)} U_j(C \cap L)\, \nu_k(\mathrm{d}L)$$

$$= \int_{G(d,k)} \frac{1}{2} \int_{G(L,k-j)} \mathbb{1}\{C \cap E \neq \{o\}\}\, \nu^L_{k-j}(\mathrm{d}E)\, \nu_k(\mathrm{d}L)$$

$$= \frac{1}{2} \int_{G(d,k-j)} \mathbb{1}\{C \cap E \neq \{o\}\}\, \nu_{k-j}(\mathrm{d}E) = U_{n-k+j}(C).$$

The integral-geometric Theorem 2.3.2 on intersections with subspaces has a counterpart dealing with projections on subspaces. If $C \in \mathcal{C}^d$ and $L \in \mathcal{L}_\bullet$, we denote by $C|L$ the image of C under orthogonal projection to L. We note that we have

$$\mathrm{cl}(C|L) = (C^\circ \cap L)^{\circ L}, \tag{2.72}$$

where $^{\circ L}$ denotes the polar with respect to L. In fact, we have $(C^\circ \cap L)^{\circ L} = (C^\circ \cap L)^\circ \cap L$, since

$$x \in (C^\circ \cap L)^{\circ L} \Leftrightarrow x \in L \wedge \langle x, y \rangle \leq 0 \ \forall y \in C^\circ \cap L$$

$$\Leftrightarrow x \in L \wedge x \in (C^\circ \cap L)^\circ.$$

Since $(C^\circ \cap L)^\circ = \mathrm{cl}(C + L^\perp)$, we obtain

$$\mathrm{cl}(C|L) = \mathrm{cl}(C + L^\perp) \cap L = (C^\circ \cap L)^\circ \cap L = (C^\circ \cap L)^{\circ L}.$$

If C is a polyhedral cone, then $C|L$ is closed, hence the closure in (2.72) can be omitted.

Theorem 2.3.4. *Let $C \in \mathcal{PC}^d$ be a polyhedral cone, and let $q \in \{1, \ldots, d-1\}$. Then*

$$\int_{G(d,q)} v_k(C|L)\, \nu_q(\mathrm{d}L) = v_k(C) \tag{2.73}$$

for $k = 0, \ldots, q - 1$, and

$$\int_{G(d,q)} v_q(C|L)\, \nu_q(\mathrm{d}L) = \sum_{k=q}^{d} v_k(C). \tag{2.74}$$

Proof. Using duality in L, we have $v_k(C|L) = v_k((C^\circ \cap L)^{\circ L}) = v_{q-k}(C^\circ \cap L)$. This shows that the function $L \mapsto v_k(C|L)$ is measurable and that, for $k \in \{0, \ldots, q-1\}$,

$$\int_{G(d,q)} v_k(C|L)\, \nu_q(\mathrm{d}L) = \int_{G(d,q)} v_{q-k}(C^\circ \cap L)\, \nu_q(\mathrm{d}L)$$

$$= v_{d-k}(C^\circ) = v_k(C),$$

where (2.48) was used. This proves (2.73). Similarly, relation (2.49) yields (2.74). □

From (2.73) and (2.68) we obtain that also the Grassmann angles are reproduced as mean values under uniform projections. For $q \in \{0, \ldots, d-1\}$ and $k \in \{0, \ldots, q-1\}$ we have

$$\int_{G(d,q)} U_k(C|L)\,\nu_q(\mathrm{d}L) = U_k(C). \tag{2.75}$$

Defining

$$W_j(C) := \int_{G(d,j)} \gamma_L(C|L)\,\nu_j(\mathrm{d}L) \tag{2.76}$$

for $j = 1, \ldots, d$, we see from (2.74) that

$$W_j(C) = \sum_{r=j}^{d} v_r(C). \tag{2.77}$$

By (2.76), the functionals W_j are increasing under set inclusion.
Relation (2.77) gives

$$v_j = W_j - W_{j+1}, \tag{2.78}$$

and (2.68) together with (2.77) yields

$$W_{j+1} = U_j + U_{j+2}. \tag{2.79}$$

For polyhedral cones, we introduce finally a general series of functionals, comprising metric information, such as the Grassmann angles, as well as combinatorial information. For $C \in \mathcal{PC}^d$, $k \in \{1, \ldots, d\}$ and $j \in \{0, \ldots, k-1\}$, we define

$$Y_{k,j}(C) := \sum_{F \in \mathcal{F}_k(C)} U_j(F). \tag{2.80}$$

Thus, in particular,

$$Y_{\dim C, j}(C) = U_j(C),$$

so that in view of (2.70) also the conic intrinsic volumes can be expressed in terms of functions $Y_{k,j}$. If $C \in \mathcal{PC}^d$ is such that the k-faces of C are not linear subspaces, then

$$Y_{k,0}(C) = \frac{1}{2}f_k(C), \tag{2.81}$$

where $f_k(C)$ denotes the number of k-dimensional faces of C. The special function $Y_{k,k-1}$ is denoted by Λ_k, thus

$$\Lambda_k(C) = \sum_{F \in \mathcal{F}_k(C)} v_k(F). \tag{2.82}$$

The function Λ_k measures the k-skeleton of the polyhedral cone C. Viewed differently, for the spherical polytope $C \cap \mathbb{S}^{d-1}$, the functional $\Lambda_k(C)$ gives the normalized total $(k-1)$-dimensional spherical volume of the union of the $(k-1)$-dimensional faces. In particular, $\Lambda_1(C)$ is the vertex number of $C \cap \mathbb{S}^{d-1}$ and, up to normalizing factors, $\Lambda_2(C)$ gives the total edge length, $\Lambda_{d-1}(C)$ the surface area, and $\Lambda_d(C)$ the volume of the spherical polytope $C \cap \mathbb{S}^{d-1}$.

Notes for Section 2.3

1. Combinatorial proofs for relations (2.37) and (2.38) were given by Mc-Mullen [130]; we have modified these proofs here. The relations themselves, in equivalent versions, are older; see Section 3.3.

2. A derivation of relation (2.38) from the Sommerville relation can be found in Amelunxen and Lotz [10].

3. As an alternative to the solid angle, Seeger and Torki [174, 175] have studied the *least partial volume* of a pointed cone $C \in \mathcal{C}^d$ with interior points, defined by

$$\mathrm{lpv}(C) := \min_{u \in \mathbb{S}^{d-1}} \lambda_d(K \cap H(u,1)).$$

More generally, Seeger [172] defines a *size index* as a real function on $\mathcal{C}^d$ (or a suitable rotation invariant subset thereof) which is continuous with respect to the metric δ_c, nonnegative, non-decreasing under set inclusion, and invariant under orthogonal transformations. He studies the properties of a number of such size indices.

4. For a given polyhedral cone, it will in general not be possible to compute the conic intrinsic volumes explicitly. However, for some explicit cones appearing in applications, this has been done successfully. Amelunxen and Bürgisser [9] (this is an abridged version of [7]) treated symmetric cones. A *symmetric cone* is a closed convex cone which is self-dual and whose automorphism group acts transitively on the interior. Godland and Kabluchko [67] were able to calculate the conic intrinsic volumes of the so-called Weyl chambers, by determining explicitly the internal and external angles of their faces.

2.4 Polyhedral Gauss–Bonnet theorems

The classical Gauss–Bonnet theorem states that the Euler characteristic of a suitable space can be obtained by integrating local curvature information. An almost trivial counterpart to this phenomenon holds for convex polytopes, if 'curvature' is defined by outer angles at vertices and integration is replaced by summation. In fact, for a polyhedron $P \in \mathcal{Q}^d$ and a point $x \in \mathbb{R}^d$ we define the *outer angle* of P at x by

$$\Gamma(P, x) := \gamma_d(N(P, x)) = v_0(\mathrm{cone}(P, x)).$$

For a polytope $P \in \mathcal{P}^d$ we then have $\Gamma(P, x) = 0$ if x is not a vertex of P, and

$$\sum_{x \in \mathrm{vert}\, P} \Gamma(P, x) = \chi(P), \qquad (2.83)$$

where $\mathrm{vert}\, P$ denotes the set of vertices of P. The latter follows immediately from the facts that the normal cones of P at its vertices cover $\mathbb{R}^d$ without overlapping, and that $\chi(P) = 1$.

That $\Gamma(P, x)$ leads to the 'correct' notion of polyhedral curvature in this context, is seen from the following facts (see [163, Sect. 4.2]). The curvature measure of a convex body K of class C^2 is given by

$$C_0(K, \beta) = \int_{\beta \cap \mathrm{bd}\, K} H_{d-1} \, \mathrm{d}\mathcal{H}^{d-1}$$

for Borel sets $\beta \subseteq \mathbb{R}^d$, where H_{d-1} is the Gauss curvature. The curvature measure $C_0(P, \cdot)$ of a polytope P is given by

$$C_0(P, \beta) = \sum_{x \in \beta \cap \mathrm{vert}\, P} \Gamma(P, x).$$

Under Hausdorff convergence of convex bodies, the curvature measure C_0 is weakly continuous.

We note that for fixed x the map $P \mapsto \Gamma(\cdot, x)$ is a valuation on $\mathcal{P}^d$, as follows from the additivity of v_0 on $\mathcal{P}C^d$. Therefore, $\Gamma(\cdot, x)$ has an additive extension to $\mathsf{U}(\mathcal{Q}_{ro}^d)$; we denote it by the same symbol. Equation (2.83) can be extended to compact general polyhedra (that is, unions of finitely many polytopes). Let $P \in \mathsf{U}(\mathcal{P}^d)$, thus P can be represented as the union of finitely many convex polytopes, say $P = \bigcup_{i=1}^{m} P_i$ with $P_i \in \mathcal{P}^d$. By the inclusion-exclusion principle,

$$\Gamma(P, x) = \sum_{r=0}^{m} (-1)^r \sum_{1 \leq i_1 < \cdots < i_r \leq m} \Gamma(P_{i_1} \cap \cdots \cap P_{i_r}, x).$$

If $\Gamma(P, x) \neq 0$, then x must be a vertex of at least one polytope $P_{i_1} \cap \cdots \cap P_{i_r}$. There are at most finitely many such vertices, hence the sum

$$\varphi(P) := \sum_{x \in P} \Gamma(P, x)$$

is finite. This defines an additive map φ on $\mathsf{U}(\mathcal{P}^d)$, and since $\varphi = \chi$ on polytopes and χ is a valuation, we deduce that

$$\sum_{x \in P} \Gamma(P, x) = \chi(P). \qquad (2.84)$$

This is the Gauss–Bonnet formula for compact general polyhedra P.

Clearly, Γ is invariant under rigid motions, in the sense that $\Gamma(gP, gx) = \Gamma(P, x)$ for each rigid motion g of $\mathbb{R}^d$, for all $P \in \mathcal{P}^d$ (and hence for all $P \in \mathsf{U}(\mathcal{P}^d)$) and all $x \in \mathbb{R}^d$.

If for $x \in \mathbb{R}^d$, $u \in \mathbb{S}^{d-1}$ and $\varepsilon > 0$ we define the halfsphere

$$S_\varepsilon(x, u) := \{x + \varepsilon v : v \in \mathbb{S}^{d-1}, \langle v, u \rangle \geq 0\},$$

then

$$\Gamma(P, x) = \chi(\{x\} \cap P) - \lim_{\varepsilon \downarrow 0} \int_{\mathbb{S}^{d-1}} \chi_s \left[S_\varepsilon(x, u) \cap P \right] \sigma(\mathrm{d}u) \qquad (2.85)$$

for $P \in \mathsf{U}(\mathcal{P}^d)$, where in the integrand χ_s denotes the Euler characteristic on finite unions of spherical polytopes, and where σ denotes the normalized spherical Lebesgue measure. For the proof, we note that $\Gamma(\cdot, x)$ thus defined is weakly additive on $\mathcal{P}^d$ and hence additive on $\mathsf{U}(\mathcal{P}^d)$, that $\Gamma(P, x) = 0$ if $x \notin P$, and that $\Gamma(P, x) = \sigma(N(P, x))$ if P is a polytope and $x \in P$. Equation (2.85) was used by Hadwiger [86] to define $\Gamma(P, x)$, and with this definition he proved the Gauss–Bonnet type formula (2.84) for compact general polyhedra P.

The following deals with two questions raised by equation (2.84). First, over which points x of P does the summation effectively extend? And second, which special role is played by the external angles?

Let $P \in \mathsf{U}(\mathcal{Q}^d)$ be a general polyhedron. For $x \in \mathbb{R}^d$, we define (extending Definition 1.3.1, with different notation)

$$\mathrm{Tan}(P, x) := \{v \in \mathbb{R}^d : [x, x + \lambda v] \subseteq P \text{ for some } \lambda > 0\}.$$

Then $\mathrm{Tan}(P, x) = \emptyset$ for $x \notin P$. For $x \in P$, the set $\mathrm{Tan}(P, x)$ is a cone (generally not convex). For any neighborhood N of x, the cone $\mathrm{Tan}(P, x)$ depends only on $P \cap N$. Since P is a finite union of polyhedra, $\mathrm{Tan}(P, x)$ is a finite union of convex polyhedral cones.

For a non-convex general polyhedron, the notion of vertex is not unambiguously defined. We use here the following definition.

Definition 2.4.1. *Let* $P \in \mathsf{U}(\mathcal{Q}^d)$. *A point* $x \in P$ *is a* geometric vertex *of* P *if the cone* $\mathrm{Tan}(P, x)$ *is not a union of parallel lines. The set of geometric vertices of* P *is denoted by* $\mathrm{vert}\, P$.

If φ is a valuation on $\mathcal{PC}^d$, then the function φ° defined by

$$\varphi^\circ(C) := \varphi(C^\circ), \quad C \in \mathcal{PC}^d,$$

and by $\varphi^\circ(\emptyset) := 0$, is a valuation on $\mathcal{PC}^d$, as follows from Theorem 1.8.1. We call it the *dual valuation* of φ.

We recall from Definition 2.2.2 that a cone angle is a simple valuation $\widehat{\alpha}$ on $\mathcal{PC}^d$ that satisfies $\widehat{\alpha}(\mathbb{R}^d) = 1$.

The following theorem provides a whole series of polyhedral Gauss–Bonnet type theorems.

Theorem 2.4.1 (and Definition). *Let $\widehat{\alpha}$ be a cone angle on $\mathcal{PC}^d$. Let $\widehat{\alpha}^\circ$ be the additive extension to $\mathsf{U}(\mathcal{PC}^d)$ of its dual valuation, and let*

$$\kappa(P, x) := \widehat{\alpha}^\circ(\mathrm{Tan}(P, x)) \tag{2.86}$$

for $P \in \mathsf{U}(\mathcal{P}^d)$ and $x \in \mathbb{R}^d$. Then

$$\sum_{x \in \mathrm{vert}\, P} \kappa(P, x) = \chi(P).$$

Proof. Let $x \in \mathbb{R}^d$ be fixed. First we note that the mapping $P \mapsto \kappa(P, x)$ is a valuation on $\mathcal{P}^d$. For the proof, let H be a hyperplane through x, let H_0 be the parallel hyperplane through o, and let H^+, H^-, H_0^+, H_0^- be the corresponding closed halfspaces. Then

$$\mathrm{Tan}(P \cap H^+, x) = \mathrm{Tan}(P, x) \cap H_0^+,$$

and similarly for H^- and H. Therefore,

$$\kappa(P \cap H^+, x) + \kappa(P \cap H^-, x) - \kappa(P \cap H, x)$$
$$= \widehat{\alpha}^\circ(\mathrm{Tan}(P \cap H^+, x)) + \widehat{\alpha}^\circ(\mathrm{Tan}(P \cap H^-, x)) - \widehat{\alpha}^\circ(\mathrm{Tan}(P \cap H, x))$$
$$= \widehat{\alpha}^\circ(\mathrm{Tan}(P, x) \cap H_0^+) + \widehat{\alpha}^\circ(\mathrm{Tan}(P, x) \cap H_0^-) - \widehat{\alpha}^\circ(\mathrm{Tan}(P, x) \cap H_0)$$
$$= \widehat{\alpha}^\circ(\mathrm{Tan}(P, x))$$
$$= \kappa(P, x).$$

Here it was used that $\widehat{\alpha}^\circ$ is weakly additive. If the hyperplane H does not pass through x, then the equality $\kappa(P \cap H^+, x) + \kappa(P \cap H^-, x) - \kappa(P \cap H, x) = \kappa(P, x)$ holds trivially. It follows that $\kappa(\cdot, x)$ is weakly additive and hence, by Theorem 1.6.4, has an additive extension to $\mathsf{U}(\mathcal{P}^d)$.

Let $P \in \mathcal{P}^d$ be a convex polytope, and let $x \in P$. Then $\mathrm{Tan}(P, x)^\circ = N(P, x)$ is the normal cone of P at x, hence $\kappa(P, x) = \widehat{\alpha}(N(P, x))$. This is different from 0 only if $\dim N(P, x) = d$, since the valuation $\widehat{\alpha}$ is simple. Thus, $\kappa(P, x) \neq 0$ only if x is a vertex of P.

Now let $P \in \mathsf{U}(\mathcal{P}^d)$ be a compact general polyhedron. As above, it follows from the inclusion-exclusion principle that $\kappa(P, x) \neq 0$ for at most finitely many points $x \in P$. Therefore, we can define

$$\psi(P) := \sum_{x \in P} \kappa(P, x) \quad \text{for } P \in \mathsf{U}(\mathcal{P}^d).$$

If P is a convex polytope, then we have seen that

$$\psi(P) = \sum_{x \in \mathrm{vert}\, P} \widehat{\alpha}(N(P, x)).$$

Since the normal cones of the vertices of P cover $\mathbb{R}^d$ without overlapping and since $\widehat{\alpha}$ is a simple and normalized valuation, this gives $\psi(P) = 1 = \chi(P)$ for $P \in \mathcal{P}^d$.

For arbitrary $P_1, P_2 \in U(\mathcal{P}^d)$ we have

$$\psi(P_1 \cup P_2) + \psi(P_1 \cap P_2) - \psi(P_1) - \psi(P_2)$$
$$= \sum_{x \in \mathbb{R}^d} [\kappa(P_1 \cup P_2, x) + \kappa(P_1 \cap P_2, x) - \kappa(P_1, x) - \kappa(P_2, x)] = 0,$$

since $\kappa(\cdot, x)$ is a valuation. Thus, ψ is additive. Since ψ and χ are both valuations and they coincide on convex polytopes, they coincide on $U(\mathcal{P}^d)$. This shows that

$$\sum_{x \in P} \kappa(P, x) = \chi(P)$$

for each compact general polyhedron P.

It remains to show that $\kappa(P, x) = 0$ if x is not a geometric vertex of P. Suppose that $P \in U(\mathcal{P}^d)$ and that $x \in P$ is not a geometric vertex of P. Then $\mathrm{Tan}(P, x)$ is a union of parallel lines, say parallel to the line L through o. The intersection $\mathrm{Tan}(P, x) \cap L^\perp$ is the union of finitely many convex polyhedral cones in $L^\perp$, say

$$\mathrm{Tan}(P, x) \cap L^\perp = C_1 \cup \cdots \cup C_m.$$

Then

$$\mathrm{Tan}(P, x) = \overline{C}_1 \cup \cdots \cup \overline{C}_m$$

with $\overline{C}_i := C_i + L$ for $i = 1, \ldots, m$. It follows that

$$\widehat{\alpha}°(\mathrm{Tan}(P, x)) = \widehat{\alpha}°(\overline{C}_1 \cup \cdots \cup \overline{C}_m)$$
$$= \sum_{r=1}^{m} (-1)^{r-1} \sum_{1 \leq i_1 < \cdots < i_r \leq m} \widehat{\alpha}°(\overline{C}_{i_1} \cap \cdots \cap \overline{C}_{i_r}).$$

Here each cone $\overline{C}_{i_1} \cap \cdots \cap \overline{C}_{i_r}$ contains the line L, hence

$$\widehat{\alpha}°(\overline{C}_{i_1} \cap \cdots \cap \overline{C}_{i_r}) = \widehat{\alpha}((\overline{C}_{i_1} \cap \cdots \cap C_{i_r})°) = 0,$$

since $\widehat{\alpha}$ is simple. This yields $\kappa(P, x) = 0$, which completes the proof.   □

It is clear that $\kappa(P, x)$, as defined by (2.86) for a given cone angle $\widehat{\alpha}$, depends only on $P \cap N$ for an arbitrary neighborhood N of x.

If we choose

$$\widehat{\alpha}(C) = \sigma(C \cap \mathbb{S}^{d-1}) \quad \text{for } C \in \mathcal{PC}^d,$$

where σ denotes the normalized spherical Lebesgue measure on $\mathbb{S}^{d-1}$, then the function κ defined in Theorem 2.4.1 coincides with Γ.

We conclude this section with a Gauss–Bonnet formula for polyhedral complexes. First we consider conic polyhedral complexes. A *conic polyhedral complex* is a finite set $\mathcal{Z}$ of polyhedral cones in $\mathcal{PC}^d$ such that each face of a cone of $\mathcal{Z}$ belongs to $\mathcal{Z}$ and that the intersection of any two cones from $\mathcal{Z}$ is a face of each of them (possibly $\{o\}$). The *carrier* $|\mathcal{Z}| = \bigcup_{C \in \mathcal{Z}} C$, or underlying space, of the conic polyhedral complex $\mathcal{Z}$ is an element of $\mathsf{U}(\mathcal{PC}^d_{ro})$. We state that for a conic polyhedral complex $\mathcal{Z}$ we have

$$\Gamma(|\mathcal{Z}|, o) = \sum_{C \in \mathcal{Z}} (-1)^{\dim C} \Gamma(C, o). \tag{2.87}$$

For the proof we first note that for a polyhedral cone $C \in \mathcal{PC}^d$ we have

$$\sum_{F \in \mathcal{F}(C)} (-1)^{\dim F} \Gamma(F, o) = \Gamma(C, o), \tag{2.88}$$

by (2.20) (with $P = C$ and $F = \{o\}$). Now let $\mathcal{Z}$ be a conic polyhedral complex. From the complex property it follows that

$$|\mathcal{Z}| = \biguplus_{C \in \mathcal{Z}} \mathrm{relint} C$$

and hence, since $\Gamma(\cdot, o)$ is a valuation on $\mathsf{U}(\mathcal{PC}^d_{ro})$,

$$\Gamma(|\mathcal{Z}|, o) = \sum_{C \in \mathcal{Z}} \Gamma(\mathrm{relint}\, C, o). \tag{2.89}$$

For an arbitrary polyhedral cone $C \in \mathcal{PC}^d$, we use (2.88), decompose a face of C into the relative interiors of its faces, and then use the local Euler relation (1.59), to obtain

$$\begin{aligned}
\Gamma(C, o) &= \sum_{F \in \mathcal{F}(C)} (-1)^{\dim F} \Gamma(F, o) \\
&= \sum_{F \in \mathcal{F}(C)} (-1)^{\dim F} \sum_{E \in \mathcal{F}(F)} \Gamma(\mathrm{relint}\, E, o) \\
&= \sum_{E \in \mathcal{F}(C)} \Gamma(\mathrm{relint}\, E, o) \sum_{E \leq F \in \mathcal{F}(C)} (-1)^{\dim F} \\
&= (-1)^{\dim C} \Gamma(\mathrm{relint}\, C, o).
\end{aligned}$$

Inserting this in (2.89), we obtain the assertion (2.87).

Now let $\mathcal{Z}$ be a polytopal complex (thus, its polyhedra are compact). For $x \in \mathbb{R}^d$ we then have

$$\Gamma(|\mathcal{Z}|, x) = \sum_{P \in \mathcal{Z}} (-1)^{\dim P} \Gamma(P, x). \tag{2.90}$$

This follows immediately from (2.87), by applying it to the induced decomposition of $\mathrm{Tan}(|\mathcal{Z}|, x)$. From (2.83) and (2.90) we now obtain that

$$\chi(|\mathcal{Z}|) = \sum_{x \in \mathrm{vert}\, \mathcal{Z}} \sum_{P \in \mathcal{Z}} (-1)^{\dim P} \Gamma(P, x), \tag{2.91}$$

where vert $\mathcal{Z} = \bigcup_{P \in \mathcal{Z}}$ vert P. This is the Gauss–Bonnet formula for polytopal complexes.

Notes for Section 2.4

1. In this section, we mainly followed [166]. In that paper, the following definition was suggested.

Definition. *A* polyhedral vertex curvature *in $\mathbb{R}^d$ is a real function K of pairs (P, x) of polytopes $P \subset \mathbb{R}^d$ and points $x \in \mathbb{R}^d$, with the following properties.*
(1) If N is a neighborhood of x, then $\mathsf{K}(P, x)$ depends only on $P \cap N$;
(2) $\mathsf{K}(P, x) \neq 0$ for only finitely many points x of P, and

$$\sum_{x \in P} \mathsf{K}(P, x) = \chi(P)$$

for each polytope P.

In this sense, the function κ constructed in Theorem 2.4.1 is a polyhedral vertex curvature. These polyhedral vertex curvatures have the property that they are translation invariant, in the sense that $\kappa(P + t, x + t) = \kappa(P, x)$ for all $t \in \mathbb{R}^d$, for $P \in \mathsf{U}(\mathcal{P}^d)$ and $x \in \mathbb{R}^d$. We recall a question from [166].

Question. Does the construction of Theorem 2.4.1 yield all translation invariant polyhedral vertex curvatures?

Also the following question is unanswered.

Question. Is Γ the only polyhedral vertex curvature that is invariant under rigid motions?

2. Let $\mathcal{Z}$ be a polytopal complex in $\mathbb{R}^d$. With the definition

$$G(\mathcal{Z}, x) = \sum_{P \in \mathcal{Z}} (-1)^{\dim P} \Gamma(P, x),$$

the Gauss–Bonnet type formula

$$\sum_{x \in \text{vert } \mathcal{Z}} G(\mathcal{Z}, x) = \chi(|\mathcal{Z}|) \tag{2.92}$$

appears in Brin [37]. He formulated it together with the fact that $G(\mathcal{Z}, x)$ depends only on $|\mathcal{Z}|$ and x, both statements without proof. Formula (2.92) appears (possibly in modified forms) at several places in the literature, see Banchoff [16], Cheeger, Müller and Schrader [45, (3.7)], Bloch [28, p. 388], Morvan [139, p. 75], always without mention that $G(\mathcal{Z}, x)$ is a quantitity of the carrier $|\mathcal{Z}|$, not depending on the particular decomposition. For some more background details, we refer to [166].

2.5 A tube formula for compact general polyhedra

The celebrated tube formula of Hermann Weyl says the following. If a compact smooth manifold M is embedded in some Euclidean space, then for sufficiently small $r > 0$ the volume of a tube of radius r about M is a polynomial in r. Its coefficients are obtained as integrals of curvature invariants (the Lipschitz–Killing curvatures) that depend only (and only locally) on the Riemannian metric induced by the embedding.

Cheeger, Müller and Schrader [45] found a counterpart to the Lipschitz–Killing curvatures for piecewise flat spaces. To establish that these were the 'correct' counterparts, they showed (citation) "that if a piecewise flat space approximates a smooth space in a suitable sense, then the corresponding curvatures are close in the sense of measures". As one of the items where the significance of the curvatures on piecewise flat spaces should be analogous to those for smooth spaces, they mentioned Weyl's tube formula. They proved such a tube formula in [46]. Parts of the methods and results of [45] were later simplified and extended by combinatorial means in the work of Budach [40], and this was further extended by Chen [49]. In the following, we shall use this more elementary approach to derive the tube formula for compact general polyhedra (that is, finite unions of polytopes), and to express its coefficients in terms of the inner metric. Angle sum relations derived earlier will play an essential role.

We begin with recalling (from [163]) the local Steiner formula for a polytope $P \in \mathcal{P}^d$. For $r \geq 0$ and for a Borel set $A \in \mathcal{B}(\mathbb{R}^d)$, the local parallel set $M_r(P, A)$ is defined by

$$M_r(P, A) := \{x \in \mathbb{R}^d : \text{dist}(x, P) \leq r, \, \Pi_P(x) \in A\}.$$

Its Lebesgue measure is given by the *local Steiner formula*

$$\mu_r(P, A) := \lambda_d(M_r(P, A)) = \frac{1}{d} \sum_{m=0}^{d} r^{d-m} \kappa_{d-m} \Phi_m(P, A), \qquad (2.93)$$

where the mth *curvature measure* $\Phi_m(P, \cdot)$ of P is defined by

$$\Phi_m(P, A) = \sum_{F \in \mathcal{F}_m(P)} \gamma(F, P) \mathcal{H}^m(F \cap A)$$

for $A \in \mathcal{B}(\mathbb{R}^d)$ and $m \in \{0, \ldots, d\}$. (Here, $\Phi_d(P, \cdot)$ does not really measure a curvature: we have $\Phi_d(P, A) = \lambda_d(P \cap A)$.) We refer to [163, Chap. 4], in particular to (4.5), (4.10), (4.19), (4.22). The normalizations are such that $\Phi_m(P, \mathbb{R}^d) = V_m(P)$, the mth intrinsic volume of P.

According to [163, Thm. 4.2.1], the mapping $P \mapsto \Phi_m(P, A)$ (for fixed $A \in \mathcal{B}(\mathbb{R}^d)$) is a valuation on $\mathcal{P}^d$. By [163, Cor. 6.2.4], it has an extension to a valuation on $U(\mathcal{P}_{ro}^d)$. We denote this extension by the same symbol. In particular, the measure $\Phi_m(P, \cdot)$ is now defined for compact general polyhedra $P \in \mathsf{U}(\mathcal{P}^d)$. It is called the mth *Lipschitz–Killing curvature measure* of P.

We turn to the counterpart of a Weyl tube formula. The usual notion of a tube at distance r around a set M requires that points at distance at most r from M have a unique nearest point in M. For a general polyhedron, this does generally not hold, even if $r > 0$ is arbitrarily small. Therefore, in this case one has to consider parallel sets with multiplicity. For a convex polytope $P \in \mathcal{P}^d$, let us first define

$$c_r(P, A, x) := \mathbb{1}[M_r(P, A)](x) = \begin{cases} 1, \text{ if } 0 \leq \operatorname{dist}(x, P) \leq r, \ \Pi_P(x) \in A, \\ 0 \text{ otherwise,} \end{cases}$$

for $x \in \mathbb{R}^d$. For $P, Q \in \mathcal{P}^d$ with $P \cup Q \in \mathcal{P}^d$ it is easy to check that

$$c_r(P \cup Q, A, x) + c_r(P \cap Q, A, x) = c_r(P, A, x) + c_r(Q, A, x).$$

Thus, $c_r(\cdot, A, x)$ is additive on $\mathcal{P}^d$ and hence has an additive extension to $\mathsf{U}(\mathcal{P}^d_{ro})$, which we denote by the same symbol. We can now consistently define

$$\mu_r(P, A) := \int_{\mathbb{R}^d} c_r(P, A, x)\,\lambda_d(dx)$$

for arbitrary bounded general ro-polyhedra $P \in \mathsf{U}(\mathcal{P}^d_{ro})$ (where the integrability follows from the inclusion-exclusion principle), and then $\mu_r(\cdot, A)$ is a valuation on $\mathsf{U}(\mathcal{P}^d_{ro})$. Now by additivity, relation (2.93) extends immediately. In particular, we have

$$\mu_r(P, A) = \frac{1}{d} \sum_{m=0}^{d} r^{d-m} \kappa_{d-m} \Phi_m(P, A) \tag{2.94}$$

for arbitrary compact general polyhedra $P \in \mathsf{U}(\mathcal{P}^d)$. This extends the local Steiner formula from polytopes to their finite unions.

While so far we have only used additive extension, our main task is now to describe the curvature measures $\Phi_m(P, \cdot)$ for $P \in \mathsf{U}(\mathcal{P}^d)$ more explicitly. Before that, we mention a more explicit representation of the function $c_r(P, A, \cdot)$. Let $B(z, \rho)$ denote the closed ball with center z and radius $\rho > 0$, and define the *index* of $P \in \mathsf{U}(\mathcal{P}^d)$ at the point $q \in \mathbb{R}^d$ with respect to $x \in \mathbb{R}^d$ by

$$j(P, q, x) := 1 - \lim_{\delta \downarrow 0} \lim_{\varepsilon \downarrow 0} \chi(P \cap B(x, \|x - q\| - \varepsilon) \cap B(q, \delta))$$

(where χ is the Euler characteristic) if $q \in P$, and $j(P, q, x) := 0$ if $q \notin P$. If P is convex, then $j(P, q, x) = 1$ if $q = \Pi_P(x)$, and $j(P, q, x) = 0$ otherwise. The function $j(\cdot, q, x)$ is additive on $\mathcal{P}^d$ and hence has an additive extension to $P \in \mathsf{U}(\mathcal{P}^d_{ro})$, denoted by the same symbol. The function c_r is then given by

$$c_r(P, A, x) = \sum_{q \in A} j(P \cap B(x, r), q, x).$$

The index $j(P, q, x)$ was introduced and used for the treatment of parallel sets with multiplicity in [161], more generally for finite unions of convex bodies.

To find explicit expressions for the Lipschitz–Killing curvature measures $\Phi_m(P, \cdot)$, we assume now that P is the carrier $|\mathcal{Z}|$ of a polytopal complex $\mathcal{Z}$. Every compact general polyhedron P can be represented in this way. While the curvature measures $\Phi_m(|\mathcal{Z}|, \cdot)$ depend only on $|\mathcal{Z}|$, the representations obtained below make use of the chosen decomposition.

We write $\mathcal{F}_m(\mathcal{Z})$ for the set of m-dimensional polytopes in $\mathcal{Z}$, and $F \leq P$ means that F is a face of P.

We fix $A \in \mathcal{B}(\mathbb{R}^d)$ and first note that for $P \in \mathcal{P}^d$ we have

$$P = \biguplus_{F \in \mathcal{F}(P)} \operatorname{relint} F.$$

By additivity, this yields

$$\Phi_m(P, A) = \sum_{F \in \mathcal{F}(P)} \Phi_m(\operatorname{relint} F, A).$$

By Lemma 1.6.2 (where $\mathcal{Q}^d$ can be replaced by $\mathcal{P}^d$), this implies

$$
\begin{aligned}
&\Phi_m(\operatorname{relint} P, A)\\
&= (-1)^{\dim P} \sum_{F \in \mathcal{F}(P)} (-1)^{\dim F} \Phi_m(F, A)\\
&= (-1)^{\dim P} \sum_{F \in \mathcal{F}(P)} (-1)^{\dim F} \sum_{E \in \mathcal{F}_m(F)} \gamma(E, F) \mathcal{H}^m(E \cap A)\\
&= (-1)^{\dim P} \sum_{E \in \mathcal{F}_m(P)} \mathcal{H}^m(E \cap A) \sum_{E \leq F \in \mathcal{F}(P)} (-1)^{\dim F} \gamma(E, F)\\
&= (-1)^{\dim P} \sum_{E \in \mathcal{F}_m(P)} \mathcal{H}^m(E \cap A)(-1)^{\dim E} \gamma(E, P)\\
&= (-1)^{\dim P - m} \Phi_m(P, A),
\end{aligned}
$$

where (2.20) was used.

Now let $\mathcal{Z}$ be a polytopal complex. Then

$$|\mathcal{Z}| = \biguplus_{P \in \mathcal{Z}} \operatorname{relint} P$$

and hence

$$
\begin{aligned}
\Phi_m(|\mathcal{Z}|, A) &= \sum_{P \in \mathcal{Z}} \Phi_m(\operatorname{relint} P, A)\\
&= \sum_{P \in \mathcal{Z}} (-1)^{\dim P - m} \Phi_m(P, A)
\end{aligned}
$$

$$= \sum_{P \in \mathcal{Z}} (-1)^{\dim P - m} \sum_{F \in \mathcal{F}_m(P)} \mathscr{H}^m(F \cap A) \gamma(F, P)$$

$$= \sum_{F \in \mathcal{F}_m(\mathcal{Z})} \mathscr{H}^m(F \cap A) \sum_{F \leq P \in \mathcal{Z}} (-1)^{\dim P - \dim F} \gamma(F, P).$$

Now we adopt the following definition from Budach [40].

Definition 2.5.1. *The function* r *defined, for* $F \in \mathcal{Z}$, *by*

$$r(F) := \sum_{F \leq P \in \mathcal{Z}} (-1)^{\dim P - \dim F} \gamma(F, P)$$

is called the Chern–Gauss–Bonnet density *of* $\mathcal{Z}$.

With this definition, the mth Lipschitz–Killing curvature measure is given by

$$\Phi_m(|\mathcal{Z}|, \cdot) = \sum_{F \in \mathcal{F}_m(\mathcal{Z})} r(F) \mathscr{H}^m(F \cap \cdot).$$

If $\mathcal{Z}$ is a simplicial complex and all its maximal simplices have the dimension n, then the number

$$\Phi_{d-m}(|\mathcal{Z}|, \mathbb{R}^d) = R^m$$

is what Cheeger, Müller and Schrader [45] call the *Lipschitz–Killing curvature* R^m of $|\mathcal{Z}|$.

The function r can be expressed in terms of inner angles. We have $\gamma(F, F) = 1$, and Theorem 2.2.6 tells us that

$$\gamma(F, P) = \sum_{j=0}^{k} (-1)^{k-j} \sigma_j(F, P)$$

with $k = \dim P - \dim F$ and

$$\sigma_j(E, G) = \sum{}^* \beta(E, F_1) \beta(F_1, F_2) \cdots \beta(F_{j-1}, G),$$

where the sum $\sum^*$ extends over all over all ordered $(j-1)$-tuples $(F_1, \ldots, F_{j-1})$ with

$$E < F_1 < F_2 < \cdots < F_{j-1} < G.$$

This gives

$$r(F) = 1 + \sum_{F < P \in \mathcal{Z}} \sum_{j=0}^{m} (-1)^j \sigma_j(F, P), \qquad (2.95)$$

with $m = \max\{\dim P : P \in \mathcal{Z}\} - \dim F$.

We derive another representation for $r(F)$, from which more conclusions can be drawn. For this, we extend the poset $(\mathcal{Z}, \leq)$ by an element $\hat{1}$ satisfying

$P < \widehat{1}$ for all $P \in \mathcal{Z}$. Let $\widehat{\mathcal{Z}} = \mathcal{Z} \cup \{\widehat{1}\}$. For the poset $(\widehat{\mathcal{Z}}, \leq)$ we use the incidence algebra $I(\widehat{\mathcal{Z}})$.

The functions δ, ζ, μ, defined as usual for the (incidence algebra of the) poset $(\widehat{\mathcal{Z}}, \leq)$, extend those for $(\mathcal{Z}, \leq)$. We extend the function β by $\beta(F, \widehat{1}) = 0$ for $F \in \mathcal{Z}$ and $\beta(\widehat{1}, \widehat{1}) = 1$. Then β is invertible. On $(\mathcal{Z}, \leq)$, the function γ satisfies

$$\gamma = \beta^{-1} * \zeta$$

by (2.26), and we use this equation to extend γ to $I(\widehat{\mathcal{Z}})$. Then γ is invertible, and

$$\gamma^{-1} = \zeta^{-1} * \beta = \mu * \beta.$$

We define

$$\nu := \frac{1}{2}(\beta + \gamma^{-1}).$$

Then $\nu(F, F) = \frac{1}{2}(\beta(F, F) + \gamma^{-1}(F, F)) = 1$ for all F, hence ν is invertible. We have

$$\nu^{-1} = 2(\beta + \gamma^{-1})^{-1} = 2(\beta + \mu * \beta)^{-1}$$
$$= 2((\delta + \mu) * \beta)^{-1} = 2\beta^{-1} * (\delta + \mu)^{-1},$$

hence

$$\beta^{-1} = \frac{1}{2}\nu^{-1} * (\delta + \mu).$$

Now we define

$$\rho := \beta^{-1} * (\zeta - \delta),$$

Then

$$\rho = \frac{1}{2}\nu^{-1} * (\delta + \mu) * (\zeta - \delta) = \frac{1}{2}\nu^{-1} * (\zeta - \mu).$$

With $\xi := \nu - \delta$ we have $\xi(F, F) = 0$ for all F, hence $\xi^k(F, G) = 0$ for large k, by Lemma 1.2.2. Now Lemma 1.2.3 gives

$$\nu^{-1} = \sum_{j=0}^{\infty} (\delta - \nu)^j,$$

(the sum is finite), thus

$$\rho = \frac{1}{2} \sum_{j=0}^{\infty} (\delta - \nu)^j * (\zeta - \mu).$$

The function ρ was introduced because we have

$$\rho(F, \widehat{1}) = (\beta^{-1} * (\zeta - \delta))(F, \widehat{1})$$
$$= \sum_{F \leq P \leq \widehat{1}} \beta^{-1}(F, P)(\zeta - \delta)(F, \widehat{1})$$

$$= \sum_{F \le P \in \mathcal{Z}} \beta^{-1}(F, P)$$

$$= r(F).$$

Therefore, with a slight change of notation (and since $(\zeta - \mu)(\widehat{1}, \widehat{1}) = 0$),

$$r(F_0) = \frac{1}{2} \sum_{j=0}^{\infty} \sum_{F_0 \le F_j \in \mathcal{Z}} (\delta - \nu)^j (F_0, F_j)(1 - \mu(F_j, \widehat{1})).$$

Since $(\delta - \nu)^0 = \delta$ and $(\delta - \nu)^j$ for $j \ge 2$ is defined by convolution, we get

$$r(F_0) = \frac{1}{2}[1 - \mu(F_0, \widehat{1})] +$$

$$+ \frac{1}{2} \sum_{j=1}^{\infty} (-1)^j \sum_{(j)} \nu(F_0, F_1)\nu(F_1, F_2) \cdots \nu(F_{j-1}, F_j)(1 - \mu(F_j, \widehat{1})),$$

where the summation in $\sum_{(j)}$ is over all $F_1, \dots, F_j \in \mathcal{Z}$ with

$$F_0 < F_1 < F_2 < \cdots < F_{j-1} < F_j.$$

By the definition of the Möbius function and by (1.62),

$$\mu(F, \widehat{1}) = - \sum_{F \le P \in \mathcal{Z}} \mu(F, P) = (-1)^{\dim F + 1} \sum_{F \le P \in \mathcal{Z}} (-1)^{\dim P}.$$

We set

$$\chi^{\perp}(F) := \frac{1}{2}[1 - \mu(F, \widehat{1})]. \tag{2.96}$$

The relation $\overline{\beta} * \gamma = \delta$, explicitly

$$\sum_{F \le G \le P} \overline{\beta}(F, G)\gamma(G, P) = \delta(F, P)$$

holds on $(\mathcal{Z}, \le)$. In fact, if $F \not\le P$, then both sides are zero, and if $F \le P$, it holds for P by (2.27). Therefore, $\gamma^{-1} = \overline{\beta}$ and thus

$$\nu(F, P) = \frac{1}{2}(\beta * \overline{\beta})(F, P) = \begin{cases} \beta(F, P), & \text{if } \dim P - \dim G \text{ is even,} \\ 0, & \text{if } \dim P - \dim G \text{ is odd.} \end{cases}$$

Thus, we can state the following theorem.

Theorem 2.5.1. *The Chern–Gauss–Bonnet density r of a polytopal complex $\mathcal{Z}$ is given by*

$$r(F_0) = \chi^{\perp}(F_0) + \sum_{j=1}^{\infty} (-1)^j \sum_{(j,2)} \beta(F_0, F_1) \cdots \beta(F_{j-1}, F_j)\chi^{\perp}(F_j) \tag{2.97}$$

for $F_0 \in \mathcal{Z}$ (with $\chi^{\perp}(F)$ defined by (2.96)), where the summation $\sum_{(j,2)}$ is over all $F_1, \ldots, F_j \in \mathcal{Z}$ with

$$F_0 < F_1 < \cdots < F_j$$

and such that $\dim F_i - \dim F_{i-1}$ is even for $i = 1, \ldots, j$.

This can be used to show that in (2.94) for a closed manifold, as in the smooth case, about half of the coefficients are zero.

Corollary 2.5.1. *Suppose that $\mathcal{Z}$ is a polytopal complex such that $|\mathcal{Z}|$ is an n-dimensional manifold without boundary. Then the Lipschitz–Killing curvature measures satisfy*

$$\Phi_m(|\mathcal{Z}|, \cdot) = 0 \quad \text{if } n - m \text{ is odd.}$$

Proof. We use a result from Stanley [184], where the proof is carried out. By definition, $2\chi^{\perp}(F) = 1 - \mu(F, \widehat{1})$, and by [184, 3.8.9],

$$\mu(F, \widehat{1}) = (-1)^{\ell(F, \widehat{1})}.$$

Here $\ell(F, \widehat{1}) = k + 1$ if

$$F_0 < F_1 < \cdots < F_k < \widehat{1}$$

is a maximal chain in the poset $(\widehat{\mathcal{Z}}, \leq)$. It follows that $\chi^{\perp}(F) = 0$ if $n - \dim F$ is odd. This yields the assertion, if we observe that in (2.97) $\dim F_j$ and $\dim F_0$ are either both even or both odd. $\qquad \square$

We remark that now also the Gauss–Bonnet formula for polytopal complexes, that is (2.91), can be given a different form. Let $\mathcal{Z}$ be a finite polytopal complex. We already know that

$$\chi(|\mathcal{Z}|) = \Phi_0(|\mathcal{Z}|, \mathbb{R}^d) = \sum_{F \in \mathcal{F}_0(\mathcal{Z})} r(F).$$

Here we can express $r(F)$ either by (2.95) or by (2.97).

Notes for Section 2.5

1. The differential geometry of Weyl's tube formula and its extensions is thoroughly treated in the monograph by Gray [76].

2. An extensive study of curvatures for piecewise flat spaces was made by Cheeger, Müller and Schrader [45]. (The approximation results proved in [45] were partly extended and simplified by Zähle [191].) Equation (2.97) appears there; see [45, (3.35)] and the comment (on page 423) that it was originally derived in Cheeger [44], by heat equation methods. Budach [40] proved part of their results by elementary combinatorial methods (and extended them by replacing internal angle and volume by more general functions). In his review (MR1029098) of this paper, Peter McMullen pointed out that some of the properties follow by standard considerations from valuation theory.

3. As mentioned in Note 4 of Section 1.7, Chen [49] has made an extensive study of angle sum relations on the more general level of characteristic functions. He treated also the relations for the Chern–Gauss–Bonnet density in this way.

3

Relations to spherical geometry

Every convex cone $C \subset \mathbb{R}^d$ determines a spherically convex subset A of the sphere $\mathbb{S}^{d-1}$, by

$$A = C \cap \mathbb{S}^{d-1}.$$

And conversely, a spherically convex subset $A \subset \mathbb{S}^{d-1}$ determines a convex cone C, by

$$C = A^\vee = \operatorname{pos} A = \{\lambda u : u \in A,\ \lambda \geq 0\}.$$

For this reason, parts of the geometry of convex sets in the sphere can be translated into results about convex cones, and conversely. It is often a matter of taste, or of convenience, which point of view one prefers. In these notes, we prefer the viewpoint of convex cones, partially because this allows us to take advantage of the linear structure of $\mathbb{R}^d$.

The present chapter develops more closely the viewpoint of spherical geometry and provides some useful details.

Sections 3.1 and 3.2 collect basic facts about spherical geometry and the relations between spherical and Euclidean geometry. Section 3.3 deals with valuations on spherical polytopes. Inequalities of isoperimetric type for subsets of the sphere are the subject of Section 3.4.

3.1 Basic facts

We call the set $A \subset \mathbb{S}^{d-1}$ *spherically convex* (or briefly *convex* if there is no danger of ambiguity) if $A^\vee$ is convex. With this definition, for example also a set consisting of a pair of antipodal points of the sphere is convex. A *spherically convex body* is the intersection of $\mathbb{S}^{d-1}$ with a closed convex cone different from $\{o\}$. The set of all spherically convex bodies in $\mathbb{S}^{d-1}$ is denoted by $\mathcal{K}_s$. The dimension of $K \in \mathcal{K}_s$ is defined by $\dim K = \dim K^\vee - 1$. We say that $K \in \mathcal{K}_s$ is *proper* if $K^\vee$ is a pointed cone, equivalently, if K is contained in some open hemisphere. (In the literature, such sets are also known as *strongly convex*.) The set of proper convex bodies in $\mathbb{S}^{d-1}$ is denoted by $\mathcal{K}_s^p$.

© The Author(s), under exclusive license to Springer Nature Switzerland AG 2022
R. Schneider, *Convex Cones*, Lecture Notes in Mathematics 2319,
https://doi.org/10.1007/978-3-031-15127-9_3

The set of spherically convex bodies contains the great subspheres. A *great subsphere* of dimension k is the intersection of $\mathbb{S}^{d-1}$ with a subspace of $\mathbb{R}^d$ of dimension $k+1$, $k \in \{0, \ldots, d-1\}$. We denote by $\mathcal{S}_k$ the set of all k-dimensional great subspheres, and we write $\mathcal{S}_\bullet := \bigcup_{k=0}^{d-1} \mathcal{S}_k$.

The set of spherically convex bodies contains also the spherical polytopes. A *spherical polytope* is the intersection of $\mathbb{S}^{d-1}$ with a polyhedral cone $C \in \mathcal{PC}_*^d$. We denote the set of spherical polytopes by $\mathcal{P}_s$ and write $\mathcal{P}_s^p := \mathcal{P}_s \cap \mathcal{K}_s^p$ for the set of spherical polytopes that are contained in some open hemisphere. A *relatively open polytope*, briefly an *ro-polytope*, in $\mathbb{S}^{d-1}$ is the relative interior (with respect to $\mathbb{S}^{d-1}$) of a polytope in $\mathcal{P}_s$. The set of relatively open polytopes in $\mathbb{S}^{d-1}$ is denoted by $\mathcal{P}_{sro}$. As in Euclidean space, we see that every polytope in $\mathcal{P}_s$ is the disjoint union of ro-polytopes. The finite unions of ro-polytopes are called *general ro-polytopes*, and the set of all general ro-polytopes (including the empty set $\emptyset$) is denoted by $\mathsf{U}(\mathcal{P}_{sro})$. This set system is an algebra in $\mathbb{S}^{d-1}$, that is, it is closed under finite unions, intersections, and complementation. Every general ro-polytope can be represented as the disjoint union of finitely many ro-polytopes.

The angular distance (also called spherical distance) of $x, y \in \mathbb{S}^{d-1}$ has been defined (in Section 1.3) by $d_a(x, y) = \arccos\langle x, y \rangle$. This defines a metric d_a on $\mathbb{S}^{d-1}$, also known as the *geodesic metric*. For the standard proof of the triangle inequality, let $x, y, z \in \mathbb{S}^{d-1}$ and $d_a(x, y) = \gamma$, $d_a(x, z) = \alpha$, $d_a(z, y) = \beta$. There are unit vectors $u, v \perp z$ such that

$$x = (\cos \alpha)z + (\sin \alpha)u, \quad y = (\cos \beta)z + (\sin \beta)v.$$

This gives

$$\cos \gamma = \langle x, y \rangle = \cos \alpha \cos \beta + \langle x, y \rangle \sin \alpha \sin \beta$$

$$\geq \cos \alpha \cos \beta - \sin \alpha \sin \beta = \cos(\alpha + \beta)$$

and hence $\gamma \leq \alpha + \beta$, thus $d_a(x, y) \leq d_a(y, z) + d_a(z, y)$.

The connection with the Euclidean distance of x and y is given by

$$\|x - y\|^2 = 2(1 - \cos d_a(x, y)). \tag{3.1}$$

Therefore, d_a and the Euclidean metric on $\mathbb{R}^d$ induce the same topology on $\mathbb{S}^{d-1}$.

The *diameter* of a nonempty set $A \subset \mathbb{S}^{d-1}$ is defined by

$$\operatorname{diam}_a A := \sup\{d_a(x, y) : x, y \in A\}.$$

It is obvious that a set $A \subset \mathbb{S}^{d-1}$ of sufficiently small diameter must be contained in some open hemisphere. The following lemma gives a sharp bound. We reproduce the elegant proof given by Böröczky and Sagmeister [31].

Lemma 3.1.1. *Let $A \subset \mathbb{S}^{d-1}$. If $\operatorname{diam}_a A < \arccos(-1/d)$, then A is contained in some open hemisphere.*

Proof. Without loss of generality, we may assume that A is closed. Let K be the convex hull of A in $\mathbb{R}^d$. Let $z := \Pi_K(o)$, the point in K nearest to o. By Carathéodory's theorem, there are points $x_1, \ldots, x_{d+1} \in K$ and numbers $\alpha_1, \ldots, \alpha_{d+1} \in [0,1]$ with $\sum_{i=1}^{d+1} \alpha_i = 1$ and $z = \sum_{i=1}^{d+1} \alpha_i x_i$. From $\operatorname{diam}_a A < \arccos(-1/d)$ it follows that $\langle x_i, x_j \rangle > -1/d$ for $i \neq j$. Using $2\alpha_i \alpha_j \leq \alpha_i^2 + \alpha_j^2$, we obtain

$$\langle z, z \rangle > \sum_{i=1}^{d+1} \alpha_i^2 - \sum_{i<j} \frac{2\alpha_i \alpha_j}{d} \geq \sum_{i=1}^{d+1} \alpha_i^2 - \sum_{i<j} \frac{\alpha_i^2 + \alpha_j^2}{d} = 0.$$

It follows that $A \subset \{x \in \mathbb{R}^d : \langle x, z \rangle > 0\}$. $\qquad\square$

The set of vertices of a regular d-simplex inscribed to $\mathbb{S}^{d-1}$ shows that the bound given in Lemma 3.1.1 cannot be increased.

We transfer to spherical geometry some notions and facts from Euclidean convex geometry. For $K \in \mathcal{K}_s$, the *polar convex body* is defined by

$$K^* := \{x \in \mathbb{S}^{d-1} : \langle x, y \rangle \leq 0 \text{ for all } y \in K\};$$

thus, $K^* = (K^\vee)^\circ \cap \mathbb{S}^{d-1}$. For a subsphere $S \in \mathcal{S}_\bullet$, we have $S^* = (\operatorname{lin} S)^\perp \cap \mathbb{S}^{d-1}$.

Next, we consider the nearest-point map. Let $K \in \mathcal{K}_s$, and let $x \in \mathbb{S}^{d-1}$ be a point with $0 \leq d_a(x, K) < \pi/2$. Then there is a unique point in K nearest to x, with respect to the metric d_a; it is denoted by $p_s(K, x)$. The obvious relations with Euclidean notions are

$$\sin d_a(x, K) = \operatorname{dist}(x, K^\vee)$$

and

$$p_s(K, x) = \frac{\Pi_{K^\vee}(x)}{\|\Pi_{K^\vee}(x)\|}. \tag{3.2}$$

We call $p_s(K, \cdot)$ the spherical *nearest-point map* or *metric projection* of K.

This leads us to a spherical interpretation of the Moreau decomposition. Let $K \in \mathcal{K}_s$. Let $x \in \mathbb{S}^{d-1} \setminus (K \cup K^*)$. By Theorem 1.3.3, the points

$$o, \ x, \ \Pi_{K^\vee}(x), \ \Pi_{(K^\vee)^\circ}(x)$$

are the vertices of a rectangle. Its conical hull intersects $\mathbb{S}^{d-1} \setminus (K \cup K^*)$ in an open great circular arc of length $\pi/2$, passing through x and with endpoints $p_s(K, x)$ and $p_s(K^*, x)$. All points of such an arc have the same nearest points in K and K^*. Thus, we can state the following.

Theorem 3.1.1. *Let $K \in \mathcal{K}_s$. Then $\mathbb{S}^{d-1} \setminus (K \cup K^*)$ is the disjoint union of a family of open great circular arcs of length $\pi/2$. If A is such an arc and $x \in A$, then the nearest points $p_s(K, x)$ and $p_s(K^*, x)$ are the endpoints of A.*

If $x \in \mathbb{S}^{d-1} \setminus (K \cup K^*)$, we also define

$$u_s(K, x) := p_s(K^*, x). \tag{3.3}$$

Using (1.6), we see that the relation to Euclidean quantities is given by

$$u(K^\vee, x) := \frac{x - \Pi_{K^\vee}(x)}{\|x - \Pi_{K^\vee}(x)\|} = \frac{\Pi_{(K^\vee)^\circ}(x)}{\|\Pi_{(K^\vee)^\circ}(x)\|} = p_s(K^*, x).$$

For $x \in \operatorname{bd} K$ (the boundary relative to $\mathbb{S}^{d-1}$), the set

$$N_s(K, x) := \{y \in K^* : \langle x, y \rangle = 0\}$$

is the set of outer unit normal vectors of K at x. A pair (x, u) with $x \in \operatorname{bd} K$ and $u \in N_s(K, x)$ is called a *support element* of K. In particular, each pair $(p_s(K, x), u_s(K, x))$ with $x \in \mathbb{S}^{d-1} \setminus (K \cup K^*)$ is a support element of K.

If $P \in \mathcal{P}_s$ is a spherical polytope and F is a face of P of dimension less than $d - 1$, then the set $N_s(P, x)$ is the same for all $x \in \operatorname{relint} F$. It is denoted by $N_s(P, F)$.

We turn to some metric and volumetric considerations. Let $\mathcal{C}p$ denote the set of nonempty compact subsets of $\mathbb{S}^{d-1}$. For $A \in \mathcal{C}p$ and $x \in \mathbb{S}^{d-1}$, the distance of x from A is defined by

$$d_a(x, A) := \min\{d_a(x, a) : a \in A\}$$

(the existence of the minimum is clear).

For $\rho \geq 0$, we define the *(outer) parallel set* of A at distance ρ by

$$A_\rho := \{x \in \mathbb{S}^{d-1} : d_a(x, A) \leq \rho\}. \tag{3.4}$$

A warning: If A is convex, then A_ρ need not be convex, even if A is contained in an open hemisphere and $\rho > 0$ is small. For example, let A, B be two great semicircles, not lying in the same circle, with common endpoints x, y. For given $\rho > 0$, we can choose a closed subarc $A' \subset A$, with endpoints different from x, y, and points $p, q \in B \setminus \{x, y\}$ such that $d_a(p, A') \leq \rho$ and $d_a(q, A') \leq \rho$. Then $p, q \in A'_\rho$, but the circular arc connecting p and q is not contained in A'_ρ, if ρ is sufficiently small, thus A'_ρ is not spherically convex.

We shall later need to know the spherical volume of a parallel set of a subsphere. For such computations, the following lemma is useful.

Lemma 3.1.2. *Let $S \in \mathcal{S}_k$, $k \in \{0, \ldots, d - 2\}$, be a subsphere, and let $f : \mathbb{S}^{d-1} \to \mathbb{R}$ be a nonnegative, measurable function. For $u \in S$, write*

$$S_u^* := \mathbb{S}^{d-1} \cap \operatorname{pos}(S^* \cup \{u\}).$$

Then

$$\int_{\mathbb{S}^{d-1}} f \, d\sigma_{d-1} = \int_S \int_{S_u^*} f(v) \sin^k d_a(v, S^*) \, \sigma_{d-k-1}(dv) \, \sigma_k(du).$$

Proof. To derive this from a corresponding orthogonal decomposition of $\mathbb{R}^d$ and its Lebesgue measure, we define

$$\bar{f}(x) := \begin{cases} \|x\|^{-d+1} f(x/\|x\|) & \text{if } 0 < \|x\| < 1, \\ 0 & \text{otherwise.} \end{cases}$$

Then, using (2.1) in $\mathbb{R}^d$ and in $L := \lim S$, we get

$$\int_{\mathbb{S}^{d-1}} f \, d\sigma_{d-1} = \int_{\mathbb{R}^d} \bar{f} \, d\lambda_d$$

$$= \int_L \int_{L^\perp} \bar{f}(x+y) \lambda_{d-k-1}(dy) \lambda_{k+1}(dx)$$

$$= \int_S \left[\int_0^1 \int_{L^\perp} \bar{f}(tu+y) \lambda_{d-k-1}(dy) t^k \, dt \right] \sigma_k(du).$$

If in the last integrand we write $tu + y = z = \tau v$ with $v \in \mathbb{S}^{d-1}$, then $t = \text{dist}(z, L^\perp) = \tau \sin d_a(v, S^*)$. For the double integral in brackets, this yields

$$\int_0^1 \int_{L^\perp} \bar{f}(tu+y) \, \text{dist}^k(tu+y, L^\perp) \lambda_{d-k-1}(dy) \, dt$$

$$= \int_{\text{pos}(L^\perp \cup \{u\})} \bar{f}(z) \, \text{dist}^k(z, L^\perp) \lambda_{d-k}(dz)$$

$$= \int_{S_u^*} \int_0^1 \bar{f}(\tau v) \tau^k \sin^k d_a(v, S^*) \tau^{d-k-1} \, d\tau \, \sigma_{d-k-1}(dv)$$

$$= \int_{S_u^*} f(v) \sin^k d_a(v, S^*) \, \sigma_{d-k-1}(dv),$$

where (2.1) was used in $L^\perp$. This yields the assertion. $\square$

We reformulate the case $k = d - 2$ of Lemma 3.1.2. Let $e \in \mathbb{S}^{d-1}$ and write $S_e := \mathbb{S}^{d-1} \cap e^\perp$. Then the lemma gives

$$\int_{\mathbb{S}^{d-1}} f \, d\sigma_{d-1} = \int_{S_e} \int_{-\pi/2}^{\pi/2} f((\cos\alpha)u + (\sin\alpha)e) \cos^{d-2}\alpha \, d\alpha \, \sigma_{d-2}(du), \quad (3.5)$$

or, alternatively,

$$\int_{\mathbb{S}^{d-1}} f \, d\sigma_{d-1} = \int_{S_e} \int_{-1}^1 f\left(te + \sqrt{1-t^2}\, u\right) (1-t^2)^{\frac{d-3}{2}} \, dt \, \sigma_{d-2}(du). \quad (3.6)$$

Now we first formulate a general lemma on decompositions of spherical integration.

Lemma 3.1.3. *Let $S \in \mathcal{S}_m$, $m \in \{0, \ldots, d-2\}$, be a subsphere. Let $f : \mathbb{S}^{d-1} \to \mathbb{R}$ be a nonnegative measurable function. Then*

$$\int_{\mathbb{S}^{d-1}} f \, d\sigma_{d-1} = \int_S \int_{S^*} \int_0^{\pi/2} f(x\cos\varphi + u\sin\varphi) \cos^m \varphi \sin^{d-m-2}\varphi$$
$$\times d\varphi \, \sigma_{d-m-2}(du) \, \sigma_m(dx).$$

Proof. Applying Lemma 3.1.2 to $\mathbb{S}^{d-1}$ and its subsphere S, we get

$$\int_{\mathbb{S}^{d-1}} f \, d\sigma_{d-1} = \int_S \left[\int_{S_x^*} \underbrace{f(v) \sin^m d_a(v, S^*)}_{F(v)} \sigma_{d-m-1}(dv) \right] \sigma_m(dx).$$

To compute the integral in brackets, Lemma 3.1.2 is applied to the sphere $\mathbb{S}^{d-1} \cap \mathrm{lin}(S^* \cup \{x\})$ and its subsphere S^*, where $x \in S$. This yields

$$\int_{S_x^*} F(v) \, \sigma_{d-m-1}(dv)$$

$$= \int_{S^*} \left[\int_{\mathbb{S}^{d-1} \cap \mathrm{lin}\{x,v\}} F(z) \sin^{d-m-2} d_a(z, \{-x, x\}) \sigma_1(dz) \right] \sigma_{d-m-2}(dv).$$

In the integrand, we write $z = x\cos\varphi + u\sin\varphi$, with $u \in S^*$. Then $d_a(z, \{-x, x\}) = \varphi$ and $d_a(z, S^*) = \pi/2 - \varphi$. Therefore

$$\int_{S_x^*} F(v) \, \sigma_{d-m-1}(dv)$$

$$= \int_{S^*} \int_0^{\pi/2} F(x\cos\varphi + u\sin\varphi) \sin^{d-m-2}\varphi \, d\varphi \, \sigma_{d-m-2}(du)$$

$$= \int_{S^*} \int_0^{\pi/2} f(x\cos\varphi + u\sin\varphi) \cos^m \varphi \sin^{d-m-2}\varphi \, d\varphi \, \sigma_{d-m-2}(du).$$

This gives the assertion.   □

Now we can compute the parallel volume of a subsphere, by applying the preceding lemma to the indicator function of the parallel set.

Lemma 3.1.4. *Let $S \in \mathcal{S}_m$, $m \in \{0, \ldots, d-2\}$, be a subsphere. The spherical volume of its parallel set at distance $0 \le \varepsilon \le \pi/2$ is given by*

$$\sigma_{d-1}(S_\varepsilon) = \omega_{d-m-1}\omega_{m+1} \int_0^\varepsilon \cos^m \varphi \sin^{d-m-2}\varphi \, d\varphi.$$

We turn to convergence. The *spherical Hausdorff distance* of two compact sets $A, B \in \mathcal{C}p$ is defined by

$$\delta_a(A, B) := \min\{\rho \ge 0 : A \subseteq B_\rho, \ B \subseteq A_\rho\}.$$

It is a well-known fact that this defines a metric on $\mathcal{C}p$, the *Hausdorff metric* δ_a. In the following, topological notions for $\mathcal{C}p$, such as convergence, refer to this metric.

The following is adapted from the Euclidean case.

Lemma 3.1.5. *If $(K_i)_{i\in\mathbb{N}}$ is a decreasing sequence in $\mathcal{C}p$ and $K := \bigcap_{j\in\mathbb{N}} K_j$, then*

$$\lim_{i\to\infty} K_i = K. \tag{3.7}$$

Proof. We have $K \in \mathcal{C}p$. Suppose that (3.7) is false. Then there exists $\varepsilon > 0$ with $K_m \subsetneqq K_\varepsilon$ for all $m \in \mathbb{N}$. Let $A_m := K_m \setminus \operatorname{int} K_\varepsilon$ for $m \in \mathbb{N}$. Since $(A_m)_{m\in\mathbb{N}}$ is a decreasing sequence in $\mathcal{C}p$, its intersection, A, is not empty. We have $A \cap K = \emptyset$, but $A_m \subseteq K_m$ for all m implies $A \subseteq K$, a contradiction. $\square$

Theorem 3.1.2. *Every sequence in $\mathcal{C}p$ has a convergent subsequence.*

Proof. By a *tessellation* of $\mathbb{S}^{d-1}$ we understand a finite set of spherical polytopes that cover $\mathbb{S}^{d-1}$ and have pairwise no common interior points. We can easily construct a sequence $(\mathcal{T}_m)_{m\in\mathbb{N}}$ of tessellations such that each polytope of $\mathcal{T}_m$ has diameter at most $1/m$.

Now let $(K_i^0)_{i\in\mathbb{N}}$ be a sequence in $\mathcal{C}p$. For $K \in \mathcal{C}p$ we denote by $A_m(K)$ the union of all polytopes in $\mathcal{T}_m$ meeting K. Since each tessellation is finite, there is a subsequence $(K_i^1)_{i\in\mathbb{N}}$ of $(K_i^0)_{i\in\mathbb{N}}$ such that $A_1(K_i^1) =: T_1$ is independent of i. Similarly, there is a subsequence $(K_i^2)_{i\in\mathbb{N}}$ of $(K_i^1)_{i\in\mathbb{N}}$ such that $A_2(K_i^2) =: T_2$ is independent of i, and so on. Thus, we obtain sequences $(T_m)_{m\in\mathbb{N}}$ and $(K_i^m)_{i\in\mathbb{N}}$ for $m \in \mathbb{N}$, where

$$T_m = A_m(K_i^m) \quad \text{for } i \in \mathbb{N} \tag{3.8}$$

is a union of polytopes of $\mathcal{T}_m$ and where

$$(K_i^k)_{i\in\mathbb{N}} \text{ is a subsequence of } (K_i^m)_{i\in\mathbb{N}} \quad \text{for } k > m. \tag{3.9}$$

Let $m, i, j \in \mathbb{N}$. By (3.8), each point $x \in K_i^m$ belongs to some polytope $P \in \mathcal{T}_m$, and P contains a point $y \in K_j^m$, hence $d_a(x, y) \le \operatorname{diam} P \le 1/m$. It follows that $\delta_a(K_i^m, K_j^m) \le 1/m$ for $m, i, j \in \mathbb{N}$, and then from (3.9) that

$$\delta_a(K_i^m, K_j^k) \le 1/m \quad \text{for } i, j \in \mathbb{N} \text{ and } k \ge m.$$

Define $K_m := K_m^m$ for $m \in \mathbb{N}$, then

$$\delta_a(K_m, K_k) \le 1/m \quad \text{for } k \ge m. \tag{3.10}$$

Now we set $B_m := \operatorname{cl} \bigcup_{i=m}^{\infty} K_i$. Then $(B_m)_{m\in\mathbb{N}}$ is a decreasing sequence in $\mathcal{C}p$, hence Lemma 3.1.5 yields $\lim_{m\to\infty} B_m = B := \bigcap_{i\in\mathbb{N}} B_i$. Therefore, for given $\varepsilon > 0$ there is $n_0 \in \mathbb{N}$ with $B_m \subseteq (B)_\varepsilon$ for $m \ge n_0$, hence $K_i \subseteq (B)_\varepsilon$ for $i \ge n_0$. By (3.10), there is $n_1 \ge n_0$ with $K_j \subseteq (K_i)_\varepsilon$ for $i, j \ge n_1$. Thus, for $i, m \geq n_1$ we have $\bigcup_{j=m}^{\infty} K_j \subseteq (K_i)_\varepsilon$ and hence $B_m \subseteq (K_i)_\varepsilon$. This implies $B \subseteq (K_i)_\varepsilon$. Both inclusions together show that $\delta_a(K_i, B) \leq \varepsilon$ for $i \geq n_1$. Thus, the sequence $(K_i)_{i \in \mathbb{N}}$, which is a subsequence of $(K_i^0)_{i \in \mathbb{N}}$, converges to B. $\square$

If a sequence in $\mathcal{K}_s$ converges to a compact set K, then it is an easy consequence of the definitions that K is spherically convex. Thus, we can formulate the following spherical version of Blaschke's selection theorem.

Theorem 3.1.3. *Every sequence in $\mathcal{K}_s$ has a subsequence that converges to some element of $\mathcal{K}_s$.*

Note for Section 3.1

1. We mention that a collection of basic results on convex sets in spherical spaces, as needed for certain applications, is found in Chapters 3 and 4 of Amelunxen [5].

3.2 The gnomonic map

For relating spherical geometry to Euclidean geometry, one can sometimes take advantage from using the gnomonic projection. To introduce it, let $e \in \mathbb{S}^{d-1}$ and define the open hemisphere with pole e by

$$\mathbb{S}_e^+ := \{u \in \mathbb{S}^{d-1} : \langle e, u \rangle > 0\}.$$

The *gnomonic projection* $g_e : \mathbb{S}_e^+ \to e^\perp$ is defined by

$$g_e(u) := \frac{u}{\langle e, u \rangle} - e \quad \text{for } u \in \mathbb{S}_e^+.$$

Let $u, v \in \mathbb{S}_e^+$ and write $g_e(u) = x$, $g_e(v) = y$. With $\langle u, v \rangle = \cos \gamma = \cos d_a(u, v)$ and $a = 1/\langle e, u \rangle$, $b = 1/\langle e, v \rangle$ we have

$$\|x - y\|^2 = (a - b)^2 + 2ab(1 - \cos \gamma) = (a - b)^2 + ab\|u - v\|^2 \geq \|u - v\|^2,$$

thus $2(1 - \cos \gamma) = \|u - v\|^2 \leq \|x - y\|^2$ and hence

$$d_a(u, v) \leq \arccos \left(1 - \frac{1}{2}\|g_e(u) - g_e(v)\|^2\right)$$

$$\leq 2\|g_e(u) - g_e(v)\| \quad \text{if } \|g_e(u) - g_e(v)\| \leq \pi/2. \qquad (3.11)$$

If $\langle e, u \rangle, \langle e, v \rangle$ are bounded away from zero, that is, if there is a constant $q \geq 1$ such that $\langle e, u \rangle^{-1}, \langle e, v \rangle^{-1} \leq q$, then

$$(a - b)^2 \leq q^4 \|u - v\|^2,$$

hence

$$\|x - y\|^2 = (a - b)^2 + ab\|u - v\|^2 \le q^4\|u - v\|^2 + q^2\|u - v\|^2$$
$$\le 2q^4\|u - v\|^2 = 4q^4(1 - \cos d_a(u, v)) \le 2q^4 d_a(u, v)^2$$

and thus

$$\|g_e(u) - g_e(v)\| \le \sqrt{2}q^2 d_a(u, v). \tag{3.12}$$

Clearly, the gnomonic projection $g_e : \mathbb{S}_e^+ \to e^\perp$ is a homeomorphism. It maps spherically convex bodies in $\mathbb{S}_e^+$ to convex bodies in $e^\perp$ and spherical polytopes in $\mathbb{S}_e^+$ to polytopes in $e^\perp$. Moreover, let $K, L \in \mathcal{K}(\mathbb{S}_e^+)$ (the set of $K \in \mathcal{K}_s$ contained in $\mathbb{S}_e^+$) and let $\overline{K}, \overline{L}$ be their images under g_e. Let B^e denote the unit ball in $e^\perp$. Then, for $0 \le \varepsilon \le \pi/2$, it follows from (3.11) that

$$\overline{K} \subseteq \overline{L} + \varepsilon B^u \Rightarrow K \subseteq L_{2\varepsilon}. \tag{3.13}$$

As a consequence, we deduce the following from the well-known Euclidean case.

Lemma 3.2.1. *Let $K \in \mathcal{K}(\mathbb{S}_u^+)$. For each sufficiently small $\varepsilon > 0$, there exists a spherical polytope P such that $P \subseteq K \subseteq P_\varepsilon$.*

If K has interior points, we can choose $P \subset \operatorname{int} K$.

The gnomonic projection is compatible with polarity, in the following sense.

Lemma 3.2.2. *Let $K \in \mathcal{K}_s$ be contained in $\mathbb{S}_e^+$. Then*

$$g_e(K)^\circ = g_{-e}(K^*), \tag{3.14}$$

where $^\circ$ refers to the space $e^\perp$.

Proof. For $y \in e^\perp$ we have

$$y \in g_e(K)^\circ \Leftrightarrow \langle y, x \rangle \le 1 \,\forall x \in g_e(K)$$
$$\Leftrightarrow \langle y, \langle e, k \rangle^{-1} k - e \rangle \le 1 \,\forall k \in K$$
$$\Leftrightarrow \langle y - e, k \rangle \le 0 \,\forall k \in K$$
$$\Leftrightarrow y - e \in (K^\vee)^\circ \quad (^\circ \text{ refers to } \mathbb{R}^d)$$
$$\Leftrightarrow y \in g_{-e}(K^*),$$

which gives the assertion. $\qquad\square$

It will later be useful to know the Jacobian of the gnomonic map.

Lemma 3.2.3. *If $A \subset \mathbb{S}_e^+$ is a Borel set, then*

$$\lambda_{d-1}(g_e(A)) = \int_A \langle e, u \rangle^{-d} \sigma_{d-1}(du).$$

Proof. We write $u \in \mathbb{S}_e^+$ in the form

$$u = (\sin \alpha)u_0 + (\cos \alpha)e = su_0 + \sqrt{1-s^2}\,e, \quad u_0 \in e^{\perp},$$

so that $\langle e, u \rangle = \sqrt{1-s^2}$ and

$$g_e(u) = ru_0 \quad \text{with } r = \frac{s}{\sqrt{1-s^2}}.$$

For the volume elements we get

$$\mathrm{d}\lambda_{d-1}(ru_0) = r^{d-2}\,\mathrm{d}r\,\sigma_{d-2}(\mathrm{d}u_0)$$

$$= \left(\frac{s}{\sqrt{1-s^2}}\right)^{d-2} \frac{1}{(1-s^2)^{3/2}}\,\mathrm{d}s\,\sigma_{d-2}(\mathrm{d}u_0)$$

and

$$\sigma_{d-1}(\mathrm{d}u) = \sin^{d-2}\alpha\,\mathrm{d}\alpha\,\sigma_{d-2}(\mathrm{d}u_0)$$

$$= s^{d-2}\frac{1}{\sqrt{1-s^2}}\,\mathrm{d}s\,\sigma_{d-2}(\mathrm{d}u_0).$$

From this, the assertion follows. □

Finally, we note the following. If $\lambda_{d-1}(g_e(A)) \neq 0$, then the centroid of $g_e(A)$ is given by

$$c(g_e(A)) = \lambda_{d-1}(g_e(A))^{-1} \int_A g_e(u)\langle e, u \rangle^{-d}\,\sigma_{d-1}(\mathrm{d}u)$$

$$= \lambda_{d-1}(g_e(A))^{-1} \int_A \left(\frac{u}{\langle e, u \rangle} - e\right) \langle e, u \rangle^{-d}\,\sigma_{d-1}(\mathrm{d}u)$$

$$= \lambda_{d-1}(g_e(A))^{-1} \int_A \langle e, u \rangle^{-(d+1)} u\,\sigma_{d-1}(\mathrm{d}u) - e. \tag{3.15}$$

3.3 Spherical and conic valuations

Every valuation φ on an intersectional family $\mathscr{S}$ of cones (not necessarily convex or closed) in $\mathbb{R}^d$ which satisfies $\varphi(\{o\}) = 0$ if $\{o\} \in \mathscr{S}$, induces a valuation φ_s on the family

$$\mathscr{S}_s := \{C \cap \mathbb{S}^{d-1} : C \in \mathscr{S}\},$$

by means of

$$\varphi_s(A) := \varphi(A^{\vee}), \quad A \in \mathscr{S}_s.$$

(Recall that we have defined $\emptyset^{\vee} = \{o\}$ and that the valuation property of φ_s requires, by definition, that $\varphi_s(\emptyset) = 0$ if $\emptyset \in \mathscr{S}_s$.) Conversely, any valuation φ_s on $\mathscr{S}_s$ induces a valuation on $\mathscr{S}$ by $\varphi(C) := \varphi_s(C \cap \mathbb{S}^{d-1})$.

In Section 1.6 we have introduced the Euler characteristic χ on the lattice $\mathsf{U}(\mathcal{CC}^d)$ of finite unions of closed convex sets in $\mathbb{R}^d$ and also on the lattice $\mathsf{U}(\mathcal{Q}_{ro}^d)$ of general ro-polyhedra. We use this now to define the spherical Euler characteristic χ_s by

$$\chi_s(K) := 1 - \chi(K^\vee), \tag{3.16}$$

if either $K \in \mathsf{U}(\mathcal{K}_s)$ or $K \in \mathsf{U}(\mathcal{P}_{sro})$. We recall that $\mathcal{P}_{sro}$ denotes the set of all relative interiors of spherically convex polytopes. Clearly, χ_s is a valuation, satisfying $\chi_s(\emptyset) = 0$.

By (1.52) we have

$$\chi_s(K) = 1 \quad \text{for } K \in \mathcal{K}_s^p$$

and

$$\chi_s(S_k) = 1 + (-1)^k \tag{3.17}$$

for a great subsphere $S_k \in \mathcal{S}_k$.

The Grassmann angles yield the *spherical quermassintegrals* by

$$U_j(A) := U_j(A^\vee)$$

(using the same notation cannot lead to ambiguities, since the domains are different). Thus, according to (2.43) and (3.17) we have

$$U_j(A) = \frac{1}{2} \int_{G(d,d-j)} \chi_s(A \cap L) \, d\nu_{d-j}(dL) \tag{3.18}$$

for $j = 0, \ldots, d-1$, if either $A \in \mathsf{U}(\mathcal{K}_s)$ or $A \in \mathsf{U}(\mathcal{P}_{sro})$. In particular,

$$U_0(A) = \frac{1}{2}\chi_s(A),$$

and for formal reasons we set $U_d(A) = 0$. For $K \in \mathcal{K}_s^p$ we have

$$U_j(K) = \frac{1}{2}\nu_{d-j}(\{L \in G(d, d-j) : L \cap K \neq \emptyset\}),$$

and for $S_k \in \mathcal{S}_k$,

$$U_j(S_k) = \begin{cases} \frac{1}{2}(1 + (-1)^{k-j}) & \text{if } j \leq k+1, \\ 0 & \text{if } j > k+1. \end{cases} \tag{3.19}$$

The *spherical intrinsic volumes* on $\mathcal{K}_s$ are defined by

$$v_j^s(K) := v_{j+1}(K^\vee) \tag{3.20}$$

for $K \in \mathcal{K}_s$ and $j = 0, \ldots, d-1$. (The upper index s is not a power, but stands for 'spherical'.)

We have chosen the notation so that v_j^s agrees with v_j as used in [170] and [98]. This notation has the advantage that the spherical intrinsic volumes

$v_0^s, \ldots, v_{d-1}^s$ on $\mathcal{K}_s$ are in strict analogy to the Euclidean intrinsic volumes of convex bodies in $\mathbb{R}^{d-1}$: for a spherical polytope $P \in \mathcal{P}_s$, we have, for $j = 0, \ldots, d-1$,

$$v_j^s(P) = \frac{1}{\omega_{j+1}} \sum_{F \in \mathcal{F}_j(P)} \sigma_j(F) \gamma(F, P), \tag{3.21}$$

where $\mathcal{F}_j(P)$ is the set of j-dimensional faces of the spherical polytope P, and $\gamma(F, P) := \gamma(F^\vee, P^\vee)$ (see [170, (6.50)], and compare this with [163, (4.23)]).

On $\mathcal{K}_s$, the functions $U_0, \ldots, U_{d-1}$ and $v_0^s, \ldots, v_{d-1}^s$ are valuations. This follows from the corresponding properties of the U_j and v_j on $\mathcal{C}^d$. We may also define

$$v_{-1}^s(K) := v_0(K^\vee),$$

but we note that $v_{-1}^s(\emptyset) = v_0(\{o\}) = 1$, so that, by our definition, v_{-1}^s is not a valuation. By (2.37) we have

$$v_{-1}^s - 1 = v_0^s + \cdots + v_{d-1}^s,$$

which is a valuation. For a subsphere S, we have

$$v_k^s(S) = \begin{cases} 1 \text{ if } \dim S = k, \\ 0 \text{ otherwise.} \end{cases}$$

Corollary 3.3.1. *The functions $v_0^s, \ldots, v_{d-1}^s$ on $\mathcal{K}_s$ are linearly independent.*

There is a different approach to the spherical intrinsic volumes, namely, in analogy to the Euclidean case, via a Steiner formula. We mention this here briefly and prove a more general result in Section 4.1. Let $K \in \mathcal{K}_s^p$ and $0 \leq \varepsilon < \pi/2$. It can be shown that the outer parallel body

$$K_\varepsilon = \{x \in \mathbb{S}^{d-1} : d_a(x, K) \leq \varepsilon\},$$

defined in (3.4), has spherical Lebesgue measure given by

$$\sigma_{d-1}(K_\varepsilon) = \omega_d v_{d-1}(K) + \sum_{j=0}^{d-2} g_{d,j}(\varepsilon) \cdot v_j^s(K) \tag{3.22}$$

with

$$g_{d,j}(\varepsilon) = \omega_{j+1} \omega_{d-j-1} \int_0^\varepsilon \cos^j \varphi \sin^{d-j-2} \varphi \, d\varphi.$$

By (3.22), the numbers $v_0^s(K), \ldots, v_{d-1}^s(K)$ are uniquely determined.

Also equation (2.42) has an interpretation in terms of spherical geometry. Due to (3.16) and (3.20), it now reads

$$\sum_{k=0}^{\lfloor \frac{d-1}{2} \rfloor} v_{2k}^s = \frac{1}{2} \chi_s. \tag{3.23}$$

If the boundary of $K \in \mathcal{K}_s$ is sufficiently smooth, then the spherical intrinsic volumes of K can be expressed as boundary integrals of elementary symmetric functions of principal curvatures in the space $\mathbb{S}^{d-1}$ of constant curvature. With this interpretation, and not restricted to convex hypersurfaces, the functionals v_j^s have appeared early (though with different notation and normalization) in differential and integral geometry of spaces of constant curvature. Equation (3.23) is then a version of the *spherical Gauss–Bonnet theorem*; see, for example, Santaló [152], formulas (1.4) and (1.5).

We continue the study of valuations in spherical space and mention first two extension results.

A function φ from $\mathcal{P}_s^p$ into an abelian group is called *weakly additive* if, after setting $\varphi(\emptyset) = 0$, it satisfies

$$\varphi(P) = \varphi(P \cap H^+) + \varphi(P \cap H^-) - \varphi(P \cap H)$$

for all $P \in \mathcal{P}_s^p$ and all hyperplanes $H \in G(d, d-1)$, where H^+, H^- are the closed halfspaces bounded by H.

A function φ from $\mathcal{K}_s^p$ into a topological (Hausdorff) vector space is called *σ-continuous* if for every decreasing sequence $(K_i)_{i \in \mathbb{N}}$ in $\mathcal{K}_s^p$ the limit relation

$$\lim_{i \to \infty} \varphi(K_i) = \varphi \left(\bigcap_{i \in \mathbb{N}} K_i \right)$$

holds. Continuity with respect to the spherical Hausdorff metric implies σ-continuity.

Theorem 3.3.1. (a) *Let φ be a function on $\mathcal{P}_s^p$ with values in an abelian group. If φ is weakly additive, then it has an extension to a valuation on the lattice $\mathsf{U}(\mathcal{P}_{sro})$.*

(b) *Let φ be a function on $\mathcal{K}_s^p$ with values in a topological vector space. If φ is weakly additive on $\mathcal{P}_s^p$ and is σ-continuous on $\mathcal{K}_s^p$, then φ has an extension to a valuation on the lattice $\mathsf{U}(\mathcal{K}_s^p)$.*

Proof. Let $e \in \mathbb{S}^{d-1}$. The gnomonic projection $g_e : \mathbb{S}_e^+ \to e^\perp$ induces a bijective correspondence between the convex bodies of $\mathcal{K}_s^p$ contained in $\mathbb{S}_e^+$ and the convex bodies in $e^\perp$, which can be identified with $\mathbb{R}^{d-1}$. Here, spherical polytopes in $\mathcal{K}_s^p$, which are contained in $\mathbb{S}_e^+$, correspond to polytopes in $e^\perp$. Defining $\overline{\varphi}(K) = \varphi(g_e^{-1}(K))$ for $g_e^{-1}(K)$ in the domain of φ, we obtain a function $\overline{\varphi}$ on $\mathcal{P}(e^\perp)$ in case (a) and on $\mathcal{K}(e^\perp)$ in case (b).

In case (a), φ is defined on $\mathcal{P}_s^p$ and is weakly additive, hence $\overline{\varphi}$ is weakly additive on $\mathcal{P}(e^\perp)$. By [163, Theorem 6.2.3], $\overline{\varphi}$ is fully additive on $\mathcal{P}(e^\perp)$. Therefore, φ satisfies (1.45) whenever $A_1, \ldots, A_m \in \mathcal{P}_s^p$ are polytopes in $\mathbb{S}_e^+$ with $A_1 \cup \cdots \cup A_m \in \mathcal{P}_s^p$. Since $e \in \mathbb{S}^{d-1}$ was arbitrary, this shows that φ is fully additive on $\mathcal{P}_s^p$. By Theorem 1.6.3, it has an extension to a valuation on $\overline{\mathsf{U}}(\mathcal{P}_s^p)$ and thus, in particular, to a valuation on $\mathsf{U}(\mathcal{P}_{sro})$.

In case (b), φ is defined on $\mathcal{K}_s^p$, is weakly additive on $\mathcal{P}_s^p$ and is σ-continuous on $\mathcal{K}_s^p$. It follows that $\overline{\varphi}$ has the corresponding properties. Now Groemer's second extension theorem shows that $\overline{\varphi}$ is fully additive. (Groemer's second extension theorem is Theorem 3 in Groemer [78]. Groemer's proof is reproduced in [170, Theorem 14.4.2].) Similarly as above, this implies that φ is fully additive on $\mathcal{K}_s^p$. Again, Theorem 1.6.3 shows that φ has an extension to a valuation on $\overline{\mathsf{U}}(\mathcal{K}_s^p)$.   □

It is a major open problem (already frequently asked) whether the well-known characterization of the Euclidean intrinsic volumes by Hadwiger (see, e.g., [163, Thm. 6.1.14]) has a spherical counterpart. More precisely, we formulate two different versions of this question. They were probably first asked, in an equivalent form, by Peter McMullen in 1974; see [80, Problem **49**].

Question. Is every real, continuous, $SO(d)$-invariant valuation on $\mathcal{K}_s$ a linear combination (with constant coefficients) of the spherical intrinsic volumes?

(We have not yet proved that the spherical intrinsic volumes are indeed continuous, but this will follow from more general results in the next chapter.)

Question. Is every real, increasing, $SO(d)$-invariant valuation on $\mathcal{K}_s^p$ a linear combination (with nonnegative coefficients) of the spherical quermassintegrals?

Here a function $\varphi : \mathcal{K}_s^p \to \mathbb{R}$ is called *increasing* if $K \subseteq M$ for $K, M \in \mathcal{K}_s^p$ implies $\varphi(K) \leq \varphi(M)$.

We can prove here only a characterization theorem for simple valuations. A valuation φ on an intersectional family of subsets of $\mathbb{S}^{d-1}$ is called *simple* if $\varphi(A) = 0$ whenever A is in the domain of φ and lies in a hyperplane through 0.

Theorem 3.3.2. *Let φ be a real simple valuation on $\mathcal{P}_s^p$ which is $SO(d)$-invariant and nonnegative. Then φ can be extended to a finite measure on the Borel sets of $\mathbb{S}^{d-1}$, which is $SO(d)$-invariant. Therefore, $\varphi = cv_{d-1}^s$ on $\mathcal{P}_s^p$, with a real constant $c \geq 0$.*

The characterization $\varphi = cv_{d-1}^s$ was first proved in Schneider [160, Theorem (6.2)]. In the following, we give a different proof, by carrying out the sketch in McMullen and Schneider [135, p. 226].

Proof. By Theorem 3.3.1, φ has an extension to a valuation (also denoted by φ) on the algebra $\mathsf{U}(\mathcal{P}_{sro})$. The idea then is to prove that φ is a premeasure on $\mathsf{U}(\mathcal{P}_{sro})$ (which is also a ring of sets) and then to apply the Carathéodory extension theorem.

Since φ vanishes on polytopes $P \in \mathcal{P}_s^p$ of dimension less than $d - 1$, the extension is still nonnegative on ro-polytopes and hence is nonnegative on $\mathsf{U}(\mathcal{P}_{sro})$. Let $A, B \in \mathsf{U}(\mathcal{P}_{sro})$ be such that $A \subset B$. Then B can be represented as a disjoint union of ro-polytopes in such a way that A is the union of a subfamily of these ro-polytopes. From the additivity of φ it follows that $\varphi(A) \leq \varphi(B)$. Thus, the extension φ is still increasing under set inclusion. We want to prove that φ is σ-additive on $\mathsf{U}(\mathcal{P}_{sro})$, and to prepare this, we show the following.

Proposition. Let $P \in \mathcal{P}_s^p$ be a $(d-1)$-polytope, and let $\varepsilon > 0$. There are a polytope Q and an ro-polytope R such that $Q \subseteq P \subseteq R$ and

$$\varphi(R) - \varepsilon < \varphi(P) < \varphi(Q) + \varepsilon.$$

For the proof, we take two vectors $v, w \in \mathbb{S}^{d-1}$, forming an angle $0 < \alpha < \pi$, and define the *open lune*

$$L(v, w) := \{u \in \mathbb{S}^{d-1} : \langle u, v \rangle < 0 < \langle u, w \rangle\}.$$

Denoting by c the value of φ at a hemisphere, it then follows from the additivity and rotation invariance of φ that for $\alpha = \pi/k$ with $k \in \mathbb{N}$ we have $\varphi(L(v, w)) = c/k$.

Now let $P \in \mathcal{P}_s^p$ be a $(d-1)$-polytope, and let $\varepsilon > 0$. Let $F_1, \ldots, F_N$ be the facets of P. We can find lunes $L_1, \ldots, L_N$ such that $\varphi(L_i) < \varepsilon/N$ and $F_i \subset L_i$ for $i = 1, \ldots, N$. It is clear that the polytope Q and the ro-polytope R can be constructed by moving the facets of P appropriately within the corresponding lunes, deleting boundary points where necessary, and using the additivity and monotonicity of φ. This proves the Proposition.

Since each $C \in \mathsf{U}(\mathcal{P}_{sro})$ is a finite union of ro-polytopes, the following is an easy consequence of the Proposition. For given $C \in \mathsf{U}(\mathcal{P}_{sro})$ and $\varepsilon > 0$, there are $B_1, B_2 \in \mathsf{U}(\mathcal{P}_{sro})$ such that

$$\mathrm{cl}\, B_1 \subset C \subset \mathrm{int}\, B_2$$

and

$$\varphi(B_2) - \varepsilon < \varphi(C) < \varphi(B_1) + \varepsilon.$$

To show now that φ is σ-additive on $\mathsf{U}(\mathcal{P}_{sro})$, let $(A_i)_{i \in \mathbb{N}}$ be a disjoint sequence in $\mathsf{U}(\mathcal{P}_{sro})$ such that $A := \bigcup_{i \in \mathbb{N}} A_i \in \mathsf{U}(\mathcal{P}_{sro})$. By the monotonicity and additivity of φ we have

$$\varphi(A) \geq \varphi\left(\bigcup_{i \in \mathbb{N}}^{k} A_i\right) = \sum_{i=1}^{k} \varphi(A_i)$$

for each $k \in \mathbb{N}$ and therefore

$$\varphi(A) \geq \sum_{i=1}^{\infty} \varphi(A_i). \tag{3.24}$$

Let $\varepsilon > 0$ be given. As shown above, there is $B \in \mathsf{U}(\mathcal{P}_{sro})$ with $\mathrm{cl}\, B \subset A$ and $\varphi(A) < \varphi(B) + \varepsilon$. Further, to each $i \in \mathbb{N}$ there is $B_i \in \mathsf{U}(\mathcal{P}_{sro})$ with

$A_i \subset \text{int } B_i$ and $\varphi(B_i) - 2^{-i}\varepsilon < \varphi(A_i)$. Since $\{B_i : i \in \mathbb{N}\}$ is an open covering of the compact set cl B, there is a number k with

$$B \subset \text{cl } B \subset \bigcup_{i=1}^{k} \text{int } B_i \subset \bigcup_{i=1}^{k} B_i.$$

By monotonicity and additivity, it follows that

$$\varphi(B) \leq \varphi\left(\bigcup_{i=1}^{k} B_i\right) \leq \sum_{i=1}^{k} \varphi(B_i),$$

hence

$$\varphi(A) - \varepsilon < \sum_{i=1}^{k} \varphi(A_i) + \sum_{i=1}^{k} 2^{-i}\varepsilon < \sum_{i=1}^{\infty} \varphi(A_i) + \varepsilon.$$

Since $\varepsilon > 0$ was arbitrary, we deduce that

$$\varphi(A) \leq \sum_{i=1}^{\infty} \varphi(A_i).$$

Together with (3.24) this shows the σ-additivity of φ.

Thus, we have shown that φ is a premeasure on the ring $\mathsf{U}(\mathcal{P}_{sro})$. By the Carathéodory extension theorem (see, e.g., [14, 1.3.10]), it has an extension to a measure (also denoted by φ) on the σ-algebra of φ^*-measurable sets, where φ^* is the outer measure derived from φ. In its definition, we can use coverings by ro-polytopes of arbitrarily small positive diameter, from which it follows that φ^* is a metric outer measure. Therefore, the σ-algebra of φ^*-measurable sets comprises the σ-algebra of Borel sets in $\mathbb{S}^{d-1}$. The extension of φ is a finite measure on the Borel sets.

It follows from the construction that the measure φ is $SO(d)$-invariant. By Theorem 2.1.2, it is a constant multiple of the spherical Lebesgue measure σ_{d-1}. In particular, this shows that for $P \in \mathcal{P}_s^p$ we have $\varphi(P) = cv_{d-1}^s(P)$. Since φ is increasing, the constant c is nonnegative. □

An answer to the following question would be a crucial step towards a spherical Hadwiger theorem.

Question. Does Theorem 3.3.2 hold with 'nonnegative' replaced by 'continuous'?

Note for Section 3.3

1. A positive solution to the spherical Hadwiger problem is known for $d = 3$; see Klain and Rota [116, Thm. 11.3.1].

3.4 Inequalities in spherical space

We have defined several series of geometrically natural functionals (in fact, valuations) on polyhedral convex cones, and we shall extend them in the next chapter to general convex cones. Motivated by the well-known Euclidean situation, there immediately arises the question for inequalities between these functionals. Very little is known, and what is known, essentially comes from inequalities of spherical geometry. The present section treats such inequalities. Their implications for inequalities between conic intrinsic volumes will briefly be discussed in Section 4.5.

We begin with a fundamental inequality of spherical geometry. Recall that for a subset $A \subset \mathbb{S}^{d-1}$ and for $\varepsilon \geq 0$ the parallel set of A at distance ε is defined by

$$A_\varepsilon = \{x \in \mathbb{S}^{d-1} : d_a(x, A) \leq \varepsilon\}.$$

A *spherical cap* is the nonempty intersection of the sphere $\mathbb{S}^{d-1}$ with a closed halfspace of $\mathbb{R}^d$.

Theorem 3.4.1. *Let $A \in \mathcal{C}p$ be a nonempty, compact subset, and let $\varepsilon \geq 0$. Let C be a spherical cap with $\sigma_{d-1}(C) = \sigma_{d-1}(A)$. Then*

$$\sigma_{d-1}(A_\varepsilon) \geq \sigma_{d-1}(C_\varepsilon). \tag{3.25}$$

Inequality (3.25) is commonly known as (one version of) the isoperimetric inequality on the sphere. This terminology is, unfortunately, a bit misleading, since the inequality does not deal with the perimeter of sets, but with the volume of their parallel sets. The great importance of this inequality for problems of measure concentration can be seen, for example, from Chapter 3 (and its Notes) in the book [12].

We shall give here a proof of inequality (3.25) by a symmetrization method which, also somewhat misleadingly, is known as 'two-point symmetrization'. With this, we follow Benyamini [26]; a short version of the proof is also given by Schechtman [155]. We add some remarks on the equality case.

We fix a point $p \in \mathbb{S}^{d-1}$, and for each hyperplane $H \subset \mathbb{R}^d$ through o not passing through p we denote by H^+ (respectively, H^-) the closed halfspace bounded by H that contains (does not contain) the point p. The map $\rho : \mathbb{S}^{d-1} \to \mathbb{S}^{d-1}$ is defined by the reflection in H (for the moment, H is fixed, so the notation will not show that ρ and T depend on H). If a set $A \subset \mathbb{S}^{d-1}$ is given, then the two-point symmetrization with respect to H transforms it into the set

$$TA := (A \cap \rho(A)) \cup [(A \cup \rho(A)) \cap H^+].$$

In other words, T pushes as much of A into H^+ as is possible without causing double covering. Writing $A_0 := A \cap \rho(A)$ we have the disjoint decomposition

$$TA = A_0 \cup \rho[(A \cap H^-) \setminus A_0] \cup [(A \cap H^+) \setminus A_0],$$

which shows that

$$\sigma_{d-1}(TA) = \sigma_{d-1}(A).$$

Let $x, y \in H^- \cap \mathbb{S}^{d-1}$. Then $\|x - y\| \leq \|x - \rho(y)\|$ by elementary geometry, hence (3.1) shows that $d_a(x, y) \leq d_a(x, \rho(y))$. The same inequality holds if $x, y \in H^+ \cap \mathbb{S}^{d-1}$. Using this, we can show that

$$(TA)_\varepsilon \subseteq T(A_\varepsilon). \tag{3.26}$$

For the proof, let $y \in (TA)_\varepsilon$ and choose $x \in TA$ such that $d_a(x, y) \leq \varepsilon$. We distinguish the following cases.

Case 1: $x \in H^+$ and $y \in H^-$. We have

$$d_a(\rho(x), y) = d_a(x, \rho(y)) \leq d_a(\rho(x), \rho(y)) = d_a(x, y) \leq \varepsilon.$$

Here, $x \in A$ or $\rho(x) \in A$, and in any case, $y \in A_\varepsilon$ and $\rho(y) \in A_\varepsilon$, so that $y \in T(A_\varepsilon)$.

Case 2: $x \in H^-$ and $y \in H^+$. The argument is as in Case 1.

Case 3: $x, y \in H^-$. Since $x \in TA \cap H^-$, both x and $\rho(x)$ are in A. We have $y \in A_\varepsilon$ and $\rho(y) \in A_\varepsilon$, hence $y \in T(A_\varepsilon)$.

Case 4: $x, y \in H^+$. Then either $x \in A$ or $\rho(x) \in A$, hence either $y \in A_\varepsilon$ or $\rho(y) \in A_\varepsilon$. Since $y \in H^+$, we have $y \in T(A_\varepsilon)$. This completes the proof of (3.26).

We prepare the proof of Theorem 3.4.1 by two Lemmas. We recall that the set $\mathcal{C}p$ of nonempty closed subsets of $\mathbb{S}^{d-1}$ has been equipped with the Hausdorff metric δ_a. The following lemmas refer to the metric space $(\mathcal{C}p, \delta_a)$.

Lemma 3.4.1. *For given $A \in \mathcal{C}p$, let $\mathcal{B}_A$ be the system of all sets $B \in \mathcal{C}p$ satisfying*

(i) *For all $\varepsilon > 0$, $\sigma_{d-1}(B_\varepsilon) \leq \sigma_{d-1}(A_\varepsilon)$,*

(ii) *$\sigma_{d-1}(B) = \sigma_{d-1}(A)$.*

Then $\mathcal{B}_A$ is closed.

Proof. Suppose that $B_j \in \mathcal{B}_A$ for $j \in \mathbb{N}$, $B \in \mathcal{C}p$ and $B_j \to B$ in the Hausdorff metric. Let $\varepsilon > 0$. For each $\delta > 0$, there exists j_0 with $B \subseteq (B_j)_\delta$ for $j \geq j_0$. For these j, we have $B_\varepsilon \subseteq (B_j)_{\delta+\varepsilon}$, hence $\sigma_{d-1}(B_\varepsilon) \leq \sigma_{d-1}((B_j)_{\delta+\varepsilon}) \leq \sigma_{d-1}(A_{\delta+\varepsilon})$. Since this holds for arbitrary $\delta > 0$, it follows that $\sigma_{d-1}(B_\varepsilon) \leq \sigma_{d-1}(A_\varepsilon)$ (since a measure is continuous from above).

From (i) it follows that $\sigma_{d-1}(B) \leq \sigma_{d-1}(A)$. For given $\delta > 0$, let j be so large that $B_j \subseteq B_\delta$. Then $\sigma_{d-1}(A) = \sigma_{d-1}(B_j) \leq \sigma_{d-1}(B_\delta)$. Since $\delta > 0$ was arbitrary, we obtain $\sigma_{d-1}(A) \leq \sigma_{d-1}(B)$ and hence (ii). We have proved that $B \in \mathcal{B}_A$. $\qquad\square$

Lemma 3.4.2. *Let $C \in \mathcal{C}p$. The function $B \mapsto \sigma_{d-1}(B \cap C)$ is upper semi-continuous on $\mathcal{C}p$.*

Proof. Let $B_j, B \in \mathcal{C}p$ and $B_j \to B$ in the Hausdorff metric. Let $\delta > 0$. Let j be so large that $B_j \subseteq B_\delta$. Let $x \in B_j \cap C$. There is some $y \in B$ so that $d_a(x, y) \leq \delta$, and then $y \in B \cap C_\delta$ and thus $x \in (B \cap C_\delta)_\delta$. It follows that $B_j \cap C \subseteq (B \cap C_\delta)_\delta$. Therefore,

$$\limsup_{j \to \infty} \sigma_{d-1}(B_j \cap C) \leq \sigma_{d-1}((B \cap C_\delta)_\delta) \quad \text{for } \delta > 0. \qquad (3.27)$$

Now,

$$\bigcap_{\delta > 0}(B \cap C_\delta)_\delta = B \cap C. \qquad (3.28)$$

Indeed, since B and C are closed, if $x \notin B \cap C$, then there is some $\varepsilon > 0$ such that either $d_a(x, B) \geq \varepsilon$ or $d_a(x, C) \geq \varepsilon$. But then $x \notin (B \cap C_\delta)_\delta$ if $\delta < \varepsilon/2$.

From (3.27) and (3.28) the assertion of the lemma follows. $\qquad \square$

Proof of Theorem 3.4.1. Let $C \subset \mathbb{S}^{d-1}$ be the closed spherical cap centered at p and satisfying $\sigma_{d-1}(C) = \sigma_{d-1}(A)$.

It follows from Lemma 3.4.2 that the function $B \mapsto \sigma_{d-1}(B \cap C)$ attains a maximum on $\mathcal{B}_A$, say at B. We state that

$$C \subseteq B. \qquad (3.29)$$

Suppose this were false. Then $\sigma_{d-1}(B \setminus C) = \sigma_{d-1}(C \setminus B) > 0$ (since $\sigma_{d-1}(B_0) = \sigma_{d-1}(A) = \sigma_{d-1}(C)$ and B and C are closed). Let $x \in B \setminus C$ be a point of density of $B \setminus C$ and $y \in C \setminus B$ a point of density of $C \setminus B$. Let H be the hyperplane through o that is orthogonal to $x - y$. Let $0 < c < 1$. There exists a number $r > 0$ such that

$$\sigma_{d-1}(B(x, r) \cap (B \setminus C)) > c\sigma_{d-1}(B(x, r)),$$
$$\sigma_{d-1}(B(y, r) \cap (C \setminus B)) > c\sigma_{d-1}(B(y, r)).$$

If we now apply the two-point symmetrization with respect to the hyperplane H, then a set of positive measure is transformed from $B \setminus C$ into C, whereas no point of $C \cap B$ is transformed to a point which is not in C (here we make use of the fact that every two-point symmetrization transforms C into itself). Therefore, the transformed set TB satisfies $\sigma_{d-1}(TB \cap C) > \sigma_{d-1}(B \cap C)$. This contradiction shows that (3.29) holds. It implies that $\sigma_{d-1}(A_\varepsilon) \geq \sigma_{d-1}(B_\varepsilon) \geq \sigma_{d-1}(C_\varepsilon)$. $\qquad \square$

In [26, p. 61], it is remarked after the proof that "caps are the only closed solutions of the isoperimetric problem". And in [155, p. 1610], where Benyamini's proof is sketched, we read: "With a bit more effort the proof above can be adjusted to show that caps are the only solutions to the isoperimetric problem in S^{n-1}." However, this is not obvious, and it seems that a proof has not been carried out.

In the special case where $K \in \mathcal{K}_s$ and K lies in some open hemisphere, we have

$$K_{\pi/2} = \mathrm{cl}(\mathbb{S}^{d-1} \setminus K^*).$$

Therefore, (3.25) implies that $\sigma_{d-1}(K^*) \le \sigma_{d-1}(C^*)$, if C is a spherical cap with $\sigma_{d-1}(K) = \sigma_{d-1}(C)$.

For this inequality, we shall give a different proof, which also yields the equality cases.

Theorem 3.4.2. *Suppose that $K \in \mathcal{K}_s$ lies in some open hemisphere. Let C be a spherical cap with $\sigma_{d-1}(K) = \sigma_{d-1}(C)$. Then*

$$\sigma_{d-1}(K^*) \le \sigma_{d-1}(C^*),$$

and equality holds if and only if K is a spherical cap.

Proof. We follow Gao, Hug and Schneider [61]. Suppose that $K \in \mathcal{K}_s$ is contained in an open hemisphere. Let $C \in \mathcal{K}_s$ be a spherical cap with $\sigma_{d-1}(K) = \sigma_{d-1}(C)$. Our goal is to show that

$$\sigma_{d-1}(K^*) \le \sigma_{d-1}(C^*),$$

with equality only if K is a spherical cap. If $\sigma_{d-1}(K) = 0$ or if $d = 2$, this is easy to see; therefore, we assume in the following that $\sigma_{d-1}(K) > 0$ and $d \ge 3$.

Let $e \in -\mathrm{int}\, K^*$ and let $g_e : \mathbb{S}_e^+ \to e^\perp$ be the gnomonic projection. Let

$$F(e) := \lambda_{d-1}(g_e(K)).$$

By Lemma 3.2.3 we have

$$F(e) = \int_K \langle e, u \rangle^{-d}\, \sigma_{d-1}(\mathrm{d}u).$$

The function $F : -\mathrm{int}\, K^* \to (0, \infty)$ defined in this way is continuous and satisfies $F(e) \to \infty$ if e approaches the boundary of $-\mathrm{int}\, K^*$. Hence, F attains a minimum at some point $e \in -\mathrm{int}\, K^*$. At this point, the derivatives of F with respect to local coordinates on $\mathbb{S}^{d-1}$ must vanish, which yields

$$\int_K \langle e, u \rangle^{-(d+1)} \langle v, u \rangle\, \sigma_{d-1}(\mathrm{d}u) = 0 \quad \text{for } v \in e^\perp.$$

Therefore, the vector

$$\int_K \langle e, u \rangle^{-(d+1)} u\, \sigma_{d-1}(\mathrm{d}u)$$

is proportional to e and hence is equal to $F(e)e$. Now it follows from (3.15) that the centroid of $g_e(K)$ in $e^\perp$ is at o. Under this condition, the Blaschke–Santaló inequality (see, e.g., [163, Sect. 10.5]), applied in $e^\perp$, says that

$$\lambda_{d-1}(g_e(K))\lambda_{d-1}(g_e(K)^\circ) \le \kappa_{d-1}^2. \tag{3.30}$$

where the polarity $^\circ$ is taken in $e^\perp$. Equality in (3.30) holds if and only if $g_e(K)$ is an ellipsoid.

We abbreviate $S_e := \mathbb{S}^{d-1} \cap e^\perp$. With $u \in \mathbb{S}_e^+$ uniquely written in the form

$$u = (\cos t)e + (\sin t)u_0, \quad u_0 \in S_e, \; t \in [0, \pi/2),$$

there is a function $\alpha : S_e \to (0, \pi/2)$ such that

$$(\cos t)e + (\sin t)u_0 \in K \Leftrightarrow 0 \leq t \leq \alpha(u_0).$$

We have

$$\sigma_{d-1}(K) = \int_{S_e} \int_0^{\alpha(u_0)} \sin^{d-2} t \, dt \, \sigma_{d-2}(du_0). \tag{3.31}$$

The function $u_0 \mapsto \tan \alpha(u_0)$ is the radial function of $g_e(K)$, hence

$$\lambda_{d-1}(g_e(K)) = \frac{1}{d-1} \int_{S_e} \tan^{d-1} \alpha(u_0) \, \sigma_{d-2}(du_0). \tag{3.32}$$

We define the functions

$$\psi(x) := \int_0^x \sin^{d-2} t \, dt, \quad x \in (0, \pi/2),$$

and

$$f(s) := \psi\left(\arctan s^{1/(d-1)}\right), \quad s \in (0, \infty).$$

Then f has an inverse function h, and ψ has an inverse function ψ^{-1}. They are related by

$$h(y) = \tan^{d-1} \psi^{-1}(y), \quad y \in \operatorname{im}(\psi).$$

Since

$$h'(y) = \frac{d-1}{\cos^d \psi^{-1}(y)},$$

the function h' is strictly increasing, hence h is strictly convex. Applying (3.32), Jensen's inequality and (3.31), we obtain

$$\frac{\lambda_{d-1}(g_e(K))}{\kappa_{d-1}} = \frac{1}{\omega_{d-1}} \int_{S_e} \tan^{d-1} \alpha(u_0) \, \sigma_{d-2}(du_0)$$

$$= \frac{1}{\omega_{d-1}} \int_{S_e} h(\psi(\alpha(u_0))) \, \sigma_{d-2}(du_0)$$

$$\geq h\left(\frac{1}{\omega_{d-1}} \int_{S_e} \psi(\alpha(u_0)) \, \sigma_{d-2}(du_0)\right)$$

$$= h\left(\frac{\sigma_{d-1}(K)}{\omega_{d-1}}\right).$$

Here equality holds if and only if the function α is constant. We have shown that

$$h\left(\frac{\sigma_{d-1}(K)}{\omega_{d-1}}\right) \leq \frac{\lambda_{d-1}(g_e(K))}{\kappa_{d-1}}, \tag{3.33}$$

with equality if and only if K is a spherical cap with center e.

Since $-e \in \operatorname{int} K^*$ and $K^* \subset \mathbb{S}_{-e}^+$, we similarly obtain

$$h\left(\frac{\sigma_{d-1}(K^*)}{\omega_{d-1}}\right) \leq \frac{\lambda_{d-1}(g_{-e}(K^*))}{\kappa_{d-1}}, \tag{3.34}$$

with equality if and only if K^* is a spherical cap with center $-e$, hence if and only if K is a spherical cap with center e. Together with (3.14), this shows that

$$h\left(\frac{\sigma_{d-1}(K^*)}{\omega_{d-1}}\right) \leq \frac{\lambda_{d-1}(g_e(K)^\circ)}{\kappa_{d-1}}, \tag{3.35}$$

with equality if and only if K is a spherical cap with center e.

Now let $C \subset \mathbb{S}^{d-1}$ be a spherical cap with center e and satisfying $\sigma_{d-1}(C) = \sigma_{d-1}(K)$. From (3.33) and its equality conditions we get

$$\frac{\lambda_{d-1}(g_e(C))}{\kappa_{d-1}} = h\left(\frac{\sigma_{d-1}(C)}{\omega_{d-1}}\right) = h\left(\frac{\sigma_{d-1}(K)}{\omega_{d-1}}\right) \leq \frac{\lambda_{d-1}(g_e(K))}{\kappa_{d-1}}. \tag{3.36}$$

From (3.35), (3.36) and the Blaschke–Santaló inequality, together with the equality conditions, we obtain

$$h\left(\frac{\sigma_{d-1}(K^*)}{\omega_{d-1}}\right) \leq \frac{\lambda_{d-1}(g_e(K)^\circ)}{\kappa_{d-1}} \leq \frac{\kappa_{d-1}}{\lambda_{d-1}(g_e(K))}$$

$$\leq \frac{\kappa_{d-1}}{\lambda_{d-1}(g_e(C))} \leq \frac{\lambda_{d-1}(g_e(C)^\circ)}{\kappa_{d-1}} = h\left(\frac{\sigma_{d-1}(C^*)}{\omega_{d-1}}\right).$$

Since the function h is strictly increasing, this shows that $\sigma_{d-1}(K^*) \leq \sigma_{d-1}(C^*)$. If equality holds here, then it holds in (3.35), which implies that K is a spherical cap. $\qquad\square$

Notes for Section 3.4

1. The proof given for Theorem 3.4.2 does not only allow to identify the equality cases, but can be strengthened to yield a stability result. For this, and its application, we refer to Hug and Reichenbacher [95].

2. A very thorough treatment of inequality (3.25), in Euclidean, spherical and hyperbolic space, together with consequences and the identification of the equality cases, was made by Schmidt [158]. A proof was sketched by Lévy [123, pp. 209–222], who emphasized the importance of this inequality for asymptotic estimates. The book [12, Chap. 3] points out that for such applications, the equality cases are irrelevant, and weaker inequalities are sufficient. A proof of (3.25) was also given by Figiel, Lindenstrauss and Milman [59, Appendix].

4

Steiner and kinematic formulas

The conic intrinsic volumes and their spherical counterparts can alternatively be introduced by means of a so-called Steiner formula. Such a formula expresses the (Euclidean, Gaussian, or spherical) volume of a parallel set of a suitable given set at a given distance ε as a function of ε, exhibiting a special form from which functionals depending only on the given set can be extracted. This approach naturally leads, by considering 'local parallel sets', to extensions, or localizations, of the intrinsic volumes in the form of measures. The intrinsic volumes and their local versions are then the subject of various formulas of integral geometry.

In the first two sections of this chapter, we introduce in this way the conic support measures, which are local versions of the conic intrinsic volumes. In large parts, we follow the article by McCoy and Tropp [129], which we localize appropriately. In particular, we obtain an extension of their 'Master Steiner formula', together with various specializations. In Section 4.1, we restrict ourselves to polyhedral cones, and the extension to general closed convex cones follows in Section 4.2.

In Section 4.3, we turn to integral geometric formulas. The kinematic formula for conic intrinsic volumes is proved in a more general version for marginal measures of the conic support measures, the conic curvature measures, first for polyhedral cones by extending an approach of Amelunxen and Lotz [10], and then for general cones.

When the kinematic formula for conic intrinsic volumes is used to compute certain intersection probabilities for random cones, the problem arises that in general the conic intrinsic volumes cannot be computed explicitly. However, good estimates in high dimensions can be obtained by employing concentration properties of these functionals, as they have been discovered and used by Amelunxen, Lotz, McCoy and Tropp [11] and McCoy and Tropp [129]. In Section 4.4 we follow, to a large extent, parts of these papers.

Once some integral geometric formulas are available, it is time to collect the known results about inequalities satisfied by the conic intrinsic volumes. This is done in Section 4.5.

© The Author(s), under exclusive license to Springer Nature Switzerland AG 2022
R. Schneider, *Convex Cones*, Lecture Notes in Mathematics 2319,
https://doi.org/10.1007/978-3-031-15127-9_4

Section 4.6 deals with some observations about the conic support measures. We show that they can be characterized by some of their most natural and simple properties. Then we strengthen their weak continuity property, by proving their Hölder continuity with respect to suitable metrics.

4.1 A general Steiner formula for polyhedral cones

As we want to localize the conic intrinsic volumes in the form of measures, we need domains for such measures that are adapted to the conical structure.

A subset $A \subseteq \mathbb{R}^d$ is called *conic* if $a \in A$ implies $\lambda a \in A$ for all $\lambda > 0$. With every subset $A \subseteq \mathbb{R}^d$ we associate the conic set

$$A^+ := \{\lambda a : a \in A, \lambda > 0\}.$$

(Recall that we have also defined $A^\vee = \{\lambda a : a \in A, \lambda \geq 0\}$.) For $A \subseteq \mathbb{S}^{d-1}$, we then have $A = A^+ \cap \mathbb{S}^{d-1}$, and the map $A \mapsto A^+$ is a bijection between the subsets of $\mathbb{S}^{d-1}$ and the conic subsets of $\mathbb{R}^d$ different from $\{o\}$.

The *conic (Borel) σ-algebra* is defined by

$$\widehat{\mathcal{B}}(\mathbb{R}^d) := \{A \in \mathcal{B}(\mathbb{R}^d) : A^+ = A\}.$$

Similarly, for $\eta \subseteq \mathbb{R}^d \times \mathbb{R}^d$, we write

$$\eta^+ := \{(\lambda x, \mu y) : (x, y) \in \eta, \lambda, \mu > 0\}.$$

We equip the Cartesian product $\mathbb{R}^d \times \mathbb{R}^d$ with the product topology and define the *biconic (Borel) σ-algebra*

$$\widehat{\mathcal{B}}(\mathbb{R}^d \times \mathbb{R}^d) := \{\eta \in \mathcal{B}(\mathbb{R}^d \times \mathbb{R}^d) : \eta^+ = \eta\},$$

(The terms 'conic σ-algebra' and 'biconic σ-algebra' were suggested in [6].)

Adapted to integration with respect to measures on the measurable space $(\mathbb{R}^d, \widehat{\mathcal{B}}(\mathbb{R}^d))$ are functions that are homogeneous of degree zero (on $\mathbb{R}^d \setminus \{o\}$). Equivalently, we may use functions on the unit sphere. For a function $f : \mathbb{S}^{d-1} \to \mathbb{R}$, we define its homogeneous extension by

$$f_h(x) := f\left(\frac{x}{\|x\|}\right) \quad \text{for } x \in \mathbb{R}^d \setminus \{o\}.$$

If f is measurable with respect to $\mathcal{B}(\mathbb{S}^{d-1})$, then f_h is measurable with respect to $\widehat{\mathcal{B}}(\mathbb{R}^d)$. That the homogeneous extension is not defined at $x = o$, is irrelevant, since we shall only integrate with respect to measures that assign measure zero to $\{o\}$. If μ is such a measure on $\widehat{\mathcal{B}}(\mathbb{R}^d)$ and if we define the measure $\overline{\mu}$ on $\mathcal{B}(\mathbb{S}^{d-1})$ by $\overline{\mu}(A) := \mu(A^+)$ for $A \in \mathcal{B}(\mathbb{S}^{d-1})$, then, for every $\overline{\mu}$-integrable function $f : \mathbb{S}^{d-1} \to \mathbb{R}$,

$$\int_{\mathbb{S}^{d-1}} f \, d\overline{\mu} = \int_{\mathbb{R}^d} f_h \, d\mu.$$

Similarly, for a function $f : \mathbb{S}^{d-1} \times \mathbb{S}^{d-1} \to \mathbb{R}$, we define its homogeneous extension by

$$f_h(x, y) := f\left(\frac{x}{\|x\|}, \frac{y}{\|y\|}\right) \quad \text{for } x, y \in \mathbb{R}^d \setminus \{o\}.$$

The measures that localize the conic intrinsic volumes of polyhedral cones can now be introduced as follows.

Definition 4.1.1. For $C \in \mathcal{PC}^d$, $\eta \in \widehat{B}(\mathbb{R}^d \times \mathbb{R}^d)$ and $k = 0, \ldots, d$, let

$$\Omega_k(C, \eta) := \mathbb{P}(\Pi_C(\mathbf{g}) \in \text{skel}_k C, \; (\Pi_C(\mathbf{g}), \Pi_{C^\circ}(\mathbf{g})) \in \eta).$$

$\Omega_0(C, \cdot), \ldots, \Omega_d(C, \cdot)$ are called the conic support measures of the polyhedral cone C.

We recall that $\mathbf{g}$ is a standard Gaussian random vector in $\mathbb{R}^d$ and that the k-skeleton $\text{skel}_k C$ was defined in (1.24) as the union of the relative interiors of the k-faces of C. If $F \in \mathcal{F}_k(C)$ and $\Pi_C(\mathbf{g}) \in \text{relint } F$, then $\Pi_{C^\circ}(\mathbf{g}) \in N(C, F)$. Thus,

$$\begin{aligned}
&\mathbb{P}(\Pi_C(\mathbf{g}) \in \text{relint } F, \; (\Pi_C(\mathbf{g}), \Pi_{C^\circ}(\mathbf{g})) \in \eta) \\
&= \mathbb{P}((\Pi_C(\mathbf{g}), \Pi_{C^\circ}(\mathbf{g})) \in ((\text{relint} F) \times N(C, F)) \cap \eta) \\
&= \gamma_d(\{z \in F + N(C, F) : (\Pi_C(z), \Pi_{C^\circ}(z)) \in \eta\}) \\
&= \gamma_d(\{x + y : x \in F, \; y \in N(C, F), \; (x, y) \in \eta\}) \\
&= \int_F \int_{N(C,F)} \mathbb{1}_\eta(x, y) \, \gamma_{\langle F \rangle^\perp}(dy) \, \gamma_{\langle F \rangle}(dx).
\end{aligned}$$

Therefore, we have the explicit representation

$$\Omega_k(C, \eta) = \sum_{F \in \mathcal{F}_k(C)} \int_F \int_{N(C,F)} \mathbb{1}_\eta(x, y) \, \gamma_{\langle F \rangle^\perp}(dy) \, \gamma_{\langle F \rangle}(dx). \tag{4.1}$$

Clearly, $\Omega_k(C, \cdot)$ is a measure, and

$$\Omega_k(C, \mathbb{R}^d \times \mathbb{R}^d) = v_k(C)$$

for $k = 0, \ldots, d$. We note the particular cases

$$\Omega_0(C, \eta) = \int_{C^\circ} \mathbb{1}_\eta(o, y) \, \gamma_d(dy) \tag{4.2}$$

and

$$\Omega_d(C,\eta) = \int_C \mathbb{1}_\eta(x,o)\,\gamma_d(dx).$$ (4.3)

Equivalently, using (2.2),

$$\Omega_k(C,\eta)$$

$$= \frac{1}{\omega_k\omega_{d-k}} \sum_{F\in\mathcal{F}_k(C)} \int_{F\cap\mathbb{S}^{d-1}} \int_{N(C,F)\cap\mathbb{S}^{d-1}} \mathbb{1}_\eta(u,v)\,\sigma_{d-k-1}(dv)\,\sigma_{k-1}(du)$$

for $k \in \{1,\dots,d-1\}$ and

$$\Omega_0(C,\eta) = \frac{1}{\omega_d}\int_{C^\circ\cap\mathbb{S}^{d-1}} \mathbb{1}_\eta(o,v)\,\sigma_{d-1}(dv),$$

$$\Omega_d(C,\eta) = \frac{1}{\omega_d}\int_{C\cap\mathbb{S}^{d-1}} \mathbb{1}_\eta(u,o)\,\sigma_{d-1}(du).$$

The conic support measures have the following duality property. If we define

$$\eta^* := \{(y,x) : (x,y)\in\eta\} \quad \text{for } \eta\in\widehat{\mathcal{B}}(\mathbb{R}^d\times\mathbb{R}^d),$$

then

$$\Omega_k(C,\eta) = \Omega_{d-k}(C^\circ,\eta^*).$$ (4.4)

This follows from (4.1), since $F\in\mathcal{F}_k(C)$ implies $G := N(C,F)\in\mathcal{F}_{d-k}(C^\circ)$ and $N(C^\circ,G) = F$.

Due to the invariance properties of the Gaussian measure, it follows from (4.1) that the functions Ω_k are $O(d)$-equivariant, in the following sense. Defining

$$\vartheta\eta := \{(\vartheta x,\vartheta y) : (x,y)\in\eta\} \quad \text{for } \eta\subseteq\mathbb{R}^d\times\mathbb{R}^d,\ \vartheta\in O(d),$$

we have

$$\Omega_k(\vartheta C,\vartheta\eta) = \Omega_k(C,\eta)$$ (4.5)

for $C\in\mathcal{PC}^d$, $\eta\in\widehat{\mathcal{B}}(\mathbb{R}^d\times\mathbb{R}^d)$, and $\vartheta\in O(d)$.

By marginalization of the conic support measures, we obtain the *conic curvature measures* of first and second kind. They are defined by

$$\Phi_k(C,A) := \Omega_k(C, A\times\mathbb{R}^d),$$

$$\Psi_k(C,A) := \Omega_k(C, \mathbb{R}^d\times A)$$

for $C\in\mathcal{PC}^d$, $A\in\widehat{\mathcal{B}}(\mathbb{R}^d)$, and $k = 0,\dots,d$. Thus,

$$\Phi_k(C,A) = \sum_{F\in\mathcal{F}_k(C)} \gamma_{\langle F\rangle}(F\cap A)\gamma_{\langle F\rangle^\perp}(N(C,F)),$$ (4.6)

$$\Psi_k(C,A) = \sum_{F\in\mathcal{F}_k(C)} \gamma_{\langle F\rangle}(F)\gamma_{\langle F\rangle^\perp}(N(C,F)\cap A).$$ (4.7)

The duality now reads

$$\Phi_k(C, A) = \Psi_{d-k}(C^\circ, A) \tag{4.8}$$

and the equivariance properties are

$$\Phi_k(\vartheta C, \vartheta A) = \Phi_k(C, A), \tag{4.9}$$

$$\Psi_k(\vartheta C, \vartheta A) = \Psi_k(C, A), \tag{4.10}$$

for $C \in \mathcal{PC}^d$, $A \in \widehat{\mathcal{B}}(\mathbb{R}^d)$, and $\vartheta \in O(d)$.

To deal with the conic curvature measures, we need a localization of the functional $v_F(C)$. For $C \in \mathcal{PC}^d$, a face F of C and a conic Borel set $A \in \widehat{\mathcal{B}}(\mathbb{R}^d)$, we define

$$\varphi_F(C, A) := \mathbb{P}(\Pi_C(\mathbf{g}) \in A \cap \operatorname{relint} F).$$

Then also

$$\varphi_F(C, A) = \gamma_d((A \cap F) + N(C, F)) = \gamma_{\langle F \rangle}(A \cap F)\gamma_{\langle F \rangle^\perp}(N(C, F)).$$

If $\dim F = k$, this can be written as

$$\varphi_F(C, A) = \Phi_k(F, A)v_{d-k}(N(C, F)). \tag{4.11}$$

We have

$$\Phi_k(C, A) = \sum_{F \in \mathcal{F}_k(C)} \varphi_F(C, A). \tag{4.12}$$

The following lemma extends Lemma 2.3.1 and follows in the same way from Lemma 1.4.4.

Lemma 4.1.1. *Let $C, D \in \mathcal{PC}^d$, $A, B \in \widehat{\mathcal{B}}(\mathbb{R}^d)$, $F \in \mathcal{F}_i(C)$, $G \in \mathcal{F}_j(D)$, and suppose that $i + j = d + k$ with $k > 0$. If $F \pitchfork G$, that is, F and G intersect transversely, then*

$$\varphi_{F \cap G}(C \cap D, A \cap B) = \Phi_k(F \cap G, A \cap B)v_{d-k}(N(C, F) + N(D, G)).$$

We turn to Steiner formulas for polyhedral cones. First we use the conic support measures to generalize (by localization) the 'Master Steiner formula' of McCoy and Tropp [129].

Let $f : \mathbb{R}^2_+ \to \mathbb{R}_+$ be a measurable function. For a cone $C \in \mathcal{C}^d$ and a Borel set $\eta \in \widehat{\mathcal{B}}(\mathbb{R}^d \times \mathbb{R}^d)$ we define

$$\varphi_f(C, \eta) := \mathbb{E}\left[f(\|\Pi_C(\mathbf{g})\|^2, \|\Pi_{C^\circ}(\mathbf{g})\|^2) \cdot \mathbb{1}_\eta(\Pi_C(\mathbf{g}), \Pi_{C^\circ}(\mathbf{g}))\right]. \tag{4.13}$$

The following theorem expresses this expectation for polyhedral cones as a linear combination of conic support measures, with coefficients depending on the function f. Special choices of this function will later lead to representations for measures of local parallel sets.

Theorem 4.1.1. *Let $C \in \mathcal{PC}^d$ be a polyhedral cone. Let $f : \mathbb{R}_+^2 \to \mathbb{R}_+$ be a nonnegative, measurable function such that $\varphi_f(C, \eta)$ is finite for all $\eta \in \widehat{\mathcal{B}}(\mathbb{R}^d \times \mathbb{R}^d)$. Then*

$$\varphi_f(C, \eta) = \sum_{k=0}^{d} I_k(f) \cdot \Omega_k(C, \eta) \tag{4.14}$$

for $\eta \in \widehat{\mathcal{B}}(\mathbb{R}^d \times \mathbb{R}^d)$, where the coefficients are given by

$$I_k(f) = \varphi_f(L_k, \mathbb{R}^d \times \mathbb{R}^d), \tag{4.15}$$

for an (arbitrary) subspace $L_k \in G(d, k)$.
Explicitly,

$$I_k(f) = \frac{\omega_k \omega_{d-k}}{\sqrt{2\pi}^d} \int_0^\infty \int_0^\infty f(r^2, s^2) e^{-\frac{1}{2}(r^2 + s^2)} r^{k-1} s^{d-k-1} \, \mathrm{d}s \, \mathrm{d}r$$

for $k = 1, \ldots, d-1$ and

$$I_0(f) = \frac{\omega_d}{\sqrt{2\pi}^d} \int_0^\infty f(0, s^2) e^{-\frac{1}{2}s^2} s^{d-1} \, \mathrm{d}s,$$

$$I_d(f) = \frac{\omega_d}{\sqrt{2\pi}^d} \int_0^\infty f(r^2, 0) e^{-\frac{1}{2}r^2} r^{d-1} \, \mathrm{d}r.$$

Proof. (We follow [129], where the global case is treated.) We abbreviate $\mathbf{u} := \Pi_C(\mathbf{g})$ and $\mathbf{v} := \Pi_{C^\circ}(\mathbf{g})$. The decomposition (1.26) shows that

$$\varphi_f(C, \eta) = \sum_{k=0}^{d} \sum_{F \in \mathcal{F}_k(C)} \mathbb{E}\left[f(\|\mathbf{u}\|^2, \|\mathbf{v}\|^2) \cdot \mathbb{1}_\eta(\mathbf{u}, \mathbf{v}) \cdot \mathbb{1}_{\mathrm{relint}\, F}(\mathbf{u}) \right].$$

Let $F \in \mathcal{F}_k(C)$ and assume, first, that $k \in \{1, \ldots, d-1\}$. Explicitly,

$$\mathbb{E}\left[f(\|\mathbf{u}\|^2, \|\mathbf{v}\|^2) \cdot \mathbb{1}_\eta(\mathbf{u}, \mathbf{v}) \cdot \mathbb{1}_{\mathrm{relint}\, F}(\mathbf{u}) \right]$$
$$= \int_{F \oplus N(C, F)} f(\|\Pi_C(z)\|^2, \|\Pi_{C^\circ} z)\|^2) \mathbb{1}_\eta(\Pi_C(z), \Pi_{C^\circ}(z)) \, \gamma_d(\mathrm{d}z).$$

The image measure of the restricted Gaussian measure $\gamma_d \llcorner F \oplus N(C, F)$ under the mapping $z \mapsto (\Pi_C(z), \Pi_{C^\circ}(z))$ is the product measure

$$\gamma_{\langle F \rangle} \llcorner F \otimes \gamma_{\langle N(C, F) \rangle} \llcorner N(C, F)$$

(cf. (2.3)). Using this together with (1.6) and (2.1), we get

$$\mathbb{E}\left[f(\|\mathbf{u}\|^2, \|\mathbf{v}\|^2) \cdot \mathbb{1}_\eta(\mathbf{u}, \mathbf{v}) \cdot \mathbb{1}_{\mathrm{relint}\, F}(\mathbf{u}) \right]$$
$$= \frac{1}{\sqrt{2\pi}^d} \int_F \int_{N(C,F)} f(\|x\|^2, \|y\|^2) \cdot \mathbb{1}_\eta(x, y) e^{-\frac{1}{2}(\|x\|^2 + \|y\|^2)} \lambda_{d-k}(\mathrm{d}y) \lambda_k(\mathrm{d}x)$$

$$= \frac{1}{\sqrt{2\pi}^d} \int_0^\infty \int_0^\infty \int_{F \cap \mathbb{S}^{d-1}} \int_{N(C,F) \cap \mathbb{S}^{d-1}} f(r^2, s^2) \cdot \mathbb{1}_\eta(u, v) e^{-\frac{1}{2}(r^2 + s^2)}$$

$$\times r^{k-1} s^{d-k-1} \sigma_{d-k-1}(dv) \, \sigma_{k-1}(du) \, ds \, dr$$

$$= \frac{I_k(f)}{\omega_k \omega_{d-k}} \int_{F \cap \mathbb{S}^{d-1}} \int_{N(C,F) \cap \mathbb{S}^{d-1}} \mathbb{1}_\eta(u, v) \, \sigma_{d-k-1}(dv) \, \sigma_{k-1}(du).$$

with

$$I_k(f) = \frac{\omega_k \omega_{d-k}}{\sqrt{2\pi}^d} \int_0^\infty \int_0^\infty f(r^2, s^2) e^{-\frac{1}{2}(r^2 + s^2)} r^{k-1} s^{d-k-1} \, ds \, dr.$$

If $k = 0$, that is, $F = \{o\}$, we obtain

$$\mathbb{E}\left[f(\|\mathbf{u}\|^2, \|\mathbf{v}\|^2) \cdot \mathbb{1}_\eta(\mathbf{u}, \mathbf{v}) \cdot \mathbb{1}_{\mathrm{relint}\, F}(\mathbf{u}) \right]$$

$$= \mathbb{E}\left[f(0, \|\mathbf{v}\|^2) \cdot \mathbb{1}_\eta(o, \mathbf{v}) \cdot \mathbb{1}\{\mathbf{u} = o\} \right]$$

$$= \frac{1}{\sqrt{2\pi}^d} \int_0^\infty \int_{C^\circ \cap \mathbb{S}^{d-1}} f(0, s^2) \cdot \mathbb{1}_\eta(o, v) e^{-\frac{1}{2}s^2} s^{d-1} \, \sigma_{d-1}(dv) \, ds$$

$$= \frac{I_0(f)}{\omega_d} \int_{C^\circ \cap \mathbb{S}^{d-1}} \mathbb{1}_\eta(o, v) \, \sigma_{d-1}(dv)$$

with

$$I_0(f) = \frac{\omega_d}{\sqrt{2\pi}^d} \int_0^\infty f(0, s^2) e^{-\frac{1}{2}s^2} s^{d-1} \, ds.$$

If $F \in \mathcal{F}_d(C)$ exists, then $\dim C = d$ and $F = C$. In this case,

$$\mathbb{E}\left[f(\|\mathbf{u}\|^2, \|\mathbf{v}\|^2) \cdot \mathbb{1}_\eta(\mathbf{u}, \mathbf{v}) \cdot \mathbb{1}_{\mathrm{relint}\, F}(\mathbf{u}) \right]$$

$$= \mathbb{E}\left[f(\|\mathbf{u}\|^2, 0) \cdot \mathbb{1}_\eta(\mathbf{u}, o) \cdot \mathbb{1}\{\mathbf{u} \in \mathrm{int}\, C\} \right]$$

$$= \frac{1}{\sqrt{2\pi}^d} \int_0^\infty \int_{C \cap \mathbb{S}^{d-1}} f(r^2, 0) \cdot \mathbb{1}_\eta(u, o) e^{-\frac{1}{2}r^2} r^{d-1} \, \sigma_{d-1}(du) \, dr$$

$$= \frac{I_d(f)}{\omega_d} \int_{C \cap \mathbb{S}^{d-1}} \mathbb{1}_\eta(u, o) \, \sigma_{d-1}(du)$$

with

$$I_d(f) = \frac{\omega_d}{\sqrt{2\pi}^d} \int_0^\infty f(r^2, 0) e^{-\frac{1}{2}r^2} r^{d-1} \, dr.$$

According to the definition of $\Omega_k(C, \eta)$, in each case this gives (4.14).

Applying (4.14) with $\eta = \mathbb{R}^d \times \mathbb{R}^d$ to a subspace $C = L_j \in G(d, j)$ and observing (2.31), we obtain (4.15). This completes the proof. $\qquad\square$

In the next two subsections we show how this general Steiner formula includes some special and more intuitive formulas. Before that, we remark (following [129]) that the coefficients $I_k(f)$ appearing in the Steiner formula (4.14) can also be interpreted in the following way.

Lemma 4.1.2. *Let $f : \mathbb{R}_+^2 \to \mathbb{R}$ be a nonnegative, measurable function such that $\varphi_f(C, \mathbb{R}^d \times \mathbb{R}^d)$ is finite. Let $\mathcal{X}_0, \ldots, \mathcal{X}_d, \mathcal{Y}_0, \ldots, \mathcal{Y}_d$ be independent real random variables such that $\mathcal{X}_k$ and $\mathcal{Y}_k$ follow the chi-square distribution with k degrees of freedom (with $\mathcal{X}_0, \mathcal{Y}_0 \equiv 0$). Then*

$$I_k(f) = \mathbb{E}\, f(\mathcal{X}_k, \mathcal{Y}_k).$$

Proof. If $L_k \in G(d, k)$ is an arbitrary k-dimensional subspace, then, by definition,

$$L_k(f) = \varphi_f(L_k, \mathbb{R}^d \times \mathbb{R}^d) = \mathbb{E}\, f\left(\|\Pi_{L_k}(\mathbf{g})\|^2, \|\Pi_{L_k^\perp}(\mathbf{g})\|^2\right).$$

Since $\mathbf{g}$ is a standard Gaussian vector on $\mathbb{R}^d = L_k \oplus L_k^\perp$, the projections $\Pi_{L_k}(\mathbf{g})$ and $\Pi_{L_k^\perp}(\mathbf{g})$ are independent standard Gaussian vectors on L_k and $L_k^\perp$, respectively. Therefore, $\|\Pi_{L_k}(\mathbf{g})\|^2$ and $\|\Pi_{L_k^\perp}(\mathbf{g})\|^2$ are independent chi-square random variables with k and $d - k$ degrees of freedom, respectively. $\qquad\square$

4.1.1 The local Gaussian Steiner formula

Let $C \in \mathcal{PC}^d$ and $\eta \in \widehat{\mathcal{B}}(\mathbb{R}^d \times \mathbb{R}^d)$. For $\lambda \geq 0$, we define the *local parallel set*

$$M_\lambda(C, \eta) := \left\{x \in \mathbb{R}^d : \operatorname{dist}(x, C) \leq \lambda,\ (\Pi_C(x), \Pi_{C^\circ}(x)) \in \eta\right\}. \qquad (4.16)$$

Due to the continuity of the functions $\operatorname{dist}(\,\cdot\,, C)$, Π_C, Π_{C°, this is a Borel set. We have

$$M_\lambda(C, \mathbb{R}^d \times \mathbb{R}^d) = C + \lambda B^d,$$

which is the usual outer parallel set of C at distance λ.

Corollary 4.1.1. *Let $C \in \mathcal{PC}^d$. The Gaussian measure of the local parallel set $M_\lambda(C, \eta)$ is given by*

$$\gamma_d(M_\lambda(C, \eta)) = \sum_{k=0}^d f_k(\lambda) \cdot \Omega_k(C, \eta), \qquad (4.17)$$

where

$$f_k(\lambda) = \frac{\omega_{d-k}}{\sqrt{2\pi}^{d-k}} \int_0^\lambda e^{-\frac{1}{2}s^2} s^{d-k-1}\, \mathrm{d}s$$

for $k = 0, \ldots, d - 1$ and $f_d \equiv 1$.

Proof. With the function f defined by

$$f(a,b) := \begin{cases} 1 \text{ if } b \leq \lambda^2, \\ 0 \text{ otherwise,} \end{cases}$$

we have

$$\varphi_f(C, \eta) = \gamma_d(M_\lambda(C, \eta)),$$

and we get

$$I_k(f) = \frac{\omega_{d-k}}{\sqrt{2\pi}^{d-k}} \int_0^\lambda e^{-\frac{1}{2}s^2} s^{d-k-1} \, \mathrm{d}s = f_k(\lambda)$$

for $k = 0, \ldots, d-1$ and $I_d(f) = 1$. The assertion follows, therefore, from Theorem 4.1.1. $\qquad\square$

With $\eta = \mathbb{R}^d \times \mathbb{R}^d$, we obtain the global *Gaussian Steiner formula*

$$\gamma_d(C + \varepsilon B^d) = \sum_{k=0}^d f_k(\varepsilon) v_k(C). \tag{4.18}$$

It determines the values $v_0(C), \ldots, v_d(C)$ uniquely, since the coefficient functions $f_0, \ldots, f_d$ are linearly independent. Thus, this is an approach to the conic intrinsic volumes, different from the one in Chapter 1, which is in strict analogy to the definition of the Euclidean intrinsic volumes via the classical Steiner formula.

Equation (4.17) can be given a reformulation in more probabilistic terms (as in McCoy and Tropp [129]). Replacing the distance by the squared distance, we have, for $\lambda \geq 0$,

$$\mathbb{P}\left(\mathrm{dist}^2(\mathbf{g}, C) \leq \lambda, \, (\Pi_C(\mathbf{g}), \Pi_{C^\circ}(\mathbf{g})) \in \eta\right)$$

$$= \gamma_d(M_{\sqrt{\lambda}}(C, \eta)) = \sum_{k=0}^d f_k\left(\sqrt{\lambda}\right) \cdot \Omega_k(C, \eta).$$

A substitution gives

$$f_k\left(\sqrt{\lambda}\right) = \frac{1}{2^{\frac{d-k}{2}}\Gamma\left(\frac{d-k}{2}\right)} \int_0^\lambda e^{-\frac{s}{2}} s^{\frac{d-k}{2}-1} \, \mathrm{d}s = \mathbb{P}(\mathcal{X}_{d-k} \leq \lambda)$$

for $k = 0, \ldots, d-1$, where $\mathcal{X}_{d-k}$ is a random variable following the chi-square distribution with $d - k$ degrees of freedom. We formulate this as a corollary.

Corollary 4.1.2. *Let $C \in \mathcal{PC}^d$ and $\eta \in \widehat{\mathcal{B}}(\mathbb{R}^d \times \mathbb{R}^d)$. If $\mathbf{g}$ is a standard Gaussian random vector in $\mathbb{R}^d$, then, for $\lambda \geq 0$,*

$$\mathbb{P}\left(\mathrm{dist}^2(\mathbf{g}, C) \leq \lambda, \, (\Pi_C(\mathbf{g}), \Pi_{C^\circ}(\mathbf{g})) \in \eta\right)$$

$$= \sum_{k=0}^d \mathbb{P}(\mathcal{X}_{d-k} \leq \lambda) \cdot \Omega_k(C, \eta), \tag{4.19}$$

where $\mathcal{X}_{d-k}$ is a random variable following the chi-square distribution with $d-k$ degrees of freedom (with the convention that $\mathcal{X}_0 \equiv 0$).

4.1.2 The local spherical Steiner formula

We consider a counterpart to (4.16), with the Euclidean distance replaced by the angular distance.

Let $C \in \mathcal{PC}_*^d$ and $\eta \in \widehat{\mathcal{B}}(\mathbb{R}^d \times \mathbb{R}^d)$. For $0 \le \lambda < \pi/2$, we define the *angular local parallel set*

$$M_\lambda^a(C, \eta) := \left\{ x \in \mathbb{R}^d : d_a(x, C) \le \lambda, \ (\Pi_C(x), \Pi_{C^\circ}(x)) \in \eta \right\}. \qquad (4.20)$$

Again, this is a Borel set. We note that $o \notin M_\lambda^a(C, \eta)$, since $d_a(o, C) = \pi/2$.

Corollary 4.1.3. *Let $C \in \mathcal{PC}_*^d$. The Gaussian measure of the angular local parallel set $M_\lambda^a(C, \eta)$, for $0 \le \lambda < \pi/2$, is given by*

$$\gamma_d(M_\lambda^a(C, \eta)) = \sum_{k=1}^d g_k(\lambda) \cdot \Omega_k(C, \eta), \qquad (4.21)$$

where

$$g_k(\lambda) = \frac{\omega_k \omega_{d-k}}{\omega_d} \int_0^\lambda \cos^{k-1} \varphi \sin^{d-k-1} \varphi \, d\varphi \qquad (4.22)$$

for $k = 1, \ldots, d-1$ and $g_d \equiv 1$.

Proof. With the function f defined by

$$f(a, b) := \begin{cases} 1 \text{ if } a \le b \tan^2 \lambda, \\ 0 \text{ otherwise,} \end{cases}$$

we have

$$\varphi_f(C, \eta) = \mathbb{P}\left(d_a(\mathbf{g}, C) \le \lambda, \ (\Pi_C(x), \Pi_{C^\circ}(x)) \in \eta\right) = \gamma_d(M_\lambda^a(C, \eta)).$$

Now (4.14) and (4.15) give

$$\gamma_d(M_\lambda^a(C, \eta)) = \sum_{k=0}^d I_k(f) \cdot \Omega_k(C, \eta)$$

with

$$I_k(f) = \varphi_f(L_k, \mathbb{R}^d \times \mathbb{R}^d),$$

where L_k is a k-dimensional subspace. First let $k \in \{1, \ldots, d-1\}$. Writing $S := L_k \cap \mathbb{S}^{d-1}$ and recalling the definition (3.4) of parallel sets on the sphere, we have,

$$I_k(f) = \gamma_d(M_\lambda^a(L_k, \mathbb{R}^d \times \mathbb{R}^d)) = \omega_d^{-1}\sigma_{d-1}(S_\lambda)$$

$$= \frac{\omega_k\omega_{d-k}}{\omega_d} \int_0^\lambda \cos^{k-1}\varphi \sin^{d-k-1}\varphi \, d\varphi,$$

by Lemma 3.1.4. Further, $I_d(f) = \varphi_f(\mathbb{R}^d, \mathbb{R}^d \times \mathbb{R}^d) = 1$.   □

Corollary 4.1.3 is the conic reformulation of the local spherical Steiner formula, as it appears in Glasauer [63, 64] and is reproduced in [170, Thm. 6.5.1]. In order to compare the two formulations, let a nonempty, closed, spherically convex set $K \subseteq \mathbb{S}^{d-1}$ and a Borel set $A \in \mathcal{B}(\mathbb{S}^{d-1} \times \mathbb{S}^{d-1})$ be given. Recall that $K^\vee = \{\lambda x : x \in K, \lambda \geq 0\}$ and $A^+ = \{(\lambda x, \mu y) : (x, y) \in A, \lambda, \mu > 0\}$. The formula given in [170, Thm. 6.5.1] can be written as

$$\omega_d\gamma_d(M_\lambda^a(K^\vee, A^+)) = \omega_d\Omega_d(K^\vee, A^+) + \sum_{k=1}^{d-1} g_{d,k-1}(\lambda) \cdot \Theta_{k-1}(K, A),$$

where $g_{d,k-1}(\lambda) = \omega_d \, g_k(\lambda)$, in our present notation. Comparison with (4.21) shows that $\Theta_{k-1}(K, A) = \Omega_k(K^\vee, A^+)$ for $k = 1, \ldots, d-1$, if (for the moment) $K^\vee$ is a polyhedral cone. We recall that the spherical intrinsic volumes which in [63] and [170] were denoted by v_j are now denoted by v_j^s and are related to the conic intrinsic volumes by $v_j^s(K) = v_{j+1}(K^\vee)$.

The following is a counterpart to Corollary 4.1.2; its global case appears in McCoy and Tropp [129].

Corollary 4.1.4. *Let $C \in \mathcal{PC}_*^d$ and $\eta \in \widehat{\mathcal{B}}(\mathbb{R}^d \times \mathbb{R}^d)$. If $\mathbf{u}$ is a uniform random vector in the sphere $\mathbb{S}^{d-1}$ (that is, with distribution σ_{d-1}/ω_d), then, for $\lambda \in [0, 1]$,*

$$\mathbb{P}\left(\mathrm{dist}^2(\mathbf{u}, C) \leq \lambda, (\Pi_C(\mathbf{u}), \Pi_{C^\circ}(\mathbf{u})) \in \eta\right)$$

$$= \sum_{k=0}^d \mathbb{P}(B_{d-k,k} \leq \lambda) \cdot \Omega_k(C, \eta), \tag{4.23}$$

where $B_{j,k}$ is a random variable following the beta distribution with parameters $j/2$ and $k/2$.

Proof. Normalizing the standard Gaussian random vector $\mathbf{g}$,

$$\mathbf{u} := \frac{\mathbf{g}}{\|\mathbf{g}\|} \quad \text{for } \mathbf{g} \neq o,$$

we obtain a random vector $\mathbf{u}$ on the unit sphere $\mathbb{S}^{d-1}$ with uniform distribution.

If $\lambda = 1$, then (using (1.5)) the left side of (4.23) is equal to

$$\mathbb{P}((\Pi_C(\mathbf{u}), \Pi_{C^\circ}(\mathbf{u})) \in \eta) = \mathbb{P}((\Pi_C(\mathbf{g}), \Pi_{C^\circ}(\mathbf{g})) \in \eta) = \varphi_1(C, \eta),$$

where $\mathbf{1}$ is the constant function with value 1, and (4.23) follows from (4.14) and (4.15), since the latter gives $I_k(\mathbf{1}) = 1$. Hence, we assume now that $0 \leq \lambda < 1$.

Elementary geometry, together with (1.6), shows that

$$\operatorname{dist}^2(\mathbf{u}, C) \leq \lambda \Leftrightarrow \|\Pi_{C^\circ}(\mathbf{g})\|^2 \leq \frac{\lambda}{1 - \lambda} \|\Pi_C(\mathbf{g})\|^2.$$

Hence,

$$\mathbb{P}\left(\operatorname{dist}^2(\mathbf{u}, C) \leq \lambda, \ (\Pi_C(\mathbf{g}), \Pi_{C^\circ}(\mathbf{g})) \in \eta\right)$$
$$= \mathbb{P}\left(\|\Pi_{C^\circ}(\mathbf{g})\|^2 \leq \lambda(1 - \lambda)^{-1}\|\Pi_C(g)\|^2, \ (\Pi_C(\mathbf{g}), \Pi_{C^\circ}(\mathbf{g})) \in \eta\right)$$
$$= \varphi_f(C, \eta)$$

with a suitable function f. Now (4.14) and (4.15) give

$$\mathbb{P}\left(\operatorname{dist}^2(\mathbf{u}, C) \leq \lambda, \ (\Pi_C(\mathbf{g}), \Pi_{C^\circ}(\mathbf{g})) \in \eta\right) = \sum_{k=0}^{d} I_k(f) \cdot \Omega_k(C, \eta)$$

with

$$I_k(f) = \varphi_f(L_k, \mathbb{R}^d \times \mathbb{R}^d) = \mathbb{P}\left(\operatorname{dist}^2(\mathbf{u}, L_k) \leq \lambda\right),$$

where L_k is a k-dimensional subspace.

For the computation of $\mathbb{P}\left(\operatorname{dist}^2(\mathbf{u}, L_k) \leq \lambda\right)$, we write $S := \mathbb{S}^{d-1} \cap L_k$ and note that $\operatorname{dist}(\mathbf{u}, L_k) = \sin d_a(\mathbf{u}, S)$, hence Lemma 3.1.4 and the substitution $\sin^2 \varphi = t$ give

$$\mathbb{P}\left(\operatorname{dist}^2(\mathbf{u}, L_k) \leq \lambda\right) = \omega_d^{-1} \sigma_{d-1}(S_{\arcsin \sqrt{\lambda}})$$

$$= \frac{\omega_{d-k}\omega_k}{\omega_d} \int_0^{\arcsin \sqrt{\lambda}} \cos^{k-1} \varphi \, \sin^{d-k-1} \varphi \, d\varphi$$

$$= \frac{\Gamma\left(\frac{d}{2}\right)}{\Gamma\left(\frac{d-k}{2}\right)\Gamma\left(\frac{k}{2}\right)} \int_0^{\lambda} t^{\frac{d-k}{2}-1}(1 - t)^{\frac{k}{2}-1} dt$$

$$= \mathbb{P}\left(B_{d-k,k} \leq \lambda\right), \tag{4.24}$$

as stated. $\qquad\qquad\qquad\qquad\qquad\qquad\qquad\qquad\qquad\qquad\qquad\qquad\square$

Notes for Section 4.1

1. Spherical support measures were introduced and studied by Glasauer [63]. The introduction of the conic support measures in Definition 4.1.1 follows the introduction of the conic intrinsic volumes in McCoy [127] and of the conic support measures in Amelunxen [6].

2. As mentioned after the proof of Corollary 4.1.3, a local Steiner formula in spherical space was proved by Glasauer [63, 64]; it was reproduced in [170,

Thm. 6.5.1]. The global case of formula (4.14) (that is, for $\eta = \mathbb{R}^d \times \mathbb{R}^d$) appeared as 'Master Steiner formula' in McCoy and Tropp [129]. Theorem 4.1.1 and the consequences in Subsections 4.1.1 and 4.1.2 are local versions of their results, as sketched in [168].

4.2 Support measures of general convex cones

Mostly we have considered polyhedral cones up to now. The fundamental notions and results can be carried over to general closed convex cones, by continuity. The present section lays the foundations for this extension.

We need to define weak convergence of measures on the biconic σ-algebra, and we reduce this to the spherical setting. For $\omega \subseteq \mathbb{S}^{d-1} \times \mathbb{S}^{d-1}$ we have defined $\omega^+ = \{(\alpha x, \beta y) : (x, y) \in \omega, \ \alpha, \beta > 0\}$. For a finite measure μ on $\widehat{\mathcal{B}}(\mathbb{R}^d \times \mathbb{R}^d)$, we define

$$\overline{\mu}(\omega) := \mu(\omega^+) \quad \text{for } \omega \in \mathcal{B}(\mathbb{S}^{d-1} \times \mathbb{S}^{d-1}).$$

Then $\overline{\mu}$ is a finite measure on $\mathcal{B}(\mathbb{S}^{d-1} \times \mathbb{S}^{d-1})$. For measures on this σ-algebra, weak convergence, denoted by $\xrightarrow{w}$, is well defined. For finite measures μ, μ_i ($i \in \mathbb{N}$) on $\widehat{\mathcal{B}}(\mathbb{R}^d \times \mathbb{R}^d)$, we now define the weak convergence $\mu_i \xrightarrow{w} \mu$ by $\overline{\mu}_i \xrightarrow{w} \overline{\mu}$. (This will only be used in the following for measures on $\widehat{\mathcal{B}}(\mathbb{R}^d \times \mathbb{R}^d)$ that assign measure zero to $\{o\}$.) The following is an easy consequence of known characterizations of weak convergence (e.g., [14, p. 176]).

Lemma 4.2.1. *For finite measures μ, μ_i on $\widehat{\mathcal{B}}(\mathbb{R}^d \times \mathbb{R}^d)$, the following are equivalent.*

(a) $\mu_i \xrightarrow{w} \mu$;

(b) *For all continuous functions $f : \mathbb{S}^{d-1} \times \mathbb{S}^{d-1} \to \mathbb{R}$,*

$$\int_{\mathbb{R}^d \times \mathbb{R}^d} f_h(x, y)\, \mu_i(d(x, y)) \to \int_{\mathbb{R}^d \times \mathbb{R}^d} f_h(x, y)\, \mu(d(x, y))$$

as $i \to \infty$;

(c) *For every open set $\eta \in \widehat{\mathcal{B}}(\mathbb{R}^d \times \mathbb{R}^d)$,*

$$\mu(\eta) \leq \liminf_{i \to \infty} \mu_i(\eta),$$

and

$$\mu(\mathbb{R}^d \times \mathbb{R}^d) = \lim_{i \to \infty} \mu_i(\mathbb{R}^d \times \mathbb{R}^d).$$

We turn to local parallel sets. The definition of the angular local parallel set makes sense for arbitrary closed convex cones of positive dimension. For $C \in \mathcal{C}_*^d$, $\eta \in \widehat{\mathcal{B}}(\mathbb{R}^d \times \mathbb{R}^d)$, and $0 \leq \lambda < \pi/2$, we define

$$M_\lambda^a(C, \eta) := \{x \in \mathbb{R}^d : d_a(x, C) \le \lambda, \ (\Pi_C(x), \Pi_{C^\circ}(x)) \in \eta\}. \tag{4.25}$$

Since this is a Borel set,

$$\mu_\lambda(C, \eta) := \gamma_d(M_\lambda^a(C, \eta)), \quad \eta \in \widehat{\mathcal{B}}(\mathbb{R}^d \times \mathbb{R}^d),$$

defines a finite measure $\mu_\lambda(C, \cdot)$ on the biconic σ-algebra $\widehat{\mathcal{B}}(\mathbb{R}^d \times \mathbb{R}^d)$. If $C \in \mathcal{PC}_*^d$, then we know from (4.21) that

$$\mu_\lambda(C, \eta) = \sum_{k=1}^d g_k(\lambda) \cdot \Omega_k(C, \eta), \tag{4.26}$$

with the coefficients $g_k(\lambda)$ given by (4.22).

We show that μ_λ depends weakly continuously on C. The counterpart to this fact for Euclidean convex bodies is [163, Thm. 4.1.1], and Glasauer [63, Hilfssatz 3.1.3] has carried this over to spherical space. We adapt his argument to the conic setting.

Lemma 4.2.2. *Let* $0 \le \lambda < \pi/2$. *Let* $C, C_i \in \mathcal{C}_*^d$ $(i \in \mathbb{N})$ *be cones with* $C_i \to C$ *for* $i \to \infty$. *Then* $\mu_\lambda(C_i, \cdot) \xrightarrow{w} \mu_\lambda(C, \cdot)$, *where* $\xrightarrow{w}$ *means weak convergence.*

Proof. We write

$$C_\lambda^a := M_\lambda^a(C, \mathbb{R}^d \times \mathbb{R}^d)$$

for the angular parallel set of the cone $C \in \mathcal{C}_*^d$ at distance λ. First we state that

$$\gamma_d(\text{bd}\, C_\lambda^a) = 0. \tag{4.27}$$

This is trivial if $\dim C < d$, hence we assume that $\dim C = d$. For the proof, we let $0 < \delta < \pi/2 - \lambda$ and choose (which is possible by Lemma 3.2.1) a polyhedral cone $P \in \mathcal{PC}^d$ with $P \subset \text{int}\, C$ and $d_a(P, C) < \delta$. Then $(\text{bd}\, C_\lambda^a) \setminus \{o\} \subset P_{\lambda+\delta} \setminus P_\lambda$, hence Corollary 4.1.3 gives

$$\gamma_d(\text{bd}\, C_\lambda^a) \le \gamma_d(P_{\lambda+\delta} \setminus P_\lambda) \le \sum_{k=1}^{d-1} \frac{\omega_k \omega_{d-k}}{\omega_d} \int_\lambda^{\lambda+\delta} \cos^{k-1}\varphi \sin^{d-k-1}\varphi \, d\varphi$$

(where $v_k(C) \le 1$ was used), which tends to zero for $\delta \to 0$.

Now let $\eta \in \widehat{\mathcal{B}}(\mathbb{R}^d \times \mathbb{R}^d)$ be an open set. Let $x \in M_\lambda^a(C, \eta)$ be a point with $d_a(C, x) < \lambda$. By Lemma 1.3.15, we have

$$\|\Pi_{C_i}(x) - \Pi_C(x)\| \le \|x\| \sqrt{10\delta_a(C_i, C)},$$

$$\|\Pi_{C_i^\circ}(x) - \Pi_{C^\circ}(x)\| \le \|x\| \sqrt{10\delta_a(C_i^\circ, C^\circ)} = \|x\| \sqrt{10\delta_a(C_i, C)}$$

for sufficiently large i (see Theorem 1.3.4). This implies that

$$\Pi_{C_i}(x) \to \Pi_C(x), \qquad \Pi_{C_i^\circ}(x) \to \Pi_{C^\circ}(x), \qquad d_a(C_i, x) \to d_a(C, x)$$

as $i \to \infty$. Hence, for almost all i we have $d_a(C_i, x) < \lambda$ and

$$(\Pi_{C_i}(x), \Pi_{C_i^\circ}(x)) \in \eta$$

and thus $x \in M_\lambda^a(C_i, \eta)$. This shows that

$$M_\lambda^a(K, \eta) \setminus \operatorname{bd} C_\lambda^a \subseteq \liminf_{i \to \infty} M_\lambda^a(C_i, \eta)$$

and hence

$$\mu_\lambda(C, \eta) = \gamma_d \left(M_\lambda^a(C, \eta) \setminus \operatorname{bd} C_\lambda^a \right) \leq \gamma_d \left(\liminf_{i \to \infty} M_\lambda^a(C_i, \eta) \right)$$

$$\leq \liminf_{i \to \infty} \gamma_d \left(M_\lambda^a(C_i, \eta) \right) = \liminf_{i \to \infty} \mu_\lambda(C_i, \eta). \tag{4.28}$$

It follows from Corollary 1.3.1 that

$$C_\lambda^a \cup (C^\circ)_{\frac{\pi}{2} - \lambda}^a = \mathbb{R}^d \quad \text{and} \quad C_\lambda^a \cap (C^\circ)_{\frac{\pi}{2} - \lambda}^a \subset \operatorname{bd} C_\lambda^a,$$

hence

$$\mu_\lambda(C, \mathbb{R}^d \times \mathbb{R}^d) = 1 - \mu_{\frac{\pi}{2} - \lambda}(C^\circ, \mathbb{R}^d \times \mathbb{R}^d).$$

Together with (4.28) this yields

$$\mu_\lambda(C, \mathbb{R}^d \times \mathbb{R}^d)$$

$$\leq \liminf_{i \to \infty} \mu_\lambda(C_i, \mathbb{R}^d \times \mathbb{R}^d) \leq \limsup_{i \to \infty} \mu_\lambda(C_i, \mathbb{R}^d \times \mathbb{R}^d)$$

$$= 1 - \liminf_{i \to \infty} \mu_{\frac{\pi}{2} - \lambda}(C_i^\circ, \mathbb{R}^d \times \mathbb{R}^d) \leq 1 - \mu_{\frac{\pi}{2} - \lambda}(C^\circ, \mathbb{R}^d \times \mathbb{R}^d)$$

$$= \mu_\lambda(C, \mathbb{R}^d \times \mathbb{R}^d)$$

and thus

$$\lim_{i \to \infty} \mu_\lambda(C_i, \mathbb{R}^d \times \mathbb{R}^d) = \mu_\lambda(C, \mathbb{R}^d \times \mathbb{R}^d).$$

Both results together prove the assertion.   □

The following theorem is deliberately formulated in analogy to [163, Thm. 4.2.1].

Theorem 4.2.1. *To every cone* $C \in \mathcal{C}_*^d$ *there exist finite positive measures* $\Omega_0(C, \cdot), \ldots, \Omega_d(C, \cdot)$ *on the biconic σ-algebra* $\widehat{\mathcal{B}}(\mathbb{R}^d \times \mathbb{R}^d)$ *such that, for every* $\eta \in \widehat{\mathcal{B}}(\mathbb{R}^d \times \mathbb{R}^d)$ *and every λ with $0 \leq \lambda < \pi/2$, the Gaussian measure of the angular local parallel set $M_\lambda^a(C, \eta)$ is given by*

$$\gamma_d(M_\lambda^a(C, \eta)) = \mu_\lambda(C, \eta) = \sum_{k=1}^d g_k(\lambda) \cdot \Omega_k(C, \eta) \tag{4.29}$$

with

$$g_k(\lambda) = \frac{\omega_k \omega_{d-k}}{\omega_d} \int_0^\lambda \cos^{k-1}\varphi \sin^{d-k-1}\varphi \, d\varphi \tag{4.30}$$

for $k = 1, \ldots, d-1$ and $g_d \equiv 1$.

The mapping $C \mapsto \Omega_k(C, \cdot)$ (from $\mathcal{C}_*^d$ into the space of finite measures on $\widehat{\mathcal{B}}(\mathbb{R}^d \times \mathbb{R}^d)$) is a weakly continuous valuation.

For each $\eta \in \widehat{\mathcal{B}}(\mathbb{R}^d \times \mathbb{R}^d)$, the function $\Omega_k(\cdot, \eta)$ (from $\mathcal{C}_*^d$ to $\mathbb{R}$) is measurable.

Proof. If $P \in \mathcal{PC}_*^d$ is a polyhedral cone, we know from Corollary 4.1.3 that

$$\mu_\lambda(P, \eta) = \sum_{k=1}^d g_k(\lambda) \cdot \Omega_k(P, \eta), \tag{4.31}$$

with the measures $\Omega_k(P, \cdot)$ given by (4.1). The functions $g_1, \ldots, g_d$ are linearly independent on $[0, \pi/2]$. This can be proved by assuming a linear relation $a_1 g_1(\lambda) + \cdots + a_d g_d(\lambda) = 0$ and then considering the values of this function and its derivatives at $\lambda = 0$. Now we state (following [63, Hilfssatz 3.1.4]) that for each $m \in \{1, \ldots, d\}$ there are numbers $\lambda_1, \ldots, \lambda_m \in (0, \pi/2)$ such that the matrix $(g_i(\lambda_j))_{\substack{i=1,\ldots,d \\ j=1,\ldots,m}}$ has rank m. For $m = 1$ this is clear. Assume it has been proved for $m \in \{1, \ldots, d-1\}$. Since $m < d$, there is a nontrivial solution $(a_1, \ldots, a_d)$ of the system $\sum_{i=1}^d a_i g_i(\lambda_r) = 0$, $r = 1, \ldots, m$. Suppose that for all $\lambda = \lambda_{m+1}$, the matrix $(g_i(\lambda_j))_{\substack{i=1,\ldots,d \\ j=1,\ldots,m+1}}$ has rank m. Then there are numbers $\alpha_1(\lambda), \ldots, \alpha_m(\lambda)$, not all zero, with $g_i(\lambda) = \sum_{r=1}^m \alpha_r(\lambda) g_i(\lambda_r)$ for $i = 1, \ldots, d$. This gives

$$\sum_{i=1}^d a_i g_i(\lambda) = \sum_{i=1}^d a_i \sum_{r=1}^m \alpha_r(\lambda) g_i(\lambda_r) = \sum_{r=1}^m \alpha_r(\lambda) \sum_{i=1}^d a_i g_i(\lambda_r) = 0,$$

which holds for all $\lambda \in (0, \pi/2)$, since $g_1, \ldots, g_d$ are linearly independent. Thus, λ_{m+1} can be chosen such that $(g_i(\lambda_j))_{\substack{i=1,\ldots,d \\ j=1,\ldots,m+1}}$ has rank $m+1$. This completes the induction. In particular, the matrix $(g_i(\lambda_j))_{\substack{i=1,\ldots,d \\ j=1,\ldots,d}}$ has an inverse. Therefore, there are numbers a_{ji} such that (4.31) can be inverted to

$$\Omega_j(P, \cdot) = \sum_{i=1}^d a_{ji} \mu_{\lambda_i}(P, \cdot) \tag{4.32}$$

for $j = 1, \ldots, d$. We use these numbers to define

$$\Omega_j(C, \cdot) = \sum_{i=1}^d a_{ji} \mu_{\lambda_i}(C, \cdot) \tag{4.33}$$

for arbitrary cones $C \in \mathcal{C}_*^d$ and for $j = 1, \ldots, d$. We supplement the definition by

$$\Omega_j(\{o\}, \cdot) := 0 \quad \text{for } j = 1, \ldots, d$$

and by

$$\Omega_0(C, \eta) = \int_{C^\circ} \mathbb{1}_\eta(o, y)\, \gamma_d(\mathrm{d}y) = \gamma_d(M^0(C, \eta)) \tag{4.34}$$

for $C \in \mathcal{C}^d$, where

$$M^0(C, \eta) := \{y \in C^\circ : (o, y) \in \eta\}.$$

By (4.33), the weak continuity of the functions μ_λ extends to the functions Ω_j, $j = 1, \ldots, d$, and then (4.31) immediately yields (4.29).

The weak continuity of Ω_0 is shown as follows. Let $(C_i)_{i \in \mathbb{N}}$ be a sequence in $\mathcal{C}^d$ converging to $C \in \mathcal{C}^d$. Let $\eta \in \widehat{\mathcal{B}}(\mathbb{R}^d \times \mathbb{R}^d)$ be open. Let $x \in M^0(C, \eta)$. Then $x \in C^\circ$ and $(o, x) \in \eta$. Since the polar mapping is continuous, we have $C_i^\circ \to C^\circ$. Therefore (as follows from [163, Thm. 1.8.8]), there exist points $x_i \in C_i^\circ$ with $x_i \to x$ for $i \to \infty$. For almost all i we have $(o, x_i) \in \eta$ and hence $x_i \in M^0(C_i, \eta)$. This shows that

$$M^0(C, \eta) \subseteq \liminf_{i \to \infty} M^0(C_i, \eta)$$

and hence

$$\Omega_0(C, \eta) \le \liminf_{i \to \infty} \Omega_0(C_i, \eta).$$

Further,

$$\Omega_0(C_i, \mathbb{R}^d \times \mathbb{R}^d) = \gamma_d(C_i^\circ) \to \gamma_d(C^\circ),$$

by the continuity of the polar mapping and Lemma 2.1.1. This proves the weak continuity of Ω_0.

Now let $C_1, C_2 \in \mathcal{C}_*^d$ be cones such that $C_1 \cup C_2$ is convex. Let $\eta \in \widehat{\mathcal{B}}(\mathbb{R}^d \times \mathbb{R}^d)$ be given. We state that

$$\mu_\lambda(C_1 \cup C_2, \eta) + \mu_\lambda(C_1 \cap C_2, \eta) = \mu_\lambda(C_1, \eta) + \mu_\lambda(C_2, \eta). \tag{4.35}$$

For the proof (adapted from [163, Thm. 4.1.3] and [63, Thm. 4.1.1]), let $x \in \mathbb{R}^d \setminus \{o\}$. Let $y := \Pi_{C_1 \cup C_2}(x)$ and assume, without loss of generality, that $y \in C_1$. Then $\Pi_{C_1 \cup C_2}(x) = \Pi_{C_1}(x)$. As in [163, Thm. 4.1.3], one shows that $\Pi_{C_1 \cap C_2}(x) = \Pi_{C_2}(x)$. This gives $d_a(x, C_1 \cup C_2) = d_a(x, C_1)$ and $d_a(x, C_1 \cap C_2) = d_a(x, C_2)$. Moreover,

$$\Pi_{(C_1 \cup C_2)^\circ}(x) = \Pi_{C_1^\circ}(x), \qquad \Pi_{(C_1 \cap C_2)^\circ}(x) = \Pi_{C_2^\circ}(x). \tag{4.36}$$

If $d_a(x, C_1) < \pi/2$ and $d_a(x, C_2) < \pi/2$, this follows from (1.6). If $x \in C_1^\circ$, then $y = o$, hence $\Pi_{C_2}(x) = o$ and thus $x \in C_2^\circ$. If $x \in C_2^\circ$, then $x \in (C_1 \cap C_2)^\circ$, hence $\Pi_{(C_1 \cap C_2)^\circ}(x) = x = \Pi_{C_2^\circ}$ and $\Pi_{(C_1 \cup C_2)^\circ}(x) = \Pi_{C_1^\circ \cap C_2^\circ}(x) = \Pi_{C_1^\circ}(x)$. Thus, (4.36) holds in each case. It follows that

$$\mathbb{1}[M_\lambda^a(C_1 \cup C_2, \eta)](x) = \mathbb{1}[M_\lambda^a(C_1, \eta)](x),$$
$$\mathbb{1}[M_\lambda^a(C_1 \cap C_2, \eta)](x) = \mathbb{1}[M_\lambda^a(C_2, \eta)](x),$$

and we obtain

$$\mathbb{1}[M_\lambda^a(C_1 \cup C_2, \eta)] + \mathbb{1}[M_\lambda^a(C_1 \cap C_2, \eta)] = \mathbb{1}[M_\lambda^a(C_1, \eta)] + \mathbb{1}[M_\lambda^a(C_2, \eta)].$$

Integration with γ_d yields the assertion (4.35).

From (4.35) and (4.33) we now obtain that each function $\Omega_j(\cdot, \eta)$, $j \in \{1, \ldots, d\}$, is a valuation.

To show also the valuation property of Ω_0, let $C_1, C_2 \in \mathcal{C}^d$ be cones such that $C_1 \cup C_2$ is convex. Then Theorem 1.3.2 says that $C_1^\circ \cup C_2^\circ$ is convex and that $(C_1 \cup C_2)^\circ = C_1^\circ \cap C_2^\circ$ and $(C_1 \cap C_2)^\circ = C_1^\circ \cup C_2^\circ$. Using this, it is easy to check that

$$\mathbb{1}[M^0(C_1 \cup C_2, \eta)] + \mathbb{1}[M^0(C_1 \cap C_2, \eta)] = \mathbb{1}[M^0(C_1, \eta)] + \mathbb{1}[M^0(C_2, \eta)].$$

Integration with γ_d gives

$$\Omega_0(C_1 \cup C_2, \eta) + \Omega_0(C_1 \cap C_2, \eta) = \Omega_0(C_1, \eta) + \Omega_0(C_2, \eta),$$

as asserted.

The proof of the last assertion, the measurability of $\Omega_k(\cdot, \eta)$, is postponed after we have stated the subsequent lemma.   □

Comment. We emphasize that now, in particular, the conic intrinsic volumes v_k, $k = 0, \ldots, d$, are defined as continuous functions on $\mathcal{C}^d$. The continuity allows us to extend to general convex cones many of the relations obtained earlier for polyhedral cones. This continuous extension will in the obvious cases be understood to have been carried out tacitly.

The following lemma is easily deduced from [170, Lemma 12.1.1], by a transformation to the spherical setting, as at the beginning of this section.

Lemma 4.2.3. *Let $(T, \mathcal{T})$ be a measurable space, and let*

$$\psi : T \times \widehat{B}(\mathbb{R}^d \times \mathbb{R}^d) \to \mathbb{R}$$

be a mapping such that $\psi(t, \cdot)$ is a measure for each $t \in T$. Suppose that the mapping

$$t \mapsto \int_{\mathbb{R}^d \times \mathbb{R}^d} f_h(x, y)\, \psi(t, \mathrm{d}(x, y)) \tag{4.37}$$

is $\mathcal{T}$-measurable for each continuous function $f : \mathbb{S}^{d-1} \times \mathbb{S}^{d-1} \to \mathbb{R}$. Then the mapping (4.37) is measurable for each nonnegative measurable function $f : \mathbb{S}^{d-1} \times \mathbb{S}^{d-1} \to \mathbb{R}$. In particular, $\psi(\cdot, \eta)$ is a $\mathcal{T}$-measurable function for every $\eta \in \widehat{\mathcal{B}}(\mathbb{R}^d \times \mathbb{R}^d)$.

The measurability of $\Omega_j(\cdot, \eta)$, for fixed $\eta \in \widehat{\mathcal{B}}(\mathbb{R}^d \times \mathbb{R}^d)$ and $j \in \{0, \ldots, d\}$, follows now from Lemma 4.2.3, by choosing $(T, \mathcal{T}) = (\mathcal{C}_*^d, \mathcal{B}(\mathcal{C}_*^d))$ and $\psi = \Omega_j$.

This yields also the measurability of $\Phi_j(\cdot, A)$, for fixed $A \in \widehat{\mathcal{B}}(\mathbb{R}^d)$ and $j \in \{0, \ldots, d\}$.

As another application of Lemma 4.2.3, we insert here a preparation for the next section. Let $C, D \in \mathcal{C}^d$, and recall that we say that C and D *touch* if $C \cap D \neq \{o\}$, but C and D can be separated by a hyperplane. We define

$$Tc(C, D) := \{\vartheta \in \mathrm{SO}(d) : C \text{ and } \vartheta D \text{ touch}\}. \tag{4.38}$$

Later (in (4.59)) we shall show that $Tc(C, D)$ has ν-measure zero. For polyhedral cones C, D, this was shown in Lemma 2.1.3.

Lemma 4.2.4. *Let* $C, D \in \mathcal{C}_*^d$ *and* $A, B \in \widehat{\mathcal{B}}(\mathbb{R}^d)$, *let* $k \in \{1, \ldots, d\}$. *The function*

$$\vartheta \mapsto \Phi_k(C \cap \vartheta D, A \cap \vartheta B), \quad \vartheta \in \mathrm{SO}(d), \tag{4.39}$$

is measurable on $\mathrm{SO}(d) \setminus Tc(C, D)$.

Proof. (We modify the proof of [170, Lemma 5.2.1].) For each $\vartheta \in \mathrm{SO}(d)$, the mapping

$$F_\vartheta : (\mathbb{R}^d, \widehat{\mathcal{B}}(\mathbb{R}^d)) \to (\mathbb{R}^d \times \mathbb{R}^d, \widehat{\mathcal{B}}(\mathbb{R}^d \times \mathbb{R}^d))$$

defined by $F_\vartheta(x) = (x, \vartheta^{-1}x)$, $x \in \mathbb{R}^d$, is measurable. Let $\psi(\vartheta, \cdot)$ denote the image measure of $\Phi_k(C \cap \vartheta D, \cdot)$ under F_ϑ. Then $\psi(\vartheta, \cdot)$ is a finite measure on $\widehat{\mathcal{B}}(\mathbb{R}^d \times \mathbb{R}^d)$, and

$$\psi(\vartheta, A \times B) = \Phi_k(C \cap \vartheta D, A \cap \vartheta B).$$

For each continuous function $f : \mathbb{S}^{d-1} \times \mathbb{S}^{d-1} \to \mathbb{R}$, we have, by the transformation formula for integrals,

$$\int_{\mathbb{R}^d \times \mathbb{R}^d} f_h \, \mathrm{d}\psi(\vartheta, \cdot) = \int_{\mathbb{R}^d} f_h(x, \vartheta^{-1}x) \, \Phi_k(C \cap \vartheta D, \mathrm{d}x).$$

By Lemma 1.3.14, the mapping $\vartheta \mapsto C \cap \vartheta D$ is continuous on $\mathrm{SO}(d) \setminus Tc(C, D)$. Let $(\vartheta_i)_{i \in \mathbb{N}}$ be a sequence in $\mathrm{SO}(d)$ converging to $\vartheta_0 \in \mathrm{SO}(d) \setminus Tc(C, D)$. Then $\vartheta_i \notin Tc(C, D)$ for all sufficiently large i. From the weak continuity of Φ_k it follows that, as $i \to \infty$,

$$\int_{\mathbb{R}^d} f_h(x, \vartheta_i^{-1}x) \, \Phi_k(C \cap \vartheta_i D, \mathrm{d}x) \to \int_{\mathbb{R}^d} f_h(x, \vartheta^{-1}x) \, \Phi_k(C \cap \vartheta D, \mathrm{d}x)$$

for each continuous (and hence uniformly continuous) function $f : \mathbb{S}^{d-1} \times \mathbb{S}^{d-1} \to \mathbb{R}$. Thus, for each such function, the mapping

$$\vartheta \mapsto \int_{\mathbb{R}^d \times \mathbb{R}^d} f_h(x, y) \, \psi(\vartheta, \mathrm{d}(x, y))$$

is continuous on $\mathrm{SO}(d) \setminus Tc(C, D)$. By Lemma 4.2.3, this implies that the mapping

$$\vartheta \mapsto \psi(\vartheta, A \times B) = \Phi_k(C \cap \vartheta D, A \cap \vartheta B)$$

is measurable on $\mathrm{SO}(d) \setminus Tc(C, D)$. $\square$

Our next aim is to extend the 'Master Steiner formula' of Theorem 4.2.1 from polyhedral cones to general convex cones (here we follow [168]). For this, we first prove the following lemma. We recall that $\varphi_f(C, \eta)$ was defined by (4.13).

Lemma 4.2.5. *Let $f : \mathbb{R}_+^2 \to \mathbb{R}_+$ be continuous and bounded. Let $C, C_i \in \mathcal{C}_*^d$ $(i \in \mathbb{N})$ be cones with $C_i \to C$ as $i \to \infty$. Then*

$$\varphi_f(C_i, \cdot) \xrightarrow{w} \varphi_f(C, \cdot),$$

where $\xrightarrow{w}$ means weak convergence.

Proof. For $x \in \mathbb{R}^d$ we have

$$\Pi_{C_i}(x) \to \Pi_C(x), \qquad \Pi_{C_i^\circ}(x) \to \Pi_{C^\circ}(x) \qquad (i \to \infty),$$

as in the proof of Lemma 4.2.2. This implies

$$\lim_{i \to \infty} f(\|\Pi_{C_i}(x)\|^2, \|\Pi_{C_i^\circ}(x)\|^2) = f(\|\Pi_C(x)\|^2, \|\Pi_{C^\circ}(x)\|^2). \tag{4.40}$$

Let $\eta \subset \mathbb{R}^d \times \mathbb{R}^d$ be open. If $(\Pi_C(x), \Pi_{C^\circ}(x)) \in \eta$, then $(\Pi_{C_i}(x), \Pi_{C_i^\circ}(x)) \in \eta$ for almost all i. Thus,

$$\mathbb{1}_\eta(\Pi_C(x), \Pi_{C^\circ}(x)) \leq \mathbb{1}_\eta(\Pi_{C_i}(x), \Pi_{C_i^\circ}(x))$$

for almost all i. It follows that

$$f\left(\|\Pi_C(x)\|^2, \|\Pi_{C^\circ}(x)\|^2\right) \mathbb{1}_\eta(\Pi_C(x), \Pi_{C^\circ}(x))$$
$$\leq \liminf_{i \to \infty} f\left(\|\Pi_{C_i}(x)\|^2, \|\Pi_{C_i^\circ}(x)\|^2\right) \mathbb{1}_\eta(\Pi_{C_i}(x), \Pi_{C_i^\circ}(x)).$$

Using Fatou's lemma, we get

$$\varphi_f(C, \eta) \leq \liminf_{i \to \infty} \varphi_f(C_i, \eta).$$

Further, (4.40) together with the dominated convergence theorem yields

$$\lim_{i \to \infty} \varphi_f(C_i, \mathbb{R}^d \times \mathbb{R}^d) = \varphi_f(C, \mathbb{R}^d \times \mathbb{R}^d).$$

The assertion follows. $\square$

Now we can prove the general local version of the 'Master Steiner formula'.

Theorem 4.2.2. *Let $C \in \mathcal{C}^d$ be a convex cone. Let $f : \mathbb{R}_+^2 \to \mathbb{R}$ be a nonnegative, measurable function such that $\varphi_f(C, \eta)$ is finite for all $\eta \in \widehat{\mathcal{B}}(\mathbb{R}^d \times \mathbb{R}^d)$. Then*

$$\varphi_f(C, \cdot) = \sum_{k=0}^{d} \varphi_f(L_k, \mathbb{R}^d \times \mathbb{R}^d) \cdot \Omega_k(C, \cdot), \tag{4.41}$$

with an arbitrary $L_k \in G(d, k)$.

Proof. Let $C \in \mathcal{C}^d$ be given. We choose a sequence of polyhedral cones $C_i \in \mathcal{PC}^d$ converging to C. By Theorem 4.1.1 we have

$$\varphi_f(C_i, \cdot) = \sum_{k=0}^{d} I_k(f) \cdot \Omega_k(C_i, \cdot)$$

for $i \in \mathbb{N}$, where $I_k(f) = \varphi_f(L_k, \mathbb{R}^d \times \mathbb{R}^d)$ with $L_k \in G(d, k)$. First let f be continuous. Then $\varphi_f(C_i, \cdot) \xrightarrow{w} \varphi_f(C, \cdot)$ by Lemma 4.2.5 and $\Omega_k(C_i, \cdot) \xrightarrow{w} \Omega_k(C, \cdot)$ by Theorem 4.2.1. Therefore,

$$\varphi_f(C, \cdot) = \sum_{k=0}^{d} I_k(f) \cdot \Omega_k(C, \cdot). \tag{4.42}$$

The following is essentially the argument of McCoy and Tropp [129]. We fix a Borel set $\eta \in \widehat{\mathcal{B}}(\mathbb{R}^d \times \mathbb{R}^d)$. Let $h : \mathbb{R}_+^2 \to \mathbb{R}$ be a bounded, continuous function. We can write $h = h^+ - h^-$ with nonnegative, bounded, continuous functions h^+, h^- and define $\varphi_h(C, \eta) = \varphi_{h^+}(C, \eta) - \varphi_{h^-}(C, \eta)$. Then

$$\varphi_h(C, \eta) = \mathbb{E}\left[h(\|\Pi_C(\mathbf{g})\|^2, \|\Pi_{C^\circ}(\mathbf{g})\|^2) \cdot \mathbb{1}_\eta(\Pi_C(\mathbf{g}), \Pi_{C^\circ}(\mathbf{g}))\right]$$

$$= \int_{B_\eta} h(\|\Pi_C(x)\|^2, \|\Pi_{C^\circ}(x)\|^2) \, \gamma_d(\mathrm{d}x)$$

$$= \int_{\mathbb{R}_+^2} h(s, t) \, \mu(\mathrm{d}(s, t)),$$

where

$$B_\eta := \{x \in \mathbb{R}^d : (\Pi_C(x), \Pi_{C^\circ}(x)) \in \eta\}$$

and where μ is the image measure of the restriction $\gamma_d \llcorner B_\eta$ under the mapping $x \mapsto (\|\Pi_C(x)\|^2, \|\Pi_{C^\circ}(x)\|^2)$. Denoting by μ_k the image measure of γ_d under the mapping $x \mapsto (\|\Pi_{L_k}(x)\|^2, \|\Pi_{L_k^\circ}(x)\|^2)$, we have

$$\sum_{k=0}^{d} \varphi_h(L_k, \mathbb{R}^d \times \mathbb{R}^d) \cdot \Omega_k(C, \eta) = \sum_{k=0}^{d} \left(\int_{\mathbb{R}_+^2} h(s, t) \, \mu_k(\mathrm{d}(s, t))\right) \Omega_k(C, \eta).$$

Therefore, (4.42) gives

$$\int_{\mathbb{R}_+^2} h \, \mathrm{d}\mu = \int_{\mathbb{R}_+^2} h \, \mathrm{d}\left(\sum_{k=0}^{d} \Omega_k(C, \eta) \mu_k\right).$$

Since this holds for all bounded, continuous real functions h on $\mathbb{R}_+^2$, it follows (e.g., from [21, Lemma 30.14]) that

$$\mu = \sum_{k=0}^{d} \Omega_k(C, \eta) \mu_k.$$

Integrating a nonnegative, measurable function f on $\mathbb{R}_+^2$ with respect to these measures gives the assertion. $\qquad\square$

The polyhedral skeleton measures

Finally in this section, we briefly introduce a more elementary series of measures. The conic support and curvature measures have been defined for general cones in $\mathcal{C}^d$. In contrast, the following series of measures is restricted to polyhedral cones.

For $C \in \mathcal{PC}^d$ and for $k \in \{1, \ldots, d\}$, we define

$$\Pi_k(C, A) := \sum_{F \in \mathcal{F}_k(C)} \gamma_{\langle F \rangle}(F \cap A) \quad \text{for } A \in \widehat{\mathcal{B}}(\mathbb{R}^d). \tag{4.43}$$

Equivalently,

$$\Pi_k(C, A) = \omega_k^{-1} \sum_{F \in \mathcal{F}_k(C)} \sigma_{k-1}(F \cap A \cap \mathbb{S}^{d-1}).$$

We have

$$\Pi_k(C, \mathbb{R}^d) = \Lambda_k(C),$$

thus the measures Π_k are the localizations of the functionals Λ_k introduced in (2.82).

We call $\Pi_k(C, \cdot)$ the kth *polyhedral skeleton measure* of C.

In preparation for a kinematic formula in Section 4.3, we consider $\Pi_k(C \cap \vartheta D, A \cap \vartheta B)$ for $C, D \in \mathcal{PC}^d$, $A, B \in \widehat{\mathcal{B}}(\mathbb{R}^d)$ and $\vartheta \in \mathrm{SO}(d)$.

It follows from Lemmas 1.4.4 and 2.1.4 (applied to all pairs of faces of C and D) that there is a set $M \subset \mathrm{SO}(d)$ with $\nu(M) = 0$ and the following property. For each $\vartheta \in \mathrm{SO}(d) \setminus M$, each k-face J of $C \cap \vartheta D$ is of the form $J = F \cap \vartheta G$ with $F \in \mathcal{F}_i(C)$, $G \in \mathcal{F}_j(D)$, $i + j = d + k$, and F and G do not touch. For these ϑ, we have

$$\Pi_k(C \cap \vartheta D, A \cap \vartheta B)$$

$$= \sum_{i+j=k+d} \sum_{F \in \mathcal{F}_i(C)} \sum_{G \in \mathcal{F}_j(D)} \gamma_{\langle F \rangle \cap \vartheta \langle G \rangle}(F \cap \vartheta G \cap A \cap \vartheta B)$$

$$= \sum_{i+j=k+d} \sum_{F \in \mathcal{F}_i(C)} \sum_{G \in \mathcal{F}_j(D)} \Phi_k(F \cap \vartheta G, A \cap \vartheta B),$$

the latter since $\dim(F \cap \vartheta G) = k$. Now the following is a consequence of Lemma 4.2.4.

Lemma 4.2.6. *Let $C, D \in \mathcal{PC}^d$ and $A, B \in \widehat{\mathcal{B}}(\mathbb{R}^d)$, let $k \in \{1, \ldots, d\}$. There is a set $M \subset \mathrm{SO}(d)$ with $\nu(M) = 0$, such that the function*

$$\vartheta \mapsto \Pi_k(C \cap \vartheta D, A \cap \vartheta B) \tag{4.44}$$

is measurable for $\vartheta \in \mathrm{SO}(d) \setminus M$.

4.3 Kinematic formulas

The main goal of this section is the proof of the following kinematic formula for the conic curvature measures of first kind.

Theorem 4.3.1. *Let $C, D \in \mathcal{C}^d$ and $A, B \in \widehat{\mathcal{B}}(\mathbb{R}^d)$. Then*

$$\int_{\mathrm{SO}(d)} \Phi_k(C \cap \vartheta D, A \cap \vartheta B) \, \nu(\mathrm{d}\vartheta) = \sum_{i=k}^{d} \Phi_i(C, A) \Phi_{d+k-i}(D, B) \qquad (4.45)$$

for $k = 1, \ldots, d$.

Since proofs of integral-geometric formulas by means of characterization theorems (as initiated by Hadwiger; see [163, Sect. 4.4] in the local case), though elegant, have their limits, proofs by direct computation are also of interest. For the global conic kinematic formula for polyhedral cones, such a proof, exploiting averaging over suitable subgroups of the rotation group, was given by Amelunxen and Lotz [10]. We extend their proof here (following [165]) to the local case. The extension from the case of polyhedral cones to general convex cones then employs standard principles, as set out in [163] and Glasauer [63].

For the proof, we shall first assume that C, D are polyhedral cones. In that case we know from Lemma 2.1.3 that the set $Tc(C, D)$ of rotations $\vartheta \in \mathrm{SO}(d)$ for which C and ϑD touch, has ν-measure zero, so that by Lemma 4.2.4 the integral (4.45) is defined.

The proof of Theorem 4.3.1 requires two preparatory results. We prove these in more general versions than immediately needed, since these can be useful in Euclidean integral geometry (for example, allowing a direct proof of [163, Lemma 4.4.4]).

For this, we require the *generalized sine function* of two subspaces $L_1, L_2 \in \mathcal{L}_\bullet$. If $\dim L_1 + \dim L_2 = m \le d$, we choose an orthonormal basis in each L_i and define $[L_1, L_2]$ as the m-dimensional volume of the parallelepiped spanned by the union of these bases. If one of the subspaces has dimension zero, then $[L_1, L_2] = 1$, by definition. Obviously, $[L_1, L_2]$ depends only on L_1 and L_2 and not on the choice of the bases. If $\dim L_1 + \dim L_2 \ge d$, we define $[L_1, L_2] = [L_1^\perp, L_2^\perp]$ (which is consistent if $\dim L_1 + \dim L_2 = d$).

If $C, D \in \mathcal{C}^d$ are cones, we set $[C, D] := [\langle C \rangle, \langle D \rangle]$. Let $f : [0, 1] \to \mathbb{R}$ be a bounded, measurable function. If $\dim C = i$, $\dim D = j$, then the constant

$$c_{ij}(f) := \int_{\mathrm{SO}(d)} f([C, \vartheta D]) \, \nu(\mathrm{d}\vartheta)$$

$$= \int_{G(d,j)} f([L_i, L]) \, \nu_j(\mathrm{d}L), \quad L_i \in G(d, i), \qquad (4.46)$$

depends only on i, j, f. This follows from the invariance properties of the measure ν and the invariance property $[\vartheta L_1, \vartheta L_2] = [L_1, L_2]$ for $\vartheta \in \mathrm{SO}(d)$. The last equality sign in (4.46) follows from (2.7).

In a special case of the kinematic formula, which we prove first, the index k of $\Phi_k(C \cap \vartheta D, \cdot)$ coincides with $\dim C + \dim D - d$.

Theorem 4.3.2. *Let $C, D \in \mathcal{PC}^d$ be polyhedral cones with $\dim C = i$, $\dim D = j$, where $i + j = d + k > d$. Let $A, B \in \hat{\mathcal{B}}(\mathbb{R}^d)$, and let $f : [0, 1] \to \mathbb{R}$ be a bounded, measurable function. Then*

$$\int_{SO(d)} \Phi_k(C \cap \vartheta D, A \cap \vartheta B) f([C, \vartheta D]) \, \nu(d\vartheta) = c_{ij}(f) \, \Phi_i(C, A) \, \Phi_j(D, B).$$
(4.47)

Proof. Since for ν-almost all $\vartheta \in SO(d)$ we have $\dim (\langle C \rangle \cap \vartheta \langle D \rangle) = k$, we get

$$\int_{SO(d)} \Phi_k(C \cap \vartheta D, A \cap \vartheta B) f([C, \vartheta D]) \, \nu(d\vartheta)$$

$$= \int_{SO(d)} \gamma_{\langle C \rangle \cap \vartheta \langle D \rangle}(C \cap \vartheta D \cap A \cap \vartheta B) f([C, \vartheta D]) \, \nu(d\vartheta)$$

$$= \int_{SO(d)} \int_{\langle C \rangle \cap \vartheta \langle D \rangle} \mathbb{1}_{C \cap A}(x) \mathbb{1}_{\vartheta(D \cap B)}(x) f([C, \vartheta D]) \, \gamma_{\langle C \rangle \cap \vartheta \langle D \rangle}(dx) \, \nu(d\vartheta).$$

We replace ϑ by $\vartheta\rho$ with $\rho \in SO_{\langle D \rangle}$ (which satisfies $\rho\langle D \rangle = \langle D \rangle$ and hence $[C, \vartheta\rho D] = [C, \vartheta D]$). This does not change the integral, hence the same holds if we integrate over all ρ with respect to the probability measure $\nu_{\langle D \rangle}$. We obtain

$$\int_{SO(d)} \Phi_k(C \cap \vartheta D, A \cap \vartheta B) f([C, \vartheta D]) \, \nu(d\vartheta)$$

$$= \int_{SO_{\langle D \rangle}} \int_{SO(d)} \int_{\langle C \rangle \cap \vartheta \langle D \rangle} \mathbb{1}_{C \cap A}(x) \mathbb{1}_{D \cap B}(\rho^{-1}\vartheta^{-1}x) f([C, \vartheta D])$$

$$\times \gamma_{\langle C \rangle \cap \vartheta \langle D \rangle}(dx) \, \nu(d\vartheta) \, \nu_{\langle D \rangle}(d\rho)$$

$$= \int_{SO(d)} \int_{\langle C \rangle \cap \vartheta \langle D \rangle} \mathbb{1}_{C \cap A}(x) \left[\int_{SO_{\langle D \rangle}} \mathbb{1}_{D \cap B}(\rho^{-1}\vartheta^{-1}x) \, \nu_{\langle D \rangle}(d\rho) \right]$$

$$\times \gamma_{\langle C \rangle \cap \vartheta \langle D \rangle}(dx) f([C, \vartheta D]) \, \nu(d\vartheta)$$

$$= \Phi_j(D, B) \int_{SO(d)} \int_{\langle C \rangle \cap \vartheta \langle D \rangle} \mathbb{1}_{C \cap A}(x) \, \gamma_{\langle C \rangle \cap \vartheta \langle D \rangle}(dx) f([C, \vartheta D]) \, \nu(d\vartheta),$$

where we have used Fubini's theorem and then (2.6) in $\langle D \rangle$, together with $\gamma_{\langle D \rangle}(D \cap B) = \Phi_j(D, B)$.

In the obtained double integral, the outer integral does not change if we replace ϑ by $\sigma\vartheta$ with $\sigma \in SO_{\langle C \rangle}$. Therefore (and since $\langle C \rangle = \sigma\langle C \rangle$, $[C, \sigma\vartheta D] = [\sigma C, \sigma\vartheta D] = [C, \vartheta D]$ and $\nu_{\langle C \rangle}(SO_{\langle C \rangle}) = 1$), we get

$$\int_{SO(d)} \Phi_k(C \cap \vartheta D, A \cap \vartheta B) f([C, \vartheta D]) \, \nu(d\vartheta)$$

$$= \Phi_j(D, B) \int_{SO_{\langle C \rangle}} \int_{SO(d)} \int_{\sigma(\langle C \rangle \cap \vartheta \langle D \rangle)} \mathbb{1}_{C \cap A}(x)$$

$$\times \gamma_{\sigma(\langle C \rangle \cap \vartheta \langle D \rangle)}(dx) f([C, \vartheta D]) \, \nu(d\vartheta) \, \nu_{\langle C \rangle}(d\sigma)$$

$$= \Phi_j(D, B) \int_{SO(d)} \int_{\langle C \rangle \cap \vartheta \langle D \rangle} \left[\int_{SO_{\langle C \rangle}} \mathbb{1}_{C \cap A}(\sigma x) \, \nu_{\langle C \rangle}(d\sigma) \right]$$

$$\times \gamma_{\langle C \rangle \cap \vartheta \langle D \rangle}(dx) f([C, \vartheta D]) \, \nu(d\vartheta)$$

$$= c_{ij}(f) \Phi_j(D, B) \Phi_i(C, A),$$

where we have used (2.6) in $\langle C \rangle$. $\qquad \qquad \square$

Another special theorem needed for the proof of the general kinematic formula is the following.

Theorem 4.3.3. *Let* $C, D \in \mathcal{PC}^d$ *be polyhedral cones with* $\dim C = i$, $\dim D = j$, *where* $i + j = d - k < d$. *Let* $f : [0, 1] \to \mathbb{R}$ *be a bounded, measurable function. Then*

$$\int_{SO(d)} v_{d-k}(C + \vartheta D) f([C, \vartheta D]) \, \nu(d\vartheta) = c_{ij}(f) v_i(C) v_j(D). \qquad (4.48)$$

Proof. It follows from Lemma 2.1.2 that $\langle C \rangle \cap \vartheta \langle D \rangle = \{o\}$ and $\dim (\langle C \rangle + \vartheta \langle D \rangle) = i + j = d - k$ for ν-almost all ϑ. Using (2.3), we get

$$v_{d-k}(C + \vartheta D) = \int_{\langle C \rangle + \vartheta \langle D \rangle} \mathbb{1}_{C + \vartheta D}(x) \, \gamma_{\langle C \rangle + \vartheta \langle D \rangle}(dx)$$

$$= \int_{\mathbb{R}^d} \mathbb{1}_{C + \vartheta D + (\langle C \rangle + \vartheta \langle D \rangle)^\perp}(x) \, \gamma_d(dx).$$

For ν-almost all ϑ, there is a unique decomposition

$$x = x_{C,\vartheta} + x_{D,\vartheta} + x_\vartheta \qquad (4.49)$$

with $x_{C,\vartheta} \in \langle C \rangle$, $x_{D,\vartheta} \in \vartheta \langle D \rangle$, $x_\vartheta \in (\langle C \rangle + \vartheta \langle D \rangle)^\perp$, which we call the ϑ-decomposition of x. By the uniqueness of the decomposition,

$$x \in C + \vartheta D + (\langle C \rangle + \vartheta \langle D \rangle)^\perp \Leftrightarrow x_{C,\vartheta} \in C \text{ and } x_{D,\vartheta} \in \vartheta D,$$

hence

$$\int_{SO(d)} v_{d-k}(C + \vartheta D) f([C, \vartheta D]) \, \nu(d\vartheta)$$

$$= \int_{SO(d)} \int_{\mathbb{R}^d} \mathbb{1}_{C + \vartheta D + (\langle C \rangle + \vartheta \langle D \rangle)^\perp}(x) \, \gamma_d(dx) f([C, \vartheta D]) \, \nu(d\vartheta)$$

$$= \int_{SO(d)} \int_{\mathbb{R}^d} \mathbb{1}_C(x_{C,\vartheta}) \mathbb{1}_{\vartheta D}(x_{D,\vartheta}) \, \gamma_d(dx) f([C, \vartheta D]) \, \nu(d\vartheta). \qquad (4.50)$$

Now let $\rho \in SO_{\langle C \rangle}$. Applying ρ to both sides of (4.49), we get

$$\rho x = \rho x_{C,\vartheta} + \rho x_{D,\vartheta} + \rho x_\vartheta$$

with $\rho x_{C,\vartheta} \in \langle C \rangle$, $\rho x_{D,\vartheta} \in \rho\vartheta\langle D \rangle$, $\rho x_\vartheta \in (\langle C \rangle + \rho\vartheta\langle D \rangle)^\perp$, because $\rho\langle C \rangle = \langle C \rangle$. On the other hand, the $\rho\vartheta$-decomposition of ρx gives

$$\rho x = (\rho x)_{C,\rho\vartheta} + (\rho x)_{D,\rho\vartheta} + (\rho x)_{\rho\vartheta}$$

with $(\rho x)_{C,\rho\vartheta} \in \langle C \rangle$, $(\rho x)_{D,\rho\vartheta} \in \rho\vartheta\langle D \rangle$, $(\rho x)_{\rho\vartheta} \in (\langle C \rangle + \rho\vartheta\langle D \rangle)^\perp$. Thus, the two decompositions are identical.

The integral (4.50) does not change if we replace ϑ by $\rho\vartheta$ and x by ρx, as follows from the invariance properties of ν and γ_d. Further, we have $[C, \rho\vartheta D] = [\rho C, \rho\vartheta D] = [C, \vartheta D]$. Therefore,

$$\int_{SO(d)} \int_{\mathbb{R}^d} \mathbb{1}_C(x_{C,\vartheta}) \mathbb{1}_{\vartheta D}(x_{D,\vartheta}) \gamma_d(dx) f([C, \vartheta D]) \nu(d\vartheta)$$

$$= \int_{SO(d)} \int_{\mathbb{R}^d} \mathbb{1}_C((\rho x)_{C,\rho\vartheta}) \mathbb{1}_{\rho\vartheta D}((\rho x)_{D,\rho\vartheta}) \gamma_d(dx) f([C, \rho\vartheta D]) \nu(d\vartheta)$$

$$= \int_{SO(d)} \int_{\mathbb{R}^d} \mathbb{1}_C(\rho x_{C,\vartheta}) \mathbb{1}_{\rho\vartheta D}(\rho x_{D,\vartheta}) \gamma_d(dx) f([C, \vartheta D]) \nu(d\vartheta)$$

$$= \int_{SO_{\langle C \rangle}} \int_{SO(d)} \int_{\mathbb{R}^d} \mathbb{1}_C(\rho x_{C,\vartheta}) \mathbb{1}_{\vartheta D}(x_{D,\vartheta}) \gamma_d(dx) f([C, \vartheta D]) \nu(d\vartheta) \nu_{\langle C \rangle}(d\rho)$$

$$= \int_{SO(d)} \int_{\mathbb{R}^d} \left[\int_{SO_{\langle C \rangle}} \mathbb{1}_C(\rho x_{C,\vartheta}) \nu_{\langle C \rangle}(d\rho) \right] \tag{4.51}$$
$$\times \mathbb{1}_D(\vartheta^{-1} x_{D,\vartheta}) \gamma_d(dx) f([C, \vartheta D]) \nu(d\vartheta)$$

$$= v_i(C) \int_{SO(d)} \int_{\mathbb{R}^d} \mathbb{1}_D(\vartheta^{-1} x_{D,\vartheta}) \gamma_d(dx) f([C, \vartheta D]) \nu(d\vartheta). \tag{4.52}$$

We have applied (2.6) in $\langle C \rangle$, which is possible since $x_{C,\vartheta} \in \langle C \rangle$ and $x_{C,\vartheta} \neq o$ for ν-almost all ϑ and γ_d-almost all x.

Let $\sigma \in SO_{\langle D \rangle}$. The integral in (4.52) does not change if we replace ϑ by $\vartheta\sigma^{-1}$. Since $\sigma^{-1}\langle D \rangle = \langle D \rangle$, we have $x_{D,\vartheta\sigma^{-1}} = x_{D,\vartheta}$. Further, $[C, \vartheta\sigma^{-1} D] = [C, \vartheta D]$. Therefore, we obtain

$$\int_{SO(d)} \int_{\mathbb{R}^d} \mathbb{1}_D(\vartheta^{-1} x_{D,\vartheta}) \gamma_d(dx) f([C, \vartheta D]) \nu(d\vartheta)$$

$$= \int_{SO_{\langle D \rangle}} \int_{SO(d)} \int_{\mathbb{R}^d} \mathbb{1}_D(\sigma\vartheta^{-1} x_{D,\vartheta\sigma^{-1}}) \gamma_d(dx) f([C, \vartheta\sigma^{-1} D]) \nu(d\vartheta) \nu_{\langle D \rangle}(d\sigma)$$

$$= \int_{SO(d)} \int_{\mathbb{R}^d} \left[\int_{SO_{\langle D \rangle}} \mathbb{1}_D(\sigma\vartheta^{-1} x_{D,\vartheta}) \nu_{\langle D \rangle}(d\sigma) \right] \gamma_d(dx) f([C, \vartheta D]) \nu(d\vartheta)$$

$$= c_{ij}(f) v_j(D), \tag{4.53}$$

where we have used (2.6) in $\langle D \rangle$, which is possible since $\vartheta^{-1}x_{D,\vartheta} \in \langle D \rangle$. The results (4.50), (4.52), (4.53) together complete the proof. $\qquad\square$

Now we are in a position to prove Theorem 4.3.1.

Proof of Theorem 4.3.1. By (4.12),

$$\Phi_k(C \cap \vartheta D, A \cap \vartheta B) = \sum_{J \in \mathcal{F}_k(C \cap \vartheta D)} \varphi_J(C \cap \vartheta D, A \cap \vartheta B).$$

By Lemmas 1.4.4 and 2.1.4 (applied to all pairs of faces of C and D), it holds for ν-almost all $\vartheta \in SO(d)$ that each k-face J of $C \cap \vartheta D$ is of the form $J = F \cap \vartheta G$ with $F \in \mathcal{F}_i(C), G \in \mathcal{F}_j(D), i+j = d+k$ (and hence $k \le i, j \le d$), and $F \pitchfork \vartheta G$, that is, F and ϑG intersect transversely. Therefore,

$$\int_{SO(d)} \Phi_k(C \cap \vartheta D, A \cap \vartheta B)\,\nu(d\vartheta)$$

$$= \sum_{i+j=k+d} \sum_{F \in \mathcal{F}_i(C)} \sum_{G \in \mathcal{F}_j(D)} \int_{SO(d)} \varphi_{F \cap \vartheta G}(C \cap \vartheta D, A \cap \vartheta B)\mathbb{1}\{F \pitchfork \vartheta G\}\,\nu(d\vartheta).$$

It suffices, therefore, to show that

$$\int_{SO(d)} \varphi_{F \cap \vartheta G}(C \cap \vartheta D, A \cap \vartheta B)\mathbb{1}\{F \pitchfork \vartheta G\}\,\nu(d\vartheta)$$

$$= \varphi_F(C, A)\varphi_G(D, B). \tag{4.54}$$

By Lemmas 2.1.4 and 4.1.1, for ν-almost all $\vartheta \in SO(d)$ we have

$$\varphi_{F \cap \vartheta G}(C \cap \vartheta G, A \cap \vartheta B)\mathbb{1}\{F \pitchfork \vartheta G\}$$

$$= \Phi_k(F \cap \vartheta G, A \cap \vartheta B)v_{d-k}(N(C, F) + \vartheta N(D, G))$$

(note that $v_{d-k}(N(C, F) + \vartheta N(D, G)) = 0$ if $\dim F \cap \vartheta G > k$). Thus, we have to prove that

$$I := \int_{SO(d)} \Phi_k(F \cap \vartheta G, A \cap \vartheta B)v_{d-k}(N(C, F) + \vartheta N(D, G))\,\nu(d\vartheta)$$

$$= \varphi_F(C, A)\varphi_G(D, B). \tag{4.55}$$

Let $F \in \mathcal{F}_i(C)$, $G \in \mathcal{F}_j(D)$ with $i + j = d + k > d$ be given. In the following, we first replace ϑ by $\rho\vartheta$ with $\rho \in SO_{\langle F \rangle}$ (noting that $N(C, F) = \rho N(C, F)$), which does not change the integral, then integrate over all ρ with respect to $\nu_{\langle F \rangle}$, and use the $SO(d)$ invariance of the v_i, and Fubini's theorem. We obtain

$$I = \int_{SO_{\langle F \rangle}} \int_{SO(d)} \Phi_k(F \cap \rho\vartheta G, A \cap \rho\vartheta B)v_{d-k}(N(C, F) + \rho\vartheta N(D, G))$$

$$\times \nu(d\vartheta)\,\nu_{\langle F \rangle}(d\rho)$$

$$= \int_{\mathrm{SO}(d)} \left[\int_{\mathrm{SO}_{\langle F \rangle}} \Phi_k(F \cap \rho \vartheta G, A \cap \rho \vartheta B) \, \nu_{\langle F \rangle}(\mathrm{d}\rho) \right]$$
$$\times v_{d-k}(N(C,F) + \vartheta N(D,G)) \, \nu(\mathrm{d}\vartheta).$$

Denoting the integral in brackets by $[\cdot]$, we have

$$[\cdot] = \int_{\mathrm{SO}_{\langle F \rangle}} \Phi_k(F \cap \rho(\vartheta G \cap \langle F \rangle), A \cap \rho(\vartheta B \cap \langle F \rangle)) \, \nu_{\langle F \rangle}(\mathrm{d}\rho).$$

If $\dim(\vartheta G \cap \langle F \rangle) = k$, we can apply (4.47) (with $f = 1$) in $\langle F \rangle$ and get

$$[\cdot] = \Phi_i(F,A) \Phi_k(\vartheta G \cap \langle F \rangle, \vartheta B).$$

If $\dim(\vartheta G \cap \langle F \rangle) < k$, this equation also holds, since both sides are zero. Thus, we obtain

$$I = \Phi_i(F,A) \int_{\mathrm{SO}(d)} \Phi_k(\vartheta G \cap \langle F \rangle, \vartheta B) v_{d-k}(N(C,F) + \vartheta N(D,G)) \, \nu(\mathrm{d}\vartheta)$$

$$= \Phi_i(F,A) \int_{\mathrm{SO}(d)} \Phi_k(G \cap \vartheta^{-1} \langle F \rangle, B) v_{d-k}(N(C,F) + \vartheta N(D,G)) \, \nu(\mathrm{d}\vartheta),$$

by the $\mathrm{SO}(d)$-equivariance of Φ_k.

The latter integral can be treated in a similar way, replacing ϑ by $\vartheta \sigma$ with $\sigma \in \mathrm{SO}_{\langle G \rangle}$ (noting that $\sigma N(D,G) = N(D,G)$ and $\sigma \langle G \rangle = \langle G \rangle$), and integrating over all σ with respect to $\nu_{\langle G \rangle}$. In this way, we obtain

$$I = \Phi_i(F,A) \int_{\mathrm{SO}_{\langle G \rangle}} \int_{\mathrm{SO}(d)} \Phi_k(\sigma G \cap \vartheta^{-1} \langle F \rangle, \sigma B) v_{d-k}(N(C,F) + \vartheta N(D,G))$$
$$\times \nu(\mathrm{d}\vartheta) \, \nu_{\langle G \rangle}(\mathrm{d}\sigma)$$

$$= \Phi_i(F,A) \int_{\mathrm{SO}(d)} \left[\int_{\mathrm{SO}_{\langle G \rangle}} \Phi_k(\sigma G \cap \vartheta^{-1} \langle F \rangle, \sigma B) \, \nu_{\langle G \rangle}(\mathrm{d}\sigma) \right]$$
$$\times v_{d-k}(N(C,F) + \vartheta N(D,G)) \, \nu(\mathrm{d}\vartheta).$$

Denoting the integral in brackets by $[\cdot]$, we have

$$[\cdot] = \int_{\mathrm{SO}_{\langle G \rangle}} \Phi_k(\vartheta^{-1} \langle F \rangle \cap \langle G \rangle \cap \sigma G, \mathbb{R}^d \cap \sigma B) \, \nu_{\langle G \rangle}(\mathrm{d}\sigma).$$

For ν-almost all ϑ we have $\dim(\vartheta^{-1} \langle F \rangle \cap \langle G \rangle) = k$ and hence can apply (4.47) in $\langle G \rangle$, to obtain

$$[\cdot] = \Phi_k(\vartheta^{-1} \langle F \rangle \cap \langle G \rangle, \mathbb{R}^d) \Phi_j(G,B) = \Phi_j(G,B).$$

We arrive at

$$I = \Phi_i(F, A)\Phi_j(G, B) \int_{SO(d)} v_{d-k}(N(C, F) + \vartheta N(D, G))\, \nu(\mathrm{d}\vartheta)$$

$$= \Phi_i(F, A)\Phi_j(G, B)v_{d-i}(N(C, F))v_{d-j}(N(D, G))$$

$$= \varphi_F(C, A)\varphi_G(D, B),$$

where we have used that $v_k(\langle F \rangle \cap \vartheta \langle G \rangle) = 1$ for ν-almost all ϑ and have finally applied (4.48) (with $f = 1$). We have shown (4.55), which finishes the proof of Theorem 4.3.1 for polyhedral cones.

The extension to general convex cones requires some intermediate considerations. First, the global case of (4.47) says that for polyhedral cones $C, D \in \mathcal{PC}^d$ we have

$$\int_{SO(d)} v_k(C \cap \vartheta D)\, \nu(\mathrm{d}\vartheta) = \sum_{i=k}^{d} v_i(C)v_{d+k-i}(D) \tag{4.56}$$

for $k = 1, \ldots, d$. The version (2.66) of the spherical Gauss–Bonnet theorem says that

$$2 \sum_{k \geq 0} v_{2k+1}(C) = 1, \quad \text{if } C \text{ is not a subspace.} \tag{4.57}$$

Assume that C and D are not both subspaces. Then, by Lemma 2.1.4, for ν-almost all $\vartheta \in SO(d)$, either $C \cap \vartheta D = \{o\}$ (in which case $v_k(C \cap \vartheta D) = 0$ for $k \geq 1$) or $C \cap \vartheta D$ is not a subspace. Hence, (4.57) and (4.56) yield

$$\int_{SO(d)} \mathbb{1}\{C \cap \vartheta D \neq \{o\}\}\, \nu(\mathrm{d}\vartheta)$$

$$= \int_{SO(d)} 2 \sum_{k \geq 0} v_{2k+1}(C \cap \vartheta D)\, \nu(\mathrm{d}\vartheta)$$

$$= 2 \sum_{k=0}^{\lfloor \frac{d-1}{2} \rfloor} \sum_{i=2k+1}^{d} v_i(C)v_{d+2k+1-i}(D). \tag{4.58}$$

Next, we recall that $Tc(C, D)$ denotes the set of all $\vartheta \in SO(d)$ for which the cones C and ϑD touch, and that $\nu(Tc(C, D)) = 0$ for polyhedral cones, by Lemma 2.1.3. Now let $C, D \in \mathcal{C}^d$ be general convex cones where, say, C is not a subspace. We can choose polyhedral cones $P_1, P_2 \in \mathcal{PC}^d$, not subspaces, with $P_1 \subseteq C \subseteq P_2$, and polyhedral cones $Q_1, Q_2 \in \mathcal{PC}^d$ with $Q_1 \subseteq D \subseteq Q_2$. Then

$$Tc(C, D) \setminus Tc(P_1, Q_1)$$

$$\subseteq \{\vartheta \in SO(d) : P_2 \cap \vartheta Q_2 \neq \{o\}\} \setminus \{\vartheta \in SO(d) : P_1 \cap \vartheta Q_1 \neq \{o\}\}.$$

Since $\nu(Tc(P_1, Q_1)) = 0$, this yields

$$\nu(Tc(C,D))$$

$$\leq 2 \sum_{k=0}^{\lfloor \frac{d-1}{2} \rfloor} \sum_{i=2k+1}^{d} v_i(P_2)v_{d+2k+1-i}(Q_2) - 2 \sum_{k=0}^{\lfloor \frac{d-1}{2} \rfloor} \sum_{i=2k+1}^{d} v_i(P_1)v_{d+2k+1-i}(Q_1).$$

Here the right-hand side can be made arbitrarily close to zero, since C and D can be approximated from the inside and from the outside by polyhedral cones and since the conic intrinsic volumes are continuous. Thus,

$$\nu(Tc(C,D)) = 0 \quad \text{for arbitrary cones } C, D \in \mathcal{C}^d. \tag{4.59}$$

We recall that we have to prove the equation

$$\int_{\mathrm{SO}(d)} \Phi_k(C \cap \vartheta D, A \cap \vartheta B)\, \nu(\mathrm{d}\vartheta) = \sum_{i=k}^{d} \Phi_i(C,A)\Phi_{d+k-i}(D,B) \tag{4.60}$$

for $C, D \in \mathcal{C}^d$ and $A, B \in \hat{\mathcal{B}}(\mathbb{R}^d)$ and for $k = 1, \ldots, d$, and that we have proved this for polyhedral cones. Now that we know that $\nu(Tc(C,D)) = 0$, Lemma 4.2.4 tells us that the integral in (4.60) is well-defined.

By a standard measure-theoretic argument (see, e.g., [170, p. 187]), the equation (4.60) is equivalent to

$$\int_{\mathrm{SO}(d)} \int_{\mathbb{R}^d} f_h(x)g_h(\vartheta^{-1}x)\, \Phi_k(C \cap \vartheta D, \mathrm{d}x)\, \nu(\mathrm{d}\vartheta)$$

$$= \sum_{i=k}^{d} \int_{\mathbb{R}^d} f_h(x)\, \Phi_i(C, \mathrm{d}x) \int_{\mathbb{R}^d} g_h(x)\, \Phi_{d+k-i}(D, \mathrm{d}x) \tag{4.61}$$

for all continuous functions $f, g : \mathbb{S}^{d-1} \to \mathbb{R}$. Relation (4.61) holds if C and D are polyhedral cones. Let $C, D \in \mathcal{C}^d$ be general cones, not both subspaces (otherwise they are polyhedral). We can choose sequences $(P_i)_{i\in\mathbb{N}}$, $(Q_i)_{i\in\mathbb{N}}$ with $P_i \to C$ and $Q_i \to D$ as $i \to \infty$. Then $P_i \cap \vartheta Q_i \to C \cap \vartheta D$ for ν-almost all ϑ. Using the weak continuity of the conic curvature measures and the dominated convergence theorem, we obtain that (4.61) and thus (4.60) hold also for the general cones C and D. This completes the proof of Theorem 4.3.1. $\qquad\square$

We note that relation (4.59), which we have proved above, extends Lemma 2.1.3 from polyhedral to general closed convex cones. For later reference, we formulate this as a lemma.

Lemma 4.3.1. *Let $C, D \in \mathcal{C}^d$ be closed convex cones. The set of all rotations $\vartheta \in \mathrm{SO}(d)$ for which C and ϑD touch has ν-measure zero.*

The following is a consequence of Lemma 4.3.1. If the sum $C + D$ of two cones $C, D \in \mathcal{C}^d$ is not closed, then we see from Theorem 1.3.1 that C and $-D$ touch. Therefore, we can state:

Lemma 4.3.2. *If $C, D \in \mathcal{C}^d$, then $C + \vartheta D$ is closed for ν-almost all $\vartheta \in SO_d$.*

By duality, namely using (4.8) and Theorem 1.3.2, we obtain from Theorem 4.3.1 the relation

$$\int_{SO(d)} \Psi_k(C + \vartheta D, A \cap \vartheta B)\, \nu(d\vartheta) = \sum_{r=0}^{k} \Psi_r(C, A)\Psi_{k-r}(D, B) \qquad (4.62)$$

for $C, D \in \mathcal{C}^d$, $A, B \in \widehat{\mathcal{B}}(\mathbb{R}^d)$ and $k = 0, \dots, d - 1$.

We repeat the global case of Theorem 4.3.1 and supplement it by some consequences.

Theorem 4.3.4. *Let $C, D \in \mathcal{C}^d$ be closed convex cones. Then*

$$\int_{SO(d)} v_k(C \cap \vartheta D)\, \nu(d\vartheta) = \sum_{i=k}^{d} v_i(C)v_{d+k-i}(D) \qquad (4.63)$$

for $k = 1, \dots, d$. Moreover,

$$\int_{SO(d)} v_0(C \cap \vartheta D)\, \nu(d\vartheta) = \sum_{i=0}^{d} \sum_{j=0}^{d-i} v_i(C)v_j(D). \qquad (4.64)$$

Dually, we have

$$\int_{SO(d)} v_k(C + \vartheta D)\, \nu(d\vartheta) = \sum_{r=0}^{k} v_r(C)v_{k-r}(D) \qquad (4.65)$$

for $k = 0, \dots, d - 1$, and

$$\int_{SO(d)} v_d(C + \vartheta D)\, \nu(d\vartheta) = \sum_{r=0}^{d} \sum_{s=d-r}^{d} v_r(C)v_s(D). \qquad (4.66)$$

Proof. Equation (4.64) follows from (4.63) by using (2.39), that is,

$$\sum_{i=0}^{d} v_i(C \cap \vartheta D) = 1,$$

applying (4.63), and then using

$$1 = \left(\sum_{i=0}^{d} v_i(C)\right)\left(\sum_{j=0}^{d} v_j(D)\right) = \sum_{i+j \le 2d} v_i(C)v_j(D).$$

Equation (4.65) is the global case of (4.62). Equation (4.66) is obtained from (4.64) by duality. $\qquad \square$

As special cases of (4.63) and (4.64) (choosing for D an m-dimensional subspace and using (2.7)) we get, for $m \in \{1, \ldots, d-1\}$, the equations

$$\int_{G(d,m)} v_k(C \cap L) \, v_m(\mathrm{d}L) = v_{d+k-m}(C) \tag{4.67}$$

for $k = 1, \ldots, d$, and

$$\int_{G(d,m)} v_0(C \cap L) \, v_m(\mathrm{d}L) = \sum_{i=0}^{d-m} v_i(C), \tag{4.68}$$

which were already proved in Theorem 2.3.2.

Special cases of (4.65) and (4.66) are

$$\int_{G(d,m)} v_k(C + L) \, v_m(\mathrm{d}L) = v_{k-m}(C) \tag{4.69}$$

for $k = 0, \ldots, d-1$, and

$$\int_{G(d,m)} v_d(C + L) \, v_m(\mathrm{d}L) = \sum_{r=d-m}^{d} v_r(C). \tag{4.70}$$

With (4.58) above, we have obtained the *conic kinematic formula*

$$\int_{\mathrm{SO}(d)} \mathbb{1}\{C \cap \vartheta D \neq \{o\}\} \nu(\mathrm{d}\vartheta) = 2 \sum_{k=0}^{\lfloor \frac{d-1}{2} \rfloor} \sum_{j=2k+1}^{d} v_j(C) v_{d+2k+1-j}(D) \tag{4.71}$$

for polyhedral cones $C, D \in \mathcal{PC}^d$, not both subspaces. Since Theorem 4.3.1 is now proved for general convex cones, also (4.71) holds for $C, D \in \mathcal{C}^d$, not both subspaces.

This formula can be used to answer the following question (as formulated in McCoy and Tropp [129, p. 518]): "When does a randomly oriented cone strike a fixed cone?" More precisely, let $C, D \subset \mathbb{R}^d$ be closed convex cones, not both subspaces. Let θ be a uniform random rotation, that is, a random element of the rotation group $\mathrm{SO}(d)$ with distribution equal to the normalized Haar measure on $\mathrm{SO}(d)$. The question asks for the probability

$$\mathbb{P}(C \cap \theta D \neq \{o\}). \tag{4.72}$$

Formula (4.71) provides the answer:

$$\mathbb{P}(C \cap \theta D \neq \{o\}) = \mathbb{E} \, \mathbb{1}\{C \cap \theta D \neq \{o\}\}$$

$$= 2 \sum_{k=0}^{\lfloor \frac{d-1}{2} \rfloor} \sum_{j=2k+1}^{d} v_j(C) v_{d+2k+1-j}(D) \tag{4.73}$$

$$= \sum_{i=1}^{d-1} (1 + (-1)^{i+1}) \sum_{j=i}^{d} v_j(C) v_{d+i-j}(D). \tag{4.74}$$

The specialization of the formula (4.73) to the case where $D = L_{d-k}$ is a subspace of codimension k can in view of (2.31) be written in the form

$$\mathbb{P}(C \cap \boldsymbol{\theta} L_{d-k} \neq \{o\}) = 2 \sum_{j=0}^{\lfloor \frac{d-1}{2} \rfloor} v_{k+2j+1}(C), \qquad (4.75)$$

where C is not a subspace. This was already obtained in Section 2.3.

Formulas (4.73) and (4.75) can be given a still more compact form. A first step is the following definition.

Definition 4.3.1. *For a cone $C \in \mathcal{C}^d$, the kth half-tail functional is defined by*

$$h_k(C) := v_k(C) + v_{k+2}(C) + \cdots = \sum_{j=0}^{\lfloor \frac{d-k}{2} \rfloor} v_{k+2j}(C)$$

and the kth tail functional by

$$t_k(C) := v_k(C) + v_{k+1}(C) + \cdots = \sum_{j=k}^{d} v_j(C),$$

for $k = 0, \ldots, d$.

Here we have used the terminology introduced in [11], but we note that these are well-known functionals, namely

$$h_{k+1}(C) = U_k(C), \qquad t_k(C) = W_k(C),$$

by (2.68) and (2.77).

Second, we write (4.73) in the form

$$\mathbb{P}(C \cap \boldsymbol{\theta} D \neq \{o\}) = 2 \sum_{k=0}^{\lfloor \frac{d-1}{2} \rfloor} \sum_{i+j=d+2k+1} v_i(C) v_j(D)$$

and recall formula (2.34) for the product cone $C \times D$, namely

$$v_m(C \times D) = \sum_{i+j=m} v_i(C) v_j(D).$$

Therefore, we may write

$$\mathbb{P}(C \cap \boldsymbol{\theta} D \neq \{o\}) = 2 \sum_{k=0}^{\lfloor \frac{d-1}{2} \rfloor} v_{d+2k+1}(C \times D) = 2h_{d+1}(C \times D).$$

We collect these formulations in the following theorem.

Theorem 4.3.5. *Let $C, D \in \mathcal{C}^d$ be closed convex cones, not both subspaces. Let $\boldsymbol{\theta}$ be a uniform random rotation in* $\mathrm{SO}(d)$. *Then*

$$\mathbb{P}(C \cap \boldsymbol{\theta} D \neq \{o\}) = 2h_{d+1}(C \times D). \tag{4.76}$$

If C is not a subspace and $L_{d-k} \in G(d, d-k)$, then

$$\mathbb{P}(C \cap \boldsymbol{\theta} L_{d-k} \neq \{o\}) = 2h_{k+1}(C). \tag{4.77}$$

Of course, with (4.77) we are back to the definition of $U_k(C)$.

For working with the half-tails and tails, the following lemma will be useful.

Lemma 4.3.3. *If the cone $C \in \mathcal{C}^d$ is not a subspace, then*

$$2h_{k+1}(C) \leq t_k(C) \leq 2h_k(C) \quad for\ k = 0, \dots, d.$$

Proof. Let $k \in \{1, \dots, d\}$ be given. Choosing subspaces $L_{d-k} \subset L_{d-k+1}$, it is clear from (4.77) that

$$2h_{k+1}(C) = \mathbb{P}(C \cap \boldsymbol{\theta} L_{d-k} \neq \{o\}) \leq \mathbb{P}(C \cap \boldsymbol{\theta} L_{d-k+1} \neq \{o\}) = 2h_k(C).$$

Since $t_k(C) = h_k(C) + h_{k+1}(C)$, the assertion follows. $\square$

A kinematic formula for polyhedral skeleton measures

Finally in this section, we note that also for the polyhedral skeleton measures, which were introduced at the end of Section 4.2, there is a kinematic formula. The following result is due to Amelunxen [6], though with a different proof, based on a characterization theorem.

Theorem 4.3.6. *Let $C, D \in \mathcal{PC}^d$ and $A, B \in \widehat{\mathcal{B}}(\mathbb{R}^d)$. Then*

$$\int_{\mathrm{SO}(d)} \Pi_k(C \cap \vartheta D, A \cap \vartheta B)\, \nu(\mathrm{d}\vartheta) = \sum_{i=k}^{d} \Pi_i(C, A)\Pi_{d+k-i}(D, B) \tag{4.78}$$

for $k = 1, \dots, d$.

Proof. The existence of the integral in (4.78) follows from Lemma 4.2.6. The argument given before that Lemma shows that

$$\int_{\mathrm{SO}(d)} \Pi_k(C \cap \vartheta D, A \cap \vartheta B)\, \nu(\mathrm{d}\vartheta)$$

$$= \sum_{i+j=k+d} \sum_{F \in \mathcal{F}_i(C)} \sum_{G \in \mathcal{F}_j(D)} \int_{\mathrm{SO}(d)} \Phi_k(F \cap \vartheta G, A \cap \vartheta B)\, \nu(\mathrm{d}\vartheta).$$

Now for $F \in \mathcal{F}_i(C)$, $G \in \mathcal{F}_j(D)$, a simplified version of the proof given above for Theorem 4.3.1 (not caring about normal cones) reads

$$\int_{\mathrm{SO}(d)} \Phi_k(F \cap \vartheta G, A \cap \vartheta B)\, \nu(\mathrm{d}\vartheta)$$

$$= \int_{\mathrm{SO}_{\langle F \rangle}} \int_{\mathrm{SO}(d)} \Phi_k(F \cap \rho\vartheta G, A \cap \rho\vartheta B)\, \nu(\mathrm{d}\vartheta)\, \nu_{\langle F \rangle}(\mathrm{d}\rho)$$

$$= \int_{\mathrm{SO}(d)} \left[\int_{\mathrm{SO}_{\langle F \rangle}} \Phi_k(F \cap \rho\vartheta G, A \cap \rho\vartheta B)\, \nu_{\langle F \rangle}(\mathrm{d}\rho) \right] \nu(\mathrm{d}\vartheta)$$

$$= \int_{\mathrm{SO}(d)} \left[\int_{\mathrm{SO}_{\langle F \rangle}} \Phi_k(F \cap \rho(\vartheta G \cap \langle F \rangle), A \cap \rho(\vartheta B \cap \langle F \rangle))\, \nu_{\langle F \rangle}(\mathrm{d}\rho) \right] \nu(\mathrm{d}\vartheta)$$

$$= \int_{\mathrm{SO}(d)} \Phi_i(F, A)\Phi_k(\vartheta G \cap \langle F \rangle, \vartheta B)\, \nu(\mathrm{d}\vartheta)$$

$$= \Phi_i(F, A) \int_{\mathrm{SO}_{\langle G \rangle}} \int_{\mathrm{SO}(d)} \Phi_k(\sigma G \cap \vartheta^{-1}\langle F \rangle, \sigma B)\, \nu(\mathrm{d}\vartheta)\, \nu_{\langle G \rangle}(\mathrm{d}\sigma)$$

$$= \Phi_i(F, A) \int_{\mathrm{SO}(d)} \left[\int_{\mathrm{SO}_{\langle G \rangle}} \Phi_k(\sigma G \cap \vartheta^{-1}\langle F \rangle, \sigma B)\, \nu_{\langle G \rangle}(\mathrm{d}\sigma) \right] \nu(\mathrm{d}\vartheta)$$

$$= \Phi_i(F, A) \int_{\mathrm{SO}(d)} \left[\int_{\mathrm{SO}_{\langle G \rangle}} \Phi_k(\vartheta^{-1}\langle F \rangle \cap \langle G \rangle \cap \sigma G, \mathbb{R}^d \cap \sigma B)\, \nu_{\langle G \rangle}(\mathrm{d}\sigma) \right] \nu(\mathrm{d}\vartheta)$$

$$= \Phi_i(F, A) \int_{\mathrm{SO}(d)} \Phi_k(\vartheta^{-1}\langle F \rangle \cap \langle G \rangle, \mathbb{R}^d)\Phi_j(G, B)\, \nu(\mathrm{d}\vartheta)$$

$$= \Phi_i(F, A)\Phi_j(G, B)$$

$$= \gamma_{\langle F \rangle}(F \cap A)\gamma_{\langle G \rangle}(G \cap B),$$

the latter because of $\dim F = i$ and $\dim G = j$. This yields

$$\int_{\mathrm{SO}(d)} \Pi_k(C \cap \vartheta D, A \cap \vartheta B)\, \nu(\mathrm{d}\vartheta)$$

$$= \sum_{i+j=k+d} \sum_{F \in \mathcal{F}_i(C)} \sum_{G \in \mathcal{F}_j(D)} \gamma_{\langle F \rangle}(F \cap A)\gamma_{\langle G \rangle}(G \cap B)$$

$$= \sum_{i+j=k+d} \Pi_i(C, A)\Pi_j(D, B),$$

as stated. □

Notes for Section 4.3

1. For smooth submanifolds of the sphere, integral geometry goes back to Santaló [153, 154]. An equivalent spherical version of Theorem 4.3.1 was proved by Glasauer [63]. A summary of this thesis appears in [64]. Glasauer gave two proofs, both based on characterization theorems, either for spherical curvature measures or for spherical support measures. Glasauer's second proof was transferred to the conic situation, and the result was expanded (by combining it with its dual version, and iteration) by Amelunxen [6].

4.4 Concentration of the conic intrinsic volumes

The conic intrinsic volumes of a subspace $L \in \mathcal{L}_\bullet$ satisfy

$$v_k(L) = \delta_{k,\dim L},$$

by (2.31), thus

$$(v_0(L), \ldots, v_d(L)) = (0, \ldots, 0, 1, 0, \ldots, 0),$$

with 1 at the place numbered by $\dim L$. It has been discovered that also for closed convex cones, the intrinsic volumes have a certain concentration property, though not as sharply as for subspaces. This concentration property is proved in the present section.

Let us briefly indicate the background why such properties are of interest. Certain random models motivated by applications lead one to ask for the probability

$$\mathbb{P}(C \cap \boldsymbol{\theta} D \neq \{o\})$$

of non-trivial intersection of a closed convex cone C with a randomly rotated cone $\boldsymbol{\theta} D$; here $\boldsymbol{\theta}$ is a uniform random rotation in $\mathrm{SO}(d)$. Formula (4.73) provides an explicit answer in terms of conic intrinsic volumes of C and D. As the latter are generally difficult to compute, estimates are of interest. Let us consider the special case where $D = L$ is a subspace. If also C is a subspace, then it is trivial that

$$\mathbb{P}(C \cap \boldsymbol{\theta} L \neq \{o\}) = \begin{cases} 0 \text{ if } \dim C + \dim L \leq d, \\ 1 \text{ if } \dim C + \dim L > d. \end{cases}$$

With a general closed convex cone C, one can associate a substitute of the dimension, denoted by $\delta(C)$, such that $\mathbb{P}(C \cap \boldsymbol{\theta} L \neq \{o\})$ is close to 0 if $\delta(C) + \dim L$ is considerably smaller than d, and $\mathbb{P}(C \cap \boldsymbol{\theta} L \neq \{o\})$ is close to 1 if $\delta(C) + \dim L$ is considerably larger than d. This can be made precise in a quantitative way; see Theorem 4.4.2. The number $\delta(C)$ is, in this sense, a threshold. A similar behavior of the probability $\mathbb{P}(C \cap \boldsymbol{\theta} D \neq \{o\})$ for another closed convex cone D is described in Theorem 4.4.3. We follow, to a large extent, parts of the papers by Amelunxen, Lotz, McCoy and Tropp [11] and McCoy and Tropp [129]. To quote from [11]: The main result "leads to accurate bounds on the probability that a randomly rotated cone shares a ray with a fixed cone".

The crucial number around which concentration takes place is given by the so-called statistical dimension. Before we introduce it, we prove the identity

$$\mathbb{E}\,\|\Pi_C(\mathbf{g})\|^2 = \sum_{k=0}^{d} k v_k(C) \tag{4.79}$$

for $C \in \mathcal{C}^d$ and the standard Gaussian random vector $\mathbf{g}$. For the proof, we note that by the 'layer cake representation' (as it is called in [125, Thm. 1.13]), we have

$$\int_{\mathbb{R}^d} \|\Pi_C(x)\|^2 \gamma_d(dx) = \int_0^\infty \gamma_d \left(\{x \in \mathbb{R}^d : \|\Pi_C(x)\|^2 > t\}\right) dt,$$

or

$$\mathbb{E}\,\|\Pi_C(\mathbf{g})\|^2 = \int_0^\infty \mathbb{P}\left(\|\Pi_C(\mathbf{g})\|^2 > t\right) dt.$$

First we assume now that C is a polyhedral cone. Then, by the Gaussian Steiner formula in the version of Corollary 4.1.2 and with $\mathcal{X}_{d-k}$ as defined there,

$$\mathbb{P}\left(\|\Pi_C(\mathbf{g})\|^2 > t\right) = \mathbb{P}\left(\text{dist}^2(\mathbf{g}, C^\circ) > t\right) = 1 - \mathbb{P}\left(\text{dist}^2(\mathbf{g}, C^\circ) \leq t\right)$$

$$= 1 - \sum_{k=0}^{d} \mathbb{P}(\mathcal{X}_{d-k} \leq t) v_k(C^\circ)$$

$$= \sum_{k=0}^{d} \mathbb{P}(\mathcal{X}_k > t) v_k(C), \tag{4.80}$$

where we have used (2.63) and (2.61). Since

$$\int_0^\infty \mathbb{P}(\mathcal{X}_k > t)\, dt = \mathbb{E}\,\mathcal{X}_k = k,$$

we obtain (4.79). If now $C \in \mathcal{C}^d$ is a general closed convex cone, we approximate it by a decreasing sequence $(C_i)_{i \in \mathbb{N}}$ of polyhedral cones. For each $i \in \mathbb{N}$ we have

$$\mathbb{E}\,\|\Pi_{C_i}(\mathbf{g})\|^2 = \sum_{k=0}^{d} k v_k(C_i).$$

Further, $\|\Pi_{C_i}(\mathbf{g})\| \to \|\Pi_C(\mathbf{g})\|$ as $i \to \infty$, and $\|\Pi_{C_i}(\mathbf{g})\| \leq \|\Pi_{C_1}(\mathbf{g})\|$ by (1.22). Now the dominated convergence theorem, together with the continuity of the conic intrinsic volumes, shows that (4.79) holds for general C.

Since the conic intrinsic volumes of a closed convex cone C satisfy

$$v_k(C) \geq 0 \quad \text{and} \quad v_0(C) + \cdots + v_d(C) = 1,$$

the following definition makes sense.

Definition 4.4.1. *The* intrinsic volume random variable $\mathbf{V}_C$ *of a cone* $C \in \mathcal{C}^d$ *is defined as a random variable with values in* $\{0, \ldots, d\}$ *and with distribution*

$$\mathbb{P}(\mathbf{V}_C = k) = v_k(C), \quad k = 0, \ldots, d.$$

By the duality relation (2.39) we have

$$\mathbb{P}(\mathbf{V}_{C^\circ} = k) = v_k(C^\circ) = v_{d-k}(C) = \mathbb{P}(\mathbf{V}_C = d - k) = \mathbb{P}\{d - \mathbf{V}_C = k\},$$

hence

$$\mathbf{V}_{C^\circ} \stackrel{d}{=} d - \mathbf{V}_C. \tag{4.81}$$

Here $\stackrel{d}{=}$ means equality in distribution. For the variances, this implies

$$\mathrm{Var}\mathbf{V}_C = \mathrm{Var}(d - \mathbf{V}_C) = \mathrm{Var}\mathbf{V}_{C^\circ}. \tag{4.82}$$

Definition 4.4.2. *The* statistical dimension *of the cone* $C \in \mathcal{C}^d$ *is the number*

$$\delta(C) := \mathbb{E}\,\mathbf{V}_C = \sum_{k=0}^{d} k v_k(C) = \mathbb{E}\,\|\Pi_C(\mathbf{g})\|^2, \tag{4.83}$$

where $\mathbf{g}$ *is a standard Gaussian random vector in* $\mathbb{R}^d$.

By (4.83), the statistical dimension inherits some properties from the conic intrinsic volumes: on $\mathcal{C}^d$, the function δ is a rotation invariant, continuous valuation. It is intrinsic, that is, $\delta(C)$ does not depend on the dimension of the space in which C is embedded. In addition, δ is increasing under set inclusion. In fact, if $C \subset D$, then $C^\circ \supset D^\circ$ and hence

$$\|\Pi_C(\mathbf{g})\|^2 = \mathrm{dist}^2(\mathbf{g}, C^\circ) \leq \mathrm{dist}^2(\mathbf{g}, D^\circ) = \|\Pi_D(\mathbf{g})\|^2.$$

The duality relation

$$\delta(C) + \delta(C^\circ) = d \tag{4.84}$$

follows from

$$\mathbb{E}\left(\|\Pi_C(\mathbf{g})\|^2 + \|\Pi_{C^\circ}(\mathbf{g})\|^2\right) = \mathbb{E}\,\|\mathbf{g}\|^2 = d$$

(cf. Thm. 1.3.3).

As a consequence of (4.84), we note that a self-dual cone $C \in \mathcal{C}^d$, that is, a cone C with $C^\circ = -C$, satisfies

$$\delta(C) = \frac{1}{2}d.$$

For a linear subspace L, it follows immediately from (2.31) and (4.83) that

$$\delta(L) = \dim L.$$

If $C \in \mathcal{C}^{d_1}$ and $D \in \mathcal{C}^{d_2}$, then

$$\delta(C \times D) = \delta(C) + \delta(D). \tag{4.85}$$

For the proof, we note that according to the convention (in Sect. 1.3) about the scalar product of $\mathbb{R}^{d_1} \times \mathbb{R}^{d_2}$, we can take an orthogonal decomposition $\mathbb{R}^{d_1+d_2} = L_1 \oplus L_2$ with $\dim L_i = d_i$ and assume that $C \subseteq L_1$, $D \subseteq L_2$; then $\delta(C \times D) = \delta(C \oplus D)$. Using Lemma 1.3.7 and (2.3), we get

$$
\begin{aligned}
\delta(C \oplus D) &= \int_{\mathbb{R}^{d_1+d_2}} \|\Pi_{C \oplus D}(x)\|^2 \, \gamma_{d_1+d_2}(dx) \\
&= \int_{L_1} \int_{L_2} \left(\|\Pi_C(y)\|^2 + \|\Pi_D(z)\|^2 \right) \gamma_{L_2}(dz) \, \gamma_{L_1}(dy) \\
&= \delta(C) + \delta(D).
\end{aligned}
$$

Similar to (4.79), we derive, first for a polyhedral cone C,

$$\mathbb{E} \|\Pi_C(\mathbf{g})\|^4 = \int_0^\infty \mathbb{P}\left(\|\Pi_C(\mathbf{g})\|^4 > t \right) dt$$

and

$$\mathbb{P}\left(\|\Pi_C(\mathbf{g})\|^4 > t \right) = \sum_{k=0}^{d} \mathbb{P}(\mathcal{X}_k > \sqrt{t}) v_k(C)$$

by (4.80), hence

$$\mathbb{E} \|\Pi_C(\mathbf{g})\|^4 = \sum_{k=0}^{d} (\mathbb{E}\, \mathcal{X}_k^2) v_k(C).$$

By approximation as above, this holds for a general cone $C \in \mathcal{C}^d$. Since $\mathbb{E}\, \mathcal{X}_k^2 = k^2 + 2k$, we obtain

$$\mathbb{E} \|\Pi_C(\mathbf{g})\|^4 = \mathbb{E}\, \mathbf{V}_C^2 + 2\delta(C).$$

This yields an expression for the variance of the random variable $\mathbf{V}_C$, namely

$$
\begin{aligned}
\operatorname{Var} \mathbf{V}_C &= \mathbb{E}\, \mathbf{V}_C^2 - (\mathbb{E}\, \mathbf{V}_C)^2 \\
&= \mathbb{E} \|\Pi_C(\mathbf{g})\|^4 - 2\delta(C) - \delta(C)^2 \\
&= \mathbb{E} \|\Pi_C(\mathbf{g})\|^4 - (\mathbb{E} \|\Pi_C(\mathbf{g})\|^2)^2 - 2\delta(C) \\
&= \operatorname{Var}(\|\Pi_C(\mathbf{g})\|^2) - 2\delta(C). \tag{4.86}
\end{aligned}
$$

This can be estimated. By Lemma 1.3.8, we have

$$\nabla \|\Pi_C(x)\|^2 = 2\Pi_C(x) \quad \text{for } x \in \mathbb{R}^d.$$

Hence, applying to the function $f(x) = \|\Pi_C(x)\|^2$ the Gaussian Poincaré inequality $\operatorname{Var} f(\mathbf{g}) \le \mathbb{E}\left(\|\nabla f(\mathbf{g})\|^2 \right)$ (see [29, Thm. 1.6.4]), we obtain

$$\operatorname{Var}(\|\Pi_C(\mathbf{g})\|^2) \le \mathbb{E}\left(4\|\Pi_C(\mathbf{g})\|^2 \right) = 4\delta(C).$$

Together with (4.86), this yields

$$\text{Var } \mathbf{V}_C \leq 2\delta(C).$$

Since $\text{Var}\mathbf{V}_C = \text{Var}\mathbf{V}_{C^\circ}$ by (4.82), this implies

$$\text{Var } \mathbf{V}_C \leq 2(\delta(C) \wedge \delta(C^\circ)), \tag{4.87}$$

where $a \wedge b := \min\{a, b\}$.

Tschebyscheff's inequality yields

$$\mathbb{P}\left(|\mathbf{V}_C - \delta(C)| > \lambda\sqrt{\delta(C)}\right) \leq \frac{\text{Var}\mathbf{V}_C}{\lambda^2\delta(C)} \leq \frac{2}{\lambda^2}$$

for $\lambda \geq 0$. The aim of the following is to obtain stronger concentration properties.

We shall need an expression for the moment generating function of the intrinsic volume random variable.

Lemma 4.4.1. *Let $C \in \mathcal{C}^d$ and $\eta \in \mathbb{R}$. Then*

$$\mathbb{E}\, e^{\eta \mathbf{V}_C} = \mathbb{E}\, e^{\xi\|\Pi_C(\mathbf{g})\|^2} \quad \text{with } \xi = \frac{1}{2}(1 - e^{-2\eta}).$$

Proof. First let C be a polyhedral cone. Let $\xi < \frac{1}{2}$. Theorem 4.1.1 with $f(a,b) = e^{\xi a}$ gives

$$\mathbb{E}\, e^{\xi\|\Pi_C(\mathbf{g})\|^2} = \sum_{k=0}^{d} \varphi_f(L_k) v_k(C)$$

with $\varphi_f(L_k) = \mathbb{E}\, e^{\xi\|\Pi_{L_k}(\mathbf{g})\|^2}$, where $L_k \in G(d, k)$. Since $\Pi_{L_k}(\mathbf{g})$ is a standard Gaussian random vector in L_k, we get $\varphi_f(L_k) = \mathbb{E}\, e^{\xi\chi_k} = (1 - 2\xi)^{-k/2}$ (the moment generating function of a chi-square variable with k degrees of freedom). Therefore, $\varphi_f(L_k) = e^{\eta k}$ and thus

$$\mathbb{E}\, e^{\xi\|\Pi_C(\mathbf{g})\|^2} = \sum_{k=0}^{d} e^{\eta k} v_k(C) = \mathbb{E}\, e^{\eta \mathbf{V}_C}.$$

The extension to general cones $C \in \mathcal{C}^d$ is achieved as above. □

Lemma 4.4.2. *Let $C \in \mathcal{C}^d$. For each real $\xi < \frac{1}{2}$,*

$$\mathbb{E}\, e^{\xi(\|\Pi_C(\mathbf{g})\|^2 - \delta(C))} \leq \exp\left(\frac{2\xi^2\delta(C)}{1 - 2\xi}\right).$$

Proof. Define

$$Z := \|\Pi_C(\mathbf{g})\|^2 - \delta(C),$$

so that Z is a random variable with mean zero, and define the moment generating function

$$m(\xi) := \mathbb{E}\, e^{\xi Z}.$$

It follows from the preceding lemma that

$$m(\xi) = e^{-\xi\delta(C)} \sum_{k=0}^{d} (\mathbb{E}\, e^{\xi \chi_k}) v_k(C) = e^{-\xi\delta(C)} \sum_{k=0}^{d} (1 - 2\xi)^{-k/2} v_k(C)$$

is finite for $\xi < \frac{1}{2}$. For $\xi < \frac{1}{2}$, the rules for the differentiation of a parameter integral allow us to conclude that

$$m'(\xi) = \mathbb{E}\left[Z e^{\xi Z}\right].$$

If $H : \mathbb{R}^d \to \mathbb{R}$ is a differentiable function, then the Gaussian logarithmic Sobolev inequality (see, e.g., [29, Thm. 1.6.1] and put $f = e^{H/2}$) says that

$$\mathbb{E}\left[H(\mathbf{g}) e^{H(\mathbf{g})}\right] - \mathbb{E}\left[e^{H(\mathbf{g})}\right] \log \mathbb{E}\left[e^{H(\mathbf{g})}\right] \le \frac{1}{2}\mathbb{E}\left[\|\nabla H(\mathbf{g})\|^2 e^{H(\mathbf{g})}\right], \quad (4.88)$$

provided that the expectations are finite. This can be applied to the function given by

$$H(x) := \xi\left[\|\Pi_C(x)\|^2 - \delta(C)\right]$$

for given $C \in \mathcal{C}^d$ and $\xi \in \mathbb{R}$, since by Lemma 1.3.8 it is differentiable. This lemma yields

$$\|\nabla H(x)\|^2 = 4\xi^2 \|\Pi_C(x)\|^2.$$

We observe that $H(\mathbf{g}) = \xi Z$ and $\|\nabla H(\mathbf{g})\|^2 = 4\xi^2(Z + \delta(C))$. If we now assume that $\xi < \frac{1}{2}$, then for this function H all expectations in (4.88) are finite, and we obtain

$$\xi\mathbb{E}\left[Z e^{\xi Z}\right] - \mathbb{E}\left[e^{\xi Z}\right] \log \mathbb{E}\left[e^{\xi Z}\right] \le 2\xi^2 \mathbb{E}\left[Z e^{\xi Z}\right] + 2\xi^2 \delta(C)\mathbb{E}\left[e^{\xi Z}\right].$$

Since $m(\xi) = \mathbb{E}\left[e^{\xi Z}\right]$ and $m'(\xi) = \mathbb{E}\left[Z e^{\xi Z}\right]$, this can be written as

$$\xi m'(\xi) - m(\xi) \log m(\xi) \le 2\xi^2 m'(\xi) + 2\delta(C)\xi^2 m(\xi) \quad \text{for } \xi < \frac{1}{2}. \quad (4.89)$$

For $\xi \ne 0$ we can divide this by $\xi^2 m(\xi)$ and write the result in the form

$$\frac{d}{dt}\left[\frac{1}{t}\log m(t)\right] \le 2\frac{d}{dt}\left[\log m(t) + 2\delta(C)t\right], \quad t < \frac{1}{2}, \, t \ne 0. \quad (4.90)$$

We can integrate this from 0 to $\xi \in (0, 1/2)$. Concerning the boundary conditions at 0, we remark that $m'(0) = 0$ and $\log m(0) = 0$ and that, using the rule of de l'Hospital together with (4.83), we obtain $\lim_{\xi \to 0} \xi^{-1} \log m(\xi) = 0$. With these boundary conditions, the integration gives

$$\frac{1}{\xi} \log m(\xi) \leq 2 \log m(\xi) + 2\delta(C)\xi, \quad 0 < \xi < \frac{1}{2},$$

which yields

$$m(\xi) \leq \exp\left(\frac{2\xi^2\delta(C)}{1 - 2\xi}\right) \tag{4.91}$$

for $0 < \xi < \frac{1}{2}$. Integrating (4.89) from $\xi < 0$ to 0, we obtain in a similar way that (4.91) holds for $\xi < 0$. For $\xi = 0$ it holds trivially. This proves the assertion of the lemma.   □

Lemma 4.4.3. *Let $C \in \mathcal{C}^d$. For each $\eta \in \mathbb{R}$,*

$$\mathbb{E}\, e^{\eta(\mathbf{V}_C - \delta(C))} \leq \exp\left(\frac{e^{2\eta} - 2\eta - 1}{2}\delta(C)\right) \tag{4.92}$$

and

$$\mathbb{E}\, e^{\eta(\mathbf{V}_C - \delta(C))} \leq \exp\left(\frac{e^{-2\eta} + 2\eta - 1}{2}\delta(C^\circ)\right). \tag{4.93}$$

Proof. Writing $\xi = \frac{1}{2}\left(1 - e^{-2\eta}\right)$ and using Lemma 4.4.1, we get

$$\mathbb{E}\, e^{\eta(\mathbf{V}_C - \delta(C))} = \mathbb{E}\, e^{\xi\|\Pi_C(\mathbf{g})\|^2}e^{-\eta\delta(C)} = e^{(\xi-\eta)\delta(C)}\mathbb{E}\, e^{\xi(\|\Pi_C(\mathbf{g})\|^2 - \delta(C))}.$$

The last term can be estimated by Lemma 4.4.2, hence

$$\mathbb{E}\, e^{\eta(\mathbf{V}_C - \delta(C))} \leq e^{(\xi-\eta)\delta(C)}\exp\left(\frac{2\xi^2\delta(C)}{1 - 2\xi}\right).$$

Since

$$\xi - \eta + \frac{2\xi^2}{1 - 2\xi} = \frac{e^{2\eta} - 2\eta - 1}{2},$$

we obtain (4.92). Since $\mathbf{V}_C \overset{d}{=} d - \mathbf{V}_{C^\circ}$ by (4.81) and $\delta(C) = d - \delta(C^\circ)$ by (4.84), we have

$$\mathbb{E}\, e^{\eta(\mathbf{V}_C - \delta(C))} = \mathbb{E}\, e^{-\eta(\mathbf{V}_{C^\circ} - \delta(C^\circ))},$$

and so (4.92) yields (4.93).   □

Theorem 4.4.1. *Let $C \in \mathcal{C}^d$. Define the function ψ by*

$$\psi(u) := (u + 1)\log(u + 1) - u \quad \text{for } u \geq -1,$$

and $\psi(u) = \infty$ for $u < -1$. Then, for all $\lambda \geq 0$,

$$\mathbb{P}\left(\mathbf{V}_C - \delta(C) \geq \lambda\right)$$
$$\leq \exp\left(-\frac{1}{2}\max\left\{\delta(C)\psi\left(\frac{\lambda}{\delta(C)}\right), \delta(C^\circ)\psi\left(\frac{-\lambda}{\delta(C^\circ)}\right)\right\}\right) \tag{4.94}$$

and

$$\mathbb{P}\left(\mathbf{V}_C - \delta(C) \leq -\lambda\right)$$
$$\leq \exp\left(-\frac{1}{2}\max\left\{\delta(C)\psi\left(\frac{-\lambda}{\delta(C)}\right), \delta(C^\circ)\psi\left(\frac{\lambda}{\delta(C^\circ)}\right)\right\}\right). \tag{4.95}$$

Proof. For any real random variable Y and for $\eta \geq 0$, we have

$$\mathbb{E}\, e^{\eta Y} = \mathbb{E}\left[e^{\eta Y} \mathbb{1}\{Y \geq 0\}\right] + \mathbb{E}\left[e^{\eta Y} \mathbb{1}\{Y < 0\}\right] \geq \mathbb{P}(Y \geq 0).$$

Hence, for $\lambda \geq 0$,

$$\mathbb{P}(\mathbf{V}_C - \delta(C) \geq \lambda) \leq e^{-\eta\lambda}\mathbb{E}\, e^{\eta(\mathbf{V}_C - \delta(C))}$$

$$\leq e^{-\eta\lambda}\exp\left(\frac{e^{2\eta} - 2\eta - 1}{2}\delta(C)\right),$$

where (4.92) was used. The function g defined by

$$g(\eta) := -\eta\lambda + \frac{\delta(C)}{2}\left(e^{2\eta} - 2\eta - 1\right) \quad \text{for } \eta \in \mathbb{R}$$

attains its minimum at the value η_0 with

$$e^{2\eta_0} = \frac{\lambda}{\delta(C)} + 1 =: u + 1,$$

hence

$$g(\eta) \geq -\frac{\delta(C)}{2}[(u+1)\log(u+1) - u] \quad \text{for } \eta \geq 0.$$

Therefore,

$$\mathbb{P}(\mathbf{V}_C - \delta(C) \geq \lambda) \leq \exp\left(-\frac{\delta(C)}{2}\psi\left(\frac{\lambda}{\delta(C)}\right)\right).$$

If we use (4.93) instead of (4.92), we obtain

$$\mathbb{P}(\mathbf{V}_C - \delta(C) \geq \lambda) \leq e^{-\eta\lambda}\exp\left(\frac{e^{-2\eta} + 2\eta - 1}{2}\delta(C^\circ)\right).$$

The function h defined by

$$h(\eta) := -\eta\lambda + \frac{\delta(C^\circ)}{2}\left(e^{-2\eta} + 2\eta - 1\right) \quad \text{for } \eta \in \mathbb{R}$$

attains its minimum at the value η_0 with

$$e^{-2\eta_0} = \frac{-\lambda}{\delta(C)} + 1,$$

provided that $\lambda < \delta(C^\circ)$. This gives

$$h(\eta) \geq -\frac{\delta(C^\circ)}{2}\psi\left(\frac{-\lambda}{\delta(C^\circ)}\right) \quad \text{for } \eta \geq 0$$

if $\lambda < \delta(C^\circ)$. By continuity, this holds also for $\lambda = \delta(C^\circ)$. If $\lambda > \delta(C^\circ)$, then $\delta(C) + \lambda > d$, hence $\mathbb{P}(\mathbf{V}_C - \delta(C) \geq \lambda) = 0$. Altogether, we get

$$\mathbb{P}(\mathbf{V}_C - \delta(C) \geq \lambda) \leq \exp\left(-\frac{\delta(C^\circ)}{2} \psi\left(\frac{-\lambda}{\delta(C^\circ)}\right)\right).$$

Both estimates together prove (4.94).

Since $\mathbf{V}_C \overset{d}{=} d - \mathbf{V}_{C^\circ}$ and $\delta(C) = d - \delta(C^\circ)$, we have

$$\mathbb{P}(\mathbf{V}_C - \delta(C) \leq -\lambda) = \mathbb{P}(\mathbf{V}_{C^\circ} - \delta(C^\circ) \geq \lambda),$$

so that (4.94) for C° yields (4.95) for C.   □

Slightly weaker forms of the estimates in Theorem 4.4.1 can be written in a simpler form. For this, we consider the function defined by

$$u \mapsto \psi(u) - \frac{u^2}{2 + 2u/3} \quad \text{for } -1 < u < \infty.$$

Its second derivative is nonnegative, hence it is a convex function. Since it vanishes at 0, together with its first derivative, it is nonnegative. Thus, $\psi(u) \geq u^2/(2+2u/3)$ for $-1 < u < \infty$. Therefore, we can state the following corollary, which is weaker than Theorem 4.4.1, but easier to handle.

Corollary 4.4.1. *For a cone $C \in \mathcal{C}^d$, define*

$$\omega(C)^2 := \delta(C) \wedge \delta(C^\circ)$$

and

$$p_C(\lambda) := \exp\left(\frac{-\lambda^2/4}{\omega(C)^2 + \lambda/3}\right).$$

Then

$$\mathbb{P}(|\mathbf{V}_C - \delta(C)| \geq \lambda) \leq p_C(\lambda) \tag{4.96}$$

for $\lambda \geq 0$.

The tail functionals appearing in the following corollary were introduced in Definition 4.3.1.

Corollary 4.4.2. *Let $C \in \mathcal{C}^d$, $m \in \{0, \ldots, d\}$ and $\lambda \geq 0$. Then*

$$m \geq \delta(C) + \lambda \;\Rightarrow\; t_m(C) \leq p_C(\lambda), \tag{4.97}$$

$$m \leq \delta(C) - \lambda \;\Rightarrow\; t_{m+1}(C) \geq 1 - p_C(\lambda). \tag{4.98}$$

Proof. Using, in this order, the definition of the random variable $\mathbf{V}_C$, the assumption $m \geq \delta(C) + \lambda$, and the definition of the tail function, we obtain

$$\mathbb{P}(\mathbf{V}_C - \delta(C) \geq \lambda) = \sum_{k \geq \delta(C)+\lambda} v_k(C) \geq \sum_{k \geq m} v_k(C) = t_m(C).$$

Now (4.96) gives (4.97). Similarly, the assumption $m \leq \delta(C) - \lambda$ yields

$$\mathbb{P}(\mathbf{V}_C - \delta(C) \leq -\lambda) = \sum_{k \leq \delta(C)-\lambda} v_k(C) \geq \sum_{k \leq m} v_k(C) = 1 - t_{m+1}(C).$$

Now (4.96) gives (4.98).   □

For the announced threshold phenomenon, we separate the case of a cone and a subspace from that of two cones.

Theorem 4.4.2. *Let* $C, L \in \mathcal{C}^d$, *where* L *is a subspace. If* $\boldsymbol{\theta}$ *is a uniform random rotation in* $\mathrm{SO}(d)$, *then the following holds for* $\lambda \geq 0$:

$$\delta(C) + \dim L \leq d - \lambda \;\Rightarrow\; \mathbb{P}(C \cap \boldsymbol{\theta} L \neq \{o\}) \leq p_C(\lambda),$$

$$\delta(C) + \dim L \geq d + \lambda \;\Rightarrow\; \mathbb{P}(C \cap \boldsymbol{\theta} L \neq \{o\}) \geq 1 - p_C(\lambda).$$

Proof. We can assume that C is not a subspace, since otherwise the assertion is trivial. We write $\dim L = d - m$.

Let $\delta(C) + \dim L \leq d - \lambda$, then $m \geq \delta(C) + \lambda$. From (4.77), Lemma 4.3.3 and (4.97) we get

$$\mathbb{P}(C \cap \boldsymbol{\theta} L_{d-m} \neq \{o\}) = 2h_{m+1}(C) \leq t_m(C) \leq p_C(\lambda).$$

Similarly, the assumption $\delta(C) + \dim L \geq d + \lambda$ yields $m \leq \delta(C) - \lambda$ and hence

$$1 - \mathbb{P}(C \cap \boldsymbol{\theta} L_{d-m} \neq \{o\}) = 1 - 2h_{m+1}C) \leq 1 - t_{m+1}(C) \leq 1 - p_C(\lambda),$$

as stated. □

Before proving a similar result for pairs of cones, we provide a counterpart to Corollary 4.4.1 for product cones.

Lemma 4.4.4. *For closed convex cones* $C, D \in \mathcal{C}^d$, *define*

$$\sigma(C, D)^2 := (\delta(C) \wedge \delta(C^\circ)) + (\delta(D) \wedge \delta(D^\circ))$$

and

$$p_{C,D}(\lambda) := \exp\left(\frac{-\lambda^2/4}{\sigma(C, D)^2 + \lambda/3}\right).$$

Then

$$\mathbb{P}(|\mathbf{V}_{C \times D} - \delta(C \times D)| \geq \lambda) \leq p_{C,D}(\lambda) \tag{4.99}$$

for $\lambda \geq 0$.

Proof. First we note that, considering the Taylor expansion of $e^{2\eta}$, it is easy to see that

$$\frac{e^{2\eta} - 2\eta - 1}{2} \leq \frac{\eta^2}{1 - 2|\eta|/3} \quad \text{for } |\eta| < \frac{3}{2}.$$

The estimates of Lemma 4.4.3 can, therefore, be combined into

$$\mathbb{E}\, e^{\eta(\mathbf{V}_C - \delta(C))} \leq \exp\left(\frac{\eta^2(\delta(C) \wedge \delta(C^\circ))}{1 - 2|\eta|/3}\right) \quad \text{for } |\eta| < \frac{3}{2}.$$

We can define the intrinsic volume random variables $\mathbf{V}_C, \mathbf{V}_D$ in such a way that they are stochastically independent (starting with a suitable simultaneous distribution). Under this assumption, we have

$$\mathbb{P}(\mathbf{V}_C + \mathbf{V}_D = k)$$

$$= \sum_{i+j=k} \mathbb{P}(\mathbf{V}_C = i, \mathbf{V}_D = j) = \sum_{i+j=k} \mathbb{P}\{\mathbf{V}_C = i\}\mathbb{P}\{\mathbf{V}_D = j\}$$

$$= \sum_{i+j=k} v_i(C)v_j(D) = v_k(C \times D) = \mathbb{P}(\mathbf{V}_{C\times D} = k),$$

where (2.34) was used. Thus,

$$\mathbf{V}_{C\times D} \overset{d}{=} \mathbf{V}_C + \mathbf{V}_D.$$

Accidentally, this shows again that $\delta(C \times D) = \delta(C) + \delta(D)$. Now the assumed independence of $\mathbf{V}_C$ and $\mathbf{V}_D$ gives

$$\mathbb{E}\, e^{\eta(\mathbf{V}_{C\times D} - \delta(C\times D))} = \mathbb{E}\, e^{\eta(\mathbf{V}_C - \delta(C))}\mathbb{E}\, e^{\eta(\mathbf{V}_D - \delta(D))}$$

$$\leq \exp\left(\frac{\eta^2\sigma(C, D)^2}{1 - 2|\eta|/3}\right)$$

for $|\eta| < 3/2$.

Using again the estimate $\mathbb{P}(Y \geq 0) \leq \mathbb{E}\, e^{\eta Y}$ for a real random variable Y and for $\eta \geq 0$, we obtain

$$\mathbb{P}(\mathbf{V}_{C\times D} - \delta(C \times D) \geq \lambda) \leq e^{-\eta\lambda}\mathbb{E}\, e^{\eta(\mathbf{V}_{C\times D} - \delta(C\times D))}$$

$$\leq e^{-\eta\lambda}\exp\left(\frac{\eta^2\sigma(C, D)^2}{1 - 2\eta/3}\right)$$

for $\eta \in [0, 3/2)$. The choice $\eta = \lambda/(2\sigma(C, D)^2 + 2\lambda/3)$ gives

$$\mathbb{P}(\mathbf{V}_{C\times D} - \delta(C \times D) \geq \lambda) \leq \exp\left(\frac{-\lambda^2/4}{\sigma(C, D)^2 + \lambda/3}\right).$$

The corresponding estimate for $\mathbb{P}(\mathbf{V}_{C\times D} - \delta(C \times D) \leq -\lambda)$ is obtained in the same way. □

Now we can state the following.

Theorem 4.4.3. *Let $C, D \in \mathcal{C}^d$. If $\boldsymbol{\theta}$ is a uniform random rotation in $\mathrm{SO}(d)$, then the following holds for $\lambda \geq 0$:*

$$\delta(C) + \delta(D) \leq d - \lambda \;\Rightarrow\; \mathbb{P}(C \cap \boldsymbol{\theta}D \neq \{o\}) \leq p_{C,D}(\lambda), \qquad (4.100)$$

$$\delta(C) + \delta(D) \geq d + \lambda \;\Rightarrow\; \mathbb{P}(C \cap \boldsymbol{\theta}D \neq \{o\}) \geq 1 - p_{C,D}(\lambda). \qquad (4.101)$$

Proof. We can assume that not both of the cones are subspaces. From (4.76) and Lemma 4.3.3 it follows that

$$\mathbb{P}(C \cap \boldsymbol{\theta}D \neq \{o\}) = 2h_{d+1}(C \times D) \leq t_d(C \times D).$$

By the definition of the random variable $\mathbf{V}_{C \times D}$ and under the assumption that $d \geq \delta(C) + \delta(D) + \lambda = \delta(C \times D) + \lambda$, we obtain

$$\mathbb{P}(\mathbf{V}_{C \times D} - \delta(C \times D) \geq \lambda) = \sum_{k \geq \delta(C \times D) + \lambda} v_k(C \times D)$$

$$\geq \sum_{k \geq d} v_k(C \times D) = t_d(C \times D).$$

Hence, by (4.99),

$$\mathbb{P}(C \cap \boldsymbol{\theta}D \neq \{o\}) \leq p_{C,D}(\lambda),$$

which is (4.100). The estimate (4.101) is obtained similarly. $\qquad\square$

Finally in this chapter, we show that the statistical dimension of a convex cone is not much different from the square of another parameter that has been used in the average case analysis of certain algorithms.

Definition 4.4.3. *The* Gaussian width *of a cone $C \in \mathcal{C}^d$ is defined by*

$$w(C) := \mathbb{E}\left[\max_{y \in C \cap \mathbb{S}^{d-1}} \langle y, \mathbf{g} \rangle\right],$$

where $\mathbf{g}$ is a random vector with distribution γ_d.

The following is taken from [11, Appendix F].

Theorem 4.4.4. *The estimates*

$$w^2(C) \leq \delta(C) \leq w^2(C) + 1 \tag{4.102}$$

hold for $C \in \mathcal{C}^d$.

Proof. Using (1.6), we get

$$\max_{y \in C \cap \mathbb{S}^{d-1}} \langle y, \mathbf{g} \rangle = \max_{y \in C \cap \mathbb{S}^{d-1}} \langle y, \Pi_C(\mathbf{g}) + \Pi_{C^\circ}(\mathbf{g}) \rangle \leq \max_{y \in C \cap \mathbb{S}^{d-1}} \langle y, \Pi_C(\mathbf{g}) \rangle$$

$$\leq \max_{y \in C \cap \mathbb{S}^{d-1}} \|y\| \|\Pi_C(\mathbf{g})\| = \|\Pi_C(\mathbf{g})\|,$$

hence (4.83) shows that

$$\delta(C) \geq \mathbb{E}\left[\left(\max_{y \in C \cap \mathbb{S}^{d-1}} \langle y, \mathbf{g} \rangle\right)^2\right].$$

Now the Cauchy–Schwarz inequality yields the left-hand inequality of (4.102).
For the right-hand inequality we note that the mapping

$$f : \mathbb{R}^d \to \mathbb{R}, \qquad f(x) := \max_{y \in C \cap \mathbb{S}^{d-1}} \langle y, x \rangle,$$

has Lipschitz constant 1. Therefore,

$$\mathbb{E}\left[f(\mathbf{g})^2\right] - w^2(C) = \mathbb{E}\left[(f(\mathbf{g}) - \mathbb{E}\,f(\mathbf{g}))^2\right] = \operatorname{Var} f(\mathbf{g}) \le 1$$

by the Gaussian Poincaré inequality (e.g., [35, Sect. 3.7]).
We use (4.83) and (1.9), together with the fact that $\max_{y \in C \cap \mathbb{S}^{d-1}} \langle y, x \rangle = 0$ for $x \in C^\circ$, to get

$$\delta(C) = \mathbb{E}\left[\|\Pi_C(\mathbf{g})\|^2\right] = \mathbb{E}\left[\left(\max_{y \in C \cap B^d} \langle y, \mathbf{g} \rangle\right)^2\right]$$

$$= \mathbb{E}\left[\left(\max_{y \in C \cap \mathbb{S}^{d-1}} \langle y, \mathbf{g} \rangle\right)^2 \mathbb{1}_{\mathbb{R}^d \setminus C^\circ}(\mathbf{g})\right] \le \mathbb{E}\left[\left(\max_{y \in C \cap \mathbb{S}^{d-1}} \langle y, \mathbf{g} \rangle\right)^2\right]$$

$$= \mathbb{E}\left[f(\mathbf{g})^2\right].$$

This yields the remaining inequality.   □

Notes for Section 4.4

1. Equation (4.79) for the statistical dimension of a cone $C \in \mathcal{C}^d$ (see Definition 4.4.2) says that

$$\mathbb{E}\,\|\Pi_C(\mathbf{g})\|^2 = \sum_{k=0}^d k v_k(C).$$

This can be seen as an analogue of a relation holding for convex bodies $K \subset \mathbb{R}^d$, namely

$$\int_{\mathbb{R}^d} e^{-\pi \operatorname{dist}(x,K)^2} \lambda_d(\mathrm{d}x) = \sum_{k=0}^d V_k(K) =: W(K),$$

where W is known as the *Wills functional* (see Hadwiger [87]). It was proved by McMullen [132] that

$$W(K) \le e^{V_1(K)},$$

and a different proof was given by Alonso–Gutiérrez, Hernández Cifre and Yepes Nicolás [4].

Question. Is there a similar inequality for conic intrinsic volumes?

2. For the intrinsic volume random variables $\mathbf{V}_C$ defined by Definition 4.4.1, Goldstein, Nourdin and Peccati [71] have proved a central limit theorem and a Berry–Esseen estimate. We quote their Theorem 1.1.

Theorem. *Let $\{d_n : n \geq 1\}$ be a sequence of nonnegative integers and let $\{C_n \subset \mathbb{R}^{d_n} : n \geq 1\}$ be a collection of nonempty closed convex cones such that $\delta_{C_n} \to \infty$, and write $\tau_{C_n}^2 = \mathrm{Var}(\mathbf{V}_{C_n})$, $n \geq 1$. For every n, let $\mathbf{g}_n \sim \mathcal{N}(0, I_{d_n})$ and write $\sigma_{C_n}^2 = \mathrm{Var}(\|\Pi_{C_n}(\mathbf{g}_n)\|^2)$, $n \geq 1$. Then the following holds:*

1. *One has that $2\delta_{C_n} \leq \sigma_{C_n}^2 \leq 4\delta_{C_n}$ for every n and, as $n \to \infty$, the sequence*

$$\frac{\|\Pi_{C_n}(\mathbf{g}_n)\|^2 - \delta_{C_n}}{\sigma_{C_n}}, \quad n \geq 1,$$

converges in distribution to a standard Gaussian random variable $N \sim \mathcal{N}(0, 1)$.

2. *If, in addition, $\liminf_{n \to \infty} \tau_{C_n}^2/\delta_{C_n} > 0$, then, as $n \to \infty$,*

$$\frac{\mathbf{V}_{C_n} - \delta_{C_n}}{\tau_{C_n}}, \quad n \geq 1,$$

also converges in distribution to $N \sim \mathcal{N}(0, 1)$, and moreover one has the Berry–Esseen estimate

$$\sup_{u \in \mathbb{R}} \left| \mathbb{P}\left[\frac{\mathbf{V}_{C_n} - \delta_{C_n}}{\tau_{C_n}} \leq u \right] - \mathbb{P}[N \leq u] \right| = O\left(\frac{1}{\sqrt{\log \delta_{C_n}}} \right).$$

4.5 Inequalities and monotonicity properties

In this section, we collect the (rather incomplete) information that is available on inequalities satisfied by the conic intrinsic volumes. Also their monotonicity properties are of interest. The latter refer to set inclusion. We say that a real function f on $\mathcal{C}^d$ is *increasing* (*strictly increasing*) if $C \subset D$ implies $f(C) \leq f(D)$ (respectively, $C \subsetneq D$ implies $f(C) < f(D)$). Similarly, *decreasing* and *strictly decreasing* are defined.

The behavior of v_d is clear. Since $v_d(C) = \gamma_d(C)$, the functional v_d is strictly increasing on d-dimensional cones. It satisfies $v_d \geq 0$, with $v_d = 0$ if and only if $\dim C < d$. For $C \in \mathcal{C}^d \setminus \{\mathbb{R}^d\}$ we have $v_d \leq 1/2$, with $v_d(C) = 1/2$ if and only if C is a halfspace.

The functional v_{d-1} is equal to the Grassmann angle U_{d-2} and hence is increasing on $\mathcal{C}^d \setminus \mathcal{L}_\bullet$. It is not strictly increasing in any dimension $d \geq 2$, as we shall see below. If $C \in \mathcal{C}^d$ satisfies $\dim C \geq d - 1$ and $C \neq \mathbb{R}^d$, we claim that $v_{d-1} > 0$. Since $\dim C \geq d - 1$, we can choose a polyhedral cone $D \subset C$ with $\dim D \geq d - 1$ and hence $v_{d-1}(D) > 0$ (as follows immediately from the definition). Since v_{d-1} is increasing, we get $v_{d-1}(C) > 0$.

By duality, it follows that v_0 is strictly decreasing on pointed cones and that $0 \leq v_0(C) \leq 1/2$ for $C \in \mathcal{C}^d \setminus \{o\}$, with $v_0(C) = 0$ if and only if $\dim \operatorname{lineal}(C) > 0$, and $v_0(C) = 1/2$ if and only if $\dim C = 1$. It follows also by duality that v_1 is decreasing on $\mathcal{C}^d \setminus \mathcal{L}_\bullet$ and that $v_1(C) \geq 0$, with equality if and only if $C = \{o\}$ or $\dim \operatorname{lineal}(C) > 1$.

For $2 \leq k \leq d - 2$, the functional v_k is neither increasing nor decreasing. To see this, we quote from [135, pp. 183–184]. Let D_j denote a half j-space. We can arrange that $D_{k-1} \subset D_k \subset D_{k+1} \subset D_{k+2}$ and v_k successively takes the values $0, 1/2, 1/2, 0$. Then we approximate the halfspaces by pointed d-dimensional polyhedral cones obeying the corresponding inclusion relations. The result follows by continuity.

To get a more intuitive picture of the monotonicity behavior of v_k, we have a look at spherical cones. The *spherical cone* $C(u, r)$ with central vector $u \in \mathbb{S}^{d-1}$ and radius $r \in [0, \pi/2]$ is defined by

$$C(u, r) := \{x \in \mathbb{R}^d : d_a(x, u) \leq r\},$$

where the angular distance d_a is defined by (1.15). In other words, $C(u, r)$ is the positive hull of the spherical cap with center u and geodesic radius r.

Lemma 4.5.1. *For $k \in \{1, \ldots, d-1\}$,*

$$v_k(C(u, r)) = \binom{d-2}{k-1} \frac{\omega_{d-1}}{\omega_k \omega_{d-k}} \sin^{k-1} r \cos^{d-k-1} r.$$

Proof. Let $0 \leq r < \pi/2$, $0 \leq \lambda$ and $r + \lambda \leq \pi/2$. By Corollary 4.1.3 we have

$$\gamma_d(C(u, r + \lambda)) = \gamma_d(M_\lambda^a(C(u, r), \mathbb{R}^d \times \mathbb{R}^d)$$

$$= v_d(C(u, r)) + \sum_{k=1}^{d-1} g_k(\lambda) \cdot v_k(C(u, r))$$

with

$$g_k(\lambda) = \frac{\omega_k \omega_{d-k}}{\omega_d} \int_0^\lambda \cos^{k-1} \varphi \sin^{d-k-1} \varphi \, d\varphi.$$

On the other hand, (2.2) and Lemma 3.1.4 (with $S := \{u, -u\}$) yield

$$\gamma_d(C(u, r + \lambda)) = \frac{1}{\omega_d} \sigma_{d-1}(C(u, r + \lambda) \cap \mathbb{S}^{d-1})$$

$$= \frac{1}{2\omega_d} \omega_{d-1} \omega_1 \int_0^{r+\lambda} \sin^{d-2} \varphi \, d\varphi$$

$$= v_d(C(u, r)) + \frac{\omega_{d-1}}{\omega_d} \int_r^{r+\lambda} \sin^{d-2} \varphi \, d\varphi.$$

Here

$$\int_r^{r+\lambda} \sin^{d-2}\varphi\,d\varphi = \int_0^\lambda \sin^{d-2}(\alpha+r)\,d\alpha$$

$$= \int_0^\lambda \binom{d-2}{i}[\sin r \cos\alpha]^i[\cos r \sin\alpha]^{d-2-i}\,d\alpha$$

$$= \sum_{k=1}^{d-1}\binom{d-2}{k-1}\sin^{k-1}r\cos^{d-k-1}r\,\frac{\omega_d}{\omega_k\omega_{d-k}}g_k(\lambda).$$

Comparison yields the assertion.   □

For $d \geq 3$ we see (as we know already) that for increasing r the conic intrinsic volume $v_1(C(u,r))$ decreases from a positive value to zero, and that $v_{d-1}(C(u,r))$ increases from zero to a positive value. For $k \in \{2,\ldots,d-2\}$, $v_k(C(u,r))$ increases from zero to its unique maximum, which is attained at

$$r_{\max} = \arctan\sqrt{\frac{k-1}{d-k-1}},$$

and then decreases to zero. Here $r_{\max}$ increases with k.

Now we deal with the extreme values of the conic intrinsic volume v_k, for arbitrary $k \in \{0,\ldots,d\}$. Trivially, $v_k \geq 0$. From (2.65) and (2.66), together with the fact that $\chi(C) = 0$ for $C \in \mathcal{C}^d \setminus \mathcal{L}_\bullet$, it follows that

$$v_k(C) \leq \frac{1}{2} \quad \text{for } C \in \mathcal{C}^d \setminus \mathcal{L}_\bullet, \ k = 0,\ldots,d.$$

We recall from (2.31) that for $C \in \mathcal{L}_\bullet$ we have

$$v_k(C) = \begin{cases} 1, & \text{if } k = \dim C, \\ 0, & \text{otherwise.} \end{cases}$$

In order to see which cones $C \in \mathcal{C}^d \setminus \mathcal{L}_\bullet$ satisfy one of the inequalities $0 \leq v_k(C) \leq 1/2$ with equality, we recall that we can write C as a direct orthogonal sum

$$C = \text{lineal}(C) \oplus C' \tag{4.103}$$

where $\text{lineal}(C) = C \cap (-C)$ is the lineality space of C and C' is a pointed cone. If $\dim C' = 1$, then C is a *halfspace* (of dimension $\dim C$). If $\dim C' = 2$, we call C a *wedge* (of dimension $\dim C$). If $\dim \text{lineal}(C) = m$ and $\dim C \geq k$, we have by (2.67) that

$$v_k(C) = v_k(\text{lineal}(C) \oplus C') = \sum_{i+j=k}\delta_{im}v_j(C') = v_{k-m}(C'). \tag{4.104}$$

First let $C \in \mathcal{PC}^d \setminus \mathcal{L}_\bullet$ be a polyhedral cone, decomposed in the form (4.103) with $\dim \text{lineal}(C) = m$. Each k-face of C is of the form $\text{lineal}(C) \oplus F'$, where F' is a face of C' of dimension $k - m$. Recalling that

$$v_k(C) = \sum_{F \in \mathcal{F}_k(C)} \beta(o, F) \gamma(F, C),$$

we see that $v_k(C) = 0$ if and only if C has no k-faces, hence if and only if $\dim C < k$ or $\dim \operatorname{lineal}(C) > k$ The extension to general cones is true, but not entirely trivial.

Theorem 4.5.1. *Let $C \in \mathcal{C}^d \setminus \mathcal{L}_\bullet$. For $k = 0, \ldots, d$, we have $0 \leq v_k(C) \leq 1/2$, where*

$$v_k(C) = 0 \Leftrightarrow \dim C < k \text{ or } \dim \operatorname{lineal}(C) > k, \tag{4.105}$$

and

$$v_k(C) = \tfrac{1}{2} \Leftrightarrow C \text{ is a halfspace of dimension } k \text{ or } k+1 \\ \text{or a } (k+1)\text{-dimensional wedge.} \tag{4.106}$$

Proof. First let $C \in \mathcal{C}^d$ be a pointed cone and let $k \in \{0, \ldots, d\}$. If $\dim C \geq k$, we state that

$$v_k C) > 0.$$

The cases $k = d$ and $k = d - 1$ already being settled, we may assume that $k \leq d - 2$. The assertion is true for $d = 2$. We use induction and assume that $d \geq 3$ and the assertion is true in dimension $d - 1$. By (2.73) (which extends to $C \in \mathcal{C}^d$ by approximation), we have

$$v_k(C) = \int_{G(d, d-1)} v_k(C|L) \, \nu_{d-1}(\mathrm{d}L).$$

If the cone $C|L$ is pointed, then $v_k(C|L) > 0$ by the induction hypothesis. Since C is pointed, there exists a hyperplane H such that $C \cap H = \{o\}$. (This holds for the hyperplanes in a full neighborhood of H in $G(d, d-1)$.) For any unit vector $e \in H$, the projection $C|e^\perp$ is a pointed cone. It follows that there is a set of positive ν_{d-1}-measure in $G(d, d-1)$ such that for L in this set we have $v_k(C|L) > 0$. From this, it follows that $v_k(C) > 0$.

The implication $\Leftarrow$ in (4.105) holds for polyhedral cones, hence by approximation for general closed convex cones. Now we show the implication $\Rightarrow$ of (4.105). Let $C \in \mathcal{C}^d$ be such that $\dim C \geq k$ and $m := \dim \operatorname{lineal}(C) \leq k$. By (4.103) and (4.104) we have $v_k(C) = v_{k-m}(C') > 0$, since $k - m \geq 0$ and C' is a pointed cone of dimension $\dim C - m \geq k - m$. This completes the proof of (4.105).

It remains to prove (4.106). Suppose that $v_k(C) = 1/2$. We may assume that $\dim C = d$, since we may work in $\operatorname{lin} C$. We may also suppose that $d \geq 3$, since the cases $d = 1, 2$, where all convex cones are polyhedral, are easily settled. If $k = d$, then $v_d(C) = \gamma_d(C) = 1/2$ implies that C is a d-dimensional halfspace. Suppose that $k = d - 1$. From $v_{d-1}(C) = 1/2$ and (2.65), (2.66) it follows that $v_{d-3}(C) = 0$ and hence, since $\dim C = d$, necessarily $\dim \operatorname{lineal}(C) \geq d - 2$. Thus, C is either a halfspace or a wedge. If

$k \leq d - 2$, then it follows from $v_k(C) = 1/2$ and (2.65), (2.66) that either $v_d(C) = 0$ or $v_{d-1}(C) = 0$. Since $\dim C = d$, either case is impossible.

The reverse direction of (4.106) follows immediately from the definition, since halfspaces and wedges are polyhedral.   □

Remark. For any $a \in (0, 1/2]$, there is a d-dimensional wedge C with $v_d(C) = a$. Thus, we have the following inequality of isoperimetric type: Among all closed convex cones C with given $v_d(C) \in (0, 1/2]$, precisely the wedges maximize the functional v_{d-1}. By considering intersections with the unit sphere, we obtain a reverse isoperimetric inequality in spherical space.

We also see now that v_{d-1} is not strictly increasing, even on d-dimensional cones: there are wedges C, D with $C \subsetneq D$ and $v_{d-1}(C) = 1/2 = v_{d-1}(D)$.

On the other hand, on pointed cones the functional v_{d-1} is strictly increasing. More precisely, let $C, D \in \mathcal{C}^d$ be cones with $C \subsetneq D$, where C is pointed. Then there is a hyperplane H such that $H \cap C = \{o\}$ and $H \cap \operatorname{int} D \neq \emptyset$. It follows that there is also a two-dimensional plane $L_0 \in G(d, 2)$ such that all $L \in G(d, 2)$ in a neighborhood of L_0 satisfy $L \cap C = \{o\}$ and $L \cap \operatorname{int} D \neq \emptyset$. It follows from (2.70) and (2.43) that $v_{d-1}(C) < v_{d-1}(D)$.

Now we turn to inequalities between different conic intrinsic volumes. First we recall that $v_d(C) = v_{d-1}^s(C \cap \mathbb{S}^{d-1})$ and $v_{d-1}(C) = v_{d-2}^s(C \cap \mathbb{S}^{d-1})$ for $C \in \mathcal{C}^d$. For $K \in \mathcal{K}_s$, the value $v_{d-1}^s(K)$ is the spherical volume of K, and v_{d-2}^s is half the spherical surface area of K. This is clear if $K \in \mathcal{P}_s$, and the general case follows by approximation. Now the isoperimetric inequality of spherical geometry (as proved, for example, by E. Schmidt [157]; see also [41]) shows that for a cone $C \in \mathcal{C}^d$ with $0 < v_d(C) \leq 1/2$ the conic intrinsic volume $v_{d-1}(C)$ is minimal if and only if the cone C is spherical, that is, $C \cap \mathbb{S}^{d-1}$ is a spherical cap.

Motivated by this observation is the following question, which was asked (in an equivalent form) in [61].

Question. Which of the functionals v_j, U_j, W_j on $\mathcal{C}^d$ have the property that they attain an extremum on the set of cones $C \in \mathcal{C}^d$ with given $v_d(C) > 0$ precisely at spherical cones?

We reformulate the answer in a special case as given by Theorem 3.4.2.

Theorem 4.5.2. *Let $C \in \mathcal{C}^d$ be a pointed cone, and let S be a spherical cone with $v_d(C) = v_d(S)$. Then*

$$U_1(C) \geq U_1(S), \tag{4.107}$$

or equivalently

$$v_d(C^\circ) \leq v_d(S^\circ). \tag{4.108}$$

Equality holds only if C is a spherical cone.

In fact, (4.108) (with the equality condition) is only a reformulation of Theorem 3.4.2, and (4.107) then follows from $U_1(C) = \frac{1}{2} - v_d(C^\circ)$ (for this, see (2.60)).

Since $U_1(C)$ is, up to a constant factor, the total invariant measure of the hyperplanes in $G(d, d-1)$ having a non-trivial intersection with C, the function U_1 can be considered as a conical counterpart to the Euclidean mean width of a convex body. Thus, inequality (4.107) may be viewed as a counterpart to the Euclidean Urysohn inequality.

Notes for Section 4.5

1. Theorem 4.5.2 was proved by Gao, Hug and Schneider [61], in two different ways: via Theorem 3.4.2 and its proof as given here, and by applying two-point symmetrization in spherical space. As noted in Section 3.4, the inequality follows also from Theorem 3.4.1, but it is not clear whether the equality case can be settled in this way.

2. Amelunxen [5] conjectured (in Conjecture 4.4.16) that the sequence of conic intrinsic volumes is log-concave.

4.6 Observations about the conic support measures

In this section, we collect some information about the conic support measures $\Omega_k(C, \cdot)$. First we note that $\Omega_k(C, \cdot)$, although defined on the whole biconic σ-algebra $\widehat{\mathcal{B}}(\mathbb{R}^d \times \mathbb{R}^d)$, is concentrated on the closed set

$$\operatorname{Nor} C := \{(x, y) \in C \times C^\circ : \langle x, y \rangle = 0\},$$

the normal bundle or set of support elements of C. This is clear for $k = 0, d$, since for general $C \in \mathcal{C}^d$ these measures can be defined by

$$\Omega_0(C, \eta) = \int_{C^\circ} \mathbb{1}_\eta(o, y) \, \gamma_d(\mathrm{d}y),$$

$$\Omega_d(C, \eta) = \int_C \mathbb{1}_\eta(x, o) \, \gamma_d(\mathrm{d}x).$$

Thus, $\Omega_0(C, \cdot)$ is concentrated on $\{(o, y) : y \in C^\circ\}$, and $\Omega_d(C, \cdot)$ is concentrated on $\{(x, o) : x \in C\}$. (Note that we consider here also the latter set as a subset of $\operatorname{Nor} C$.) For $k \in \{1, \dots, d-1\}$ and $C \in \mathcal{C}_*^d$, the concentration of $\Omega_k(C, \cdot)$ on $\operatorname{Nor} C$ follows from (4.29) and the definition (4.25) of the angular local parallel set, together with (1.7).

Next, we prove a characterization of the conic support measures, which is due to Glasauer [63, Thm. 4.2.1]. We translate his result into the conical setting and modify it slightly (since, in our case, also $\{(x, o) : x \in C\} \subset \operatorname{Nor} C$). We point out that the valuation property is not needed in this characterization.

Theorem 4.6.1. *Let* $\psi : \mathcal{C}^d \times \widehat{B}(\mathbb{R}^d \times \mathbb{R}^d)$ *be a mapping with the following properties:*

(a) *For all* $C \in \mathcal{C}^d$, $\psi(C, \cdot)$ *is a finite signed measure, concentrated on* $\operatorname{Nor} C$,

(b) *For all* $(C, \eta) \in \mathcal{C}^d \times \widehat{B}(\mathbb{R}^d \times \mathbb{R}^d)$,

$$\psi(\vartheta C, \vartheta \eta) = \psi(C, \eta) \quad \text{for all } \vartheta \in \operatorname{SO}(d),$$

(c) *If* $C, D \in \mathcal{PC}^d$ *and* $\eta \in \widehat{B}(\mathbb{R}^d \times \mathbb{R}^d)$ *is such that* $\eta \cap \operatorname{Nor} C = \eta \cap \operatorname{Nor} D$, *then*

$$\psi(C, \eta) = \psi(D, \eta).$$

(d) *If* $C_i, C \in \mathcal{C}^d$ *and* $C_i \to C$ *in the metric* δ_c, *then* $\psi(C_i, \cdot) \overset{w}{\to} \psi(C, \cdot)$.

Then there are constants $c_0, \ldots, c_d$ *such that*

$$\psi(C, \cdot) = \sum_{k=0}^{d} c_k \Omega_k(C, \cdot) \quad \text{for all } C \in \mathcal{C}^d.$$

Proof. First we consider a k-dimensional subspace $L \in G(d, k)$, $k \in \{0, \ldots, d\}$. Let $B \in \widehat{B}(L^\perp)$ be given. Then $\psi(L, \cdot \times B)$ is a finite signed measure on $\widehat{B}(L)$, which is invariant under SO_L, by condition (b). By Corollary 2.1.2, there is a constant $c(L, B)$ such that

$$\psi(L, A \times B) = c(L, B)\gamma_L(A) \quad \text{for } A \in \widehat{B}(L).$$

Now we fix $A \in \widehat{B}(L)$ and let $B \in \widehat{B}(L^\perp)$ vary. Then we find that $c(L, \cdot)$ is a finite signed measure on $\widehat{B}(L^\perp)$, which is invariant under $\operatorname{SO}_{L^\perp}$. This yields $c(L, B) = c(L)\gamma_{L^\perp}(B)$ for $B \in \widehat{B}(L^\perp)$. Since $c(\vartheta L) = c(L)$ for $\vartheta \in \operatorname{SO}(L)$, the constant $c(L)$ depends only on $\dim L$ and hence can be denoted by c_k. Since, by condition (a), $\psi(L, \cdot)$ is concentrated on $L \times L^\perp$, we have found that

$$\psi(L, A \times B) = c_k \gamma_L(A \cap L)\gamma_{L^\perp}(B \cap L^\perp)$$

for arbitrary $A, B \in \widehat{B}(\mathbb{R}^d)$.

Now let $C \in \mathcal{PC}^d$ and $A, B \in \widehat{B}(\mathbb{R}^d)$. We have

$$(A \times B) \cap \operatorname{Nor} C = \bigcup_{k=0}^{d} \bigcup_{F \in \mathcal{F}_k(C)} (A \cap \operatorname{relint} F) \times (B \cap N(C, F)),$$

where the right-hand side is a disjoint union. Since

$$(A \cap \operatorname{relint} F) \times (B \cap N(P, F)) \subset \operatorname{Nor} \langle F \rangle \cap \operatorname{Nor} C,$$

condition (c) yields

$$\psi(C, (A \cap \operatorname{relint} F) \times (B \cap N(P, F))$$
$$= \psi(\langle F \rangle, (A \cap \operatorname{relint} F) \times (B \cap N(P, F))$$
$$= c_k \gamma_{\langle F \rangle}(A \cap F)\gamma_{\langle F \rangle^\perp}(B \cap N(P, F)).$$

Since $\psi(C, \cdot)$ is concentrated on $\operatorname{Nor} C$, we obtain

$$\psi(C, A \times B) = \sum_{k=0}^{d} c_k \sum_{F \in \mathcal{F}_k(C)} \gamma_{\langle F \rangle}(A \cap F) \gamma_{\langle F \rangle^{\perp}}(B \cap N(P, F))$$

$$= \sum_{k=0}^{d} c_k \Omega_k(C, A \times B).$$

We have shown that the finite measures

$$\psi(C, \cdot) \quad \text{and} \quad \sum_{k=0}^{d} c_k \Omega_k(C, \cdot)$$

agree on all sets $A \times B$ with $A, B \in \widehat{\mathcal{B}}(\mathbb{R}^d)$. These sets form an intersection stable generator of the σ-algebra $\widehat{\mathcal{B}}(\mathbb{R}^d \times \mathbb{R}^d)$. Hence, the two measures coincide on this σ-algebra (see, e.g. [55, Thm. 5.6]).

Finally, we employ condition (d) to extend the result from polyhedral cones to all of $\mathcal{C}^d$. □

In Section 4.2 we have shown that the conic support measures of convex cones are weakly continuous. We now improve this result (following [168]), by establishing Hölder continuity with respect to a metric which metrizes the weak convergence.

We use the standard Euclidean norm $\| \cdot \|$ on $\mathbb{R}^d \times \mathbb{R}^d$, which is thus also defined on $\mathbb{S}^{d-1} \times \mathbb{S}^{d-1}$. For a bounded real function f on $\mathbb{S}^{d-1} \times \mathbb{S}^{d-1}$, let

$$\|f\|_L := \sup_{a \neq b} \frac{|f(a) - f(b)|}{\|a - b\|}, \qquad \|f\|_\infty := \sup_a |f(a)|.$$

We set $\mathcal{F}_{bL} := \{f : \mathbb{S}^{d-1} \times \mathbb{S}^{d-1} \to \mathbb{R} : \|f\|_L \leq 1, \|f\|_\infty \leq 1\}$. The functions in $\mathcal{F}_{bL}$ are integrable with respect to every finite Borel measure on $\mathbb{S}^{d-1} \times \mathbb{S}^{d-1}$. The *bounded Lipschitz distance* of finite Borel measures μ, ν on $\mathbb{S}^{d-1} \times \mathbb{S}^{d-1}$ is defined by

$$d_{bL}(\mu, \nu) := \sup \left\{ \left| \int_{\mathbb{S}^{d-1} \times \mathbb{S}^{d-1}} f \, d\mu - \int_{\mathbb{S}^{d-1} \times \mathbb{S}^{d-1}} f \, d\nu \right| : f \in \mathcal{F}_{bL} \right\}.$$

This gives a metric d_{bL} on the finite Borel measures on $\mathbb{S}^{d-1} \times \mathbb{S}^{d-1}$. Convergence with respect to this metric is equivalent to weak convergence (see, e.g., [54, Sect. 11.3]).

For finite measures μ, ν on the biconic σ-algebra $\widehat{\mathcal{B}}(\mathbb{R}^d \times \mathbb{R}^d)$, we define

$$d_{bL}(\mu, \nu) := \sup \left\{ \left| \int_{\mathbb{R}^d \times \mathbb{R}^d} f_h \, d\mu - \int_{\mathbb{R}^d \times \mathbb{R}^d} f_h \, d\nu \right| : f \in \mathcal{F}_{bL} \right\}.$$

This yields a metric d_{bL} which metrizes the weak convergence of finite measures on the biconic σ-algebra.

The assumption $\delta_a(C, D) \le 1$ in the following theorem is made for convenience. It could be replaced by $\delta_a(C, D) \le \pi/2 - \varepsilon$, with any $\varepsilon > 0$; then the constant c would also depend on ε.

Theorem 4.6.2. *Let $C, D \in \mathcal{C}_*^d$ be convex cones, and suppose that $\delta_a(C, D) \le 1$. Then*
$$d_{bL}(\Omega_i(C, \cdot), \Omega_i(D, \cdot)) \le c\, \delta_a(C, D)^{1/2}$$
for $i \in \{1, \ldots, d-1\}$, where c is a constant depending only on the dimension d.

Proof. Let $C \in \mathcal{C}_*^d$, $\eta \in \widehat{\mathcal{B}}(\mathbb{R}^d \times \mathbb{R}^d)$, and $0 \le \lambda < \pi/2$. We define
$$M^\lambda(C, \eta) := \{x \in \mathbb{R}^d \setminus \{o\} : 0 < d_a(x, C) \le \lambda, \, (\Pi_C(x), \Pi_{C^\circ}(x)) \in \eta\},$$
$$C^\lambda := C_\lambda^a \setminus C = \{x \in \mathbb{R}^d \setminus \{o\} : 0 < d_a(x, C) \le \lambda\},$$
$$\mu^\lambda(C, \eta) := \gamma_d(M^\lambda(C, \eta)).$$
Further, we define $F_\lambda : C^\lambda \to \mathbb{R}^d \times \mathbb{R}^d$ by $F_\lambda(x) := (\Pi_C(x), \Pi_{C^\circ}(x))$ for $x \in C^\lambda$. Then F_λ is continuous, and for $\eta \in \widehat{\mathcal{B}}(\mathbb{R}^d \times \mathbb{R}^d)$ we have $\gamma_d(F_\lambda^{-1}(\eta)) = \mu^\lambda(C, \eta)$. For any $\mu^\lambda(C, \cdot)$-integrable, homogeneous function f_h, we therefore get
$$\int_{\mathbb{R}^d \times \mathbb{R}^d} f_h \, d\mu^\lambda(C, \cdot) = \int_{C^\lambda} f_h \circ (\Pi_C, \Pi_{C^\circ}) \, d\gamma_d.$$

Let $D \in \mathcal{C}_*^d$ be such that $\delta_a(C, D) \le 1$. Let $f : \mathbb{S}^{d-1} \times \mathbb{S}^{d-1} \to \mathbb{R}$ be a function with $\|f\|_L \le 1$ and $\|f\|_\infty \le 1$. Then

$$\left| \int_{\mathbb{R}^d \times \mathbb{R}^d} f_h \, d\mu^\lambda(C, \cdot) - \int_{\mathbb{R}^d \times \mathbb{R}^d} f_h \, d\mu^\lambda(D, \cdot) \right|$$
$$= \left| \int_{C^\lambda} f_h \circ (\Pi_C, \Pi_{C^\circ}) \, d\gamma_d - \int_{D^\lambda} f_h \circ (\Pi_D, \Pi_{D^\circ}) \, d\gamma_d \right|$$
$$\le \int_{C^\lambda \cap D^\lambda} |f_h \circ (\Pi_C, \Pi_{C^\circ}) - f_h \circ (\Pi_D, \Pi_{D^\circ})| \, d\gamma_d$$
$$+ \int_{C^\lambda \setminus D^\lambda} |f_h \circ (\Pi_C, \Pi_{C^\circ})| \, d\gamma_d + \int_{D^\lambda \setminus C^\lambda} |f_h \circ (\Pi_D, \Pi_{D^\circ})| \, d\gamma_d$$
$$\le \int_{C^\lambda \cap D^\lambda} |f_h \circ (\Pi_C, \Pi_{C^\circ}) - f_h \circ (\Pi_D, \Pi_{D^\circ})| \, d\gamma_d$$
$$+ \gamma_d(C^\lambda \setminus D^\lambda) + \gamma_d(D^\lambda \setminus C^\lambda).$$

Let $x \in C^\lambda \cup D^\lambda$, $x \ne o$, and write
$$\frac{\Pi_C(x)}{\|\Pi_C(x)\|} = u, \quad \frac{\Pi_{C^\circ}(x)}{\|\Pi_{C^\circ}(x)\|} = u^\circ, \quad \frac{\Pi_D(x)}{\|\Pi_D(x)\|} = v, \quad \frac{\Pi_{D^\circ}(x)}{\|\Pi_{D^\circ}(x)\|} = v^\circ.$$

The homogeneity of f_h and the Lipschitz property of f lead to

$$|f_h \circ (\Pi_C, \Pi_{C^\circ}) - f_h \circ (\Pi_D, \Pi_{D^\circ})|\,(x)$$
$$= |f_h(\Pi_C(x), \Pi_{C^\circ}(x)) - f_h(\Pi_D(x), \Pi_{D^\circ}(x))|$$
$$= |f(u, u^\circ) - f(v, v^\circ)|$$
$$\leq \|(u, u^\circ) - (v, v^\circ)\|$$
$$\leq \|u - v\| + \|u^\circ - v^\circ\|$$
$$= 2\sin\frac{1}{2}d_a(u, v) + 2\sin\frac{1}{2}d_a(u^\circ, v^\circ)$$
$$= 2\sin\frac{1}{2}d_a(\Pi_C(x), \Pi_D(x)) + 2\sin\frac{1}{2}d_a(\Pi_{C^\circ}(x), \Pi_{D^\circ}(x)),$$

where (1.16) was used. By Lemma 1.3.16,

$$2\sin^2\frac{1}{2}d_a(\Pi_C(x), \Pi_D(x)) = 1 - \cos d_a(\Pi_C(x), \Pi_D(x)) \leq c_\varepsilon \delta_a(C, D),$$

with $\varepsilon = (\pi/2) - 1$. By Theorem 1.3.4, we have $\delta_a(C^\circ, D^\circ) = \delta_a(C, D)$, hence also

$$2\sin^2\frac{1}{2}d_a(\Pi_{C^\circ}(x), \Pi_{D^\circ}(x)) \leq c_\varepsilon \delta_a(C, D).$$

This yields

$$\int_{C^\lambda \cap D^\lambda} |f_h \circ (\Pi_C, \Pi_{C^\circ}) - f_h \circ (\Pi_D, \Pi_{D^\circ})|\,\mathrm{d}\gamma_d \leq b\delta_a(C, D)^{1/2}$$

with $b = 2\sqrt{2c_\varepsilon}$.

We abbreviate $\delta_a(C, D) =: \delta$. To estimate $\gamma_d(C^\lambda \setminus D^\lambda)$, let $x \in C^\lambda \setminus D^\lambda$. Then $x \in C_\lambda^a \setminus C$ and $x \notin D_\lambda^a \setminus D$. If $x \in D$, then $d_a(x, C) \leq \delta$, hence $x \in C_\delta^a \setminus C$. If $x \notin D$, then $x \notin D_\lambda^a$, but $x \in C_\lambda^a \subseteq (D_\delta^a)_\lambda \subseteq D_{\lambda+\delta}^a$, thus $x \in D_{\lambda+\delta}^a \setminus D_\lambda^a$. It follows that

$$C^\lambda \setminus D^\lambda \subseteq (C_\delta^a \setminus C) \cup (D_{\lambda+\delta}^a \setminus D_\lambda^a).$$

Therefore,

$$\gamma_d(C^\lambda \setminus D^\lambda) \leq \gamma_d(C_\delta^a) - \gamma_d(C) + \gamma_d(D_{\lambda+\delta}^a) - \gamma_d(D_\lambda^a).$$

By (4.29),

$$\gamma_d(C_\delta^a) - \gamma_d(C) = \sum_{k=1}^{d-1} g_k(\delta)v_k(C) \leq \sum_{k=1}^{d-1}\frac{\omega_k\omega_{d-k}}{\omega_d} \cdot \delta =: s(d)\delta,$$

where (4.22) and $v_k \leq 1$ were observed. Similarly, $\gamma_d(D_{\lambda+\delta}^a) - \gamma_d(D_\lambda^a) \leq s(d)\delta$. Here C and D can be interchanged, and we obtain

$$\gamma_d(C^\lambda \setminus D^\lambda) + \gamma_d(D^\lambda \setminus C^\lambda) \leq 4s(d)\delta.$$

We have obtained

$$\left| \int_{\mathbb{R}^d \times \mathbb{R}^d} f_h \, d\mu^\lambda(C, \cdot) - \int_{\mathbb{R}^d \times \mathbb{R}^d} f_h \, d\mu^\lambda(D, \cdot) \right|$$

$$\leq b\delta_a(C, D)^{1/2} + 4s(d)\delta_a(C, D).$$

This holds for all functions $f : \mathbb{S}^{d-1} \times \mathbb{S}^{d-1} \to \mathbb{R}$ with $\|f\|_L \leq 1$ and $\|f\|_\infty \leq 1$. Since $\delta_a(C, D) \leq 1$, we obtain

$$d_{bL}(\mu^\lambda(C, \cdot), \mu^\lambda(D, \cdot)) \leq b_1 \delta_a(C, d)^{1/2}$$

with a constant b_1 depending only on d.

By (4.29) and since $M_\lambda^a(C, \eta) \cap C = \emptyset$, we have

$$\mu^\lambda(C, \eta) = \sum_{k=1}^{d-1} g_k(\lambda) \cdot \Omega_k(C, \eta).$$

Since the coefficient functions $g_1, \ldots, g_{d-1}$ are linearly independent on $[0, 1]$, we can choose numbers $0 < \lambda_1 < \cdots < \lambda_{d-1} < 1$ and numbers a_{ij}, depending only on i and j, such that

$$\Omega_i(C, \cdot) = \sum_{i=1}^{d-1} a_{ij} \mu^{\lambda_j}(C, \cdot) \quad \text{for } i = 1, \ldots, d-1$$

(see the proof of Theorem 4.2.1). Using the definition of the bounded Lipschitz metric, we obtain

$$d_{bL}(\Omega_i(C, \cdot), \Omega_i(D, \cdot)) \leq \sum_{i=1}^{d-1} |a_{ij}| d_{bL}\left(\mu^{\lambda_j}(C, \cdot), \mu^{\lambda_j}(D, \cdot)\right)$$

$$\leq c \, \delta_a(C, D)^{1/2},$$

which completes the proof. $\qquad\square$

Notes for Section 4.6

1. The Hölder continuity of conic support measures treated in this section is analogous to the case of support measures of convex bodies, which was considered in [97].

2. The support measures and their marginal measures are indispensable in integral geometry if one studies distance integrals and contact probabilities. For treatments of the Euclidean case and hints to the literature, we refer to [163, Thms. 4.4.7, 4.4.11, Note 12 for Sect. 4.4], [170, Sect. 8.5], [171, Sect. 4]. Similar investigations in spherical space were briefly undertaken by Glasauer [63, Chap. 7]. This can be carried over to the conical case.

Central hyperplane arrangements and induced cones

A *central hyperplane arrangement* $\mathcal{A}$ is a non-empty finite set of hyperplanes through o (also called 'linear hyperplanes') in $\mathbb{R}^d$. (For an introduction to hyperplane arrangements, we refer, e.g., to Stanley [183].) Such a family $\mathcal{A}$ induces a tessellation of $\mathbb{R}^d$ into polyhedral cones: the closures of the components of $\mathbb{R}^d \setminus \bigcup_{H \in \mathcal{A}} H$ are the d-dimensional cones of this tessellation. This chapter deals with the polyhedral cones induced by a central hyperplane arrangement. Of course, every polyhedral cone can be generated as a cone in such a tessellation. The special viewpoint of this chapter is to consider, in one way or the other, the totality of the polyhedral cones generated by a central hyperplane arrangement. We shall collect some results pertaining to these cones.

For a central hyperplane arrangement $\mathcal{A}$ in $\mathbb{R}^d$, we denote by $\mathcal{F}_d(\mathcal{A})$ the set of d-dimensional cones it generates, and we write

$$\mathcal{F}_k(\mathcal{A}) := \bigcup_{C \in \mathcal{F}_d(\mathcal{A})} \mathcal{F}_k(C)$$

for the set of all k-faces of the cones of $\mathcal{F}_d(\mathcal{A})$. The elements of $\mathcal{F}_k(\mathcal{A})$ are also called the k-*faces* of the tessellation induced (or generated) by $\mathcal{A}$. We write $f_k(\mathcal{A})$ for the number of these k-faces. The arrangement $\mathcal{A}$ is *in general position* if any $k \leq d$ hyperplanes of $\mathcal{A}$ have an intersection of dimension $d-k$.

We mention one historical result already here in the introduction. If the central hyperplane arrangement $\mathcal{A}$ contains n hyperplanes and is in general position, then it was proved by Steiner [185] for $d \leq 3$ and by Schläfli [156] in general that

$$f_d(\mathcal{A}) = C(n,d) := 2 \sum_{j=0}^{d-1} \binom{n-1}{j}. \tag{5.1}$$

The simple proof can be found, e.g., in [170, Lemma 8.2.1].

Theorems 5.1.1 and 5.1.2 below can be considered as extensions of formula (5.1), to central hyperplane arrangements which are not necessarily in general position, and to numbers of lower-dimensional faces. The main aim of Section 5.1 will then be to obtain a generalization of a formula of Klivans and Swartz, yielding the sum of the kth conic intrinsic volumes of all faces $F \in \mathcal{F}_j(\mathcal{A})$. The gist of the matter is that this sum depends only on the combinatorics of $\mathcal{A}$, although the conic intrinsic volumes are metric quantities.

Section 5.2 reproduces a beautiful probabilistic application of central arrangements, due to Kabluchko, Vysotsky and Zaporozhets [113]. These authors determine the probability that the convex hull of a certain n-step random walk in $\mathbb{R}^d$ contains the origin, expressing the result in terms of conic intrinsic volumes of certain Weyl chambers, which can be determined explicitly.

The subsequent sections of this chapter deal with random cones generated by random central hyperplane arrangements. This topic was initiated a long time ago by Cover and Efron [50]. Their work is expanded considerably in Sections 5.3–5.5.

Isotropic random cones for which expected conic intrinsic volumes can be calculated allow more general versions of intersection probabilities based on kinematic formulas. This is briefly set out in Section 5.6.

5.1 The Klivans–Swartz formula

Let $\mathcal{A}$ be a central hyperplane arrangement, briefly a central arrangement, in $\mathbb{R}^d$. In this section, we are interested in the conic intrinsic volumes of its induced cones, in particular in their sum over all cones. Trivially,

$$\sum_{C \in \mathcal{F}_d(\mathcal{A})} v_d(C) = 1, \qquad \sum_{C \in \mathcal{F}_d(\mathcal{A})} v_{d-1}(C) = n,$$

where n is the number of hyperplanes in $\mathcal{A}$. Less trivially, and perhaps surprisingly, each sum $\sum_{C \in \mathcal{F}_d(\mathcal{A})} v_k(C)$ depends only on the combinatorics of the central arrangement $\mathcal{A}$. The precise character of this dependence is the content of the Klivans–Swartz formula. It can be extended to a formula for $\sum_{C \in \mathcal{F}_j(\mathcal{A})} v_k(C)$, $j \in \{1, \dots, d\}$. We shall derive a combinatorial identity, from which this formula can be deduced by integration. This requires some well-established combinatorial notions, which we now recall.

Definition 5.1.1. *The* intersection poset $\mathcal{L}(\mathcal{A})$ *of the central arrangement $\mathcal{A}$ is the set of all intersections of collections of hyperplanes from $\mathcal{A}$ (including $\mathbb{R}^d$, interpreted as the intersection of the empty collection), together with the partial order $\leq$ defined by reverse inclusion.*

Thus, for $S, T \in \mathcal{L}(\mathcal{A})$ the order relation $S \leq T$ is defined by $S \supseteq T$. We write $S < T$ for $S \leq T$ and $S \neq T$. The poset $\mathcal{L}(\mathcal{A})$ is graded by the rank defined by $\mathrm{rk}(S) = d - \dim S$. Moreover, $\mathcal{L}(\mathcal{A})$ is a lattice, with minimal element $\mathbb{R}^d$ and maximal element $\hat{1}$, the intersection of the hyperplanes in $\mathcal{A}$. The central arrangement $\mathcal{A}$ is called *essential* if $\hat{1} = \{o\}$.

We repeat the definition of the Möbius function for $\mathcal{L}(\mathcal{A})$.

Definition 5.1.2. *The* Möbius function *of the poset* $\mathcal{L}(\mathcal{A})$ *is defined by*

$$\mu(S, S) = 1 \ \text{for all } S \in \mathcal{L}(\mathcal{A}),$$

$$\mu(S, T) = - \sum_{S \leq L < T} \mu(S, L) \ \text{for all } S, T \in \mathcal{L}(\mathcal{A}) \text{ with } S < T.$$

Further, one defines $\mu(S, T) = 0$ *for* $S, T \in \mathcal{L}(\mathcal{A})$ *with* $S \not\leq T$, *and* $\mu(S) = \mu(\mathbb{R}^d, S)$ *for* $S \in \mathcal{L}(\mathcal{A})$.

Definition 5.1.3. *For the central arrangement* $\mathcal{A}$, *the* characteristic polynomial *is defined by*

$$\chi_{\mathcal{A}}(t) = \sum_{S \in \mathcal{L}(\mathcal{A})} \mu(S) \, t^{\dim S}.$$

For example, if $\mathcal{A}$ consists of $n \geq d$ hyperplanes in general position, then its characteristic polynomial is given by

$$\chi_{\mathcal{A}}(t) = \sum_{j=0}^{d-1} (-1)^j \binom{n}{j} t^{d-j} + (-1)^d \binom{n-1}{d-1}. \tag{5.2}$$

Proof. For $S \in \mathcal{L}(\mathcal{A})$ with $\mathrm{rk}(S) < d$ we state that

$$\mu(S) = (-1)^{\mathrm{rk}(S)} \tag{5.3}$$

if $\mathcal{A}$ is in general position. We prove this by induction. If $\mathrm{rk}(S) = 0$, then $S = \mathbb{R}^d$ and hence $\mu(S) = 1$, by definition. Let $0 < s \leq d-1$ and suppose that (5.3) has been proved for all S with $\mathrm{rk}(S) < s$. Let $\mathrm{rk}(S) = s$. Then $S = H_{i_1} \cap \cdots \cap H_{i_s}$ with unique indices $1 \leq i_1 < \cdots < i_s \leq n$. The number of elements $T \in \mathcal{L}(\mathcal{A})$ of rank $j < s$ with $T < S$ is given by $\binom{s}{j}$. It follows that

$$\mu(S) = - \sum_{j=0}^{s-1} \binom{s}{j} (-1)^j = (-1)^s,$$

which finishes the induction.

Finally, the number of elements in $\mathcal{L}(\mathcal{A})$ of rank $j < d$ is given by $\binom{n}{j}$, hence

$$\mu(\{o\}) = - \sum_{j=0}^{d-1} \binom{n}{j} (-1)^j = (-1)^d \binom{n-1}{d-1}. \tag{5.4}$$

This yields the assertion (5.2). $\qquad\square$

We mention that the characteristic polynomial has also a representation not involving the Möbius function. By Whitney's theorem (see Stanley [183, Thm. 2.4]),

$$\chi_{\mathcal{A}}(t) = \sum_{\mathcal{B} \subseteq \mathcal{A}} (-1)^{|\mathcal{B}|} t^{d - \mathrm{rk}(\mathcal{B})}.$$

Here the sum extends over all subsets $\mathcal{B}$ of $\mathcal{A}$, $|\mathcal{B}|$ denotes the number of elements of $\mathcal{B}$, and $\mathrm{rk}(\mathcal{B})$ is the codimension of the intersection of the elements of $\mathcal{B}$.

We extend some of the previous definitions.

Definition 5.1.4. *For $j \in \{0, \ldots, d\}$, let*

$$\mathcal{L}_j(\mathcal{A}) = \{L \in \mathcal{L}(\mathcal{A}) : \dim L = j\}.$$

The number of elements of $\mathcal{L}_j(\mathcal{A})$ is denoted by

$$\ell_j(\mathcal{A}) = |\mathcal{L}_j(\mathcal{A})|.$$

For a linear subspace $E \subseteq \mathbb{R}^d$,

$$\mathcal{A}^E := \{H \cap E : H \in \mathcal{A}, \ E \not\subseteq H\}.$$

Thus, $\mathcal{A}^E$ is a central hyperplane arrangement relative to E.

The following is a well-known result due to Zaslavsky [192]. It can be extended (with a modification of (5.6)) to more general hyperplane arrangements, but we consider here only central ones. For the reader's convenience, we reproduce (essentially) the proof given in [183, Sect. 2.2].

Theorem 5.1.1. *If $\mathcal{A}$ is a central hyperplane arrangement in $\mathbb{R}^d$, then*

$$f_d(\mathcal{A}) = (-1)^d \chi_{\mathcal{A}}(-1). \tag{5.5}$$

Further,

$$\chi_{\mathcal{A}}(1) = 0. \tag{5.6}$$

Proof. Since $\mathbb{R}^d$ is the disjoint union of the relative interiors of the faces of the tessellation induced by $\mathcal{A}$, the Euler characteristic of $\mathbb{R}^d$ can be written as

$$(-1)^d = \chi(\mathbb{R}^d) = \sum_{k=0}^{d} \sum_{F \in \mathcal{F}_k(\mathcal{A})} \chi(\mathrm{relint}\,F) = \sum_{k=0}^{d} (-1)^k f_k(\mathcal{A}),$$

where (1.55) was used. Each k-face of $\mathcal{A}$ lies in precisely one subspace $E \in \mathcal{L}_k(\mathcal{A})$ and is a k-face of the arrangement $\mathcal{A}^E$, hence

$$f_k(\mathcal{A}) = \sum_{E \in \mathcal{L}_k(\mathcal{A})} f_k(\mathcal{A}^E).$$

Thus,

$$(-1)^d = \sum_{k=0}^{d} (-1)^k \sum_{E \in \mathcal{L}_k(\mathcal{A})} f_k(\mathcal{A}^E) = \sum_{E \in \mathcal{L}(\mathcal{A})} (-1)^{\dim E} f_{\dim E}(\mathcal{A}^E).$$

Here $\mathbb{R}^d$ can be replaced by any $S \in \mathcal{L}(\mathcal{A})$, which yields

$$(-1)^{\dim S} = \sum_{E \in \mathcal{L}(\mathcal{A}),\, S \leq E} (-1)^{\dim E} f_{\dim E}(\mathcal{A}^E).$$

The Möbius inversion formula (1.2) gives

$$(-1)^{\dim S} f_{\dim S}(\mathcal{A}^S) = \sum_{E \in \mathcal{L}(\mathcal{A}),\, S \leq E} (-1)^{\dim E} \mu(S, E)$$

for all $S \in \mathcal{L}(\mathcal{A})$. The choice $S = \mathbb{R}^d$ results in

$$(-1)^d f_d(\mathcal{A}) = \sum_{E \in \mathcal{L}(\mathcal{A})} (-1)^{\dim E} \mu(E) = \chi_{\mathcal{A}}(-1),$$

which is (5.5).

For the proof of (5.6), we set $\dim \widehat{1} =: q$ and denote by $\delta_{i,j}$ the Kronecker symbol. For $S \in \mathcal{L}(\mathcal{A})$ we have

$$\sum_{L \in \mathcal{L}(\mathcal{A}),\, S \leq L} \delta_{q, \dim L} = 1,$$

hence the Möbius inversion formula gives

$$\delta_{q, \dim S} = \sum_{L \in \mathcal{L}(\mathcal{A})} \mu(S, L).$$

The choice $S = \mathbb{R}^d$ yields (5.6). $\square$

Equation (5.5) can also be written as

$$f_d(\mathcal{A}) = \sum_{L \in \mathcal{L}(\mathcal{A})} (-1)^{\mathrm{rk}(L)} \mu(L).$$

For an extension of (5.5) to the number of j-faces, the characteristic polynomial must be generalized.

Definition 5.1.5. *For $j \in \{0, \ldots, d\}$, the jth-level characteristic polynomial of $\mathcal{A}$ is defined by*

$$\chi_{\mathcal{A},j}(t) = \sum_{S \in \mathcal{L}_j(\mathcal{A})} \sum_{T \in \mathcal{L}(\mathcal{A})} \mu(S, T) t^{\dim T} = \sum_{m=0}^{j} a_{jm} t^m. \tag{5.7}$$

For example,

$$\chi_{\mathcal{A},0}(t) = \ell_0(\mathcal{A}), \qquad \chi_{\mathcal{A},1}(t) = \ell_1(\mathcal{A})(t - \ell_0(\mathcal{A})), \tag{5.8}$$

as follows directly from the definitions. Clearly, $\chi_{\mathcal{A},d} = \chi_{\mathcal{A}}$.

The following theorem shows that in order to determine the number of j-faces of the tessellation generated by $\mathcal{A}$, one need not know its full combinatorial structure, but only the intersection poset of $\mathcal{A}$.

Theorem 5.1.2. *If $\mathcal{A}$ is a central hyperplane arrangement in $\mathbb{R}^d$, then*

$$f_j(\mathcal{A}) = (-1)^j \chi_{\mathcal{A},j}(-1) \tag{5.9}$$

for $j = 0, \ldots, d$.

Proof. Each j-face of $\mathcal{A}$ is contained in an element of $\mathcal{L}_j(\mathcal{A})$, hence the set of all j-faces of $\mathcal{A}$ is the union of the sets of j-faces of all $\mathcal{A}^S$, $S \in \mathcal{L}_j(\mathcal{A})$. Therefore,

$$f_j(\mathcal{A}) = \sum_{S \in \mathcal{L}_j(\mathcal{A})} f_j(\mathcal{A}^S).$$

Equation (5.5), applied to $\mathcal{A}^S$, gives

$$f_j(\mathcal{A}^S) = (-1)^{\dim S} \chi_{\mathcal{A}^S}(-1).$$

By the definition of the characteristic polynomial,

$$\chi_{\mathcal{A}^S}(t) = \sum_{T \in \mathcal{L}(\mathcal{A}^S)} \mu_{\mathcal{A}^S}(S,T) t^{\dim T},$$

where $\mu_{\mathcal{A}^S}$ is the Möbius function of the poset $\mathcal{A}^S$. This poset is isomorphic to the poset $\{T \in \mathcal{L}(\mathcal{A}) : S \leq T\}$ with reverse inclusion. Therefore,

$$\chi_{\mathcal{A}^S}(t) = \sum_{T \in \mathcal{L}(\mathcal{A}),\, S \leq T} \mu(S,T) t^{\dim T},$$

by the definition of the Möbius function. This gives

$$f_j(\mathcal{A}) = \sum_{S \in \mathcal{L}_j(\mathcal{A})} (-1)^{\dim S} \sum_{T \in \mathcal{L}(\mathcal{A}),\, S \leq T} \mu(S,T)(-1)^{\dim T}$$

$$= (-1)^j \chi_{\mathcal{A},j}(-1),$$

as stated.   □

Recall that subspaces $S, T \subset \mathbb{R}^d$ are said to be in general position if

$$\dim(S \cap T) = \max\{0, \dim S + \dim T - d\},$$

equivalently,

$$S + T = \mathbb{R}^d \quad \text{or} \quad S \cap T = \{o\}.$$

Definition 5.1.6. *A subspace $L \subset \mathbb{R}^d$ is in general position with respect to the central arrangement $\mathcal{A}$ if it is in general position with respect to each element of $\mathcal{L}(\mathcal{A})$.*

The following lemma is modeled after an argument of Amelunxen and Lotz [10].

Lemma 5.1.1. *Let $q \in \{1, \ldots, d-1\}$, let $E \in G(d, d-q)$ be a subspace which is in general position with respect to $\mathcal{A}$. Let μ^E be the Möbius function of $\mathcal{L}(\mathcal{A}^E)$. Then, for $S, T \in \mathcal{L}(\mathcal{A})$,*

$$\mu^E(S \cap E, T \cap E) = \mu(S, T) \quad \text{if } \dim S, \dim T \geq q + 1. \tag{5.10}$$

Let $q + 1 \leq j \leq d$. If

$$\chi_{\mathcal{A},j}(t) = \sum_{i=0}^{j} a_{ji} t^i, \tag{5.11}$$

then

$$\chi_{\mathcal{A}^E, j-q}(t) = \sum_{i=0}^{q} a_{ji} + \sum_{i=q+1}^{j} a_{ji} t^{i-q}. \tag{5.12}$$

If $q = 1$, then

$$f_{j-1}(\mathcal{A}^E) = f_j(\mathcal{A}) - (-1)^j 2\chi_{\mathcal{A},j}(0). \tag{5.13}$$

Proof. Let $S, T \in \mathcal{L}(\mathcal{A})$ be subspaces with $\dim S, \dim T \geq q + 1$. Suppose that $T \cap E \subseteq S \cap E$. We want to show that this implies $T \subseteq S$.

Since $\dim T + \dim E \geq d$ and E and T are in general position, we have

$$\dim(T + E) = d.$$

From

$$\dim(T \cap E) + \dim(T + E) = \dim T + \dim E$$

it then follows that

$$\dim(T \cap E) = \dim T - q \geq 1. \tag{5.14}$$

Since $T \cap E \subseteq S \cap E$, we have

$$\dim(S \cap T \cap E) = \dim(T \cap E) \geq 1. \tag{5.15}$$

Since E and $S \cap T$ are in general position, this shows that

$$\dim((S \cap T) + E) = d.$$

As above, this implies that

$$\dim((S \cap T) \cap E) = \dim(S \cap T) - q. \tag{5.16}$$

From (5.16), (5.15), (5.14) we get

$$\dim(S \cap T) = \dim(S \cap T \cap E) + q = \dim(T \cap E) + q = \dim T.$$

This implies $T \subseteq S$. Thus, for $S, T \in \mathcal{L}(\mathcal{A})$ with $\dim S, \dim T \geq q + 1$, we have

$$T \cap E \subseteq S \cap E \Leftrightarrow T \subseteq S.$$

It follows that the map $S \mapsto S \cap E$ is a bijection between $\mathcal{L}_j(\mathcal{A})$ and $\mathcal{L}_{j-q}(\mathcal{A}^E)$, for all $j \geq q + 1$. It preserves the partial orders of $\mathcal{L}(\mathcal{A})$ and $\mathcal{L}(\mathcal{A}^E)$. Relation (5.10) now follows from the recursive definition of the Möbius function.

Recall that $\chi_{\mathcal{A},j}(t) = \sum_{i=0}^{j} a_{ji} t^i$ with

$$a_{ji} = \sum_{S \in \mathcal{L}_j(\mathcal{A})} \sum_{T \in \mathcal{L}_i(\mathcal{A})} \mu(S, T).$$

Let $j \geq q + 1$. We write the $(j - q)$th-level characteristic polynomial of $\mathcal{A}^E$ in the form

$$\chi_{\mathcal{A}^E, j-q}(t) = \sum_{r=0}^{j-q} a_{j-q,r}^E t^r.$$

In the following, we use that the elements of $\mathcal{L}_m(\mathcal{A}^E)$ with $m \geq 1$ are precisely the intersections $S \cap E$ with $S \in \mathcal{L}_{m+q}(\mathcal{A})$. For $r \geq 1$, we get

$$\begin{aligned}
a_{j-q,r}^E &= \sum_{S' \in \mathcal{L}_{j-q}(\mathcal{A}^E)} \sum_{T' \in \mathcal{L}_r(\mathcal{A}^E)} \mu^E(S', T') \\
&= \sum_{S \in \mathcal{L}_j(\mathcal{A})} \sum_{T \in \mathcal{L}_{r+q}(\mathcal{A})} \mu^E(S \cap E, T \cap E) \\
&= \sum_{S \in \mathcal{L}_j(\mathcal{A})} \sum_{T \in \mathcal{L}_{r+q}(\mathcal{A})} \mu(S, T) = a_{j,r+q}.
\end{aligned}$$

Thus, we have

$$\chi_{\mathcal{A}^E, j-q}(t) = a_{j-q,0}^E + a_{j,q+1} t + a_{j,q+2} t^2 + \ldots \tag{5.17}$$

To determine $a_{j-q,0}^E$, we use the recursive definition of the Möbius function and recall that $\mu(S, M) = 0$ if $S \not\supseteq M$, to obtain

$$\begin{aligned}
a_{j-q,0}^E &= \sum_{S' \in \mathcal{L}_{j-q}(\mathcal{A}^E)} \mu^E(S', \{o\}) \\
&= - \sum_{S' \in \mathcal{L}_{j-q}(\mathcal{A}^E)} \sum_{M' \in \mathcal{L}(\mathcal{A}^E), M' \neq \{o\}} \mu^E(S', M') \\
&= - \sum_{S \in \mathcal{L}_j(\mathcal{A})} \sum_{M \in \mathcal{L}(\mathcal{A}), \dim M \geq q+1} \mu^E(S \cap E, M \cap E) \\
&= - \sum_{S \in \mathcal{L}_j(\mathcal{A})} \sum_{M \in \mathcal{L}(\mathcal{A}), \dim M \geq q+1} \mu(S, M).
\end{aligned}$$

This can be related to

$$a_{j0} = \sum_{S \in \mathcal{L}_j(\mathcal{A})} \mu(S, \{o\}) = - \sum_{S \in \mathcal{L}(\mathcal{A})} \sum_{M \in \mathcal{L}(\mathcal{A}), \dim M \geq 1} \mu(S, M)$$

$$= a_{j-q,0}^E - \sum_{r=1}^{q} \sum_{S \in \mathcal{L}(\mathcal{A})} \sum_{M \in \mathcal{L}_r(\mathcal{A})} \mu(S, M)$$

$$= a_{j-q,0}^E - \sum_{r=1}^{q} a_{jr}.$$

Together with (5.17) this yields (5.12).

Now let $q = 1$. Immediately from (5.11) and (5.12) we get

$$\chi_{\mathcal{A}^E, j-1}(-1) = 2\chi_{\mathcal{A}, j}(0) - \chi_{\mathcal{A}, j}(-1).$$

Zaslavsky's formula (5.9) gives

$$f_{j-1}(\mathcal{A}^E) = (-1)^{j-1}\chi_{\mathcal{A}^E, j-1}(-1) = (-1)^{j-1}[2\chi_{\mathcal{A}, j}(0) - \chi_{\mathcal{A}, j}(-1)]$$
$$= f_j(\mathcal{A}) - (-1)^j 2\chi_{\mathcal{A}, j}(0),$$

which is (5.13).   □

We can now prove the following.

Theorem 5.1.3. *Let $\mathcal{A}$ be a central hyperplane arrangement in $\mathbb{R}^d$, and let*

$$\chi_{\mathcal{A}, j}(t) = \sum_{i=0}^{j} a_{ji} t^i$$

be its jth-level characteristic polynomial. Let $j \in \{1, \ldots, d\}$. Let $E \subset \mathbb{R}^d$ be a subspace.

If $\dim E = 1$ and $E^\perp$ is in general position with respect to $\mathcal{L}(\mathcal{A})$, then

$$\sum_{F \in \mathcal{F}_j(\mathcal{A})} \mathbb{1}\{E \cap F^\circ \neq \{o\}\} = 2(-1)^j a_{j0}. \tag{5.18}$$

If $\dim E = d - k$ with $k \in \{1, \ldots, j-1\}$ and E is in general position with respect to $\mathcal{L}(\mathcal{A})$, then

$$\sum_{F \in \mathcal{F}_j(\mathcal{A})} \mathbb{1}\{E \cap F \neq \{o\}\} = (-1)^{j-k} \left[\sum_{i=0}^{k} a_{ji} + \sum_{i=k+1}^{j} a_{ji}(-1)^{i-k} \right]. \tag{5.19}$$

Proof. First let $\dim E = 1$, and suppose that $E^\perp$ is in general position with respect to $\mathcal{L}(\mathcal{A})$.

Let $j = 1$. If $\ell_0(\mathcal{A}) \neq 0$, then F° is a halfspace for each $F \in \mathcal{F}_1(\mathcal{A})$, hence $\sum_{F \in \mathcal{F}_1(\mathcal{A})} \mathbb{1}\{E \cap F^\circ \neq \{o\}\} = f_1(\mathcal{A}) = 2\ell_1(\mathcal{A}) = -2a_{10}$ by (5.8). If $\ell_0(\mathcal{A}) = 0$, then F° is a hyperplane for each $F \in \mathcal{F}_1(\mathcal{A})$, hence $\sum_{F \in \mathcal{F}_1(\mathcal{A})} \mathbb{1}\{E \cap F^\circ \neq \{o\}\} = 0$ and $a_{10} = -\ell_0(\mathcal{A})\ell_1(\mathcal{A}) = 0$. Thus, (5.18) holds if $j = 1$.

Now let $j \in \{2, \ldots, d\}$. Let $H \in G(d, d-1)$ be a hyperplane which is in general position with respect to $\mathcal{A}$. By this general position, it follows from Lemma 1.3.10 that $H \cap F = \{o\} \Leftrightarrow H^\perp \cap F^\circ \neq \{o\}$ for $F \in \mathcal{F}_j(\mathcal{A})$ and therefore

$$f_{j-1}(\mathcal{A}^H) = \sum_{F \in \mathcal{F}_j(\mathcal{A})} \mathbb{1}\{H \cap F \neq \{o\}\} = \sum_{F \in \mathcal{F}_j(\mathcal{A})} [1 - \mathbb{1}\{H \cap F = \{o\}\}]$$

$$= f_j(\mathcal{A}) - \sum_{F \in \mathcal{F}_j(\mathcal{A})} \mathbb{1}\{H^\perp \cap F^\circ \neq \{o\}\}.$$

Together with (5.13), this yields (5.18) (with $E = H^\perp$).

We turn to the proof of (5.19). We have

$$\sum_{F \in \mathcal{F}_j(\mathcal{A})} \mathbb{1}\{E \cap F \neq \{o\}\} = f_{j-k}(\mathcal{A}^E) = (-1)^{j-k} \chi_{\mathcal{A}^E, j-k}(-1),$$

by Zaslavsky's formula (5.9), applied to $\mathcal{A}^E$. This together with (5.12) yields (5.19).   □

Now we formulate the extended Klivans–Swartz formula. For $j = d$, it was proved by Klivans and Swartz [118] (after a special case had been conjectured by Drton and Klivans [53]); a different, short proof was given by Kabluchko, Vysotsky and Zaporozhets [113]. The extension to dimensions $j < d$ is due to Amelunxen and Lotz [10]. We extend here the approach of [113], following [164].

Theorem 5.1.4. *If $\mathcal{A}$ is a central hyperplane arrangement, then*

$$\sum_{F \in \mathcal{F}_j(\mathcal{A})} v_k(F) = (-1)^{j-k} a_{jk} \tag{5.20}$$

for $j \in \{0, \ldots, d\}$ and $k \in \{0, \ldots, j\}$, where a_{jk} is a coefficient of the jth-level characteristic polynomial of $\mathcal{A}$, as defined by (5.7).

The essence of (5.20) is that the left side, which involves the metric functionals v_k, depends only on the partial order of $\mathcal{L}(\mathcal{A})$ and thus is a combinatorial quantity.

Proof. From the combinatorial result of Theorem 5.1.3, formula (5.20) can be obtained by integration, observing that the assumptions on general position are satisfied almost everywhere, due to Lemma 2.1.2.

First let $k = 0$. Integration of (5.18) over $G(d, 1)$ with the measure ν_1 gives

$$2(-1)^j a_{j0} = \sum_{F \in \mathcal{F}_j(\mathcal{A})} \mathbb{1}\{L \cap F^\circ \neq \{o\}\}$$

$$= \sum_{F \in \mathcal{F}_j(\mathcal{A})} \int_{G(d,1)} \mathbb{1}\{L \cap F^\circ \neq \{o\}\} \nu_1(dL)$$

$$= \sum_{F \in \mathcal{F}_j(\mathcal{A})} 2U_{d-1}(F^\circ) = \sum_{F \in \mathcal{F}_j(\mathcal{A})} 2v_d(F^\circ) = \sum_{F \in \mathcal{F}_j(\mathcal{A})} 2v_0(F),$$

where we have used (2.43), (2.70), (2.61). This is (5.20) for $k = 0$.

For $k \geq 1$, we use (2.70), (2.43), (5.19) and obtain

$$\sum_{F \in \mathcal{F}_j(\mathcal{A})} v_k(F)$$

$$= \sum_{F \in \mathcal{F}_j(\mathcal{A})} [U_{k-1}(F) - U_{k+1}(F)]$$

$$= \sum_{F \in \mathcal{F}_j(\mathcal{A})} \left[\frac{1}{2} \int_{G(d,d-k+1)} \mathbb{1}\{L \cap F \neq \{o\}\} \nu_{d-k+1}(dL) \right.$$

$$\left. - \frac{1}{2} \int_{G(d,d-k-1)} \mathbb{1}\{L \cap F \neq \{o\}\} \nu_{d-k-1}(dL) \right]$$

$$= \frac{1}{2}(-1)^{j-k-1} \left[\sum_{i=0}^{k-1} a_{ji} + \sum_{i=k}^{j} a_{ji}(-1)^{i-k+1} - \sum_{i=0}^{k+1} a_{ji} - \sum_{i=k+2}^{j} a_{ji}(-1)^{i-k-1} \right]$$

$$= (-1)^{j-k} a_{jk},$$

which is (5.20) for $k \geq 1$. $\square$

For use in the next section, we conveniently reformulate a special case of (5.19).

Corollary 5.1.1. *Let $\mathcal{A}$ be a central hyperplane arrangement in $\mathbb{R}^d$, let $L \in G(d, d-k)$ with $k \in \{1, \ldots, d-1\}$ be a subspace which is in general position with respect to $\mathcal{A}$. If the characteristic polynomial of $\mathcal{A}$ is written in the form*

$$\chi_\mathcal{A}(t) = \sum_{m=0}^{d} (-1)^{d-m} b_m t^m,$$

then

$$\sum_{C \in \mathcal{F}_d(\mathcal{A})} \mathbb{1}\{L \cap C \neq \{o\}\} = 2(b_{k+1} + b_{k+3} + \ldots).$$

Proof. First we remark that (5.6) implies that

$$b_0 + b_2 + \cdots = b_1 + b_3 + \ldots \tag{5.21}$$

From (5.19) we have

$$\sum_{C \in \mathcal{F}_d(\mathcal{A})} \mathbb{1}\{L \cap C \neq \{o\}\} = (-1)^k \sum_{i=0}^{k} (-1)^i b_i + \sum_{i=k+1}^{d} b_i.$$

If k is even, this is equal to

$$\sum b_i - 2 \sum_{i \text{ odd}}^{k} b_i = \sum_{i \text{ even}} b_i + \sum_{i \text{ odd}} b_i - 2 \sum_{i \text{ odd}}^{k} b_i = 2 \sum_{i \geq k+1, \, i \text{ odd},} b_i$$

by (5.21). If k is odd, the argument is the same, with 'even' and 'odd' interchanged.   □

5.2 Absorption probabilities via central arrangements

The topic of this section is a remarkable application of central arrangements to the determination of a certain absorption probability. It is due to Kabluchko, Vysotsky and Zaporozhets [113].

To introduce the subject, namely applications of central arrangements to absorption probabilities, let us first recall (e.g., from [170, 8.2.1]) a classical example, due to Wendel [190]. Let $X_1, \ldots, X_n$ be random points in $\mathbb{R}^d$. We say that they absorb the origin if $o \in \mathrm{conv}\{X_1, \ldots, X_n\}$, and the probability of this event is an example of an absorption probability.

Let ϕ be a distribution on $\mathbb{R}^d$ which is symmetric with respect to o and assigns measure zero to each hyperplane through o. Let $X_1, \ldots, X_n$ be stochastically independent random points with distribution ϕ. We ask (equivalently) for the probability

$$P_{d,n} := \mathbb{P}(o \notin \mathrm{conv}\{X_1, \ldots, X_n\})$$
$$= \mathbb{P}(\mathrm{pos}\{X_1, \ldots, X_n\} \neq \mathbb{R}^d).$$

Here we have used that almost surely

$$o \in \mathrm{conv}\{X_1, \ldots, X_n\} \Leftrightarrow o \in \mathrm{int}\,\mathrm{conv}\{X_1, \ldots, X_n\}$$
$$\Leftrightarrow \mathrm{pos}\{X_1, \ldots, X_n\} = \mathbb{R}^d,$$

where the last equivalence follows, e.g., from [163, Thm. 1.1.14]).

Defining $g(x_1, \ldots, x_n) = 1$ if there is an open halfspace with o in the boundary that contains $x_1, \ldots, x_n$, and $g(x_1, \ldots, x_n) = 0$ otherwise, it follows from the assumptions on ϕ that

$$P_{d,n} = \frac{1}{2^n} \int_{(\mathbb{R}^d)^n} \sum_{\epsilon_i \in \{1,-1\}} g(\epsilon_1 x_1, \ldots, \epsilon_n x_n)\, \phi^n(\mathrm{d}(x_1, \ldots, x_n)).$$

Let $(x_1, \ldots, x_n)$ be an n-tuple for which any d or fewer entries are linearly independent (only such tuples contribute to the integral). Let $\mathcal{A}(x_1, \ldots, x_n)$ be the set of hyperplanes through o with normal vectors $x_1, \ldots, x_n$. Then

$$g(\epsilon_1 x_1, \ldots, \epsilon_n x_n) = 1 \Leftrightarrow \exists y \in \mathbb{R}^d : \langle \epsilon_i x_i, y \rangle > 0 \ \forall i = 1, \ldots, n,$$

which is equivalent to

$$\bigcap_{i=1}^n \{y \in \mathbb{R}^d : \langle \epsilon_i x_i, y \rangle > 0\} \neq \emptyset.$$

This shows that the n-tuples $(\epsilon_1, \ldots, \epsilon_n)$ with $g(\epsilon_1 x_1, \ldots, \epsilon_n x_n) = 1$ are in one-to-one correspondence with the d-dimensional cones of the tessellation induced by the central arrangement $\mathcal{A}(x_1, \ldots, x_n)$. Schläfli's result (5.1) now shows that

$$P_{d,n} = \frac{C(n,d)}{2^n}.$$

The numbers $P_{d,n}$ will play an important role in some later sections. Therefore, we introduce a name for them.

Definition 5.2.1. *The numbers*

$$P_{d,n} = \mathbb{P}(\text{pos}\{X_1, \ldots, X_n\} \neq \mathbb{R}^d) = \frac{C(n,d)}{2^n} = \frac{1}{2^{n-1}} \sum_{i=0}^{d-1} \binom{n-1}{i}, \quad (5.22)$$

defined for $d, n \in \mathbb{N}$, are called the Wendel probabilities.

In the following, we give a more sophisticated example of an absorption probability, taken from [113]. The authors deal with an n-step random walk in $\mathbb{R}^d$, that is, a sequence $(S_1, \ldots, S_n)$, where $S_k = X_1 + \cdots + X_k$ $(k = 1, \ldots, n)$ and $X_1, \ldots, X_n$, called the *increments* of the random walk, are i.i.d. random vectors in $\mathbb{R}^d$.

Theorem 5.2.1. *Let $(S_1, \ldots, S_n)$ (with $n > d$) be a random walk in $\mathbb{R}^d$ with i.i.d. increments $X_1, \ldots, X_n$. Suppose that the distribution of X_i is symmetric with respect to o and assigns measure zero to each hyperplane through o. Then*

$$\mathbb{P}(o \in \text{conv}\{S_1, \ldots, S_n\}) = \frac{2}{2^n n!}(B(n, d+1) + B(n, d+3) + \ldots),$$

where the numbers $B(n, k)$ are the coefficients of the polynomial

$$(t+1)(t+3) \cdots (t+2n-1) = \sum_{k=0}^n B(n,k) t^k. \quad (5.23)$$

From (5.23) with $t = 1$ and $t = -1$ it follows that

$$B(n, 0) + B(n, 2) + \cdots = B(n, 1) + B(n, 3) + \cdots = 2^{n-1} n!.$$

Therefore, the complementary probability to that in Theorem 5.2.1 is given by

$$\mathbb{P}(o \notin \operatorname{conv}\{S_1, \ldots, S_n\}) = \frac{2}{2^n n!} (B(n, d-1) + B(n, d-3) + \ldots).$$

We see that, as in Wendel's theorem, the absorption probability is independent of the underlying distribution, as long as this is symmetric and satisfies some technical assumption. As shown in [113], even the independence assumption is not necessary. Under the additional assumption on ϕ that it assigns measure zero to each affine hyperplane, it suffices to assume that the tuple $(X_1, \ldots, X_n)$ is *symmetrically exchangeable*, which means that the common distribution of $X_1, \ldots, X_n$ is invariant under permutations and under multiplication of the vectors by -1. But here we treat only a special case and restrict ourselves to the assumptions made in Theorem 5.2.1.

The proof of Theorem 5.2.1 uses Weyl chambers in $\mathbb{R}^n$. A *Weyl chamber* of type B_n in $\mathbb{R}^n$ is any of the $2^n n!$ open cones in $\mathbb{R}^n$ of the form

$$\{(x_1, \ldots, x_n) \in \mathbb{R}^n : 0 < \epsilon_1 x_{\sigma(1)} < \epsilon_2 x_{\sigma(2)} < \cdots < \epsilon_n x_{\sigma(n)}\},$$

where σ is a permutation of $(1, \ldots, n)$ and $\epsilon_i \in \{-1, 1\}$. These Weyl chambers are the open cones induced by the central arrangement $\mathcal{A}_W$ of the n^2 hyperplanes in $\mathbb{R}^n$ with equations $x_i = 0$ or $x_i \pm x_j = 0$, $i < j$. The characteristic polynomial of this central arrangement is known (see [183, Sect. 5.1]), namely

$$\chi_{\mathcal{A}_W}(t) = (t-1)(t-3) \cdots (t - (2n-1)) = \sum_{k=0}^{n} (-1)^{n-k} B(n, k) t^k.$$

The coefficients $B(n, k)$ can be expressed in terms of conic intrinsic volumes. In fact, the Klivans–Swartz formula (Theorem 5.1.4 for $d = n$ and with $j = n$) gives

$$\sum_{C \in \mathcal{F}_n(\mathcal{A}_W)} v_k(C) = B(n, k).$$

Since the elements of $\mathcal{F}_n(\mathcal{A}_W)$ are all congruent to one of them, say $\operatorname{cl} W$, and there are $2^n n!$ of them, we get

$$B(n, k) = 2^n n! \, v_k(\operatorname{cl} W).$$

Theorem 5.2.1 is a consequence of Lemmas 5.2.1 and 5.2.3, which we are now going to prove.

Lemma 5.2.1. *Let $L \in G(n, n-d)$ with $d \in \{1, \ldots, n-1\}$ be a subspace of $\mathbb{R}^n$ which is in general position with respect to $\mathcal{A}_W$. Then*

$$N_{n,d} := \sum_{C \in \mathcal{F}_n(\mathcal{A}_W)} \mathbb{1}\{L \cap C \neq \{o\}\} = 2[B(n, d+1) + B(n, d+3) + \ldots].$$

This follows from Corollary 5.1.1, with (d, k) replaced by (n, d).

In the following, when we consider the random matrix A with columns $X_1, \ldots, X_n$, we regard A also as a random linear operator $\mathsf{A} : \mathbb{R}^n \to \mathbb{R}^d$. The kernel $\ker \mathsf{A}$ of this operator is a random subspace of $\mathbb{R}^n$.

Lemma 5.2.2. *Let A be the random matrix with columns $X_1, \ldots, X_n$, where $X_1, \ldots, X_n$ are i.i.d. random vectors in $\mathbb{R}^d$ with distribution ϕ. Let $S \in G(n, k)$ with $k \in \{1, \ldots, n-1\}$. Then almost surely $\dim \ker \mathsf{A} = n - d$, and S and $\ker \mathsf{A}$ are in general position.*

Proof. Almost surely, any d or less of the vectors $X_1, \ldots, X_n$ are linearly independent. This follows from the case $n = d$, and the latter we prove by induction. Define the event $\mathcal{E}_j := \{\dim \operatorname{lin} \{X_1, \ldots, X_j\} = j\}$ for $j = 1, \ldots, d$. We prove that $\mathbb{P}(\mathcal{E}_j) = 1$. For $j = 1$ this is clear, and we assume that $j \in \{2, \ldots, d-1\}$ and the assertion has been proved for smaller indices. We have

$$\mathbb{P}(\mathcal{E}_j) = \int_{\mathbb{R}^d} \cdots \int_{\mathbb{R}^d} \mathbb{1}\{\dim \operatorname{lin} \{x_1, \ldots, x_j\} = j\} \, \phi(\mathrm{d}x_j) \cdots \phi(\mathrm{d}x_1).$$

When evaluating the inner integal over $\mathbb{R}^d$, with integration variable x_j, we note that for ϕ^{j-1}-almost all $(x_1, \ldots, x_{j-1})$ we have $\dim \operatorname{lin} \{x_1, \ldots, x_{j-1}\} = j - 1$, by the induction hypothesis. For such a fixed tuple $(x_1, \ldots, x_{j-1})$, we have $\dim \operatorname{lin} \{x_1, \ldots, x_j\} \neq j$ only if x_j varies in a fixed subspace of dimension less than d. Therefore, the integral is equal to one. This completes the induction. The case $j = d$ proves the assertion. It follows that almost surely A has rank d and hence $\dim \ker \mathsf{A} = n - d$.

We have

$$\ker \mathsf{A} = \{(\beta_1, \ldots, \beta_n) \in \mathbb{R}^n : \beta_1 X_1 + \cdots + \beta_n X_n = o\}.$$

Now let $S \in G(n, k)$. We have to show that almost surely

$$\dim (S \cap \ker \mathsf{A}) = \begin{cases} k - d & \text{if } k > d, \\ 0 & \text{if } k \leq d. \end{cases}$$

For a fixed rotation ϑ of $\mathbb{R}^d$, also the image measure of ϕ under ϑ has the property that it associates measure zero to each hyperplane through o. Therefore, we may assume without loss of generality that

$$S = \{(\beta_1, \ldots, \beta_n) \in \mathbb{R}^n : \beta_{k+1} = \cdots = \beta_n = 0\}.$$

First we assume now that $k \leq d$. Suppose that $\dim (S \cap \ker \mathsf{A}) > 0$. Then there exists $(\beta_1, \ldots, \beta_n) \in S \cap \ker \mathsf{A}$ different from the zero vector. We have $\beta_{k+1} = \cdots = \beta_n = 0$, hence $\beta_1 X_1 + \cdots + \beta_k X_k = 0$. Thus, $X_1, \ldots, X_k$ are linearly dependent. Since $k \leq d$, this happens only with probability zero.

Now suppose that $k > d$. Assume that $m := \dim (S \cap \ker \mathsf{A}) \neq k - d$, then $m > k - d$. There are m linearly independent vectors $(\beta_1^{(i)}, \ldots, \beta_k^{(i)}, 0, \ldots, 0) \in S \cap \ker \mathsf{A}$, $i = 1, \ldots, m$. From

$$\beta_1^{(i)} X_1 + \cdots + \beta_k^{(i)} X_k = 0, \quad i = 1, \ldots, m > k - d,$$

it follows that the matrix with columns $X_1, \ldots, X_k$ has rank less than d. This happens only with probability zero. $\qquad\square$

Lemma 5.2.3. *Under the assumptions of Theorem 5.2.1,*

$$\mathbb{P}(o \in \mathrm{conv}\{S_1, \ldots, S_n\}) = \frac{N_{n,d}}{2^n n!}.$$

Proof. We denote by $(e_1, \ldots, e_n)$ the standard basis of $\mathbb{R}^n$ and consider vectors of $\mathbb{R}^n$ as coordinate tuples with respect to this basis. Let $g_{\sigma,\epsilon} : \mathbb{R}^n \to \mathbb{R}^n$ be the orthogonal transformation defined by

$$g_{\sigma,\epsilon}(e_k) := \epsilon_k e_{\sigma(k)}, \quad k = 1, \ldots, n,$$

where σ is a permutation of $(1, \ldots, n)$ and $\epsilon = (\epsilon_1, \ldots, \epsilon_n) \in \{-1, 1\}^n$. These transformations form the reflection group B_n, a subgroup of the orthogonal group $O(n)$. The Weyl chambers of B_n are the images of the open convex cone

$$W := \{(x_1, \ldots, x_n) \in \mathbb{R}^n : 0 < x_1 < x_2 < \cdots < x_n\}$$

under B_n. Let A be the random matrix with colums $X_1, \ldots, X_n$ (given as in Theorem 5.2.1). We state that

$$\mathbb{P}(o \in \mathrm{conv}\{S_1, \ldots, S_n\}) = \mathbb{P}(\ker \mathsf{A} \cap g(\mathrm{cl}\, W) \neq \{o\}) \quad \text{for } g \in B_n. \quad (5.24)$$

For the proof, we consider the event

$$\mathcal{E}_g := \{\ker \mathsf{A} \cap g(\mathrm{cl}\, W) \neq \{o\}\} = \{\ker (\mathsf{A}g) \cap \mathrm{cl}\, W \neq \{o\}\}$$

for $g = g_{\epsilon,\sigma}$. Since $(\mathsf{A}g)(e_k) = \mathsf{A}(ge_k)$ for $k = 1, \ldots, n$, the columns of the matrix $\mathsf{A}g$ with $g = g_{\epsilon,\sigma}$ are given by $\epsilon_1 X_{\sigma(1)}, \ldots, \epsilon_n X_{\sigma(n)}$, hence

$$\mathcal{E}_g = \{\exists (x_1, \ldots, x_n) \in \mathrm{cl}\, W \setminus \{o\} : x_1 \epsilon_1 X_{\sigma(1)} + \cdots + x_n \epsilon_n X_{\sigma(n)} = o\}.$$

Write

$$\mathbb{R}_+^n := \{x \in \mathbb{R}^n : x_i \geq 0 \text{ for } i = 1, \ldots, n\}$$

for the nonnegative orthant in $\mathbb{R}^n$. There is a bijection between

$$x = (x_1, \ldots, x_n) \in \mathrm{cl}\, W \setminus \{o\}$$

and $y = (y_1, \ldots, y_n) \in \mathbb{R}_+^n \setminus \{o\}$, given by

$$x_1 = y_1, \ x_2 = y_1 + y_2, \ldots, x_n = y_1 + \cdots + y_n.$$

Therefore,

$$\mathcal{E}_g = \{\exists x \in \mathrm{cl}\, W \setminus \{o\} : x_1 \epsilon_1 X_{\sigma(1)} + \cdots + x_n \epsilon_n X_{\sigma(n)} = o\}$$
$$= \{\exists y \in \mathbb{R}_+^n \setminus \{o\} : y_1 \epsilon_1 X_{\sigma(1)} + (y_1 + y_2)\epsilon_2 X_{\sigma(2)} + \ldots$$
$$+ (y_1 + \cdots + y_n)\epsilon_n X_{\sigma(n)} = o\}$$
$$= \{\exists y \in \mathbb{R}_+^n \setminus \{o\} : y_1(\epsilon_1 X_{\sigma(1)} + \cdots + \epsilon_n X_{\sigma(n)}) +$$
$$+ y_2(\epsilon_2 X_{\sigma(2)} + \cdots + \epsilon_n X_{\sigma(n)}) + \cdots + y_n \epsilon_n X_{\sigma(n)} = o\}.$$

By the assumption on the distribution of $X_1, \ldots, X_n$, the n-tuple

$$(\epsilon_1 X_{\sigma(1)} + \cdots + \epsilon_n X_{\sigma(n)}, \epsilon_2 X_{\sigma(2)} + \cdots + \epsilon_n X_{\sigma(n)}, \ldots, \epsilon_n X_{\sigma(n)})$$

has the same distribution as

$$(S_n, S_{n-1}, \ldots, S_1).$$

Therefore,

$$\mathbb{P}(\mathcal{E}_g) = \mathbb{P}(\exists y \in \mathbb{R}_+^n \setminus \{o\} : y_1 S_n + y_2 S_{n-1} + \cdots + y_n S_1 = o)$$
$$= \mathbb{P}(o \in \mathrm{conv}\{S_1, \ldots, S_n\})$$

is independent of g, and this proves the assertion (5.24)

Summing relation (5.24) over all $g \in B_n$ and observing that B_n has $2^n n!$ elements, we obtain

$$\mathbb{P}(o \in \mathrm{conv}\{S_1, \ldots, S_n\}) = \frac{1}{2^n n!} \sum_{g \in B_n} \mathbb{P}(\ker \mathsf{A} \cap g(\mathrm{cl}\, W) \neq \{o\}).$$

With probability one (if $n > d$) we have $\dim \ker \mathsf{A} = n - d$, and $\ker \mathsf{A}$ is in general position with respect to $\mathcal{A}_W$, by Lemma 5.2.2. Therefore, with probability one we have

$$\sum_{g \in B_n} \mathbf{1}\{\ker \mathsf{A} \cap g(\mathrm{cl}\, W) \neq \{o\}\} = N_{n,d}$$

by Lemma 5.2.1 and hence

$$\sum_{g \in B_n} \mathbb{P}(\ker \mathsf{A} \cap g(\mathrm{cl}\, W) \neq \{o\}) = \mathbb{E}\, N_{n,d} = N_{n,d}.$$

This yields the assertion of the lemma.   □

As said, the proof of Theorem 5.2.1 follows immediately from Lemmas 5.2.1 and 5.2.3.

Notes for Section 5.2

1. The approach of Kabluchko, Vysotsky and Zaporozhets [113] yields more. The following is a quotation from their abstract. "We obtain analogous distribution-free results for Weyl chambers of type A_{n-1} (yielding the probability of absorption of the origin by the convex hull of a generic random walk bridge), type D_n, and direct products of Weyl chambers (yielding the absorption probability for the joint convex hull of several random walks or bridges). The simplest case of products of the form $B_1 \times \cdots \times B_1$ recovers Wendel's formula ... for the probability that the convex hull of an i.i.d. multidimensional sample chosen from a centrally symmetric distribution does not contain the origin." (end of quotation)

The authors of [113] also deduce several asymptotic results, for large n as well as for increasing dimensions.

2. The conic intrinsic volumes of the Weyl chambers of types A_{n-1}, B_n, D_n (of which here type B_n was needed) were determined in a different way by Godland and Kabluchko [67]. These authors did not employ the Klivans–Swartz formula, but used the representation

$$v_k(C) = \sum_{F \in \mathcal{F}_k(C)} \beta(o, F)\gamma(F, C)$$

and were able to find explicit formulas for the external and internal angles.

3. The investigation of the random walk $(S_1, \ldots, S_n)$ appearing in Theorem 5.2.1 is continued in Kabluchko, Vysotsky and Zaporozhets [114]. Here, the authors consider the convex hull $C_n := \mathrm{conv}\{S_0, \ldots, S_n\}$ (where $S_0 := o$) and obtain explicit formulas for $\mathbb{E} f_k(C_n)$. Further, they compute the probability that for given indices $0 \le i_1 < \cdots < i_{k+1} \le n$, the points $S_{i_1}, \ldots, S_{i_{k+1}}$ are the vertices of a k-face of C_n.

5.3 Random cones generated by central arrangements

Let $\mathcal{H}_1, \ldots, \mathcal{H}_n$ be i.i.d. random hyperplanes through o (where the underlying distribution satisfies some mild conditions). They induce a tessellation of $\mathbb{R}^d$ into convex polyhedral cones. Choosing one of the d-dimensional cones of this tessellation at random (with equal chances) we obtain a random cone S_n. This random cone can also be generated in a different way. Let $X_1, \ldots, X_n$ be i.i.d. random points in $\mathbb{R}^d \setminus \{o\}$ (again with a suitable distribution). Let C_n be their positive hull, under the condition that this positive hull is not all of $\mathbb{R}^d$. Then C_n is a random cone. If we define the hyperplanes $\mathcal{H}_1, \ldots, \mathcal{H}_n$ in such a way that $\mathcal{H}_i = X_i^\perp$, then S_n has the same distribution as the polar cone of C_n. These two types of random cones were introduced and studied by Cover and Efron [50]. In this section, we resume and continue their investigation, mainly following Hug and Schneider [98]. We shall determine the expectations of various functionals of these random cones. The fact that the results are essentially independent of the distribution reveals their mainly combinatorial nature.

Hyperplanes $H_1, \ldots, H_n \in G(d, d-1)$ are said to be *in general position* if any $k \le d$ of them have an intersection of dimension $d-k$. Vectors $x_1, \ldots, x_n \in \mathbb{R}^d$ are *in general position* if any d or fewer of them are linearly independent. Thus, the hyperplanes $x_1^\perp, \ldots, x_n^\perp$ (where $x_i \ne o$) are in general position if and only if $x_1, \ldots, x_n$ are in general position. For a vector $x \in \mathbb{R}^d$, we write

$$x^- := \{y \in \mathbb{R}^d : \langle y, x \rangle \le 0\}.$$

If $x_1, \ldots, x_n \in \mathbb{R}^d$ are in general position, then

$$\bigcap_{i=1}^n x_i^- \ne \{o\} \Leftrightarrow \dim \bigcap_{i=1}^n x_i^- = d. \tag{5.25}$$

To prove the non-trivial direction $\Rightarrow$ of (5.25), suppose that $C := \bigcap_{i=1}^n x_i^-$ satisfies $0 < k := \dim C < d$. Let $L_k := \lim C$. Choose $o \ne p \in \operatorname{relint} C$ and define $I := \{i \in \{1, \ldots, n\} : p \in x_i^\perp\}$. Then $I \ne \emptyset$, since $k < d$. Moreover, $p \in \operatorname{int} x_j^-$ for $j \in \{1, \ldots, n\} \setminus I$, and $C \subseteq \bigcap_{i \in I} x_i^\perp$. From the latter, we have $L_k \subseteq \bigcap_{i \in I} x_i^\perp \subseteq \bigcap_{i \in I} x_i^-$. Since $p \in \operatorname{int} x_j^-$ for $j \in \{1, \ldots, n\} \setminus I$, we also have $\bigcap_{i \in I} x_i^- \subseteq L_k$ and thus $L_k = \bigcap_{i \in I} x_i^\perp = \bigcap_{i \in I} x_i^-$. This implies that $L_k^\perp = \operatorname{pos}\{x_i : i \in I\}$ and hence $|I| \ge d - k + 1$. But $|I| = d - k$ by the assumption of general position, a contradiction.

Let $H_1, \ldots, H_n \in G(d, d-1)$ be hyperplanes through o in general position. They induce a tessellation of $\mathbb{R}^d$ into d-dimensional polyhedral cones; we denote it now by $\mathcal{T}_n = \mathcal{T}_n(H_1, \ldots, H_n)$. We call $\mathcal{T}_n$ a *conical tessellation* of $\mathbb{R}^d$. We write $\mathcal{F}_k(H_1, \ldots, H_n)$ for the set of k-faces of $\mathcal{T}_n$, that is, for the set of k-faces of the d-cones in $\mathcal{T}_n$. We also write $f_k(\mathcal{T}_n) = |\mathcal{F}_k(H_1, \ldots, H_n)|$ for the number of k-faces of $\mathcal{T}_n$. As long as the dimension d is fixed, we do not exhibit it in the notation.

For a hyperplane $H \in G(d, d-1)$, we denote by H^- one of the two closed halfspaces bounded by H. It follows from (5.25) that the d-dimensional cones of the tessellation $\mathcal{T}_n$ induced by $H_1, \ldots, H_n$ (in general position) are precisely the cones different from $\{o\}$ of the form

$$\bigcap_{i=1}^n \epsilon_i H_i^-, \quad \epsilon_i \in \{-1, 1\}.$$

We call these cones the *Schläfli cones* induced by $H_1, \ldots, H_n$, because, as already mentioned, Schläfli has shown that there are exactly

$$C(n, d) := 2 \sum_{r=0}^{d-1} \binom{n-1}{r} \tag{5.26}$$

of them. For formal reasons, we supplement definition (5.26) by $C(0, d) := 1$ and $C(n, d) := 0$ for $n < 0$.

With $H_1, \ldots, H_n$ as above, each choice of $d - k$ indices $1 \le i_1 < \cdots < i_{d-k} \le n$, where $k \in \{1, \ldots, d\}$, determines a k-dimensional subspace $L =$

$H_{i_1} \cap \cdots \cap H_{i_{d-k}}$ (= $\mathbb{R}^d$ if $k = d$). For $i \in \{1, \ldots, n\} \setminus \{i_1, \ldots, i_{d-k}\}$, the intersections of L with the hyperplanes H_i are in general position in L and hence determine $C(n-d+k, k)$ Schläfli cones with respect to L. Each of these is a k-face of the tessellation $\mathcal{T}_n$, and each k-face of $\mathcal{T}_n$ is obtained in this way. Therefore, the total number of k-faces of the tessellation $\mathcal{T}_n$ is given by

$$f_k(\mathcal{T}_n) = \binom{n}{d-k} C(n-d+k, k) =: C(n, d, k). \tag{5.27}$$

In particular, $f_k(\mathcal{T}_n) = 1$ if $n = d - k$ and $f_k(\mathcal{T}_n) = 0$ if $n < d - k$.

We turn to random cones. Let ϕ be a probability measure on $\mathbb{R}^d$, which is even (that is, symmetric with respect to o) and which is zero on every hyperplane through o.

Definition 5.3.1. *Let $n \in \mathbb{N}$, and let $X_1, \ldots, X_n$ be independent random vectors in $\mathbb{R}^d$ with distribution ϕ. The*

$$(\phi, n)\text{-}Cover\text{-}Efron \text{ cone } C_n$$

is the random cone defined as the positive hull of $X_1, \ldots, X_n$ under the condition that this is different from $\mathbb{R}^d$.

We recall the Wendel probability (see Definition 5.2.1)

$$P_{d,n} = \mathbb{P}\left(\text{pos}\{X_1, \ldots, X_n\} \neq \mathbb{R}^d\right) = \frac{C(n, d)}{2^n}. \tag{5.28}$$

Thus, C_n is a random cone with distribution given by

$$\mathbb{P}(C_n \in B) \tag{5.29}$$
$$= \frac{1}{P_{d,n}} \int_{(\mathbb{R}^d)^n} \mathbb{1}\{\text{pos}\{x_1, \ldots, x_n\} \in B \setminus \{\mathbb{R}^d\}\} \, \phi^n(\mathrm{d}(x_1, \ldots, x_n))$$

for $B \in \mathcal{B}(\mathcal{P}\mathcal{C}^d)$.

A second way of defining random cones, starting from $X_1, \ldots, X_n$, is to consider the tessellation induced by $X_1^\perp, \ldots, X_n^\perp$. For this, let ϕ^* denote the image measure of the measure ϕ under the mapping $x \mapsto x^\perp$ from $\mathbb{R}^d \setminus \{o\}$ to the Grassmannian $G(d, d-1)$. A measure on $G(d, d-1)$ is obtained in this way from a measure ϕ, with the properties specified above, if and only if it assigns measure zero to any set of hyperplanes in $G(d, d-1)$ containing a fixed line. Independent random hyperplanes $\mathcal{H}_1, \ldots, \mathcal{H}_n$ with distribution ϕ^* are almost surely in general position.

Definition 5.3.2. *Let $n \in \mathbb{N}$, and let $\mathcal{H}_1, \ldots, \mathcal{H}_n$ be independent random hyperplanes in $G(d, d-1)$ with distribution ϕ^*. The*

$$(\phi^*, n)\text{-}Schläfli \text{ cone } S_n$$

is obtained by choosing at random (with equal chances) one of the Schläfli cones induced by $\mathcal{H}_1, \ldots, \mathcal{H}_n$.

More formally, the Schläfli cone S_n is defined as the random cone with distribution

$$\mathbb{P}(S_n \in B) \tag{5.30}$$

$$= \int_{G(d,d-1)^n} \frac{1}{C(n,d)} \sum_{C \in \mathcal{F}_d(H_1,\ldots,H_n)} \mathbb{1}_B(C) \, \phi^{*n}(\mathrm{d}(H_1,\ldots,H_n))$$

for $B \in \mathcal{B}(\mathcal{PC}^d)$.

Theorem 5.3.1. *Let ϕ and ϕ^* be defined as above, let $n \in \mathbb{N}$. The (ϕ, n)-Cover–Efron cone C_n and the polar cone S_n° of the (ϕ^*, n)-Schläfli cone S_n are stochastically equivalent,*

$$C_n \overset{d}{=} S_n^\circ. \tag{5.31}$$

Proof. We have $\mathbb{P}(C_n = \mathbb{R}^d) = 0 = \mathbb{P}(S_n^\circ = \mathbb{R}^d)$. Thus, we need only consider Borel sets $B \in \mathcal{PC}^d$ with $\mathbb{R}^d \notin B$. Using the symmetry of ϕ, we first rewrite equation (5.29) and then use (5.28) and the relation

$$\mathrm{pos}\{x_1,\ldots,x_n\} = \left(\bigcap_{i=1}^n x_i^-\right)^\circ. \tag{5.32}$$

For $B \in \mathcal{B}(\mathcal{PC}^d)$ with $\mathbb{R}^d \notin B$ we thus obtain

$$\mathbb{P}(C_n \in B)$$

$$= \frac{1}{P_{d,n}} \int_{(\mathbb{R}^d)^n} \frac{1}{2^n} \sum_{\epsilon_i = \pm 1} \mathbb{1}_B(\mathrm{pos}\{\epsilon_1 x_1, \ldots, \epsilon_n x_n\}) \, \phi^n(\mathrm{d}(x_1,\ldots,x_n))$$

$$= \int_{(\mathbb{R}^d)^n} \frac{1}{C(n,d)} \sum_{\epsilon_i = \pm 1} \mathbb{1}_B\left(\left(\bigcap_{i=1}^n \epsilon_i x_i^-\right)^\circ\right) \phi^n(\mathrm{d}(x_1,\ldots,x_n))$$

$$= \int_{(\mathbb{R}^d)^n} \frac{1}{C(n,d)} \sum_{C \in \mathcal{F}_d(x_1^\perp,\ldots,x_n^\perp)} \mathbb{1}_B(C^\circ) \, \phi^n(\mathrm{d}(x_1,\ldots,x_n))$$

$$= \int_{G(d,d-1)^n} \frac{1}{C(n,d)} \sum_{C \in \mathcal{F}_d(H_1,\ldots,H_n)} \mathbb{1}_B(C^\circ) \, \phi^{*n}(\mathrm{d}(H_1,\ldots,H_n))$$

$$= \mathbb{P}(S_n^\circ \in B),$$

by (5.30). This completes the proof. $\qquad\square$

Our goal is to determine for the random Schläfli cones the expectations of the functionals $Y_{k,j}$, defined by (2.80). This includes the determination of expected face numbers, Grassmann angles and conic intrinsic volumes for random Schläfli cones and hence also for Cover–Efron cones.

We adapt to conical tessellations a method first used by Miles [136, Chap. 11]. It consists in defining, by means of combinatorial selection procedures, different weighted random cones, which can then be used to yield results on expectations. Let $H_1, \ldots, H_n \in G(d, d-1)$ be hyperplanes in general position, and let $L \in G(d, k)$, with $k \in \{1, \ldots, d\}$, be a k-dimensional linear subspace in general position with respect to $H_1, \ldots, H_n$. This means that the intersections $H_1 \cap L, \ldots, H_n \cap L$ are $(k-1)$-dimensional subspaces of L which are in general position in L. Let $j \in \{1, \ldots, k\}$. The tessellation $\mathcal{T}_{L,n}$ induced in L by the intersections $H_1 \cap L, \ldots, H_n \cap L$, has $C(n, k, j)$ faces of dimension j, by (5.27). If $n < k - j$, then clearly $C(n, k, j) = 0$. We shall tacitly make use of the following lemma, which is an immediate consequence of general position.

Lemma 5.3.1. *Let $j \geq 1$. To each j-face F_j of $\mathcal{T}_{L,n}$, there exists a unique $(d - k + j)$-face F of the tessellation $\mathcal{T}_n$ induced by $H_1, \ldots, H_n$, such that $F_j = F \cap L$.*

Conversely, if $F \in \mathcal{F}_{d-k+j}(\mathcal{T}_n)$ and $F \cap L \neq \{o\}$, then $F \cap L$ is a j-face of $\mathcal{T}_{L,n}$.

We assume now that $n \geq k - j$. We choose one of the j-faces of $\mathcal{T}_{L,n}$ at random (with equal chances) and denote it by F_j. Then $F_j = L \cap F$ with a unique face $F \in \mathcal{F}_{d-k+j}(\mathcal{T}_n)$. The face F_j is contained in 2^{k-j} Schläfli cones of $\mathcal{T}_{L,n}$ and thus in 2^{k-j} Schläfli cones of $\mathcal{T}_n$. These are precisely the Schläfli cones of $\mathcal{T}_n$ that contain F. We select one of them at random (with equal chances) and call it $S^{[k,j]}(H_1, \ldots, H_n, L)$.

This is now applied to random hyperplanes. We assume that the probability measure ϕ^* on $G(d, d-1)$ is defined as above, and we let $\mathcal{H}_1, \ldots, \mathcal{H}_n$ be independent random hyperplanes with distribution ϕ^*. Let $k \in \{1, \ldots, d\}$, $j \in \{1, \ldots, k\}$, and $n > k - j$. Further, let $\mathcal{L} \in G(d, k)$ be a random subspace, independent of $\mathcal{H}_1, \ldots, \mathcal{H}_n$, with distribution ν_k, the rotation invariant probability measure on $G(d, k)$. We may assume, since this happens with probability one, that $\mathcal{H}_1, \ldots, \mathcal{H}_n$ and $\mathcal{L}$ are in general position. We define

$$S_n^{[k,j]} := S^{[k,j]}(\mathcal{H}_1, \ldots, \mathcal{H}_n, \mathcal{L}). \tag{5.33}$$

Thus, $S_n^{[k,j]}$ is a random polyhedral cone with distribution given by

$$\mathbb{P}(S_n^{[k,j]} \in B) \tag{5.34}$$

$$= \int_{G(d,d-1)^n} \int_{G(d,k)} \frac{1}{C(n,k,j)} \sum_{\substack{F \in \mathcal{F}_{d-k+j}(H_1,\ldots,H_n) \\ F \cap L \neq \{o\}}} \frac{1}{2^{k-j}}$$

$$\times \sum_{\substack{C \in \mathcal{F}_d(H_1,\ldots,H_n) \\ C \supseteq F}} \mathbb{1}_B(C)\, \nu_k(\mathrm{d}L)\, \phi^*(\mathrm{d}(H_1, \ldots, H_n))$$

for $B \in \mathcal{B}(\mathcal{PC}^d)$.

In the following computation, we abbreviate the n-tuple $(H_1, \ldots, H_n)$ by H_n. Since $n > k - j$, any $F \in \mathcal{F}_{d-k+j}(\mathsf{H}_n)$ is almost surely not a linear subspace. Therefore, the inner integral in (5.34), without the combinatorial factors, can with the use of (1.9.1) and (1.9.33) be written in the form

$$\int_{G(d,k)} \sum_{F \in \mathcal{F}_{d-k+j}(\mathsf{H}_n)} \mathbb{1}\{F \cap L \neq \{o\}\} \sum_{C \in \mathcal{F}_d(\mathsf{H}_n)} \mathbb{1}\{F \subseteq C\} \mathbb{1}_B(C) \, \nu_k(\mathrm{d}L)$$

$$= \sum_{C \in \mathcal{F}_d(\mathsf{H}_n)} \mathbb{1}_B(C) \sum_{F \in \mathcal{F}_{d-k+j}(\mathsf{H}_n)} \mathbb{1}\{F \subseteq C\} \int_{G(d,k)} \mathbb{1}\{F \cap L \neq \{o\}\} \nu_k(\mathrm{d}L)$$

$$= \sum_{C \in \mathcal{F}_d(\mathsf{H}_n)} \mathbb{1}_B(C) \sum_{F \in \mathcal{F}_{d-k+j}(\mathsf{H}_n)} \mathbb{1}\{F \subseteq C\} 2U_{d-k}(F)$$

$$= 2 \sum_{C \in \mathcal{F}_d(\mathsf{H}_n)} \mathbb{1}_B(C) Y_{d-k+j,d-k}(C).$$

As a result, we obtain that

$$\mathbb{P}(S_n^{[k,j]} \in B) \tag{5.35}$$

$$= \frac{2}{2^{k-j} C(n,k,j)} \int_{G(d,d-1)^n} \sum_{C \in \mathcal{F}_d(\mathsf{H}_n)} \mathbb{1}_B(C) Y_{d-k+j,d-k}(C) \, \phi^{*n}(\mathrm{d}\mathsf{H}_n).$$

Let g be any nonnegative, measurable function on $\mathcal{PC}^d$. From (5.30) and (5.35), both extended to expectations, we get

$$\mathbb{E} g\left(S_n^{[k,j]}\right) = \frac{2C(n,d)}{2^{k-j} C(n,k,j)} \mathbb{E}\left(gY_{d-k+j,d-k}\right)(S_n). \tag{5.36}$$

The choice $g = 1$ yields the following theorem.

Theorem 5.3.2. *The expectations of the functions $Y_{r,s}$ of the (ϕ^*, n)-Schläfli cone S_n are given by*

$$\mathbb{E} Y_{d-k+j,d-k}(S_n) = \frac{2^{k-j} C(n,k,j)}{2C(n,d)} \tag{5.37}$$

for $1 \leq j \leq k \leq d$ and $n > k - j$.

As a consequence, we can write (5.36) in the form

$$\mathbb{E} g\left(S_n^{[k,j]}\right) = \frac{\mathbb{E}\left(gY_{d-k+j,d-k}\right)(S_n)}{\mathbb{E} Y_{d-k+j,d-k}(S_n)}. \tag{5.38}$$

This can be interpreted as saying that the distribution of $S_n^{[k,j]}$ is obtained from the distribution of the random Schläfli cone S_n by weighting it with the function $Y_{d-k+j,d-k}$.

We emphasize the special case $k = j = 1$. Here the described procedure is equivalent to choosing a uniform random point in $\mathbb{S}^{d-1}$, independent of $\mathcal{H}_1, \ldots, \mathcal{H}_n$, and taking for $S_n^{[1,1]}$ the Schläfli cone containing it. The weight function is $Y_{d,d-1} = v_d$, and (5.38) specializes to

$$\mathbb{E}\, g\left(S_n^{[1,1]}\right) = \frac{\mathbb{E}\,(gv_d)(S_n)}{\mathbb{E}\,v_d(S_n)} = C(n,d)\mathbb{E}\,(gv_d)(S_n). \tag{5.39}$$

We list some special cases and consequences of (5.37). The results of the following Corollary were already obtained by Cover and Efron [50].

Corollary 5.3.1. *For* $k \in \{1, \ldots, d\}$,

$$\mathbb{E}\,f_k(S_n) = \frac{2^{d-k}\binom{n}{d-k}C(n-d+k,k)}{C(n,d)}, \tag{5.40}$$

and for $k \in \{0, \ldots, d-1\}$,

$$\mathbb{E}\,f_k(C_n) = \frac{2^k\binom{n}{k}C(n-k,d-k)}{C(n,d)}. \tag{5.41}$$

Equation (5.40) is obtained from (5.37) by choosing $k = d$ and then replacing j by k (and observing (5.27), if $n > d - k$). For $n = d - k$, both sides of (5.40) are equal to 1, and for $n < d - k$ both sides are zero. Equation (5.41) is obtained by using the duality relation (5.31), together with $f_k(S_n) = f_{d-k}(S_n^\circ)$.

The following corollary yields the expectations of the Grassmann angles.

Corollary 5.3.2. *For* $k \in \{0, \ldots, d-1\}$,

$$\mathbb{E}\,U_k(S_n) = \frac{C(n,d-k)}{2C(n,d)}, \tag{5.42}$$

and for $k \in \{1, \ldots, d-1\}$,

$$\mathbb{E}\,U_k(C_n) = \frac{C(n,d) - C(n,k)}{2C(n,d)}. \tag{5.43}$$

Equation (5.42) is obtained by replacing k and j in (5.37) both by $d-k$. If $n \leq d-k$, then both sides of the equation are equal to $1/2$. Since C_n is almost surely a pointed cone, the dualities (2.47) and (5.31) yield (5.43), where both sides are zero if $n < k$.

Next we consider the expectations of the conic intrinsic volumes.

Corollary 5.3.3. *We have*

$$\mathbb{E}\,v_j(S_n) = \begin{cases} \binom{n}{d-j}C(n,d)^{-1}, & j = 1, \ldots, d, \\ \binom{n-1}{d-1}C(n,d)^{-1}, & j = 0, \end{cases} \tag{5.44}$$

and

$$\mathbb{E}\, v_j(C_n) = \begin{cases} \binom{n}{j} C(n,d)^{-1}, & j = 0,\ldots,d-1, \\ \binom{n-1}{d-1} C(n,d)^{-1}, & j = d, \end{cases} \tag{5.45}$$

For $j = 1,\ldots,d$, relation (5.44) follows from (2.70) and (5.42); for $j = 0$ it follows from (5.43), (2.61) and (5.31). The duality relations (2.61) and (5.31) then yield (5.45). We mention that (5.44) can also be deduced easily from the Klivans–Swartz formula.

Now we turn to the skeleton measures Λ_k defined by (2.82).

Theorem 5.3.3. *For $k = 1,\ldots,d$,*

$$\mathbb{E}\, \Lambda_k(S_n) = \frac{2^{d-k} \binom{n}{d-k}}{C(n,d)}, \tag{5.46}$$

and for $k = 1,\ldots,d-1$,

$$\mathbb{E}\, \Lambda_k(C_n) = \frac{\binom{n}{k} C(n-k,d-k)}{C(n,d)}. \tag{5.47}$$

Proof. Relation (5.46) is the special case of (5.37) where k is replaced by $d-k+1$ and where $j = 1$. Here we use that for $n > d-k$, the k-faces of S_n are not subspaces. For $n \le d-k$, the equation is obviously true.

To prove (5.47), we extend and complete some arguments given in [50]. Let $k \in \{1,\ldots,d-1\}$. We can assume that $n \ge k$. By (5.30) and (5.31),

$$\mathbb{E}\, \Lambda_k(C_n) = \mathbb{E}\, \Lambda_k(S_n^\circ) = \int_{G(d,d-1)^n} \frac{1}{C(n,d)} \sum_{C \in \mathcal{F}_d(\mathsf{H}_n)} \Lambda_k(C^\circ)\, \phi^{*n}(\mathrm{d}\mathsf{H}_n).$$

Here, again, $\mathsf{H}_n = (H_1,\ldots,H_n)$ with hyperplanes $H_1,\ldots,H_n \in G(d,d-1)$, which we can assume to be in general position. Let $F \in \mathcal{F}_{d-k}(\mathsf{H}_n)$. There are indices $1 \le i_1 < \cdots < i_k \le n$ such that

$$F \subseteq L_{i_1,\ldots,i_k} := H_{i_1} \cap \cdots \cap H_{i_k}.$$

Let $\mathcal{C}_F$ be the set of Schläfli cones $C \in \mathcal{F}_d(\mathsf{H}_n)$ with $F \subset C$. Let u_j be a normal vector of H_{i_j}, $j = 1,\ldots,k$. Then the cones $C \in \mathcal{C}_F$ are in one-to-one correspondence with the choices $\epsilon_1,\ldots,\epsilon_k \in \{-1,1\}$ such that

$$C \subseteq \bigcap_{j=1}^{k} \epsilon_j u_j^-.$$

The face of C° conjugate to F is given by

$$\widehat{F}_C = \mathrm{pos}\{\epsilon_1 u_1,\ldots,\epsilon_k u_k\}.$$

It follows that the faces $\widehat{F}_C$ with $C \in \mathcal{C}_F$ form a tiling of $L_{i_1,\ldots,i_k}^\perp$, and therefore

$$\sum_{C \in \mathcal{F}_d(\mathsf{H}_n)} \mathbb{1}\{F \subset C\} v_k(\widehat{F}_C) = 1. \tag{5.48}$$

The faces $F \in \mathcal{F}_{d-k}(\mathsf{H}_n)$ with $F \subset L_{i_1,\dots,i_k}$ are the Schläfli cones of the tessellation induced by $L_{i_1,\dots,i_k}$, hence there are precisely $C(n-k, d-k)$ such cones. Now we obtain, using (5.48),

$$\sum_{C \in \mathcal{F}_d(\mathsf{H}_n)} \Lambda_k(C^\circ)$$

$$= \sum_{C \in \mathcal{F}_d(\mathsf{H}_n)} \sum_{G \in \mathcal{F}_k(C^\circ)} v_k(G)$$

$$= \sum_{C \in \mathcal{F}_d(\mathsf{H}_n)} \sum_{F \in \mathcal{F}_{d-k}(C)} v_k(\widehat{F}_C)$$

$$= \sum_{F \in \mathcal{F}_{d-k}(\mathsf{H}_n)} \sum_{C \in \mathcal{F}_d(\mathsf{H}_n)} \mathbb{1}\{F \subset C\} v_k(\widehat{F}_C)$$

$$= \sum_{1 \le i_1 < \dots < i_k \le n} \sum_{F \in \mathcal{F}_{d-k}(\mathsf{H}_n)} \mathbb{1}\{F \subset L_{i_1,\dots,i_k}\} \sum_{C \in \mathcal{F}_d(\mathsf{H}_n)} \mathbb{1}\{F \subset C\} v_k(\widehat{F}_C)$$

$$= \binom{n}{k} C(n-k, d-k),$$

which yields (5.47). $\square$

We add a remark on the expected number of k-faces of the Cover–Efron cone C_n. Formula (5.41) can be written as

$$\frac{\mathbb{E} f_k(C_n)}{\binom{n}{k}} = 2^k \frac{C(n-k, d-k)}{C(n, d)}. \tag{5.49}$$

There is a good reason for considering the quotient on the left: the polyhedral cone C_n can have at most $\binom{n}{k}$ faces of dimension k. If this maximum number is reached, the cone C_n is called k-neighborly. Therefore, the larger the quotient in (5.49) is, the more the random cone C_n tends to be k-neighborly. One might expect that increasing n decreases this tendency. This is correct, but not entirely trivial. A proof will be given in Theorem 6.5.1.

Notes for Section 5.3

1. As Cover and Efron [50] have observed, it follows from (5.40) that

$$\lim_{n \to \infty} \mathbb{E} f_k(S_n) = 2^{d-k} \binom{d-1}{k-1}.$$

This is equal to the number of $(k-1)$-faces of a $(d-1)$-cube.

We remark that for the typical cell Z of a stationary Poisson hyperplane tessellation in $\mathbb{R}^{d-1}$ we have

$$\mathbb{E} f_{k-1}(Z) = 2^{d-k} \binom{d-1}{k-1}.$$

This follows (in a more general version) from [170, Thm. 10.3.1].

Kabluchko, Temesvari and Thäle [111] have found a way to prove a weak convergence result, leading from $S_n \cap \mathbb{S}^{d-1}$ (as $n \to \infty$) to the typical cell of a stationary, isotropic Poisson hyperplane process in $\mathbb{R}^{d-1}$.

2. Godland and Kabluchko [66] have found analogues of the random Schläfli cone for which similarly explicit expectation formulas can be proved. Let $y_1, \ldots, y_n \in \mathbb{R}^d$ be vectors with $n \geq d$, so that

$$(y_i + y_j)^{\perp}, \ (y_i - y_j)^{\perp} \ (1 \leq i < j \leq n), \quad y_i^{\perp} \ (1 \leq i \leq n)$$

are hyperplanes. The following definition is motivated by the definition of Weyl chambers in $\mathbb{R}^n$. The hyperplanes above induce a tessellation $\mathcal{W}^B(y_1, \ldots, y_n)$ of $\mathbb{R}^d$ into polyhedral cones. Under a suitable assumption (which can be interpreted as a kind of general position), it is proved in [66] that the number of d-cones in $\mathcal{W}^B(y_1, \ldots, y_n)$ is given by

$$D^B(n, d) = 2(B(n, n-d+1) + B(n, n-d+3) + \ldots),$$

where the numbers $B(n, k)$ are those defined by (5.23) (and $B(n, k) := 0$ if $k \notin \{0, \ldots, n\}$). The formula is deduced from a more general result, which yields the number of k-faces of $\mathcal{W}^B(y_1, \ldots, y_n)$, each counted with the multiplicity given by the number of d-cones of which it is a face.

Now let $Y_1, \ldots, Y_n$ with $n \geq d$ be random vectors in $\mathbb{R}^d$. Under a mild assumption on their distribution, the general position assumption mentioned above holds almost surely. The *random Weyl cone* $\mathcal{D}_n^B$ of type B_n is obtained by picking at random, with equal chances, one of the d-cones from $\mathcal{W}^B(Y_1, \ldots, Y_n)$. A main result of Godland and Kabluchko is the expectation formula

$$\mathbb{E} Y_{d-k+j, d-k}(\mathcal{D}_n^B) = \frac{2^{k-j} \binom{n}{k-j} D^B(n-k+j, j)}{2 D^B(n, d)} \frac{n!}{(n-k+j)!}$$

for $1 \leq j \leq k \leq d$. This includes formulas for the expectations of $f_j(\mathcal{D}_n^B)$, $U_j(\mathcal{D}_n^B)$, and $v_j(\mathcal{D}_n^B)$.

In a similar way, for pairwise distinct vectors $y_1, \ldots, y_n \in \mathbb{R}^d$ with $n \geq d+1$, the hyperplanes

$$(y_i - y_j)^{\perp} \ (1 \leq i < j \leq n)$$

induce a tessellation $\mathcal{W}^A(y_1, \ldots, y_n)$ of $\mathbb{R}^d$ into polyhedral cones. Again under a suitable assumption of general position, it is proved in [66] that the number of d-cones in $\mathcal{W}^A(y_1, \ldots, y_n)$ is given by

$$D^A(n, d) = 2 \left(\begin{bmatrix} n \\ n - d + 1 \end{bmatrix} + \begin{bmatrix} n \\ n - d + 3 \end{bmatrix} + \cdots \right).$$

Here $\begin{bmatrix} n \\ k \end{bmatrix}$ are the Stirling numbers of the first kind, which can be defined by

$$t(t + 1) \cdots (t + n - 1) = \sum_{k=1}^{n} \begin{bmatrix} n \\ k \end{bmatrix} t^k$$

and $\begin{bmatrix} n \\ k \end{bmatrix} := 0$ for $k \notin \{1, \ldots, n\}$.

Let $Y_1, \ldots, Y_n$ with $n \geq d+1$ be random vectors in $\mathbb{R}^d$. Under a suitable assumption on their distribution, the crucial general position assumption holds almost surely. The *random Weyl cone* $\mathcal{D}_n^A$ of type A_{n-1} is obtained by picking at random, with equal chances, one of the d-cones from $\mathcal{W}^A(Y_1, \ldots, Y_n)$. Godland and Kabluchko showed that

$$\mathbb{E} Y_{d-k+j, d-k}(\mathcal{D}_n^A) = \frac{\binom{n-1}{k-j} D^A(n - k + j, j)}{2 D^A(n, d)} \frac{n!}{(n - k + j)!}$$

for $1 \leq j \leq k \leq d$.

2. Godland and Kabluchko [68] have discovered further classes of polyhedral random cones for which, similarly as in the previous note, explicit expressions for various expectations can be obtained. These are, for example, the cones of the form $\mathrm{pos}\{S_1, \ldots, S_n\}$, where $S_k = X_1 + \cdots + X_k$ for $k = 1, \ldots, n$ and $X_1, \ldots, X_n$ are random vectors, with suitable assumptions on their distribution. Together with the cones of the previous note, and by dualizing and conditioning to be non-degenerate, one obtains several classes of random cones for which explicit results can be deduced.

5.4 Volume weighted Schläfli cones

Given a random tessellation with an invariance property, one has different possibilities of choosing a typical cell. One may either take the cell containing a given fixed point, or one may select randomly, with equal chances, one of the cells (the procedure must be suitably modified for an infinite tessellation). For a stationary hyperplane tessellation in Euclidean space, the first procedure leads to the zero cell (also known as Crofton cell), and the second to the typical cell. If translations are ignored, it turns out that the zero cell can be considered as the volume weighted version of the typical cell. (Matheron [126], for example, distinguished between the 'volume law' and the 'number law'.) In our present context, namely tessellations of $\mathbb{R}^d$ into polyhedral cones by n i.i.d. random central hyperplanes, the notion of the Schläfli cone corresponds to the second procedure. In this section, which again mainly follows Hug and Schneider [98], we consider the counterpart to the first procedure.

Assumption. In this section, we consider only the isotropic case, thus we assume that the distribution of the considered random hyperplanes is rotation invariant.

Recall that ν_{d-1} denotes the unique rotation invariant probability measure on the Grassmannian $G(d, d-1)$. Most of the subsequent results require this special distribution for the considered random hyperplanes, instead of the general distribution ϕ^* of the previous sections. We choose a fixed vector $e \in \mathbb{S}^{d-1}$.

Definition 5.4.1. *Let $\mathcal{H}_1, \ldots, \mathcal{H}_n$ be independent random hyperplanes with distribution ν_{d-1}. We denote by*

$$S_n^e$$

the almost surely unique cell of the tessellation induced by $\mathcal{H}_1, \ldots, \mathcal{H}_n$ that contains the vector e.

If $e \notin H \in G(d, d-1)$, we denote by H^e the closed halfspace bounded by H that contains e.

The following lemma relates the distribution of S_n^e to that of the (ν_{d-1}, n)-Schläfli cone S_n.

Lemma 5.4.1. *Let h be a nonnegative measurable function on $\mathcal{PC}^d$ which is invariant under rotations. Then*

$$\mathbb{E}\, h(S_n^e) = C(n, d)\, \mathbb{E}\, (h\nu_d)(S_n).$$

Proof. In the following, we denote by ν the invariant probability measure on the rotation group $SO(d)$, and we make use of the fact that

$$\int_{SO(d)} g(\vartheta e)\, \nu(d\vartheta) = \frac{1}{\omega_d} \int_{\mathbb{S}^{d-1}} g(u)\, \sigma_{d-1}(du)$$

for every nonnegative measurable function g on $\mathbb{S}^{d-1}$. Using the rotation invariance of the function h and of the probability distribution ν_{d-1}, we obtain, with $\vartheta \in SO(d)$,

$$\mathbb{E}\, h(S_n^e) = \mathbb{E} \sum_{C \in \mathcal{F}_d(\mathcal{H}_1, \ldots, \mathcal{H}_n)} h(C) \mathbb{1}_{\operatorname{int} C}(e)$$

$$= \mathbb{E} \sum_{C \in \mathcal{F}_d(\mathcal{H}_1, \ldots, \mathcal{H}_n)} h(C) \mathbb{1}_{\operatorname{int} C}(\vartheta e)$$

$$= \mathbb{E} \int_{SO(d)} \sum_{C \in \mathcal{F}_d(\mathcal{H}_1, \ldots, \mathcal{H}_n)} h(C) \mathbb{1}_{\operatorname{int} C}(\vartheta e)\, \nu(d\vartheta)$$

$$= \frac{1}{\omega_d} \mathbb{E} \int_{\mathbb{S}^{d-1}} \sum_{C \in \mathcal{F}_d(\mathcal{H}_1, \ldots, \mathcal{H}_n)} h(C) \mathbb{1}_{\operatorname{int} C}(u)\, \sigma_{d-1}(du)$$

$$= \frac{1}{\omega_d} \mathbb{E} \sum_{C \in \mathcal{F}_d(\mathcal{H}_1, \dots, \mathcal{H}_n)} h(C) \sigma_{d-1}(C \cap \mathbb{S}^{d-1})$$

$$= C(n, d) \, \mathbb{E} \, (h v_d)(S_n),$$

by (5.30) (extended to expectations) with $\phi^* = \nu_{d-1}$. □

A comparison with (5.39) shows that, for rotation invariant functions h,

$$\mathbb{E} \, h(S_n^e) = \mathbb{E} \, h \left(S_n^{[1,1]} \right).$$

This, or Lemma 5.4.1, shows that, if rotations are ignored, then S_n^e can be interpreted as the volume weighted version of the Schläfli cone (the latter taken with respect to the rotation invariant measure ν_{d-1}).

Aiming at evaluating expectations of geometric functionals for the random cone S_n^e, we start with v_d, the solid angle (or spherical volume of $S_n^e \cap \mathbb{S}^{d-1}$).

Let $K \subset \mathbb{S}^{d-1}$ be a closed, spherically convex set containing e. Writing $u \in \mathbb{S}^{d-1}$ in the form

$$u = te + \sqrt{1 - t^2} \, \overline{u} \quad \text{with } \overline{u} \in e^\perp \cap \mathbb{S}^{d-1},$$

we have

$$\sigma_{d-1}(K) = \int_{e^\perp \cap \mathbb{S}^{d-1}} \int_{\cos \rho(K, \overline{u})}^{1} (1 - t^2)^{\frac{d-3}{2}} \, dt \, \sigma_{d-2}(d\overline{u}) \qquad (5.50)$$

where

$$\rho(K, \overline{u}) = \max\{\rho \in [0, \pi] : (\cos \rho)e + (\sin \rho)\overline{u} \in K\}, \quad \overline{u} \in e^\perp \cap \mathbb{S}^{d-1}.$$

Let $Z_n^e := S_n^e \cap \mathbb{S}^{d-1}$. For fixed $\overline{u} \in e^\perp \cap \mathbb{S}^{d-1}$, the distribution function of the random variable $\rho(Z_n^e, \overline{u})$ is given by

$$F(x) = \mathbb{P}(\rho(Z_n^e, \overline{u}) < x) = 1 - \left(1 - \frac{x}{n}\right)^n,$$

since $\rho(Z_n^e, \overline{u}) > x$ holds if and only if none of the random hyperplanes $\mathcal{H}_1, \dots, \mathcal{H}_n$ intersects the geodesic arc connecting e and $(\cos \rho)e + (\sin \rho)\overline{u}$. Let

$$G(x) := \int_{\cos x}^{1} (1 - t^2)^{\frac{d-3}{2}} \, dt = \int_0^x \sin^{d-2} \alpha \, d\alpha \qquad \text{for } x \in [0, \pi].$$

From (5.50) we have $G(\pi) = \omega_d / \omega_{d-1}$. Since the distribution of the random variable $\rho(Z_n^e, \overline{u})$ does not depend $\overline{u}$, we obtain

$$\mathbb{E} \, \sigma_{d-1}(Z_n^e) = \mathbb{E} \int_{e^\perp \cap \mathbb{S}^{d-1}} \int_{\cos \rho(Z_n^e, \overline{u})}^{1} (1 - t^2)^{\frac{d-3}{2}} \, dt \, \sigma_{d-2}(d\overline{u})$$

$$= \omega_{d-1} \mathbb{E}\, G(\rho(Z_n^e, \overline{u}))$$

$$= \omega_{d-1} \int_0^\pi G(x) F'(x)\, dx$$

$$= \omega_{d-1} \left[G(\pi) - \int_0^\pi G'(x) F(x)\, dx \right]$$

$$= \omega_{d-1} \left[\frac{\omega_d}{\omega_{d-1}} - \int_0^\pi \sin^{d-2} x \left(1 - \left(1 - \frac{x}{\pi} \right)^n \right) dx \right]$$

$$= \omega_{d-1} \int_0^\pi \left(1 - \frac{x}{\pi} \right)^n \sin^{d-2} x\, dx.$$

After using the binomial theorem, the integral can be evaluated by using recursion formulas and known definite integrals; e.g., see [77, p. 117]. We abbreviate

$$\theta(n, d) := \frac{\omega_{d-1}}{\omega_d} \int_0^\pi \left(1 - \frac{x}{\pi} \right)^n \sin^{d-2} x\, dx, \quad n \in \mathbb{N}_0. \qquad (5.51)$$

For formal reasons, we supplement this by $\theta(n, d) := 0$ for $n < 0$. Since $v_d(S_n^e) = \sigma_{d-1}(Z_n^e)/\omega_d$, we can write the result as

$$\mathbb{E}\, v_d(S_n^e) = \theta(n, d). \qquad (5.52)$$

This can be extended to the skeleton volumes, as follows.

Theorem 5.4.1. *For $k \in \{0, \ldots, d-1\}$,*

$$\mathbb{E}\, \Lambda_{d-k}(S_n^e) = \binom{n}{k} \theta(n - k, d). \qquad (5.53)$$

Proof. First we note that for $A \in \mathcal{B}(\mathbb{S}^{d-1})$ and $k \in \{1, \ldots, d-1\}$ we have

$$\int_{G(d, d-1)} \sigma_{d-k-1}(A \cap H_1 \cap \cdots \cap H_k)\, \nu_{d-1}^k(d(H_1, \ldots, H_k))$$

$$= \frac{\omega_{d-k}}{\omega_d} \sigma_{d-1}(A). \qquad (5.54)$$

In fact, the left-hand side of (5.54), as a function of A, is a finite measure. Due to the rotation invariance of ν_{d-1} and σ_{d-1}, it is rotation invariant and hence is a constant multiple of σ_{d-1}. The choice $A = \mathbb{S}^{d-1}$ then reveals the factor.

Now let $k \in \{0, \ldots, d-1\}$. We can assume that $n \geq k$, since otherwise both sides of (5.53) are zero. Almost surely, each $(d-k)$-face of S_n^e is the intersection of S_n^e with exactly k of the random hyperplanes $\mathcal{H}_1, \ldots, \mathcal{H}_n$. Conversely, each intersection of k distinct hyperplanes from $\mathcal{H}_1, \ldots, \mathcal{H}_n$ a.s. intersects S_n^e either in a $(d-k)$-face or in $\{o\}$. With these prerequisites, we compute

$$\mathbb{E}\,\Lambda_{d-k}(S_n^e)$$

$$= \mathbb{E} \sum_{1 \le i_1 < \cdots < i_k \le n} v_{d-k}(S_n^e \cap \mathcal{H}_{i_1} \cap \cdots \cap \mathcal{H}_{i_k})$$

$$= \sum_{1 \le i_1 < \cdots < i_k \le n} \mathbb{E}\, v_{d-k}(\mathcal{H}_1^e \cap \cdots \cap \mathcal{H}_n^e \cap \mathcal{H}_{i_1} \cap \cdots \cap \mathcal{H}_{i_k})$$

$$= \binom{n}{k} \mathbb{E}\, v_{d-k}(\mathcal{H}_{k+1}^e \cap \cdots \cap \mathcal{H}_n^e \cap \mathcal{H}_1 \cap \cdots \cap \mathcal{H}_k)$$

$$= \binom{n}{k} \int_{G(d,d-1)^{n-k}} \int_{G(d,d-1)^k} v_{d-k}(H_{k+1}^e \cap \cdots \cap H_n^e \cap H_1 \cap \cdots \cap H_k)$$

$$\times \nu_{d-1}^k(\mathrm{d}(H_1, \ldots H_k))\, \nu_{d-1}^{n-k}(\mathrm{d}(H_{k+1}, \ldots, H_n)).$$

If $n = k$, the outer integration does not appear, and $\mathcal{H}_{k+1}^e \cap \cdots \cap \mathcal{H}_n^e$ has to be interpreted as $\mathbb{R}^d$. By the initial remark, the inner integral is equal to

$$\frac{1}{\omega_d} \sigma_{d-1} \left(H_{k+1}^e \cap \cdots \cap H_n^e \cap \mathbb{S}^{d-1} \right),$$

hence we obtain

$$\mathbb{E}\,\Lambda_{d-k}(S_n^e) = \binom{n}{k} \mathbb{E}\, v_d(S_{n-k}^e).$$

Together with (5.52), this gives the assertion.   □

Lemma 5.4.1 and equation (5.53) open the way to determine some second order moments for the (ν_{d-1}, n)-Schläfli cone S_n: choosing $f = \Lambda_{d-k}$ in Lemma 5.4.1, we obtain

$$\mathbb{E}\,(\Lambda_{d-k} v_d)(S_n) = \frac{\binom{n}{k}\theta(n-k,d)}{C(n,d)}, \tag{5.55}$$

in particular,

$$\mathbb{E}\, v_d(S_n)^2 = \frac{\theta(n,d)}{C(n,d)}. \tag{5.56}$$

The case $k = d - 1$ of (5.53) yields combinatorial information, namely $\mathbb{E}\, f_1(S_n^e)$ (the expected vertex number of $S_n^e \cap \mathbb{S}^{d-1}$): for $n \ge d$,

$$\mathbb{E}\, f_1(S_n^e) = 2\binom{n}{d-1}\theta(n-d+1,d). \tag{5.57}$$

For the other face numbers, we can only show that their expectations are related to the expectations of the Grassmann angles (and thus to those of the conic intrinsic volumes).

Theorem 5.4.2. *Let* $k \in \{0, \ldots, d-1\}$ *and* $n > k$. *Then*

$$\mathbb{E}\, f_{d-k}(S_n^e) = 2\binom{n}{k} \mathbb{E}\, U_k(S_{n-k}^e). \tag{5.58}$$

Proof. Similarly as in the proof of Theorem 5.4.1, we obtain

$$\mathbb{E}\, f_{d-k}(S_n^e)$$

$$= \mathbb{E} \sum_{1 \le i_1 < \cdots < i_k \le n} \mathbb{1}\{S_n^e \cap \mathcal{H}_{i_1} \cap \cdots \cap \mathcal{H}_{i_k} \ne \{o\}\}$$

$$= \binom{n}{k} \int_{G(d,d-1)^{n-k}} \int_{G(d,d-1)^k} \mathbb{1}\{H_{k+1}^e \cap \cdots \cap H_n^e \cap H_1 \cap \cdots \cap H_k \ne \{o\}\}$$

$$\times \nu_{d-1}^k(\mathrm{d}(H_1, \ldots H_k))\, \nu_{d-1}^{n-k}(\mathrm{d}(H_{k+1}, \ldots, H_n)).$$

Let $G(d, d-1)_*^k$ denote the set of all k-tuples of $(d-1)$-dimensional linear subspaces with linearly independent normal vectors. The image measure of ν_{d-1}^k under the mapping $(H_1, \ldots, H_k) \mapsto H_1 \cap \cdots \cap H_k$ from $G(d, d-1)_*^k$ to $G(d, d-k)$ is the invariant measure ν_k, hence

$$\int_{G(d,d-1)^k} \mathbb{1}\{C \cap H_1 \cap \cdots \cap H_k \ne \{o\}\} \nu_{d-1}^k(\mathrm{d}(H_1, \ldots H_k)) = 2U_k(C)$$

for $C = H_{k+1}^e \cap \cdots \cap H_n^e$ and ν_{d-1}^{n-k} almost all $(H_{k+1}, \ldots, H_n) \in G(d, d-1)^{n-k}$. Inserting this, we obtain the assertion. $\qquad\square$

Notes for Section 5.4

1. Equation (5.55) can be extended to

$$\mathbb{E}\,(\Lambda_s \Lambda_r)(S_n)$$

$$= \frac{\binom{n}{d-s}}{C(n,d)} \sum_{p=\max\{r,s\}}^{d} 2^{d-p} \binom{d-s}{d-p} \binom{n-d+s}{p-r} \theta(n-d-p+r+s, p)$$

for $r, s = 1, \ldots, d$. This was proved by Hug and Schneider [98].

5.5 Typical faces

Given a random conical tessellation as in the previous sections, we can also consider typical faces of a lower dimension. This leads to new aspects, partially due to the fact that a face F with $1 \le \dim F \le d-1$ has a direction. A thorough investigation of faces in random conical tessellations (in the equivalent version of tessellations of the sphere by great hyperspheres) was undertaken by Kabluchko and Thäle [110]. In this section, we shall present a selection of their results.

We recall that for hyperplanes $H_1, \ldots, H_n \in G(d, d-1)$ in general position we denote by $\mathcal{F}_k(H_1, \ldots, H_n) = \mathcal{F}_k(\mathcal{T}_n)$ the set of k-dimensional faces of the conical tessellation $\mathcal{T}_n = \mathcal{T}_n(H_1, \ldots, H_n)$ induced by the hyperplanes, for $k = 1, \ldots, d$. By $f_k(\mathcal{T}_n)$ we denote the number of k-faces of the tessellation $\mathcal{T}_n$. We assume in the following that $n \geq d - 1$, so that $f_k(\mathcal{T}_n) > 0$ for $k \geq 1$. We recall from (5.27) that

$$f_k(\mathcal{T}_n) = C(n, d, k) = \binom{n}{d - k} C(n - d + k, k). \qquad (5.59)$$

The probability measure ϕ^* on $G(d, d - 1)$ is as in Section 5.3, thus it assigns measure zero to the set of hyperplanes in the Grassmannian $G(d, d-1)$ containing a fixed line.

Definition 5.5.1. *Let $k \in \{1, \ldots, d\}$, $n \geq d$, and let $\mathcal{T}_n$ be the tessellation induced by n independent random hyperplanes with distribution ϕ^*. The typical k-face $S_{n,d}^{(k)}$ of $\mathcal{T}_n$ is obtained by choosing at random (with equal chances) one of the faces from $\mathcal{F}_k(\mathcal{T}_n)$.*

In other words, $S_{n,d}^{(k)}$ is the random cone with distribution given by

$$\mathbb{P}\left(S_{n,d}^{(k)} \in B\right) \qquad (5.60)$$

$$= \int_{G(d,d-1)^n} \frac{1}{C(n, d, k)} \sum_{F \in \mathcal{F}_k(H_1, \ldots, H_n)} \mathbf{1}_B(F)\, \phi^{*n}(\mathrm{d}(H_1, \ldots, H_n))$$

for $B \in \mathcal{B}(\mathcal{PC}^d)$. Note that $S_{n,d}^{(d)} = S_n$ gives the (ϕ^*, n)-Schläfli cone again.

To represent this distribution also in a different way, we first define, for $k \in \{1, \ldots, d - 1\}$, a probability measure ψ_k on $G(d, k)$ by

$$\psi_k(A) := \int_{G(d,d-1)^{d-k}} \mathbf{1}\{H_1 \cap \cdots \cap H_{d-k} \in A\}\, \phi^{*d-k}(\mathrm{d}(H_1, \ldots, H_{d-k}))$$

for $A \in \mathcal{B}(G(d, k))$. Then we extend, for a given subspace $L \in G(d, m)$, some of the definitions in previous sections. For hyperplanes $H_1, \ldots, H_n \in G(d, d - 1)$ in general position, and in general position to L, we denote by $\mathcal{F}_k(H_1, \ldots, H_n; L)$ the set of k-dimensional faces of the tessellation in L induced by $H_1 \cap L, \ldots, H_n \cap L$.

Definition 5.5.2. *Let $n \in \mathbb{N}$, and let $\mathcal{H}_1, \ldots, \mathcal{H}_n$ be independent random hyperplanes in $G(d, d - 1)$ with distribution ϕ^*, let $L \in G(d, k)$. The*

$$(\phi^*, n, L)\text{-Schläfli cone } S_{n,L}$$

is obtained by choosing at random (with equal chances) one of the intersections of L with the d-cones induced by $\mathcal{H}_1, \ldots, \mathcal{H}_n$.

Thus, the random cone $S_{n,L}$ has the distribution

$$\mathbb{P}(S_{n,L} \in B) \tag{5.61}$$

$$= \int_{G(d,d-1)^n} \frac{1}{C(n,k)} \sum_{F \in \mathcal{F}_k(H_1,\ldots,H_n;L)} \mathbb{1}_B(F)\, \phi^{*n}(\mathrm{d}(H_1,\ldots,H_n))$$

for $B \in \mathcal{B}(\mathcal{PC}^d)$.

Theorem 5.5.1. *Let* $k \in \{1,\ldots,d\}$ *and* $n \ge d$. *Let* $S_{n,d}^{(k)}$ *be as in Definition 5.5.1. Let* $h : \mathcal{C}^d \to \mathbb{R}$ *be a nonnegative, measurable function. Then*

$$\mathbb{E}\, h\left(S_{n,d}^{(k)}\right) = \int_{G(d,k)} \mathbb{E}\, h(S_{n-d+k,L})\, \psi_k(\mathrm{d}L),$$

where $S_{n-d+k,L}$ *is the* $(\phi^*, n-d+k, L)$-*Schläfli cone.*

Proof. By (5.60),

$$\mathbb{E}\, h\left(S_{n,d}^{(k)}\right)$$

$$= \int_{G(d,d-1)^n} \frac{1}{C(n,d,k)} \sum_{F \in \mathcal{F}_k(H_1,\ldots,H_n)} h(F)\, \phi^{*n}(\mathrm{d}(H_1,\ldots,H_n))$$

$$= \int_{G(d,d-1)^n} \frac{1}{C(n,d,k)} \sum_{1 \le i_1 < \cdots < i_{d-k} \le n} \sum_{\substack{F \in \mathcal{F}_k(H_1,\ldots,H_n) \\ F \subseteq H_{i_1} \cap \cdots \cap H_{i_{d-k}}}} h(F)$$

$$\times \phi^{*n}(\mathrm{d}(H_1,\ldots,H_n))$$

$$= \int_{G(d,d-1)^{d-k}} \frac{1}{C(n-d+k,k)} \int_{G(d,d-1)^{n-d+k}}$$

$$\sum_{F \in \mathcal{F}_k(H_{d-k+1},\ldots,H_n;H_1\cap\cdots\cap H_{d-k})} h(F)\, \phi^{*n-d+k}(\mathrm{d}(H_{d-k+1},\ldots,H_n))$$

$$\times \phi^{*d-k}(\mathrm{d}(H_1,\ldots,H_{d-k})),$$

where we have used (5.59) and the fact that a face $F \in \mathcal{F}_k(H_1,\ldots,H_n)$ satisfying $F \subseteq H_1\cap\cdots\cap H_{d-k}$ is a face of $\mathcal{F}_k(H_{d-k+1},\ldots,H_n; H_1\cap\cdots\cap H_{d-k})$. Now we apply the definition of the measure ψ_k and the representation (5.61) of the distribution of $S_{n,L}$, to obtain

$$\mathbb{E}\, h\left(S_{n,d}^{(k)}\right)$$

$$= \int_{G(d,k)} \frac{1}{C(n-d+k,k)} \int_{G(d,d-1)^{n-d+k}}$$

$$\sum_{F \in \mathcal{F}_k(H_{d-k+1},\ldots,H_n;L)} h(F)\, \phi^{*n-d+k}(\mathrm{d}(H_{d-k+1},\ldots,H_n))\, \psi_k(\mathrm{d}L)$$

$$= \int_{G(d,k)} \mathbb{E}\, h(S_{n-d+k,L})\, \psi_k(\mathrm{d}L),$$

as stated.   □

In the isotropic case, the preceding result can be simplified. We fix a k-dimensional subspace $E_k \in G(d,k)$.

Corollary 5.5.1. *Let* $k \in \{1,\ldots,d\}$ *and* $n \geq d$. *Suppose that* $S_{n,d}^{(k)}$ *is the typical* k*-face of the tessellation induced by* n *independent, isotropic random hyperplanes in* $\mathbb{R}^d$. *Let* $h : \mathcal{C}^d \to \mathbb{R}$ *be a rotation invariant, non-negative, measurable function. Then*

$$\mathbb{E}\, h\left(S_{n,d}^{(k)}\right) = \mathbb{E}\, h(S_{n-d+k,E_k}).$$

Proof. For given $L \in G(d,k)$, there exists a rotation $\rho \in \mathrm{SO}(d)$ with $\rho E_k = L$. By the rotation invariance of the function h and of the considered tessellation, we obtain

$$\mathbb{E}\, h(S_{n-d+k,L}) = \mathbb{E}\, h(\rho^{-1}S_{n-d+k,L}) = \mathbb{E}\, h(S_{n-d+k,\rho^{-1}L}) = \mathbb{E}\, h(S_{n-d+k,E_k}).$$

Now Theorem 5.5.1 gives the assertion.   □

Back to a general distribution ϕ^*, we can determine the expected face numbers and the expected Grassmann angles of the random cone $S_{n,d}^{(k)}$. Of course, by (2.54) this yields also the expected conic intrinsic volumes.

Theorem 5.5.2. *Let* $k \in \{1,\ldots,d\}$ *and* $n \geq d$. *Let* $S_{n,d}^{(k)}$ *be as in Definition 5.5.1. Then*

$$\mathbb{E}\, f_\ell\left(S_{n,d}^{(k)}\right) = \frac{2^{k-\ell}\binom{n-d+k}{k-\ell}C(n-d+\ell,\ell)}{C(n-d+k,k)}$$

for $\ell = 1,\ldots,k$. *Further,*

$$\mathbb{E}\, U_\ell\left(S_{n,d}^{(k)}\right) = \frac{C(n-d+k,k-\ell)}{2C(n-d+k,k)}$$

for $\ell = 0,\ldots,k-1$.

Proof. By Theorem 5.5.1, we have

$$\mathbb{E}\, f_\ell\left(S_{n,d}^{(k)}\right) = \int_{G(d,k)} \mathbb{E}\, f_\ell(S_{n-d+k,L})\, \psi_k(\mathrm{d}L),$$

where $S_{n-d+k,L}$ is the $(\phi^*, n-d+k, L)$-Schläfli cone. Let $L \in G(d,k)$ be fixed. Let ϕ_L^* be the image measure of the ϕ^*-almost everywhere defined map

$$G(d,d-1) \to G(L,k-1), \quad H \mapsto H \cap L,$$

where $G(L, k-1)$ is the Grassmannian of hyperplanes through o with respect to L. Then $S_{n-d+k,L}$ has the same distribution as the $(\phi_L^*, n-d+k)$-Schläfli cone in L. Therefore (5.40), with (d, n, k) replaced by $(k, n-d+k, \ell)$, yields

$$\mathbb{E}\, f_\ell(S_{n-d+k,L}) = \frac{2^{k-\ell}\binom{n-d+k}{k-\ell}C(n-d+\ell,\ell)}{C(n-d+k,k)}.$$

This gives the assertion, since $\psi_k(G(d, k)) = 1$.

The expectation $\mathbb{E}\, U_\ell\left(S_{n,d}^{(k)}\right)$ is found in the same way, using Corollary 5.3.2. □

To introduce weighted typical faces, we associate with any n hyperplanes $H_1, \ldots, H_n \in G(d, d-1)$ in general position the spherical $(k-1)$-skeleton

$$ss_{k-1}(H_1, \ldots, H_n) := \mathbb{S}^{d-1} \cap \bigcup_{1 \leq i_1 < \cdots < i_{d-k} \leq n} H_{i_1} \cap \cdots \cap H_{i_{d-k}}$$

and the probability measure

$$\eta_{k-1}(H_1, \ldots, H_n)$$
$$:= (\mathscr{H}^{k-1} \llcorner ss_{k-1}(H_1, \ldots, H_n)) / \mathscr{H}^{k-1}(ss_{k-1}(H_1, \ldots, H_n)).$$

Here $\mathscr{H}^{k-1}$ denotes the $(k-1)$-dimensional Hausdorff measure. For a vector $v \in ss_{k-1}(H_1, \ldots, H_n)$ which lies in the relative interior of a unique face of $\mathcal{F}_k(H_1, \ldots, H_n)$, we denote this face by $F_{k,v}(H_1, \ldots, H_n) = F_{k,v}(\mathcal{T}_n)$, where $\mathcal{T}_n$ is the tessellation induced by $H_1, \ldots, H_n$. Thus, the face $F_{k,v}(H_1, \ldots, H_n)$ is defined for $\eta_{k-1}(H_1, \ldots, H_n)$-almost all $v \in ss_{k-1}(H_1, \ldots, H_n)$.

Definition 5.5.3. *Let $k \in \{1, \ldots, d\}$, $n \geq d-k$, and let $\mathcal{T}_n$ be the tessellation induced by n independent random hyperplanes with distribution ϕ^*. The weighted typical k-face of $\mathcal{T}_n$ is defined by*

$$W_{n,d}^{(k)} := F_{k,\mathbf{v}}(\mathcal{T}_n),$$

where $\mathbf{v}$ is a random point with distribution $\eta_{k-1}(\mathcal{H}_1, \ldots, \mathcal{H}_n)$, independent of $\mathcal{T}_n$.

Thus, the distribution of $W_{n,d}^{(k)}$ is given by

$$\mathbb{P}\left(W_{n,d}^{(k)} \in B\right) := \int_{G(d,d-1)^n} \int_{ss_{k-1}(H_1, \ldots, H_n)} \mathbb{1}_B(F_{k,v}(H_1, \ldots, H_n))$$
$$\times \eta_{k-1}(H_1, \ldots, H_n)(dv)\, \phi^{*n}(d(H_1, \ldots, H_n)) \quad (5.62)$$

for $B \in \mathcal{B}(\mathcal{PC}^d)$.

Also the distribution of the weighted typical face can be represented in a different way, by means of intersections with full-dimensional random cells.

Theorem 5.5.3. *Let $k \in \{1, \ldots, d\}$ and $n \geq d$. Let $h : C^d \to \mathbb{R}$ be a non-negative, measurable function. Let $\mathcal{H}_1, \ldots, \mathcal{H}_{n-d+k}$ be independent random hyperplanes with distribution ϕ^*. Then*

$$\mathbb{E}\, h\left(W_n^{(k)}\right)$$

$$= \frac{1}{\omega_k} \int_{G(d,k)} \int_{L \cap \mathbb{S}^{d-1}} \mathbb{E}\, h(F_{d,v}(\mathcal{H}_1, \ldots, \mathcal{H}_{n-d+k}) \cap L)\, \mathcal{H}^{k-1}(\mathrm{d}v)\, \psi_k(\mathrm{d}L).$$

Proof. First we remark that

$$\mathcal{H}^{k-1}(ss_{k-1}(H_1, \ldots, H_n)) = \binom{n}{d-k} \omega_k.$$

In the following computations, ss_{k-1}, $F_{k,v}$ and η_{k-1} have the arguments $H_1, \ldots, H_n$, which we do not write down. Using (5.62), we obtain

$$\mathbb{E}\, h\left(W_{n,d}^{(k)}\right)$$

$$= \int_{G(d,d-1)^n} \int_{ss_{k-1}} h(F_{k,v})\, \eta_{k-1}(\mathrm{d}v)\, \phi^{*n}(\mathrm{d}(H_1, \ldots, H_n))$$

$$= \int_{G(d,d-1)^n} \frac{1}{\binom{n}{d-k}\omega_k} \int_{ss_{k-1}} h(F_{k,v})\, \mathcal{H}^{k-1}(\mathrm{d}v)\, \phi^{*n}(\mathrm{d}(H_1, \ldots, H_n)).$$

Since ss_{k-1} is the union of the $\binom{n}{d-k}$ sets $S_{i_1} \cap \cdots \cap S_{i_{d-k}}$, where $S_i := \mathbb{S}^{d-1} \cap H_i$, and each integral $\int_{S_{i_1} \cap \cdots \cap S_{i_{d-k}}}$ contributes the same to the total integral, we obtain

$$\mathbb{E}\, h\left(W_{n,d}^{(k)}\right) \qquad\qquad\qquad\qquad\qquad\qquad (5.63)$$

$$= \int_{G(d,d-1)^{d-k}} \frac{1}{\omega_k} \int_{S_1 \cap \cdots \cap S_{d-k}} \int_{G(d,d-1)^{n-d+k}} h(F_{k,v})$$

$$\times \phi^{*n-d+k}(\mathrm{d}(H_{d-k+1}, \ldots, H_n))\, \mathcal{H}^{k-1}(\mathrm{d}v)\, \phi^{*d-k}(\mathrm{d}(H_1, \ldots, H_{d-k})).$$

For fixed $H_1, \ldots, H_{d-k}$ with $v \in L := H_1 \cap \cdots \cap H_{d-k}$, we have

$$F_{k,v}(H_1, \ldots, H_n) = F_{d,v}(H_{d-k+1}, \ldots, H_n) \cap L.$$

This, together with the definition of the measure ψ_k, gives

$$\mathbb{E}\, h\left(W_{n,d}^{(k)}\right)$$

$$= \frac{1}{\omega_k} \int_{G(d,k)} \int_{L \cap \mathbb{S}^{d-1}} \int_{G(d,d-1)^{n-d+k}} h(F_{d,v}(H_{d-k+1}, \ldots, H_n) \cap L)$$

$$\times \phi^{*n-d+k}(\mathrm{d}(H_{d-k+1}, \ldots, H_n))\, \mathcal{H}^{k-1}(\mathrm{d}v)\, \psi_k(\mathrm{d}L)$$

$$= \frac{1}{\omega_k} \int_{G(d,k)} \int_{L \cap \mathbb{S}^{d-1}} \mathbb{E}\, h(F_{d,v}(\mathcal{H}_1, \ldots, \mathcal{H}_{n-d+k}) \cap L)\, \mathcal{H}^{k-1}(\mathrm{d}v)\, \psi_k(\mathrm{d}L),$$

which completes the proof.   □

In the isotropic case, the situation simplifies again. We fix a vector $e \in \mathbb{S}^{d-1}$ and a k-dimensional subspace $E_k \in G(d,k)$ with $e \in E_k$.

Corollary 5.5.2. *Let $k \in \{1, \ldots, d\}$ and $n \geq d$. Suppose that $W_{n,d}^{(k)}$ is the weighted typical k-face of the tessellation induced by n independent, isotropic random hyperplanes in $\mathbb{R}^d$. Let $h : C^d \to \mathbb{R}$ be a rotation invariant, non-negative, measurable function. Let $\mathcal{H}_1, \ldots, \mathcal{H}_{n-d+k}$ be independent random hyperplanes with the isotropic distribution on $G(d, d-1)$. Then*

$$\mathbb{E}\, h\left(W_{n,d}^{(k)}\right) = \mathbb{E}\, h(F_{d,e}(\mathcal{H}_1, \ldots, \mathcal{H}_{n-d+k}) \cap E_k).$$

Proof. Theorem 5.5.3 says that

$$\mathbb{E}\, h\left(W_{n,d}^{(k)}\right)$$

$$= \frac{1}{\omega_k} \int_{G(d,k)} \int_{L \cap \mathbb{S}^{d-1}} \mathbb{E}\, h(F_{d,v}(\mathcal{H}_1, \ldots, \mathcal{H}_{n-d+k}) \cap L)\, \mathscr{H}^{k-1}(\mathrm{d}v)\, \psi_k(\mathrm{d}L).$$

In the isotropic case, the integrand is constant. In fact, for given $L \in G(d,k)$ and $v \in L \cap \mathbb{S}^{d-1}$, there is a rotation $\rho \in SO(d)$ with $\rho E_k = L$ and $\rho e = v$ (since any rotation σ with $\sigma E_k = L$ satisfies $\sigma e \in L$). The rotation invariance of the function h and that of the distribution of $\mathcal{H}_1, \ldots, \mathcal{H}_{n-d+k}$ now give

$$\mathbb{E}\, h(F_{d,v}(\mathcal{H}_1, \ldots, \mathcal{H}_{n-d+k}) \cap L)$$
$$= \mathbb{E}\, h(\rho^{-1}[F_{d,v}(\mathcal{H}_1, \ldots, \mathcal{H}_{n-d+k}) \cap L])$$
$$= \mathbb{E}\, h(F_{d,\rho^{-1}v}(\rho^{-1}\mathcal{H}_1, \ldots, \rho^{-1}\mathcal{H}_{n-d+k}) \cap \rho^{-1}L)$$
$$= \mathbb{E}\, h(F_{d,e}(\rho^{-1}\mathcal{H}_1, \ldots, \rho^{-1}\mathcal{H}_{n-d+k}) \cap E_k)$$
$$= \mathbb{E}\, h(F_{d,e}(\mathcal{H}_1, \ldots, \mathcal{H}_{n-d+k}) \cap E_k).$$

When this is inserted, the assertion is obtained.   □

We return to the case of a general distribution ϕ^*. The following theorem, which is similar to Lemma 5.4.1, shows that the distribution of $W_{n,d}^{(k)}$ is the weighted distribution of $S_{n,d}^{(k)}$. The weight is the solid angle, or the normalized $(k-1)$-dimensional spherical measure of the intersection with the unit sphere.

Theorem 5.5.4. *Let $k \in \{1, \ldots, d\}$ and $n \geq d$. Let $S_{n,d}^{(k)}$ be the typical k-face and $W_{n,d}^{(k)}$ the weighted typical k-face of the tessellation induced by n independent random hyperplanes with distribution ϕ^*. Let $h : C^d \to \mathbb{R}$ be a nonnegative, measurable function. Then*

$$\mathbb{E}\, h\left(W_{n,d}^{(k)}\right) = \frac{\mathbb{E}\, (hv_k)\left(S_{n,d}^{(k)}\right)}{\mathbb{E}\, v_k\left(S_{n,d}^{(k)}\right)}.$$

Proof. We use (5.60) and then argue as in the proof of Theorem 5.5.1, to obtain

$$\mathbb{E}\,(hv_k)\left(S_{n,d}^{(k)}\right)$$

$$= \int_{G(d,d-1)} \frac{1}{C(n,d,k)} \sum_{F\in\mathcal{F}_k(H_1,\ldots,H_n)} h(F)v_k(F)\,\phi^{*n}(\mathrm{d}(H_1,\ldots,H_n))$$

$$= \frac{\binom{n}{d-k}}{C(n,d,k)} \int_{G(d,d-1)^{d-k}} \int_{G(d,d-1)^{n-d+k}}$$

$$\sum_{F\in\mathcal{F}_k(H_{d-k+1},\ldots,H_n;H_1\cap\cdots\cap H_{d-k})} h(F)v_k(F)$$

$$\times \phi^{*n-d+k}(\mathrm{d}(H_{d-k+1},\ldots,H_n))\,\phi^{*d-k}(\mathrm{d}(H_1,\ldots,H_d))$$

$$= \frac{\binom{n}{d-k}}{C(n,d,k)\omega_k} \int_{G(d,d-1)^{d-k}} \int_{S_1\cap\cdots\cap S_{d-k}} \int_{G(d,d-1)^{n-d+k}}$$

$$\sum_{F\in\mathcal{F}_k(H_{d-k+1},\ldots,H_n;H_1\cap\cdots\cap H_{d-k})} h(F)\mathbf{1}\{v\in F\}$$

$$\times \phi^{*n-d+k}(\mathrm{d}(H_{d-k+1},\ldots,H_n))\,\mathscr{H}^{k-1}(\mathrm{d}v)\,\phi^{*d-k}(\mathrm{d}(H_1,\ldots,H_{d-k})),$$

where $S_i := \mathbb{S}^{d-1}\cap H_i$, as in the proof of Theorem 5.5.3. For ϕ^{*d-k}-almost all $(H_1,\ldots,H_{d-k})$ and ϕ^{*n-d+k}-almost all $(H_{d-k+1},\ldots,H_n)$, the relation

$$\mathbf{1}\{v\in F\} = 1$$

is satisfied for precisely one face $F\in\mathcal{F}_k(H_{d-k+1},\ldots,H_n;H_1\cap\cdots\cap H_{d-k})$, which has been denoted by $F_{k,v} = F_{k,v}(H_1,\ldots,H_n)$. Therefore, we get

$$\mathbb{E}\,(hv_k)\left(S_{n,d}^{(k)}\right)$$

$$= \frac{\binom{n}{d-k}}{C(n,d,k)\omega_k} \int_{G(d,d-1)^{d-k}} \int_{S_1\cap\cdots\cap S_{d-k}} \int_{G(d,d-1)^{n-d+k}} h(F_{k,v})$$

$$\times \phi^{*n-d+k}(\mathrm{d}(H_{d-k+1},\ldots,H_n))\,\mathscr{H}^{k-1}(\mathrm{d}v)\,\phi^{*d-k}(\mathrm{d}(H_1,\ldots,H_d)).$$

Applying this with $h\equiv 1$, we obtain

$$\mathbb{E}\,v_k\left(S_{n,d}^{(k)}\right) = \frac{\binom{n}{d-k}}{C(n,d,k)}.$$

Division yields

$$\frac{1}{\mathbb{E}\,v_k\left(S_{n,d}^{(k)}\right)}\,\mathbb{E}\,(hv_k)\left(S_{n,d}^{(k)}\right)$$

$$= \int_{G(d,d-1)^{d-k}} \frac{1}{\omega_k} \int_{S_1 \cap \cdots \cap S_{d-k}} \int_{G(d,d-1)^{n-d+k}} h(F_{k,v})$$

$$\times \phi^{*n-d+k}(\mathrm{d}(H_{d-k+1},\ldots,H_n)) \, \mathcal{H}^{k-1}(\mathrm{d}v) \, \phi^{*d-k}(\mathrm{d}(H_1,\ldots,H_{d-k}))$$

$$= \mathbb{E}\, h(W_n^{(k)})$$

by (5.63). This was the assertion. $\qquad\square$

We recall that the random cone S_n^e was introduced in Section 5.4 as the a.s. unique cone induced by n independent isotropic random hyperplanes that contains the fixed vector e. We observe the following.

Lemma 5.5.1. *Let h be a nonnegative, measurable and rotation invariant function on $\mathcal{PC}^d$. Then, in the isotropic case,*

$$\mathbb{E}\, h(S_n^e) = \mathbb{E}\, h\left(W_{n,d}^{(d)}\right).$$

Proof. By Lemma 5.4.1,

$$\mathbb{E}\, h(S_n^e) = C(n,d)\mathbb{E}\,(hv_d)(S_n).$$

By Theorem 5.5.4,

$$\mathbb{E}\, h\left(W_{n,d}^{(d)}\right) = \frac{\mathbb{E}\,(hv_d)(S_n)}{\mathbb{E}\, v_d(S_n)}.$$

Since $\mathbb{E}\, v_d(S_n) = C(n,d)^{-1}$ by (5.44), the assertion follows. $\qquad\square$

We can now extend Theorem 5.4.2 to lower-dimensional weighted faces.

Theorem 5.5.5. *Let $n \geq d+1$, $k \in \{1,\ldots,d\}$, $\ell \in \{0,\ldots,k-1\}$. Let $W_{n,d}^{(k)}$ be the weighted typical k-face induced by n independent isotropic random hyperplanes. Then*

$$\mathbb{E}\, f_{k-\ell}\left(W_{n,d}^{(k)}\right) = 2\binom{n-d+k}{\ell}\mathbb{E}\, U_\ell\left(W_{n-\ell,d}^{(k)}\right).$$

Proof. Corollary 5.5.2 (with the notation used there) gives

$$\mathbb{E}\, f_{k-\ell}\left(W_{n,d}^{(k)}\right) = \mathbb{E}\, f_{k-\ell}(F_{d,e}(\mathcal{H}_1,\ldots,\mathcal{H}_{n-d+k}) \cap E_k)$$

$$= \mathbb{E}\, f_{k-\ell}\left(W_{n-d+k,k}^{(k)}\right).$$

Using Lemma 5.5.1 and (5.58), we obtain

$$\mathbb{E}\, f_{k-\ell}\left(W_{n,d}^{(k)}\right) = 2\binom{n-d+k}{\ell}\mathbb{E}\, U_\ell\left(W_{n-d+k-\ell,k}^{(k)}\right).$$

Again by Corollary 5.5.2, we have

$$\mathbb{E}\, U_\ell\left(W^{(k)}_{n-\ell,d}\right) = \mathbb{E}\, U_\ell(F_{d,e}(\mathcal{H}_1,\dots,\mathcal{H}_{n-\ell-d+k}\cap E_k)$$

$$= \mathbb{E}\, U_\ell\left(W^{(k)}_{n-d+k-l,k}\right).$$

This gives the assertion.   □

We close this section with a remark on directions. The typical k-face $S^{(k)}_{n,d}$ and the weighted typical k-face $W^{(k)}_{n,d}$ have, somewhat surprisingly, the same directional distribution. The *direction* $D(C)$ of a cone $C \in \mathcal{C}^d$ is just its linear hull,

$$D(C) := \lim C.$$

Theorem 5.5.6. *Let* $k \in \{1,\dots,d\}$ *and* $n \ge d$. *Let* $S^{(k)}_{n,d}$ *be the typical k-face and* $W^{(k)}_{n,d}$ *the weighted typical k-face of the tessellation induced by n independent random hyperplanes with distribution* ϕ^*. *Then*

$$\mathbb{P}\left(D\left(S^{(k)}_{n,d}\right) \in A\right) = \mathbb{P}\left(D\left(W^{(k)}_{n,d}\right) \in A\right) = \psi_k(A)$$

for $A \in \mathcal{B}(G(d,k))$.

Proof. By Theorem 5.5.1, with $h(C) = \mathbb{1}\{D(C) \in A\}$,

$$\mathbb{P}\left(D\left(S^{(k)}_{n,d}\right) \in A\right) = \int_{D(d,k)} \mathbb{P}\left(D\left(S_{n-d+k,L}\right) \in A\right) \psi_k(dL)$$

$$= \int_{G(d,k)} \mathbb{1}\{L \in A\}\, \psi_k(dL)$$

$$= \psi_k(A).$$

Similarly, Theorem 5.5.3 gives

$$\mathbb{P}\left(D\left(W^{(k)}_{n,d}\right) \in A\right)$$

$$= \frac{1}{\omega_k}\int_{G(d,k)}\int_{L\cap\mathbb{S}^{d-1}} \mathbb{P}\left(D(F_{d,v}(\mathcal{H}_1,\dots,\mathcal{H}_{n-d+k})\cap L) \in A\right)$$

$$\times \mathscr{H}^{k-1}(dv)\,\psi_k(dL)$$

$$= \frac{1}{\omega_k}\int_{G(d,k)}\int_{L\cap\mathbb{S}^{d-1}} \mathbb{1}\{L \in A\}\,\mathscr{H}^{k-1}(dv)\,\psi_k(dL)$$

$$= \psi_k(A).$$

This completes the proof.   □

Notes for Section 5.5

1. Kabluchko and Thäle [110] have more results (on spherical tessellations, which can be re-interpreted in terms of conical tessellations). For example, as a counterpart to Theorem 5.5.2, they determine the expectations $\mathbb{E} f_\ell\left(W_{n,d}^{(k)}\right)$ and $\mathbb{E} U_\ell\left(W_{n,d}^{(k)}\right)$, but only in the isotropic case. This is based on earlier deeper work and is much less elementary. They also (see Section 5.1) ask the following question (in an equivalent version):

Question. Is it true, in the isotropic case, that

$$\mathbb{E} f_\ell\left(W_{n,d}^{(k)}\right) > \mathbb{E} f_\ell\left(S_{n,d}^{(k)}\right),$$

whenever $n \geq d+1$ and $\ell \in \{1,\ldots,k-1\}$?

2. Kabluchko and Thäle [110] also determine, for the typical k-face and for the weighted typical k-face, the regular conditional distribution, given the direction. Here the 'direction' of a face of a polyhedral cone is just the linear hull of the cone.

5.6 Intersections of random cones

Having defined Schläfli cones and Cover–Efron cones, it is time to modify an earlier question. We recall that the following was quoted in Section 4.3: "When does a randomly oriented cone strike a fixed cone?" A more precise version of this question reads as follows. Given are two closed convex cones $C, D \in \mathcal{C}^d$, not both subspaces, and a uniform random rotation θ in $\mathrm{SO}(d)$. We ask for the probability

$$\mathbb{P}(C \cap \theta D \neq \{o\}). \tag{5.64}$$

The conic kinematic formula provides the answer

$$\mathbb{P}(C \cap \theta D \neq \{o\}) = 2 \sum_{k=0}^{\lfloor \frac{d-1}{2} \rfloor} \sum_{j=2k+1}^{d} v_j(C) v_{d+2k+1-j}(D),$$

see (4.73).

Here the randomness of the cone θD comes only from the random rotation θ, which is applied to the fixed cone D. The question arises whether a probability of type (5.64) can also be determined for more general random cones, where also the shape is random and not only the position. Positive answers are possible for random cones with an isotropic distribution.

Theorem 5.6.1. *Let $C \in \mathcal{C}^d$ be a closed convex cone, not a subspace. Let D be a random cone in $\mathcal{C}^d$, whose distribution is isotropic, that is, invariant under $\mathrm{SO}(d)$. Then*

$$\mathbb{P}(C \cap \mathbf{D} \neq \{o\}) = 2 \sum_{k=0}^{\lfloor \frac{d-1}{2} \rfloor} \sum_{j=2k+1}^{d} v_j(C) \, \mathbb{E} \, v_{d+2k+1-j}(\mathbf{D}).$$

Proof. Let $\mathbb{Q}$ be the distribution of $\mathbf{D}$. For each rotation $\vartheta \in \mathrm{SO}(d)$, the random cones $\mathbf{D}$ and $\vartheta \mathbf{D}$ have the same distribution, since $\mathbb{Q}$ is invariant under rotations. Therefore,

$$\mathbb{P}(C \cap \mathbf{D} \neq \{o\}) = \int_{\mathcal{C}^d} \mathbb{1}\{C \cap \vartheta D \neq \{o\}\} \, \mathbb{Q}(\mathrm{d}D).$$

We can integrate this over all ϑ with the invariant probability measure on the rotation group and then apply Fubini's theorem, to obtain

$$\mathbb{P}(C \cap \mathbf{D} \neq \{o\}) = \int_{\mathrm{SO}(d)} \int_{\mathcal{C}^d} \mathbb{1}\{C \cap \vartheta D \neq \{o\}\} \, \mathbb{Q}(\mathrm{d}D) \, \nu(\mathrm{d}\vartheta)$$

$$= \int_{\mathcal{C}^d} \int_{\mathrm{SO}(d)} \mathbb{1}\{C \cap \vartheta D \neq \{o\}\} \, \nu(\mathrm{d}\vartheta) \, \mathbb{Q}(\mathrm{d}D)$$

$$= \int_{\mathcal{C}^d} 2 \sum_{k=0}^{\lfloor \frac{d-1}{2} \rfloor} \sum_{j=2k+1}^{d} v_j(C) v_{d+2k+1-j}(D) \, \mathbb{Q}(\mathrm{d}D)$$

$$= 2 \sum_{k=0}^{\lfloor \frac{d-1}{2} \rfloor} \sum_{j=2k+1}^{d} v_j(C) \, \mathbb{E} \, v_{d+2k+1-j}(\mathbf{D}),$$

where the kinematic formula (4.71) was used. $\qquad \square$

In particular, for the isotropic random Schläfli cone S_n, we know the expected intrinsic volumes by (5.44), hence we obtain the following result.

Theorem 5.6.2. *Let $C \in \mathcal{C}^d$ be a closed convex cone, not a subspace. The isotropic random Schläfli cone S_n satisfies*

$$\mathbb{P}(C \cap S_n \neq \{o\}) = \frac{2}{C(n,d)} \sum_{j=1}^{n} \sum_{k=0}^{\lfloor \frac{j-1}{2} \rfloor} \binom{n}{j-1-2k} v_j(C).$$

Using (5.45), we obtain similar explicit results for the isotropic random Cover–Efron cone.

We can also ask for nontrivial intersections of two independent isotropic random cones.

Theorem 5.6.3. *If $\mathbf{C}$, $\mathbf{D}$ are stochastically independent isotropic random cones in $\mathcal{C}^d$, then*

$$\mathbb{P}(\mathbf{C} \cap \mathbf{D} \neq \{o\}) = 2 \sum_{k=0}^{\lfloor \frac{d-1}{2} \rfloor} \sum_{j=2k+1}^{d} \mathbb{E} \, v_j(\mathbf{C}) \, \mathbb{E} \, v_{d+2k+1-j}(\mathbf{D}).$$

Proof. Let $\mathbb{T}$ and $\mathbb{Q}$ be the distributions of $\mathbf{C}$ and $\mathbf{D}$, respectively. The independence of $\mathbf{C}$ and $\mathbf{D}$ allows us to write

$$\mathbb{P}(\mathbf{C} \cap \mathbf{D} \neq \{o\}) = \int_{\mathcal{C}^d} \int_{\mathcal{C}^d} \mathbb{1}\{C \cap D \neq \{o\}\} \, \mathbb{Q}(dC) \, \mathbb{T}(dC)$$

$$= \int_{\mathcal{C}^d} \mathbb{P}(C \cap \mathbf{D} \neq \{o\}) \, \mathbb{T}(dC)$$

$$= \int_{\mathcal{C}^d} 2 \sum_{k=0}^{\lfloor \frac{d-1}{2} \rfloor} \sum_{j=2k+1}^{d} v_j(C) \, \mathbb{E} \, v_{d+2k+1-j}(D) \, \mathbb{T}(dC)$$

$$= 2 \sum_{k=0}^{\lfloor \frac{d-1}{2} \rfloor} \sum_{j=2k+1}^{d} \mathbb{E} \, v_j(\mathbf{C}) \, \mathbb{E} \, v_{d+2k+1-j}(\mathbf{D}),$$

as stated.   □

Again, we can apply this to isotropic random cones for which the expectations of the conic intrinsic volumes can be determined.

Notes for Section 5.6

1. The results of this section were noted in [167]. The essential properties of the random cones yielding explicit results are their isotropy and the fact that there are formulas for the expectations of their conic intrinsic volumes. This holds also for the weighted faces considered in Section 5.5. For these, theorems corresponding to Theorem 5.6.2 were proved by Kabluchko and Thäle [110].

6

Miscellanea on random cones

A natural way to generate a random cone is to apply a random linear map to a fixed cone. If the distribution of the random linear mapping allows it, it may be possible to obtain explicit results for some expected geometric functionals of the random cone. The brief Section 6.1 deals with uniform random orthogonal projections of polyhedral cones (or general convex polyhedra). Section 6.2 treats images of general convex cones under linear maps defined by Gaussian matrices.

The starting point of Section 6.4 is a quite general linear image of a special cone, the nonnegative orthant in a higher-dimensional space. This is equivalent to a simple way of generating a random polyhedral cone, namely by taking the positive hull of a fixed number of stochastically independent random vectors with a reasonable distribution. Here we are interested in expected face numbers when the dimension tends to infinity. The investigation of random cones in high dimensions is initiated in Section 6.3 and is continued in Section 6.5, where Cover–Efron cones are studied from this point of view. Section 6.6 deals with random cones in halfspaces.

6.1 Random projections

A simple way to generate a random cone is to apply a random linear map to a fixed cone. We recall here an older result about random projections of polyhedra. It can, in particular, be applied to polyhedral cones. The following may also be viewed as an application of Grassmann angles.

We consider face numbers of projections of polyhedra. Let $L \in G(d, n)$ be a linear subspace, and let $\Pi_L : \mathbb{R}^d \to L$ be the orthogonal projection to L. We also write $\Pi_L(\cdot) = \cdot | L$. The kernel of Π_L is the orthogonal complement $L^\perp$. Let $P \in \mathcal{Q}^d$ be a polyhedron, and let $F \in \mathcal{F}_k(P)$, with $k < \dim L$.

The subspace L is said to be *in general position* with respect to the polyhedron P if L is in general position with respect to $\langle F \rangle$ for all faces F of P.

R. Schneider, *Convex Cones*, Lecture Notes in Mathematics 2319,
https://doi.org/10.1007/978-3-031-15127-9_6

Under this condition, the following lemma describes the faces of the projection $P|L$.

Lemma 6.1.1. *Let $P \in \mathcal{Q}^d$ be a d-dimensional polyhedron, let $L \in G(d, n)$ with $n \in \{1, \ldots, d-1\}$ be a subspace which is in general position with respect to P. If the face $F \in \mathcal{F}_k(P)$ with $k < n$ satisfies*

$$L^\perp \cap A(F, P) = \{o\}, \tag{6.1}$$

then $F|L \in \mathcal{F}_k(P|L)$. Each k-face of $P|L$ is obtained in this way.

Proof. Let $F \in \mathcal{F}_k(P)$ be a face satisfying (6.1). Since $L^\perp \cap \langle F \rangle = \{o\}$, the space $\langle F \rangle$ is mapped under Π_L to a k-dimensional subspace, thus $\dim F|L = k$.

Suppose that the normal cone $N(P, F)$ satisfies $L \cap \operatorname{relint} N(P, F) = \emptyset$. Then $L^\perp \cap N(P, F)^\circ \subsetneq \operatorname{lineal}(N(P, F)^\circ)$ by Lemma 1.3.9, thus $L^\perp \cap A(F, P) \subsetneq \operatorname{lineal}(A(F, P)) = \langle F \rangle$, a contradiction. Hence, there is a unit vector $u \in L \cap \operatorname{relint} N(P, F)$. The supporting hyperplane H of P with outer normal vector u satisfies $H \cap P = F$. From $u \in L$ it follows that $L^\perp + F \subset H$. Therefore, $H|L$ is a supporting hyperplane to $P|L$ in L with $(H|L) \cap (P|L) = F|L$. We conclude that $F|L$ is a k-face of $P|L$.

Conversely, suppose that $G \in \mathcal{F}_k(P|L)$. Since $k < \dim L$, in L there is a supporting hyperplane to $P|L$ that contains G. Its pre-image H with respect to Π_L is a supporting hyperplane of P. Let $F := H \cap P$; then $F|L = G$. Since $L^\perp \cap A(F, P) = \{o\}$, we have $\dim F = k$. □

As above, let $P \in \mathcal{Q}^d$ be a d-dimensional polyhedron, and let $L \in G(d, n)$ be a linear subspace of dimension $n \in \{1, \ldots, d-1\}$. Let $k \in \{0, \ldots, n-1\}$ and $F \in \mathcal{F}_k(P)$. If L is in general position with respect to P, then Lemma 6.1.1 shows that

$$F|L \text{ is a } k\text{-face of } P|L \Leftrightarrow A(F, P) \cap L^\perp = \{o\}.$$

Now let $\mathbf{L}$ be a uniform random element of $G(d, n)$ (that is, with distribution ν_n, the rotation invariant probability measure on $G(d, n)$). Almost surely, $\mathbf{L}$ is in general position with respect to P, hence

$$\mathbb{P}(F|\mathbf{L} \text{ is a } k\text{-face of } P|\mathbf{L}) = \mathbb{P}(A(F, P) \cap \mathbf{L}^\perp = \{o\}).$$

As a side remark, we can formulate the following consequence of Theorem 4.4.2 (as suggested in [11, p. 265]). Here $\lambda \geq 0$ is arbitrary, and $p_C(\lambda)$ (for $C \in \mathcal{C}^d$) is defined in Corollary 4.4.1. We have

$$\delta(A(F, P)) \leq n - \lambda \Rightarrow \mathbb{P}(F|\mathbf{L} \text{ is a } k\text{-face of } P|\mathbf{L}) \leq p_{A(F,P)}(\lambda),$$

$$\delta(A(F, P)) \geq n + \lambda \Rightarrow \mathbb{P}(F|\mathbf{L} \text{ is a } k\text{-face of } P|\mathbf{L}) \geq 1 - p_{A(F,P)}(\lambda).$$

Now we turn to the total number of k-dimensional faces of the projection $P|L$. It is given by

$$f_k(P|L) = \sum_{F \in \mathcal{F}_k(P)} \mathbb{1}\{L^\perp \cap A(F,P) = \{o\}\}.$$

Under the same assumptions on the random subspace $\mathbf{L}$ as above, we obtain, using Lemma 2.1.2 and the definition (2.43) of Grassmann angles, that

$$\mathbb{E}\, f_k(P|\mathbf{L}) = \sum_{F \in \mathcal{F}_k(P)} \int_{G(d,n)} \mathbb{1}\{L^\perp \cap A(F,P) = \{o\}\}\, \nu_n(\mathrm{d}L)$$

$$= \sum_{F \in \mathcal{F}_k(P)} (1 - 2U_n(A(F,P))).$$

Applying either (2.59) or (2.60), we obtain

$$\mathbb{E}\, f_k(P|\mathbf{L}) = f_k(P) - \sum_{F \in \mathcal{F}_k(P)} 2\sum_{s \geq 0} \upsilon_{n+2s+1}(A(F,P))$$

$$= \sum_{F \in \mathcal{F}_k(P)} 2\sum_{s \geq 0} \upsilon_{n-2s-1}(A(F,P)).$$

Now,

$$\upsilon_m(A(F,P)) = \sum_{G \in \mathcal{F}_m(A(F,P))} \beta(o,G)\gamma(G,A(F,P)).$$

The faces $G \in \mathcal{F}_m(A(F,P))$ are in one-to-one correspondence with the faces $G' \in \mathcal{F}_m(P)$ containing F, such that $\beta(o,G) = \beta(F,G')$ and $\gamma(G,A(F,P)) = \gamma(G',P)$. Therefore, we get the following theorem.

Theorem 6.1.1. *If* $\mathbf{L}$ *is an* n*-dimensional random linear subspace of* $\mathbb{R}^d$ *with uniform distribution (given by the measure* ν_n *on* $G(d,n)$*) and if* $P \in \mathcal{Q}^d$ *is a polyhedron, then*

$$\mathbb{E}\, f_k(P|\mathbf{L}) = f_k(P) - 2\sum_{s \geq 0} \sum_{G \in \mathcal{F}_{n+1+2s}(P)} \gamma(G,P) \sum_{F \in \mathcal{F}_k(G)} \beta(F,G)$$

$$= 2\sum_{s \geq 0} \sum_{G \in \mathcal{F}_{n-1-2s}(P)} \gamma(G,P) \sum_{F \in \mathcal{F}_k(G)} \beta(F,G).$$

Notes for Section 6.1

1. The last formulas were proved by Affentranger and Schneider [2], and independently by Vershik and Sporyshev [187]. These papers derived different asymptotic conclusions.

6.2 Gaussian images of cones

In this section, we follow Götze, Kabluchko and Zaporozhets [72] and present some results about images of convex cones under Gaussian random linear mappings from $\mathbb{R}^d$ to $\mathbb{R}^k$, where $k \in \mathbb{N}$. We can assume that $\mathbb{R}^d$ and $\mathbb{R}^k$ are equipped with fixed orthonormal bases and then denote a linear mapping $\mathsf{A} : \mathbb{R}^d \to \mathbb{R}^k$ and its matrix with respect to the given bases by the same symbol.

A random $k \times d$ matrix $\mathbf{G}$ is called a *standard Gaussian random matrix* if its entries are independent standard normal random variables. First let X be a $k \times 1$ standard Gaussian random matrix (a row vector). Then X has a d-dimensional normal distribution, with expectation vector o and covariance matrix the identity matrix. Let R be a fixed orthogonal $k \times k$ matrix. Then $\mathsf{R}X$ also has a d-dimensional normal distribution, with expectation vector o and covariance matrix the identity matrix. Its entries are pairwise uncorrelated and hence independent. Therefore, $\mathsf{R}X$ has the same distribution as X. If now $\mathbf{G}$ is a standard Gaussian random $k \times d$ matrix, it follows that the entries of $\mathsf{R}\mathbf{G}$ are independent standard normal random variables, thus $\mathsf{R}\mathbf{G}$ has the same distribution as $\mathbf{G}$. Let Let S be a fixed orthogonal $d \times d$ matrix. Then $(\mathbf{G}\mathsf{S})^\top = \mathsf{S}^\top \mathbf{G}^\top$, and $\mathsf{S}^\top \mathbf{G}^\top$ has the same distribution as $\mathbf{G}^\top$, hence $\mathbf{G}\mathsf{S}$ has the same distribution as $\mathbf{G}$.

Again let $\mathbf{G}$ be a standard Gaussian random $k \times d$ matrix, where now $k \leq d$. Let $\ker \mathbf{G}$ denote the kernel of $\mathbf{G}$. Then $\ker \mathbf{G}$ is a random subspace, a.s. of dimension $d - k$, and from the preceding observations it follows that its distribution is invariant under rotations. Therefore, if $\mathbf{L}_{d-k}$ denotes a $(d-k)$-dimensional uniform random subspace (a random element of $G(d, d-k)$ with distribution given by the normalized Haar measure), then

$$\ker \mathbf{G} \overset{d}{=} \mathbf{L}_{d-k}. \tag{6.2}$$

Before stating the first result, we establish a preparatory lemma.

Lemma 6.2.1. *Let $C \in \mathcal{C}^d$, let $k \in \mathbb{N}$, let $m := \min\{k, \dim C\}$, and let $\mathbf{G}$ be a $k \times d$ standard Gaussian random matrix. Then*

$$\mathbb{P}(\dim \mathbf{G}C = m) = 1.$$

Proof. Trivially, $\dim \mathbf{G}C \leq m$. Let $v_1, \ldots, v_m \in C$ be linearly independent vectors, and write $L_m := \lin\{v_1, \ldots, v_m\}$. If $\dim \mathbf{G}C \leq m - 1$, then $\mathbf{G}v_1, \ldots, \mathbf{G}v_m$ are linearly dependent, hence there are numbers $c_1, \ldots, c_m \in \mathbb{R}$, not all zero, such that

$$o = \sum_{i=1}^m c_i \mathbf{G}v_i = \mathbf{G}\sum_{i=1}^m c_i v_i.$$

This shows that $L_m \cap \ker \mathbf{G} \neq \{o\}$. Therefore,

$$\begin{aligned}
\mathbb{P}(\dim \mathbf{G}C \leq m-1) &\leq \mathbb{P}\big(L_m \cap \ker \mathbf{G} \neq \{o\}\big) \\
&\leq \mathbb{P}\big(L_m \cap \mathbf{L}_{d-k} \neq \{o\}\big) \\
&= 0,
\end{aligned}$$

where (6.2) was used and then Lemma 2.1.2, together with the fact that $m \leq k$. □

Now we can prove the main results of this section.

Theorem 6.2.1. *Let $C \in \mathcal{C}^d \setminus \mathcal{L}_\bullet$, let $k \in \mathbb{N}$, and let $\mathbf{G}$ be a standard Gaussian random $k \times d$ matrix. Then*

$$\mathbb{P}(o \in \mathrm{int} \mathbf{G} C) = 2U_k(C).$$

Proof. We have to show that

$$\mathbb{P}(\mathbf{G} C = \mathbb{R}^k) = 2U_k(C). \tag{6.3}$$

If $\dim C < k$, then both sides of (6.3) are zero, by Lemma 6.2.1 and (2.44). Hence, we assume that $\dim C \geq k$. By Lemma 6.2.1, we have $\mathbb{P}(\dim \mathbf{G} C = k) = 1$, hence $\mathbb{P}(\mathrm{lin}(\mathbf{G} C) = \mathbb{R}^k) = 1$. Using this, and then Lemma 1.3.11, (6.2), and Lemma 4.3.1, we obtain

$$\mathbb{P}(\mathbf{G} C = \mathbb{R}^k) = \mathbb{P}(\mathbf{G} C = \mathrm{lin}(\mathbf{G} C))$$
$$= \mathbb{P}((\mathrm{relint}\, C) \cap \ker \mathbf{G} \neq \emptyset)$$
$$= \mathbb{P}((\mathrm{relint}\, C) \cap \mathbf{L}_{d-k} \neq \emptyset)$$
$$= \mathbb{P}(C \cap \mathbf{L}_{d-k} \neq \{o\})$$
$$= 2U_k(C),$$

where we have finally used the assumption that C is not a subspace. This completes the proof. □

Theorem 6.2.2. *Let $C \in \mathcal{C}^d$, $k \in \mathbb{N}$, $m := \min\{k, \dim C\}$, and let $\mathbf{G}$ be a $k \times d$ standard Gaussian random matrix. Then*

$$\mathbb{E}\, U_j(\mathbf{G} C) = U_j(C)$$

for $j \in \{0, \ldots, m-1\}$.

Proof. First we recall that $\mathbf{G}$ and $\mathsf{R}_1 \mathbf{G} \mathsf{R}_2$ have the same distribution if $\mathsf{R}_1, \mathsf{R}_2$ are fixed orthogonal matrices, where R_1 is a $k \times k$-matrix and R_2 a $d \times d$-matrix. We refer to this fact in the following as the invariance of $\mathbf{G}$.

We can assume that $\dim C = d$. In fact, if $\dim C = \ell < d$, we can assume, by the invariance of $\mathbf{G}$, that $C \subseteq \mathrm{lin}\{e_1, \ldots, e_\ell\}$, where $(e_1, \ldots, e_d)$ is the standard basis of $\mathbb{R}^d$. Since the restriction of $\mathbf{G}$ to $\mathrm{lin}\{e_1, \ldots, e_\ell\}$ is also a standard Gaussian random matrix, we can replace $\mathbb{R}^d$ by the space $\mathrm{lin}\{e_1, \ldots, e_\ell\}$, where C has full dimension. In the following, we assume that $\dim C = d$. By Lemma 6.2.1, $\dim \mathbf{G} C = m$ almost surely.

Let $j \in \{0, \ldots, m-1\}$. Since $\mathbf{G} C \subseteq \mathbb{R}^k$ and $j < k$, we see from (2.43) and Lemma 4.3.1 that

$$U_j(\mathbf{G}C) = \frac{1}{2} \int_{G(k,k-j)} \mathbb{1}\{L \cap \operatorname{relint}(\mathbf{G}C) \neq \emptyset\} \, \nu_{k-j}(\mathrm{d}L).$$

With a fixed subspace $E_{k-j} \in G(k, k-j)$, we can write this, by (2.7), as

$$U_j(\mathbf{G}C) = \frac{1}{2} \int_{\mathrm{SO}(k)} \mathbb{1}\{\vartheta E_{k-j} \cap \operatorname{relint}(\mathbf{G}C) \neq \emptyset\} \, \nu_{(k)}(\mathrm{d}\vartheta),$$

where $\nu_{(k)}$ denotes the normalized Haar measure on $\mathrm{SO}(k)$. Therefore,

$$\mathbb{E}\, U_j(\mathbf{G}C) = \frac{1}{2} \int_{\mathrm{SO}(k)} \mathbb{P}(\vartheta E_{k-j} \cap \operatorname{relint}(\mathbf{G}C) \neq \emptyset) \, \nu_{(k)}(\mathrm{d}\vartheta).$$

By the invariance of $\mathbf{G}$, we have, for fixed $\vartheta \in \mathrm{SO}(k)$,

$$\mathbb{P}(\vartheta E_{k-j} \cap \operatorname{relint}(\mathbf{G}C) \neq \emptyset) = \mathbb{P}(E_{k-j} \cap \operatorname{relint}(\vartheta^{-1}\mathbf{G}C) \neq \emptyset)$$
$$= \mathbb{P}(E_{k-j} \cap \operatorname{relint}(\mathbf{G}C) \neq \emptyset),$$

thus

$$\mathbb{E}\, U_j(\mathbf{G}C) = \frac{1}{2}\mathbb{P}(E_{k-j} \cap \operatorname{relint}(\mathbf{G}C) \neq \emptyset). \tag{6.4}$$

By $\mathbf{G}^{-1}E_{k-j}$ we denote the linear subspace of $\mathbb{R}^d$ that is the preimage of E_{k-j} under $\mathbf{G}$. We state that

$$\mathbf{G}^{-1}E_{k-j} \cap \operatorname{int} C \neq \emptyset \;\Rightarrow\; E_{k-j} \cap \operatorname{relint}(\mathbf{G}C) \neq \emptyset \tag{6.5}$$

and that

$$\mathbb{P}(E_{k-j} \cap \operatorname{relint}(\mathbf{G}C) \neq \emptyset \wedge \mathbf{G}^{-1}E_{k-j} \cap \operatorname{int} C = \emptyset) = 0. \tag{6.6}$$

Both relations together yield

$$\mathbb{P}(E_{k-j} \cap \operatorname{relint}(\mathbf{G}C) \neq \emptyset) = \mathbb{P}(\mathbf{G}^{-1}E_{k-j} \cap \operatorname{int} C \neq \emptyset)$$

and hence, by (6.4), that

$$\mathbb{E}\, U_j(\mathbf{G}C) = \frac{1}{2}\mathbb{P}(\mathbf{G}^{-1}E_{k-j} \cap \operatorname{int} C \neq \emptyset). \tag{6.7}$$

When this is proved, we can complete the proof as follows. We note that $j \leq m - 1 < \min\{k, d\}$. Almost surely, the matrix $\mathbf{G}$ has full rank, and $\dim \mathbf{G}^{-1}E_{k-j} = d - j$. By the invariance of $\mathbf{G}$, the distribution of $\mathbf{G}^{-1}E_{k-j}$ is invariant under rotations, hence $\mathbf{G}^{-1}E_{k-j}$ has the same distribution as a uniform random subspace $\mathbf{L}_{d-j}$ in $G(d, d-j)$. It follows from (6.7) that

$$\mathbb{E}\, U_j(\mathbf{G}C) = \frac{1}{2}\mathbb{P}(\mathbf{L}_{d-j} \cap \operatorname{int} C \neq \emptyset) = U_j(C),$$

where again (2.43) and Lemma 4.3.1 were used. This is the assertion.

It remains to prove (6.5) and (6.6). Suppose that $\mathbf{G}^{-1}E_{k-j} \cap \operatorname{int} C \neq \emptyset$ holds. Then there exists $x \in \operatorname{int} C$ such that $\mathbf{G}x \in E_{k-j}$. In $\mathbb{R}^d$, there exists an open neighborhood U of x with $U \subset C$. We have $\mathbf{G}x \in \mathbf{G}C$, and the image $\mathbf{G}U$ is relatively open in $\mathbf{G}\operatorname{lin} C = \operatorname{lin}(\mathbf{G}C)$. It follows that $\mathbf{G}x \in E_{k-j} \cap \operatorname{relint}(\mathbf{G}C)$. This proves (6.5).

To prove (6.6), consider the event

$$\mathcal{E} := \{E_{k-j} \cap \operatorname{relint}(\mathbf{G}C) \neq \emptyset \wedge \mathbf{G}^{-1}E_{k-j} \cap \operatorname{int} C = \emptyset\}.$$

We cannot have $E_{k-j} \cap \operatorname{relint}(\mathbf{G}C) = \{o\}$, since this would imply that the cone $\mathbf{G}C$ is a subspace, of dimension $\leq j$, which is a contradiction, by Lemma 6.2.1. Hence, we can choose $y \in E_{k-j} \cap \operatorname{relint}(\mathbf{G}C)$ with $y \neq o$. Then there exists $x \in C$ with $y = \mathbf{G}x$ and $x \neq o$. Since $x \in \mathbf{G}^{-1}E_{k-j}$, we have $\mathbf{G}^{-1}E_{k-j} \cap C \neq \{o\}$. We have proved that

$$\mathcal{E} \subset \{\mathbf{G}^{-1}E_{k-j} \cap C \neq \emptyset \wedge \mathbf{G}^{-1}E_{k-j} \cap \operatorname{int} C = \emptyset\}.$$

As remarked above, $\mathbf{G}^{-1}E_{k-j}$ has the same distribution as a uniform random subspace $\mathbf{L}_{d-j}$ in $G(d, d-j)$. Therefore, it follows from Lemma 4.3.1 that $\mathbb{P}(\mathcal{E}) = 0$. This proves (6.6) and thus completes the proof. $\qquad\square$

6.3 Wendel probabilities in high dimensions

We assume that ϕ is a probability distribution on $\mathbb{R}^d$ that is symmetric with respect to o and assigns measure zero to each hyperplane through o. We consider $n \geq d$ independent random vectors $X_1, \ldots, X_n$ with distribution ϕ. Their positive hull

$$D_n = \operatorname{pos}\{X_1, \ldots, X_n\}$$

is a polyhedral random cone. Perhaps the most basic question one can ask about this cone is whether it is different from $\mathbb{R}^d$. If it is, then with probability one it is a pointed cone. The probability of this event is given by the Wendel probability

$$P_{d,n} = \mathbb{P}\left(\operatorname{pos}\{X_1, \ldots, X_n\} \neq \mathbb{R}^d\right),$$

introduced in Section 5.2 and given explicitly by

$$P_{d,n} = \frac{1}{2^{n-1}} \sum_{i=0}^{d-1} \binom{n-1}{i}.$$

It is useful in the following that the number $P_{d,n}$ has a different probabilistic interpretation. In fact, let ξ_n denote a random variable with binomial distribution $B(n, 1/2)$, so that

$$\mathbb{P}(\xi_n = k) = \frac{1}{2^n}\binom{n}{k} \quad \text{for } k \in \mathbb{N}_0.$$

Then

$$P_{d,n} = \mathbb{P}(\xi_{n-1} \leq d-1). \tag{6.8}$$

Below, this will allow us to make use of large deviation estimates and the central limit theorem.

We shall now be interested in the asymptotic behavior of the Wendel probabilities if the dimension d tends to infinity and the number n grows with it, in a linearly coordinated way; we will then denote it by N. (The dependence of the number N on the dimension d will usually not be shown by the notation.) The results of this section are taken from Hug and Schneider [100].

Theorem 6.3.1. *Let the integer N depend on d such that*

$$\frac{d}{N} \to \delta \quad as \ d \to \infty,$$

with a number $\delta \in [0,1]$. Then the Wendel probabilities satisfy

$$\lim_{d \to \infty} P_{d,N} = \begin{cases} 1 \ if \ 1/2 < \delta \leq 1, \\ 0 \ if \ 0 \leq \delta < 1/2. \end{cases}$$

Proof. Let $\delta > \frac{1}{2}$. If d is sufficiently large, we have $\frac{d}{N-1} - \frac{1}{2} \geq \frac{1}{\sqrt{2}}\left(\delta - \frac{1}{2}\right) > 0$. Therefore, an estimate of Okamoto [143, Theorem 2(i)] for random variables with binomial distributions yields

$$\mathbb{P}(\xi_{N-1} \geq d) = \mathbb{P}\left(\frac{\xi_{N-1}}{N-1} - \frac{1}{2} \geq \frac{d}{N-1} - \frac{1}{2}\right)$$
$$\leq \exp\left(-2\left(\frac{d}{N-1} - \frac{1}{2}\right)^2 (N-1)\right).$$

It follows that

$$P_{d,N} = \mathbb{P}(\xi_{N-1} \leq d-1) \geq 1 - \exp\left(-\left(\delta - \frac{1}{2}\right)^2 (N-1)\right) \to 1$$

as $d \to \infty$.

Let $\delta < \frac{1}{2}$. If d is sufficiently large, we have $\frac{1}{2} - \frac{d-1}{N-1} \geq \frac{1}{\sqrt{2}}\left(\frac{1}{2} - \delta\right) > 0$. Therefore,

$$\mathbb{P}(\xi_{N-1} \leq d-1) = \mathbb{P}\left(\frac{\xi_{N-1}}{N-1} - \frac{1}{2} \leq -\left(\frac{1}{2} - \frac{d-1}{N-1}\right)\right)$$
$$\leq \exp\left(-2\left(\frac{d-1}{N-1} - \frac{1}{2}\right)^2 (N-1)\right)$$

by [143, Theorem 2(ii)]. It follows that

$$P_{d,N} = \mathbb{P}(\xi_{N-1} \le d-1) \le \exp\left(-\left(\delta - \frac{1}{2}\right)^2 (N-1)\right) \to 0$$

as $d \to \infty$. $\qquad\qquad\qquad\qquad\qquad\qquad\qquad\qquad\qquad\qquad\qquad\qquad$ $\square$

We have here a very simple example of a threshold phenomenon in stochastic geometry: when the parameter δ passes the value $1/2$ (where the value of N is around $2d$), then the limit of $P_{d,N}$ changes abruptly from 0 to 1. The question arises, what happens at $\delta = 1/2$. We note that $\frac{d}{N} \to \frac{1}{2}$ is equivalent to $\frac{N-2d}{d} \to 0$. It turns out that the order of this convergence is relevant for the limit behavior of the Wendel probability. This is subsumed in the follow- · ing theorem. Here $M(d) \sim cd^\alpha$ means that $\lim_{d\to\infty} M(d)/d^\alpha = c$. Further, Φ denotes the distribution function of the standard normal distribution.

Theorem 6.3.2. *Let c be a real constant.*
(a) *If*
$$N - 2d \sim cd^{1/2}, \tag{6.9}$$
then
$$\lim_{d\to\infty} P_{d,N} = \Phi\left(-c/\sqrt{2}\right) \in (0,1).$$

(b) *If*
$$N - 2d \ge cd^\alpha \quad with \quad \frac{1}{2} < \alpha < 1 \ and \ c > 0, \tag{6.10}$$
then
$$\lim_{d\to\infty} P_{d,N} = 0.$$

(c) *If*
$$-cd^\alpha \le N - 2d \le cd^\alpha \quad with \ 0 < \alpha < \frac{1}{2} \ and \ c \ge 0, \tag{6.11}$$
then
$$\lim_{d\to\infty} P_{d,N} = \frac{1}{2}.$$

(d) *If*
$$N - 2d \le cd^\alpha \quad with \quad \frac{1}{2} < \alpha < 1 \ and \ c < 0, \tag{6.12}$$
then
$$\lim_{d\to\infty} P_{d,N} = 1.$$

Proof. (a) Let (6.9) be satisfied. The proof is based on the central limit theorem. By ξ_n^* we denote the standardized version of the binomial random variable ξ_n, that is,

$$\xi_n^* = \frac{\xi_n - \mathbb{E}(\xi_n)}{\sqrt{\mathbb{V}(\xi_n)}}, \quad where \ \mathbb{E}(\xi_n) = \frac{n}{2}, \ \mathbb{V}(\xi_n) = \frac{n}{4}. \tag{6.13}$$

Then (6.8) implies that

$$P_{d,N} = \mathbb{P}\left(\xi_{N-1}^* \le \frac{2d - N - 1}{\sqrt{N-1}}\right)$$

$$= \Phi\left(\frac{2d - N - 1}{\sqrt{N-1}}\right) + O\left(\frac{1}{\sqrt{N-1}}\right),$$

by the Berry–Esseen Theorem (see, e.g., Shiryaev [181, p. 426]). From (6.9) it follows immediately that

$$\frac{2d - N - 1}{\sqrt{N-1}} \to -\frac{c}{\sqrt{2}} \quad \text{as } d \to \infty,$$

· so that the assertion follows.

(b) Suppose that (6.10) holds. For any given number $c_1 \in \mathbb{R}$ we can choose a sequence $(N_1(d))_{d \in \mathbb{N}}$ of integers such that $N_1 = N_1(d)$ satisfies

$$N_1 - 2d \sim c_1 d^{1/2}$$

and $N_1 < N$ for sufficiently large d. From (a) it follows that $\lim_{d\to\infty} P_{d,N_1} = \Phi(-c_1/\sqrt{2})$. From the (trivial) monotonicity $P_{d,N_1} > P_{d,N}$ we deduce that

$$\limsup_{d\to\infty} P_{d,N} \le \Phi(-c_1/\sqrt{2}).$$

Since $\Phi(-c_1/\sqrt{2})$ can be any number in $(0,1)$, it follows that we must have $\lim_{d\to\infty} P_{d,N} = 0$.

(c) Now let (6.11) be satisfied. Then we have

$$\frac{|N - 2d|}{\sqrt{d}} \le cd^{\alpha - \frac{1}{2}} \to 0$$

as $d \to \infty$. Hence $N - 2d \sim 0\sqrt{d}$ and the assertion follows from (a), since

$$\lim_{d\to\infty} P_{d,N} = \Phi(0) = \frac{1}{2}.$$

(d) Suppose that (6.12) holds. Since $(N - 2d)/d^{1/2} \to -\infty$, for any given number $c_1 \in \mathbb{R}$ we can chose a sequence $(N_1(d))_{d \in \mathbb{N}}$ of integers such that $N_1 = N_1(d)$ satisfies $(N_1 - 2d)/d^{1/2} \to c_1$ and $N_1 > N$ for sufficiently large d. From (a) it follows that

$$\liminf_{d\to\infty} P_{d,N} \ge \Phi\left(-c_1/\sqrt{2}\right).$$

Since this holds for all $c_1 \in \mathbb{R}$, we get $\lim_{d\to\infty} P_{d,N} = 1$. $\qquad\square$

As a consequence, for $1/2 < \alpha < 1$ we can choose $N = N(d)$ in such a way that $\frac{d}{N} \to \frac{1}{2}$,

$$N - 2d = O(d^\alpha),$$

and

$$\liminf_{d\to\infty} P_{d,N} = 0, \qquad \limsup_{d\to\infty} P_{d,N} = 1.$$

This follows by choosing a sequence $(N(d))_{d \in \mathbb{N}}$ with suitable subsequences.

6.4 Donoho–Tanner cones in high dimensions

This section is motivated by an investigation of Donoho and Tanner [52], where random cones have shown up. These authors related sparse solutions of underdetermined linear systems to face counting in linear images of the nonnegative orthant. They considered the problem

$$Ax = b, \quad x_i \geq 0. \tag{6.14}$$

Here A is a real $d \times N$ matrix, where $d < N$, and $b \in \mathbb{R}^d$ is a given vector. For $x \in \mathbb{R}^N$, we write $x = (x_1, \ldots, x_N)^t$, where $x_1, \ldots, x_N$ are the coordinates of x with respect to the standard basis. Thus, one asks for nonnegative solutions of the underdetermined linear system given by the matrix A (and the vector b). A solution vector x of (6.14) is called k-*sparse* if it has at most k non-zero coordinates. Although the system (6.14) is underdetermined, it may happen for $k < d$ that a k-sparse solution is the unique (nonnegative) solution, and this can be very useful (we refer to [52] for hints to the importance of k-sparse solutions in various practical settings). In the canonical way, A is also viewed as a linear map from $\mathbb{R}^N$ to $\mathbb{R}^d$. By

$$\mathbb{R}_+^N := \{x \in \mathbb{R}^N : x_i \geq 0 \text{ for } i = 1, \ldots, N\}$$

we denote the nonnegative orthant of $\mathbb{R}^N$. Let $x \in \mathbb{R}_+^N$ be a k-sparse vector. Then x is contained in some k-face of the polyhedral cone $\mathbb{R}_+^N$. Donoho and Tanner related the investigation of the uniqueness of k-sparse solutions of (6.14) to the investigation of k-faces F of $\mathbb{R}_+^N$ for which AF is a face of $A\mathbb{R}_+^N$.

They did this for random matrices A, so that the expectations $\mathbb{E} f_k(A\mathbb{R}_+^N)$ became of interest; here f_k denotes the number of k-faces. About the distribution of the random matrix A, they assumed that $A\Pi$ and A have the same distribution for each $N \times N$ signed permutation matrix Π, together with some assumption on general position. We note that for the columns of A, considered as vectors of $\mathbb{R}^d$, this means that they are symmetrically exchangeable, as defined after Theorem 5.2.1. As there, we consider here only the case where the columns are i.i.d. random vectors with a distribution ϕ that is symmetric with respect to o and assigns measure zero to each hyperplane through o. We note that if $X_1, \ldots, X_N$ are the columns of A, then the image $A\mathbb{R}_+^N$ is the positive hull of $X_1, \ldots, X_N$.

We do not further pursue the connection to sparse vectors, but consider the preceding as a motivation for studying the random cones appearing here. The following definition should be compared to Definition 5.3.1. As before, let ϕ be a probability distribution on $\mathbb{R}^d$ that is symmetric with respect to o and assigns measure zero to each hyperplane through o.

Definition 6.4.1. Let $n \in \mathbb{N}$, and let $X_1, \ldots, X_n$ be independent random vectors in $\mathbb{R}^d$ with distribution ϕ. The

$$(\phi, n)\text{-}Donoho\text{-}Tanner \text{ cone } D_n$$

is the random cone defined as the positive hull of $X_1, \ldots, X_n$.

The cones D_n are precisely the random cones considered in the previous section. We point out that the difference to the Cover–Efron cones C_n is that the latter are defined by $\mathrm{pos}\{X_1,\dots,X_n\}$ under the condition that $\mathrm{pos}\{X_1,\dots,X_n\} \neq \mathbb{R}^d$. Thus, the distributions are distinctly different.

First we determine the expected face numbers of the random cone D_n. We recall the Wendel probabilities

$$P_{d,n} := \mathbb{P}\left(\mathrm{pos}\{X_1,\dots,X_n\} \neq \mathbb{R}^d\right) \tag{6.15}$$

$$= \frac{C(n,d)}{2^n} = \frac{1}{2^{n-1}} \sum_{i=0}^{d-1} \binom{n-1}{d}.$$

For the Cover–Efron cone C_n we know from (5.41) that

$$\mathbb{E}\, f_k(C_n) = \frac{2^k \binom{n}{k} C(n-k, d-k)}{C(n,d)}.$$

Each k-face of C_n is the positive hull of some k vectors from $\{X_1,\dots,X_n\}$, and there are $\binom{n}{k}$ choices. Therefore, we consider the quotient $f_k(C_n)/\binom{n}{k}$. The previous expectation can now be expressed in terms of the Wendel probabilities, namely

$$\frac{\mathbb{E}\, f_k(C_n)}{\binom{n}{k}} = \frac{P_{d-k,n-k}}{P_{d,n}}. \tag{6.16}$$

We deduce that for the random cone D_n we have

$$\frac{\mathbb{E}\, f_k(D_n)}{\binom{n}{k}} = P_{d-k,n-k} \tag{6.17}$$

for $k \in \{1,\dots,d-1\}$. This follows immediately from (6.16) and the fact that $f_k(D_n) = 0$ if $D_n = \mathbb{R}^d$.

For the quotient (6.17), Donoho and Tanner [52] have established a threshold phenomenon. They defined

$$\rho_W(\delta) := \max\{0, 2 - \delta^{-1}\} \quad \text{for } 0 < \delta < 1 \tag{6.18}$$

and proved the following theorem.

Theorem 6.4.1. *Let $0 < \delta < 1$ and $0 \le \rho < 1$ be given. Let $k < d < N$ be integers (depending on d) such that*

$$\frac{d}{N} \to \delta, \qquad \frac{k}{d} \to \rho \qquad \text{as } d \to \infty.$$

Then

$$\lim_{d \to \infty} \frac{\mathbb{E}\, f_k(D_N)}{\binom{N}{k}} = \begin{cases} 1 \text{ if } \rho < \rho_W(\delta), \\ 0 \text{ if } \rho > \rho_W(\delta). \end{cases}$$

In view of (6.17), the previous section allows us to give a short proof of this theorem. In fact, under the assumptions of Theorem 6.4.1, we have $d - k \to \infty$ as $d \to \infty$, since $\rho < 1$. Further, the limit

$$\lim_{d \to \infty} \frac{d - k}{N - k} = \frac{1 - \rho}{1/\delta - \rho}$$

is larger (smaller) than $1/2$ if $\rho < \rho_W(\delta)$ (respectively, $\rho > \rho_W(\delta)$). Hence, Theorem 6.3.1 implies Theorem 6.4.1. This argument is taken from [100], as is Theorem 6.4.2 below.

To decide what happens in Theorem 6.4.1 if $\rho = \rho_W(\delta)$, we distinguish two cases:

(i) $\delta < 1/2$ and $\rho = 0$.

(ii) $\delta \geq 1/2$ and $\rho = 2 - 1/\delta$.

As $d \to \infty$, we have

$$\frac{d - k}{N - k} \to \begin{cases} \delta & \text{in case (i),} \\ \frac{1}{2} & \text{in case (ii).} \end{cases}$$

Moreover, we also see that there is a positive constant a such that

$$d^\alpha \sim a^\alpha (d - k)^\alpha \quad \text{for } 0 < \alpha < 1.$$

Therefore, in view of (6.17) and of $N - 2d + k = (N - k) - 2(d - k)$, the following theorem follows by applying Theorems 6.3.1 and 6.3.2 to $P_{d-k,N-k}$ instead of $P_{d,N}$.

Theorem 6.4.2. *Let $0 < \delta < 1$. Let $k < d < N$ be integers (depending on d) such that*

$$\frac{d}{N} \to \delta, \quad \frac{k}{d} \to \rho = \rho_W(\delta) = \begin{cases} 0 & \text{if } \delta \leq 1/2, \\ 2 - \frac{1}{\delta} & \text{if } \delta \geq 1/2, \end{cases} \quad \text{as } d \to \infty.$$

If $\delta < 1/2$, then

$$\lim_{d \to \infty} \frac{\mathbb{E} f_k(D_N)}{\binom{N}{k}} = 0. \tag{6.19}$$

Let $\delta \geq 1/2$.

(a) If

$$N - 2d + k \sim c d^{1/2}, \tag{6.20}$$

then

$$\lim_{d \to \infty} \frac{\mathbb{E} f_k(D_N)}{\binom{N}{k}} = \Phi\left(-c/\sqrt{2}\right) \in (0, 1).$$

(b) If

$$N - 2d + k \geq c d^\alpha \quad \text{with } \frac{1}{2} < \alpha < 1 \text{ and } c > 0, \tag{6.21}$$

then

$$\lim_{d \to \infty} \frac{\mathbb{E} f_k(D_N)}{\binom{N}{k}} = 0.$$

(c) *If*

$$-cd^\alpha \le N - 2d + k \le cd^\alpha \quad \text{with } 0 < \alpha < \frac{1}{2} \text{ and } c \ge 0, \tag{6.22}$$

then

$$\lim_{d \to \infty} \frac{\mathbb{E} f_k(D_N)}{\binom{N}{k}} = \frac{1}{2}.$$

(d) *If*

$$N - 2d + k \le cd^\alpha \quad \text{with } \frac{1}{2} < \alpha < 1 \text{ and } c < 0, \tag{6.23}$$

then

$$\lim_{d \to \infty} \frac{\mathbb{E} f_k(D_N)}{\binom{N}{k}} = 1.$$

Theorem 6.4.1 deals with the quotient $\mathbb{E} f_k(D_N)/\binom{N}{k}$. Similar results can be obtained for the difference $\binom{N}{k} - \mathbb{E} f_k(D_N)$ if $\delta > 1/2$, at the cost of a smaller threshold. First we introduce this threshold. For this, let

$$H(x) := -x \log x - (1 - x) \log(1 - x) \quad \text{for } 0 \le x \le 1;$$

thus H is the binary entropy function for base e. Following [52] (with different notation), we define

$$G(\delta, \rho) := H(\delta) + \delta H(\rho) - (1 - \delta\rho) \log 2 \quad \text{for } \delta, \rho \in (0, 1). \tag{6.24}$$

Writing $G_\delta(x) := G(\delta, x)$, we have

$$G'_\delta(x) = \delta \log \left(\frac{2(1 - x)}{x} \right),$$

hence $G'_\delta(x) > 0$ for $x \in (0, 2/3)$ and $G'_\delta(x) < 0$ for $x \in (2/3, 1)$. From $1/2 < \delta < 1$ it follows that $G_\delta(0) < 0$ and $G_\delta(1) > 0$, hence there is a unique number $x_0 \in [0, 1]$ with $G_\delta(x_0) = 0$. We set

$$x_0 =: \rho_S(\delta). \tag{6.25}$$

Further, we define $g(\delta) := G_\delta(2 - \delta^{-1})$ for $\delta \in [1/2, 1]$ and note that $g''(\delta) < 0$ for $\delta \in (1/2, 1)$, moreover $g(1/2) = g(1) = 0$. Therefore, $g(x) > 0$ for $x \in (1/2, 1)$, hence $G_\delta(2 - \delta^{-1}) > 0$, and it follows that $\rho_S(\delta) < \rho_W(\delta)$.

Preparing the next theorem, we state a lemma on binomial coefficients.

Lemma 6.4.1. *If $m, n \in \mathbb{N}$ and $2m < n + 1$, then*

$$\sum_{i=0}^{m} \binom{n}{i} < \binom{n}{m} \frac{n - m + 1}{n - 2m + 1}.$$

Proof. From the assumptions it follows that $n \geq 2$. For $i \in \{0, \ldots, m-1\}$ we have

$$\frac{i+1}{n-i} = \frac{n+1}{n-i} - 1 \leq \frac{n+1}{n-(m-1)} - 1 = \frac{m}{n-m+1} =: q < 1$$

and hence

$$\frac{\binom{n}{i}}{\binom{n}{m}} = \frac{m}{n-m+1} \cdots \frac{i+1}{n-i} \leq q^{m-i}.$$

This gives

$$\sum_{i=0}^{m} \binom{n}{i} \leq \binom{n}{m} \sum_{i=0}^{m} q^{m-i} < \binom{n}{m} \frac{1}{1-q} = \binom{n}{m} \frac{n-m+1}{n-2m+1},$$

as stated. $\qquad\square$

Below, we say that a function h depending on the dimension d remains in an interval I if $h(d) \in I$ for all d. An interval is called positive if its bounds are positive numbers.

Theorem 6.4.3. *Let $1/2 < \delta < 1$ and $0 < \rho < \rho_W(\delta) = 2 - 1/\delta$ be given. Let $k < d < N$ be integers (depending on d) such that*

$$\frac{d}{N} =: \delta_d \to \delta, \qquad \frac{k}{d} =: \rho_d \to \rho \qquad as \; d \to \infty.$$

Then

$$\binom{N}{k} - \mathbb{E}\, f_k(D_N) = \frac{1}{N} e^{NG(\delta_d, \rho_d)} \cdot h(d), \tag{6.26}$$

where h remains in a bounded positive interval.
In particular,

$$\lim_{d \to \infty} \left[\binom{N}{k} - \mathbb{E}\, f_k(D_N) \right] = \begin{cases} 0 & if \; \rho < \rho_S(\delta), \\ \infty & if \; \rho > \rho_S(\delta). \end{cases}$$

If $\rho = \rho_S(\delta)$, then the numbers N and k can be chosen such that

$$\liminf_{d \to \infty} \left[\binom{N}{k} - \mathbb{E}\, f_k(D_N) \right] = 0 \quad and \quad \limsup_{d \to \infty} \left[\binom{N}{k} - \mathbb{E}\, f_k(D_N) \right] = \infty.$$

Proof. From (6.17) and the obvious relation $P_{m,M} + P_{M-m,M} = 1$ we have

$$\binom{N}{k} - \mathbb{E}\, f_k(D_N) = \binom{N}{k}[1 - P_{d-k,N-k}] = \binom{N}{k}P_{N-d,N-k}, \qquad (6.27)$$

hence

$$\binom{N}{k} - \mathbb{E}\, f_k(D_N) = \binom{N}{k}\frac{1}{2^{N-k-1}}\sum_{i=0}^{N-d-1}\binom{N-k-1}{i}. \qquad (6.28)$$

We abbreviate $\delta_d\rho_d =: \tau_d$ and use Stirling's approximation to obtain

$$\binom{N}{k} = \frac{1}{\sqrt{2\pi\tau_d(1-\tau_d)}}\cdot\frac{1}{\sqrt{N}}\cdot e^{NH(\tau_d)}\cdot e^{\frac{\theta_1}{12N}}, \qquad (6.29)$$

with a number θ_1 (depending on d) in a fixed interval independent of d. Next, we write

$$\frac{1}{2^{N-k-1}} = 2e^{-N(1-\tau_d)\log 2}. \qquad (6.30)$$

Lemma 6.4.1 gives

$$\binom{n}{m} \le \sum_{i=0}^{m}\binom{n}{i} \le \binom{n}{m}\frac{n-m+1}{n-2m+1},$$

if $2m < n+1$. With $m = N-d-1$ and $n = N-k-1$ we have $2m < n$ for sufficiently large d, since

$$\frac{N-d-1}{N-k-1} < \frac{N-d}{N-k} =: \sigma_d = \frac{1-\delta_d}{1-\tau_d} \to \frac{1-\delta}{1-\delta\rho} < \frac{1}{2}.$$

Since

$$\frac{n-m+1}{n-2m+1} = \frac{d-k+1}{2d-N-k+2} \to \frac{1-\rho}{\rho_W(\delta)-\rho}$$

as $d \to \infty$ and

$$\binom{N-k-1}{N-d-1} = \binom{N-k}{N-d}\sigma_d,$$

we see that the third factor in (6.28) satisfies

$$\sum_{i=0}^{N-d-1}\binom{N-k-1}{i} = \binom{N-k}{N-d}\cdot h_1(d),$$

where h_1 remains in a bounded positive interval. Stirling's approximation yields

$$\binom{N-k}{N-d} = \frac{1}{\sqrt{2\pi\sigma_d(1-\sigma_d)}}\cdot\frac{1}{\sqrt{N-k}}\cdot e^{(N-k)H(\sigma_d)}\cdot e^{\frac{\theta_2}{12(N-k)}} \qquad (6.31)$$

with a number θ_2 in a fixed interval independent of d. We combine (6.29), (6.30), (6.31) and note that each of the expressions

$$\frac{e^{\theta_1/12N}}{\sqrt{2\pi\tau_d(1-\tau_d)}}, \qquad \frac{e^{\theta_2/12(N-k)}}{\sqrt{2\pi\sigma_d(1-\sigma_d)}}, \qquad \sqrt{\frac{N}{N-k}}$$

has a positive finite limit as $d \to \infty$. It follows that

$$\binom{N}{k} - \mathbb{E}\, f_k(D_N) = \frac{1}{N} e^{N[H(\tau_d)+(1-\tau_d)H(\sigma_d)-(1-\tau_d)\log 2]} \cdot h(d),$$

where h remains in a bounded positive interval. An elementary calculation gives

$$H(\tau_d) + (1 - \tau_d)H(\sigma_d) - (1 - \tau_d)\log 2 = G(\delta_d, \rho_d), \qquad (6.32)$$

so that relation (6.26) follows.

To prove the remaining statements, suppose first that $\rho < \rho_S(\delta)$. Since $G(\delta_d, \rho_d) \to G(\delta, \rho) < 0$, there is a constant c such that $G(\delta_d, \rho_d) \le c < 0$ for all large d. Hence, (6.26) implies that $\lim_{d\to\infty} \left[\binom{N}{k} - \mathbb{E}\, f_k(D_N) \right] = 0$. This was proved in a different way in [52].

Second, suppose that $\rho > \rho_S(\delta)$ (which implies that $G(\delta, \rho) > 0$). The assertion that $\lim_{d\to\infty} \left[\binom{N}{k} - \mathbb{E}\, f_k(D_N) \right] = \infty$ is obtained similarly as above.

Third, suppose that $\rho = \rho_S(\delta)$. With the given $\delta \in (1/2, 1)$, define $N = N(d)$ by

$$N := \lfloor d/\delta \rfloor, \qquad \text{then } 0 \le \delta_d - \delta < \frac{1}{N}.$$

The function $G(\delta, \cdot)$ is equal to 0 at ρ and is strictly increasing in a neighborhood of ρ. Since $0 < \left((-1)^d \frac{1}{2} + 1 \right) N^{-1} \cdot \log N \to 0$ as $d \to \infty$, for sufficiently large d there is a unique number $\rho(d) \in (\rho, 1)$ such that

$$G(\delta, \rho(d)) = \left((-1)^d \frac{1}{2} + 1 \right) \cdot \frac{\log N}{N}.$$

We define the number $k = k(d)$ by

$$k := \lfloor \rho(d)d \rfloor, \qquad \text{then } 0 \le \rho(d) - \rho_d < \frac{2}{N},$$

since we can assume that $d/N > 1/2$ if d is sufficiently large.

The function G is differentiable in a neighborhood of (δ, ρ), hence there are positive constants c_1, c_2, c_3, independent of d, such that for sufficiently large d we have

$$|G(\delta_d, \rho_d) - G(\delta, \rho(d))| \le |G(\delta_d, \rho_d) - G(\delta, \rho_d)| + |G(\delta, \rho_d) - G(\delta, \rho(d))|$$
$$\le c_1 |\delta_d - \delta| + c_2 |\rho_d - \rho(d)| \le \frac{c_3}{N}.$$

Thus, if we write

$$G(\delta_d, \rho_d) - G(\delta, \rho(d)) = \frac{g_2}{N},$$

then g_2 is a function which remains in a bounded interval (independent of d), and

$$\frac{1}{N} e^{NG(\delta_d, \rho_d)} = e^{g_2} \cdot \begin{cases} \sqrt{N}, & \text{if } d \text{ is even,} \\ \sqrt{N}^{-1}, & \text{if } d \text{ is odd,} \end{cases} \tag{6.33}$$

From (6.26) and (6.33), we obtain the remaining assertion. □

Under the assumptions of Theorem 6.4.3, one can draw the conclusion, due to Donoho and Tanner [52], that

$$\lim_{d \to \infty} \mathbb{P}\left(f_k(D_N) = \binom{N}{k} \right) = 1 \quad \text{if } \rho < \rho_S(\delta). \tag{6.34}$$

In fact, let D_N be the positive hull of the random vectors $X_1, \ldots, X_N$, and let $k \in \{1, \ldots, d-1\}$. Let $\left\{ M_j : j = 1, \ldots, \binom{N}{k} \right\}$ be the set of all k-element subsets of $\{X_1, \ldots, X_N\}$. Then $\mathbb{P}(\text{pos } M_j \in \mathcal{F}_k(D_N))$ is independent of j, hence

$$\binom{N}{k} \mathbb{P}\left(\text{pos } M_1 \in \mathcal{F}_k(D_N))\right) = \sum_j \mathbb{P}(\text{pos } M_j \in \mathcal{F}_k(D_N))$$

$$= \mathbb{E} \sum_j \mathbb{1}\{\text{pos } M_j \in \mathcal{F}_k(D_N)\}$$

$$= \mathbb{E} f_k(D_N).$$

Therefore,

$$\mathbb{P}\left(\text{pos } M_1 \notin \mathcal{F}_k(D_N)\right) = 1 - \frac{\mathbb{E} f_k(D_N)}{\binom{N}{k}}.$$

By Boole's inequality,

$$\mathbb{P}\left(\text{pos } M_j \notin \mathcal{F}_k(D_N) \text{ for some } j\right) \leq \binom{N}{k} \left[1 - \frac{\mathbb{E} f_k(D_N)}{\binom{N}{k}} \right],$$

and thus

$$\mathbb{P}\left(f_k(D_N) = \binom{N}{k} \right) \geq 1 - \left[\binom{N}{k} - \mathbb{E} f_k(D_N) \right].$$

Now (6.34) follows from Theorem 6.4.3.

Finally in this section, we apply one of the previous results to a special class of polytopes, the zonotopes. We start with some terminology.

Definition 6.4.2. Let $H_1, \ldots, H_n \in G(d, d-1)$. The fan induced by the hyperplanes $H_1, \ldots, H_n$ is the set of all k-faces, $k = 0, \ldots, d$, of the tessellation induced by $H_1, \ldots, H_n$.

Let $H_1, \ldots, H_n \in G(d, d-1)$. For each $i \in \{1, \ldots, n\}$, we choose a closed segment S_i (of positive length) orthogonal to H_i. The Minkowski sum

$$Z_n = S_1 + \cdots + S_n$$

is known as a *zonotope*. We call it a *zonotope associated with* $H_1, \ldots, H_n$.

Generally, the *normal fan* of a polytope $P \in \mathcal{P}^d$ is the set

$$\{N(P, F) : F \in \mathcal{F}(P)\}$$

of the normal cones of all faces of P.

Lemma 6.4.2. *The normal fan of a zonotope* Z_n *associated with* $H_1, \ldots, H_n$ *is the fan induced by* $H_1, \ldots, H_n$.

Proof. Each point of $x \in Z_n$ is of the form $x = x_1 + \cdots + x_n$ with $x_i \in S_i$ for $i = 1, \ldots, n$. By [163, Thm. 2.2.1] we have

$$N(Z_n, x) = \bigcap_{i=1}^{n} N(S_i, x_i).$$

If $x_i \in \operatorname{relint} S_i$, then $N(S_i, x_i) = H_i$. If x_i is an endpoint of the segment S_i, then $N(S_i, x_i)$ is one the closed halfspaces bounded by H_i. Therefore, $N(Z_n, x)$ is an intersection of hyperplanes from $H_1, \ldots, H_n$ and of closed halfspaces bounded by these hyperplanes, hence $N(Z_n, x)$ is a face of the tessellation induced by $H_1, \ldots, H_n$. Thus, it belongs to the fan induced by $H_1, \ldots, H_n$.

Conversely, let F belong to this fan. Then it is a k-face of the tessellation induced by $H_1, \ldots, H_n$, for some $k \in \{0, \ldots, d\}$. Hence, there are different numbers $i_1, \ldots, i_{d-k} \in \{1, \ldots, n\}$ and signs $\epsilon_i \in \{-1, 1\}$ with

$$F = H_{i_1} \cap \cdots \cap H_{i_{d-k}} \cap \bigcap_{i \in I} H_i^{\epsilon_i},$$

where $I := \{1, \ldots, n\} \setminus \{i_1, \ldots, i_{d-k}\}$ and H^{-1}, H^1 are the closed halfspaces bounded by H. If we choose $x_{i_j} \in \operatorname{relint} S_{i_j}$ for $j = 1, \ldots, d-k$ and, for $i \in I$, let x_i be the endpoint of S_i for which $N(S_i, x_i) = H_i^{\epsilon_i}$, then the point $x := x_{i_1} + \cdots + x_{i_{d-k}} + \sum_{i \in I} x_i$ satisfies $N(Z_n, x) = F$. $\quad\square$

Now let $H_1, \ldots, H_n \in G(d, d-1)$ be in general position. Then also nondegenerate segments S_i orthogonal to H_i, $i = 1, \ldots, n$, are said to be in general position. Let $Z_n = S_1 + \cdots + S_n$, as above. It follows from Lemma 6.4.2 that there is a one-to-one correspondence between the k-faces of Z_n and the $(d-k)$-faces of the tessellation induced by $H_1, \ldots, H_n$. Therefore, it follows from (5.27) and (5.22) that

$$f_k(Z_n) = \binom{n}{k} 2^{n-k} P_{d-k, n-k}.$$

We note that

$$\binom{n}{k} 2^{n-k} = f_k(I^n)$$

is the number of k-faces of the cube I^n in $\mathbb{R}^n$. Thus, we have

$$\frac{f_k(Z_n)}{f_k(I^n)} = P_{d-k,n-k} = 1 - P_{n-d,n-k}.$$

This is the same number as it appeared in (6.17). Therefore, the proof of Theorem 6.4.1 yields also the following result. We recall that $\rho_W(\delta)$ was defined in (6.18).

Theorem 6.4.4. *Let $0 < \delta, \rho < 1$ be given. Let $k < d < N$ be integers such that*

$$\frac{k}{d} \to \rho, \qquad \frac{d}{N} \to \delta \qquad as\ d \to \infty.$$

For each d and N, let Z_N be a sum of N segments in $\mathbb{R}^d$ in general position. Then

$$\lim_{d \to \infty} \frac{f_k(Z_N)}{f_k(I^N)} = \begin{cases} 1 & if\ \rho < \rho_W(\delta), \\ 0 & if\ \rho > \rho_W(\delta). \end{cases}$$

This can also be interpreted in a different way. Let I^N be the standard unit cube in $\mathbb{R}^N$. Let $\mathsf{A} : \mathbb{R}^N \to \mathbb{R}^d$ (where $N > d$) be a linear map which is generic in the sense that the vectors of the standard orthonormal basis of $\mathbb{R}^N$ are mapped to vectors in general position in $\mathbb{R}^d$. If this is the case, then $\mathsf{A}I^N$ is the sum of segments in general position. Thus, under the assumptions of Theorem 6.4.4 we have

$$\lim_{d \to \infty} \frac{f_k(\mathsf{A}I^N)}{f_k(I^N)} = \begin{cases} 1 & if\ \rho < \rho_W(\delta), \\ 0 & if\ \rho > \rho_W(\delta). \end{cases}$$

This result appears in Donoho and Tanner [52].

Notes for Section 6.4

1. We refer to Donoho and Tanner [52] for a thorough discussion of some of the previous results and their possible applications.

6.5 Cover–Efron cones in high dimensions

For the basic geometric functionals of the Cover–Efron cones C_n, such as face numbers, Grassmann angles and conic intrinsic volumes, we have obtained in Section 5.3 simple explicit formulas for their expectations. This enables us to study their behavior in high dimensions, and to describe threshold phenomena in the style of Section 6.4. Such a goal requires more effort than for the random cones D_n, since the expressions for the expectations now have denominators.

In this section, where we follow Hug and Schneider [99], we study the face numbers of the Cover–Efron cone C_n in $\mathbb{R}^d$, when the dimension d tends to infinity and the number n increases with d in a suitable way. We shall see that it makes a difference in the nature of the threshold phenomenon whether or not also the dimension of the considered faces tends to infinity.

We repeat that the definition of the Cover–Efron cone C_n is based on a distribution ϕ on $\mathbb{R}^d$ which is symmetric with respect to o and assigns measure zero to each hyperplane through o. Then C_n is defined as the positive hull of independent random vectors $X_1, \ldots, X_n$ with distribution ϕ, under the condition that this positive hull is not all of $\mathbb{R}^d$.

We recall from (5.41) that

$$\frac{\mathbb{E} f_k(C_n)}{\binom{n}{k}} = 2^k \frac{C(n-k, d-k)}{C(n,d)} = \frac{P_{d-k, n-k}}{P_{d,n}}, \tag{6.35}$$

which involves the Wendel probabilities

$$P_{d,n} = \mathbb{P}(\mathrm{pos}\{X_1, \ldots, X_n\} \neq \mathbb{R}^d),$$

with $X_1, \ldots, X_n$ as above. According to (5.22), the explicit values are

$$P_{d,n} = \frac{C(n,d)}{2^n} = \frac{1}{2^{n-1}} \sum_{i=0}^{d-1} \binom{n-1}{i}.$$

Denoting, as in Section 6.3, by ξ_n a random variable which has the Bernoulli distribution with parameters n and $1/2$, we have

$$\frac{\mathbb{E} f_k(C_n)}{\binom{n}{k}} = \frac{\mathbb{P}(\xi_{n-k-1} \leq d-k-1)}{\mathbb{P}(\xi_{n-1} \leq d-1)}. \tag{6.36}$$

First we show a monotonicity property of the quotient (6.36).

Theorem 6.5.1. *Let $n > d > k$. Then*

$$\frac{\mathbb{E} f_k(C_{n+1})}{\binom{n+1}{k}} < \frac{\mathbb{E} f_k(C_n)}{\binom{n}{k}}.$$

Proof. Let $d, n, k \in \mathbb{N}$ with $n > d > k$ be given. We define

$$s_j := \sum_{i=0}^{d-j-2} \binom{n-j-1}{i}$$

for $j \in \{0, \ldots, d-2\}$. We claim that

$$\frac{s_{j+1}}{\binom{n-j-2}{d-j-2}} < \frac{s_j}{\binom{n-j-1}{d-j-1}} \tag{6.37}$$

(where, by definition, an empty sum is zero). For the proof, we use

$$\sum_{i=0}^{p} \binom{M+1}{i} = 2\sum_{i=0}^{p-1} \binom{M}{i} + \binom{M}{p}, \qquad (6.38)$$

with $M = n - j - 2$ and $p = d - j - 2$. We get

$$s_j = \sum_{i=0}^{d-j-2} \binom{n-j-1}{i} = 2\sum_{i=0}^{d-j-3} \binom{n-j-2}{i} + \binom{n-j-2}{d-j-2}$$

$$= 2s_{j+1} + \binom{n-j-2}{d-j-2}.$$

Thus, (6.37) is equivalent to

$$\frac{\binom{n-j-1}{d-j-1}}{\binom{n-j-2}{d-j-2}} s_{j+1} < 2s_{j+1} + \binom{n-j-2}{d-j-2},$$

which is equivalent to

$$\frac{n-2d+j+1}{d-j-1} s_{j+1} < \binom{n-j-2}{d-j-2}.$$

If $n - 2d + j \leq -1$, then this holds trivially. Hence, we assume now that $n - 2d + j \geq 0$. Then the previous relation is equivalent to

$$\frac{s_{j+1}}{\binom{n-j-2}{d-j-2}} < \frac{d-j-1}{n-2d+j+1}. \qquad (6.39)$$

If $d-j-2 \geq 1$, we can use Lemma 6.4.1 (with n replaced by $n = n-j-2$ and $m = d-j-3$; the assumption of the lemma is satisfied since $n - 2d + j \geq 0$). Observing that

$$\binom{n}{m} = \binom{n}{m+1} \frac{m+1}{n-m},$$

we obtain

$$\frac{s_{j+1}}{\binom{n-j-2}{d-j-2}} \leq \frac{d-j-2}{n-d+1} \cdot \frac{n-d+2}{n-2d+j+5} \leq \frac{d-j-2}{n-d+1} \cdot \frac{n-d+2}{n-2d+j+1}.$$

If $d - j - 2 = 0$, then this inequality holds trivially, because then $s_{j+1} = 0$. Since $n - 2d + j \geq 0$, the right side is strictly smaller than the right side of (6.39). This proves (6.37).

By (6.35) we have

$$g(n) := \frac{\mathbb{E} f_k(C_n)}{\binom{n}{k}} = 2^k \frac{C(n-k, d-k)}{C(n,d)} = 2^k \frac{\sum_{i=0}^{d-k-1} \binom{n-k-1}{i}}{\sum_{i=0}^{d-1} \binom{n-1}{i}}$$

$$= 2^k \frac{s_k + \binom{n-k-1}{d-k-1}}{s_0 + \binom{n-1}{d-1}}.$$

Using (6.38) again, we obtain

$$g(n+1) = 2^k \frac{\sum_{i=0}^{d-k-1} \binom{n-k}{i}}{\sum_{i=0}^{d-1} \binom{n}{i}} = 2^k \frac{2\sum_{i=0}^{d-k-2} \binom{n-k-1}{i} + \binom{n-k-1}{d-k-1}}{2\sum_{i=0}^{d-2} \binom{n-1}{i} + \binom{n-1}{d-1}}$$

$$= 2^k \frac{2s_k + \binom{n-k-1}{d-k-1}}{2s_0 + \binom{n-1}{d-1}}.$$

Therefore, $g(n+1) < g(n)$ holds if and only if

$$\frac{s_k}{\binom{n-k-1}{d-k-1}} < \frac{s_0}{\binom{n-1}{d-1}}. \tag{6.40}$$

But this follows from the inequality (6.37) by induction.   □

In the following investigations, the number n will depend on d, and is then denoted by N. To simplify the notation, we introduce:

Convention. In the following, C_N denotes a Cover–Efron cone in $\mathbb{R}^d$. The integers $N = N(d) \geq d$, and possibly $k = k(d) \leq d - 1$, depend on the dimension d, but we shall not indicate the dimension d in the notation.

We consider the quotient $f_k(C_N)/\binom{N}{k}$, since C_N can have at most $\binom{N}{k}$ faces of dimension k. If this quotient is equal to one, this means that the polyhedral cone C_N is k-neighborly, that is, the positive hull of any k of its generating rays is a k-face of the cone. Our first theorem provides assumptions on the growth of N and k under which the expectation of this quotient tends to one as $d \to \infty$.

Theorem 6.5.2. *Let $k = k(d) \in \{1, \ldots, d\}$ be such that*

$$\frac{k}{\sqrt{N}} \to 0 \ as \ d \to \infty.$$

If $\sqrt{N}\left(\frac{d}{N} - \frac{1}{2}\right)$ is bounded from below, then

$$\lim_{d\to\infty} \frac{\mathbb{E} f_k(C_N)}{\binom{N}{k}} = 1.$$

Proof. The proof is based on the central limit theorem. By ξ_n^* we denote again the standardized version of the binomial random variable ξ_n, see (6.13). Then (6.36) implies that

$$\frac{\mathbb{E} f_k(C_N)}{\binom{N}{k}} = \frac{\mathbb{P}\left(\xi_{N-k-1}^* \leq \frac{2d-N-k-1}{\sqrt{N-k-1}}\right)}{\mathbb{P}\left(\xi_{N-1}^* \leq \frac{2d-N-1}{\sqrt{N-1}}\right)}$$

$$= \frac{\Phi\left(\frac{2d-N-k-1}{\sqrt{N-k-1}}\right) + O\left(\frac{1}{\sqrt{N-k-1}}\right)}{\Phi\left(\frac{2d-N-1}{\sqrt{N-1}}\right) + O\left(\frac{1}{\sqrt{N-1}}\right)},$$

by the Berry–Esseen Theorem (see, e.g., Shiryaev [181, p. 426]), where Φ is the distribution function of the standard normal distribution. To compare the arguments of this function in the numerator and the denominator, we write

$$\frac{2d - N - k - 1}{\sqrt{N - k - 1}} = 2\sqrt{N}\left(\frac{d}{N} - \frac{1}{2}\right)\frac{1}{\sqrt{1 - \frac{k+1}{N}}} - \frac{k+1}{\sqrt{N - k - 1}}.$$

For sufficiently large d we have $\frac{k+1}{N} \le \frac{1}{2}$ and hence

$$1 + \frac{1}{2}\frac{k+1}{N} \le \frac{1}{\sqrt{1 - \frac{k+1}{N}}} \le 1 + 2\frac{k+1}{N},$$

thus

$$\frac{1}{\sqrt{1 - \frac{k+1}{N}}} = 1 + \theta\frac{k+1}{2N}$$

with $\theta \in [1, 4]$ (depending on d). Therefore,

$$\frac{2d - N - k - 1}{\sqrt{N - k - 1}} = 2\sqrt{N}\left(\frac{d}{N} - \frac{1}{2}\right) + \theta\frac{k+1}{2N}2\sqrt{N}\left(\frac{d}{N} - \frac{1}{2}\right) - \frac{k+1}{\sqrt{N - k - 1}}$$

$$= a(d) + b_k(d)$$

with

$$a(d) := 2\sqrt{N}\left(\frac{d}{N} - \frac{1}{2}\right), \qquad b_k(d) := \theta\frac{k+1}{\sqrt{N}}\left(\frac{d}{N} - \frac{1}{2}\right) - \frac{k+1}{\sqrt{N - k - 1}}.$$

It follows from the assumptions that $a(d) \ge a$ with some constant a and $b_k(d) \to 0$ as $d \to \infty$.

Using the mean value theorem, we arrive at

$$\frac{\mathbb{E}\, f_k(C_N)}{\binom{N}{k}} = \frac{\Phi(a(d) + b_k(d)) + O\left(\frac{1}{\sqrt{N-k-1}}\right)}{\Phi(a(d) + b_0(d)) + O\left(\frac{1}{\sqrt{N-1}}\right)}$$

$$= \frac{\Phi(a(d)) + b_k(d) \cdot \Phi'(z_k) + O\left(\frac{1}{\sqrt{N-k-1}}\right)}{\Phi(a(d)) + b_0(d) \cdot \Phi'(z_0) + O\left(\frac{1}{\sqrt{N-1}}\right)} \qquad (6.41)$$

with intermediate values $z_k, z_0 \in \mathbb{R}$. Since the derivative Φ' is bounded, further $b_k(d), b_0(d) \to 0$ as $d \to \infty$, and $\Phi(a(d)) \ge \Phi(a) > 0$, we conclude that the quotient tends to 1 as $d \to \infty$. $\qquad\square$

We prepare the proof of the next theorem by some results on binomial coefficients. First we remark that, under the assumption $2m < n + 1$, Lemma 6.4.1 (together with a trivial lower bound) implies that

$$\binom{n}{m} \le \sum_{i=0}^{m} \binom{n}{i} \le \binom{n}{m} \frac{n-m+1}{n-2m+1}.$$

This can also be written as

$$\frac{m+1}{n-m} \le \frac{1}{\binom{n}{m+1}} \sum_{i=0}^{m} \binom{n}{i} \le \frac{m+1}{n-m} \cdot \frac{n-m+1}{n-2m+1}.$$

If $N > 2d - 3$, we can use this with $n = N - 1$ and $m = d - 2$, to obtain

$$\frac{d-1}{N-d+1} \le \frac{1}{\binom{N-1}{d-1}} \sum_{j=0}^{d-2} \binom{N-1}{j} \le \frac{d-1}{N-d+1} \cdot \frac{N-d+2}{N-2d+4}. \tag{6.42}$$

The following lemma gives better lower bounds.

Lemma 6.5.1. *Let $n, m, p \in \mathbb{N}$, with $2m \le n+1$ and $2 \le p \le m$. Then*

$$\frac{1}{\binom{n}{m+1}} \sum_{i=0}^{m} \binom{n}{i} \ge \frac{m-p+1}{n-2m+2p-1} \left(1 - \left(\frac{m-p+1}{n-m+p} \right)^{p+1} \right). \tag{6.43}$$

Proof. First we observe that $m \le (n+1)/2 \le n$, and $m \le n - 1$ if $n \ge 2$. Hence, if $2 \le p \le m$, then $m \le n - 1$, further $n - 2m + 2p - 1 \ge 2$ and $(n+1)/(m+1) > 1$. We set

$$q := \frac{m-p+1}{n-m+p},$$

then $0 < q < 1$. For $j \in \{m-p, \ldots, m\}$ we have

$$\frac{\binom{n}{j}}{\binom{n}{m+1}} = \left(\frac{n+1}{m+1} - 1 \right)^{-1} \cdots \left(\frac{n+1}{j+1} - 1 \right)^{-1} \ge \left(\frac{n+1}{j+1} - 1 \right)^{-(m+1-j)}$$

$$\ge \left(\frac{m-p+1}{n-m+p} \right)^{m+1-j} = q^{m+1-j}$$

and hence

$$\frac{1}{\binom{n}{m+1}} \sum_{j=m-p}^{m} \binom{n}{j} \ge q \sum_{r=0}^{p} q^r = \frac{q}{1-q} \left(1 - q^{p+1} \right),$$

which yields (6.43). $\qquad\square$

Lemma 6.5.2. *If $d/N \to \delta$ as $d \to \infty$, with $0 \le \delta < 1/2$, then*

$$\lim_{d \to \infty} \frac{1}{\binom{N-1}{d-1}} \sum_{j=0}^{d-2} \binom{N-1}{j} = \frac{\delta}{1-2\delta}.$$

Proof. Let $d/N \to \delta$ as $d \to \infty$, where $0 \le \delta < 1/2$. We write $N = \alpha d$, where α depends on d and satisfies $\alpha \to 1/\delta$ as $d \to \infty$. If $\delta = 0$, this means that $\alpha \to \infty$. We assume that d is so large that $\alpha > 2$. From (6.42) we have

$$\frac{1}{\binom{N-1}{d-1}} \sum_{j=0}^{d-2} \binom{N-1}{j} \le \frac{d-1}{(\alpha-1)d+1} \cdot \frac{(\alpha-1)d+2}{(\alpha-2)d+4}.$$

We conclude that

$$\limsup_{d \to \infty} \frac{1}{\binom{N-1}{d-1}} \sum_{j=0}^{d-2} \binom{N-1}{j} \le \frac{\delta}{1-2\delta}. \tag{6.44}$$

For $2 \le p \le d-2$, Lemma 6.5.1 provides the lower bound

$$\frac{1}{\binom{N-1}{d-1}} \sum_{j=0}^{d-2} \binom{N-1}{j} \ge \frac{d-p}{N-2d+2p+2} \left(1 - \left(\frac{d-p-1}{N-d+p+1}\right)^{p+1}\right).$$

From this, we obtain

$$\liminf_{d \to \infty} \frac{1}{\binom{N-1}{d-1}} \sum_{j=0}^{d-2} \binom{N-1}{j} \ge \frac{\delta}{1-2\delta} \left(1 - \left(\frac{\delta}{1-\delta}\right)^{p+1}\right)$$

for each fixed $p \ge 2$. Letting $p \to \infty$, we find that

$$\liminf_{d \to \infty} \frac{1}{\binom{N-1}{d-1}} \sum_{j=0}^{d-2} \binom{N-1}{j} \ge \frac{\delta}{1-2\delta}.$$

Together with (6.44), this completes the proof. □

Now we are in a position to prove the following theorem.

Theorem 6.5.3. *Suppose that*

$$\frac{d}{N} \to \delta \quad as \ d \to \infty,$$

with a number $\delta \in [0,1]$. Let $k \in \mathbb{N}$ be fixed. Then the Cover–Efron cone C_N satisfies

$$\lim_{d \to \infty} \frac{\mathbb{E} f_k(C_N)}{\binom{N}{k}} = \begin{cases} 1 & if \ 1/2 \le \delta \le 1, \\ (2\delta)^k & if \ 0 \le \delta \le 1/2. \end{cases}$$

Proof. The proof has three parts: (a) $1/2 < \delta \le 1$, (b) $0 \le \delta < 1/2$, and (c) $\delta = 1/2$. This third step is necessary, since none of the other two proofs extends to $\delta = 1/2$.

(a) If $\frac{d}{N} \to \delta > \frac{1}{2}$, then $\frac{d}{N} - \frac{1}{2} > 0$ for almost all d, hence $\mathbb{E} f_k(C_N)/\binom{N}{k} \to 1$ follows from Theorem 6.5.2.

(b) Suppose that $\frac{d}{N} \to \delta$ as $d \to \infty$, where $0 \le \delta < \frac{1}{2}$. We write (6.35) in the form

$$\frac{\mathbb{E} f_k(C_N)}{\binom{N}{k}} = 2^k \frac{\binom{N-k-1}{d-k-1} \left[1 + \binom{N-k-1}{d-k-1}^{-1} \sum_{i=0}^{d-k-2} \binom{N-k-1}{i}\right]}{\binom{N-1}{d-1} \left[1 + \binom{N-1}{d-1}^{-1} \sum_{i=0}^{d-2} \binom{N-1}{i}\right]}. \qquad (6.45)$$

We do this since

$$\frac{\binom{N-k-1}{d-k-1}}{\binom{N-1}{d-1}} = \frac{d-1}{N-1} \cdots \frac{d-k}{N-k} \to \delta^k \quad \text{as } d \to \infty.$$

Since also $(d-k)/(N-k) \to \delta$, we deduce from Lemma 6.5.2 that the terms in brackets in the numerator and denominator of (6.45) tend to the same finite limit. It follows that

$$\lim_{d \to \infty} \frac{\mathbb{E} f_k(C_N)}{\binom{N}{k}} = (2\delta)^k,$$

as stated.

(c) Now we assume that $\frac{d}{N} \to \frac{1}{2}$ as $d \to \infty$. First we note that if $N < 2d$ for all d, then

$$\lim_{d \to \infty} \frac{\mathbb{E} f_k(C_N)}{\binom{N}{k}} = 1 \qquad (6.46)$$

follows from Theorem 6.5.2. On the other hand, assume that $N \ge 2d$ for all d. Let $0 < \delta_1 < 1/2$ and choose integers $N_1 = N_1(d)$ such that

$$\frac{d}{N_1} \to \delta_1 \quad \text{as } d \to \infty.$$

Then $d/N \ge d/N_1$ and hence $N \le N_1$ for sufficiently large d. Theorem 6.5.1 therefore gives

$$\frac{\mathbb{E} f_k(C_N)}{\binom{N}{k}} \ge \frac{\mathbb{E} f_k(C_{N_1})}{\binom{N_1}{k}},$$

which implies that

$$1 \ge \liminf_{d \to \infty} \frac{\mathbb{E} f_k(C_N)}{\binom{N}{k}} \ge \liminf_{d \to \infty} \frac{\mathbb{E} f_k(C_{N_1})}{\binom{N_1}{k}} = (2\delta_1)^k.$$

Since this holds for all $\delta_1 < 1/2$, relation (6.46) follows.

Finally we consider an arbitrary sequence $(N(d))_{d \in \mathbb{N}}$ with $\frac{d}{N} \to \frac{1}{2}$. The subsequence for which $N < 2d$ and the subsequence for which $N \ge 2d$ both satisfy (6.46), hence this holds also for the sequence itself. □

A more abrupt change than in Theorem 6.5.3 occurs when also the face dimension k tends to infinity. In the following theorem, $k = k(d)$ depends on d, but we omit the argument d in the notation. As it turns out, the threshold is the same as in Theorem 6.4.1, defined by (6.18), although the considered random cones have different distributions.

Theorem 6.5.4. *Let $0 < \delta < 1$ and $0 \leq \rho < 1$ be given, and define the function ρ_W by*

$$\rho_W(\delta) := \max\{0, 2 - \delta^{-1}\}.$$

Let $k(d) = k < d < N = N(d)$ be integers such that

$$\frac{d}{N} \to \delta, \qquad \frac{k}{d} \to \rho \qquad \text{as } d \to \infty.$$

Then

$$\lim_{d \to \infty} \frac{\mathbb{E} f_k(C_N)}{\binom{N}{k}} = \begin{cases} 1, & \text{if } \rho < \rho_W(\delta), \\ 0, & \text{if } \rho > \rho_W(\delta). \end{cases}$$

Proof. First let $0 \leq \rho < \rho_W(\delta)$. The representation given by (6.36),

$$\frac{\mathbb{E} f_k(C_N)}{\binom{N}{k}} = \frac{\mathbb{P}(\xi_{N-k-1} \leq d - k - 1)}{\mathbb{P}(\xi_{N-1} \leq d - 1)}, \tag{6.47}$$

allows us to use a probabilistic argument. Since ξ_n is in distribution equal to the sum of n i.i.d. Bernoulli random variables with parameter $1/2$, the weak law of large numbers gives

$$\lim_{n \to \infty} \mathbb{P}\left(\xi_n > n\left(\frac{1}{2} + \varepsilon\right)\right) = 0$$

for each $\varepsilon > 0$. If now $\frac{d}{N} \to \delta > \frac{1}{2}$ and $\frac{k}{d} \to \rho < 2 - \delta^{-1}$, then

$$\lim_{d \to \infty} \frac{d - k - 1}{N - k - 1} = \frac{1 - \rho}{1/\delta - \rho} > \frac{1}{2} \quad \text{and} \quad \lim_{d \to \infty} \frac{d - 1}{N - 1} = \delta > \frac{1}{2}.$$

Therefore,

$$\lim_{d \to \infty} \mathbb{P}(\xi_{N-k-1} \leq d - k - 1) = 1, \qquad \lim_{d \to \infty} \mathbb{P}(\xi_{N-1} \leq d - 1) = 1$$

and hence

$$\lim_{d \to \infty} \frac{\mathbb{E} f_k(C_N)}{\binom{N}{k}} = 1,$$

as stated.

Now we assume that $\rho > \rho_W(\delta)$. We use (6.45) and show first that for increasing d the terms in brackets remain between two positive constants. Define

$$\frac{d}{N} =: \delta_d, \qquad \frac{k}{d} =: \rho_d,$$

then $\delta_d \to \delta$ and $\rho_d \to \rho$ as $d \to \infty$. For sufficiently large d (which we assume in the following), we then have $\rho_d > 2 - \delta_d^{-1}$, which implies $N - 2d + k > 0$. Therefore, we can apply (6.42) to the normalized sum in the numerator of (6.45). This yields

$$\frac{d-k-1}{N-d+1} \le \binom{N-k-1}{d-k-1}^{-1} \sum_{i=0}^{d-k-2} \binom{N-k-1}{i}$$

$$\le \frac{d-k-1}{N-d+1} \cdot \frac{N-d+2}{N-2d+k+4}.$$

Here,

$$\lim_{d\to\infty} \frac{d-k-1}{N-d+1} = \frac{\delta(1-\rho)}{1-\delta}, \qquad \lim_{d\to\infty} \frac{N-d+2}{N-2d+k+4} = \frac{1-\delta}{1-2\delta+\rho\delta},$$

where the last denominator is positive. It follows that there are positive constants c_1, c_2, independent of d, such that

$$c_1 \le \binom{N-k-1}{d-k-1}^{-1} \sum_{i=0}^{d-k-2} \binom{N-k-1}{i} \le c_2 \qquad (6.48)$$

for all sufficiently large d.

It remains to determine the asymptotic behavior, as $d \to \infty$, of

$$2^k \frac{\binom{N-k-1}{d-k-1}}{\binom{N-1}{d-1}} = 2^k \frac{N}{d} \frac{d-k}{N-k} \frac{\binom{d}{k}}{\binom{N}{k}},$$

appearing in (6.45). Here,

$$\lim_{d\to\infty} \frac{N}{d} \frac{d-k}{N-k} = \frac{1-\rho}{1-\rho\delta}.$$

For the remaining term, we use the Stirling formula

$$n! = \sqrt{2\pi n}\, e^{-n} n^n e^{\theta/12n}, \quad 0 < \theta < 1. \qquad (6.49)$$

It gives

$$2^k \frac{\binom{d}{k}}{\binom{N}{k}} = 2^{\rho_d\delta_d N} \sqrt{\frac{1-\rho_d\delta_d}{1-\rho_d}} \cdot \frac{(\delta_d N)^{\delta_d N}(N-\rho_d\delta_d N)^{N-\rho_d\delta_d N}}{(\delta_d N-\rho_d\delta_d N)^{\delta_d N-\rho_d\delta_d N} N^N} \cdot e^{\varphi/12N}$$

$$= e^{\varphi/12N} \sqrt{\frac{1-\rho_d\delta_d}{1-\rho_d}} H(\delta_d,\rho_d)^N, \qquad (6.50)$$

where φ is contained in a fixed interval independent of d, and

$$H(a,b) := \frac{(2a)^{ab}(1-ab)^{1-ab}}{(1-b)^{a(1-b)}}, \quad a \in (0,1),\, b < 1.$$

We define

$$g(a) := H\left(a, 2-a^{-1}\right) = 2a^a(1-a)^{1-a} \qquad \text{for } a \in (0,1). \qquad (6.51)$$

Note that for $a, b \in (0,1)$ we have $b > 2 - a^{-1}$ if and only if $a < 1/(2-b)$. Let $H_a(b) := H(a,b)$. Differentiation yields

$$H_a'(b) = a \log \left(\frac{2a(1-b)}{1-ab} \right) H(a,b).$$

Hence $H_a'(b) < 0$ for $b > 2 - a^{-1}$, since

$$\frac{2a(1-b)}{1-ab} < 1 \Leftrightarrow b > 2 - a^{-1}.$$

If $a \leq 1/2$, then

$$H(a,b) < H(a,0) = 1$$

for $b \in (0,1)$. On the other hand, if $a > 1/2$ and $b > 2 - a^{-1}$, we have

$$H(a,b) < H\left(a, 2 - a^{-1}\right) = 2a^a (1-a)^{1-a}. \tag{6.52}$$

Since the function g defined by (6.51) satisfies $g(1/2) = 1$ and $g'(a) > 0$ for $1/2 < a < 1$, we also have

$$H(a, 2 - a^{-1}) > 1 \quad \text{if } a > 1/2. \tag{6.53}$$

Now we distinguish two cases.

(a) Let $\delta \leq 1/2$. Then (6.45), (6.48) and (6.50) yield

$$\limsup_{d \to \infty} \frac{\mathbb{E} f_k(C_N)}{\binom{N}{k}} \leq \frac{1-\rho}{1-\rho\delta} \frac{\sqrt{1-\rho\delta}}{\sqrt{1-\rho}} \limsup_{d \to \infty} H(\delta_d, \rho_d)^N \frac{1+c_2}{1}$$

$$\leq (1+c_2) \limsup_{d \to \infty} c_3^N = 0,$$

where $H(\delta_d, \rho_d) \leq c_3 < 1$, since $H(\delta_d, \rho_d) \to H(\delta, \rho) < H(\delta, 0) = 1$. Here and below, c_i are positive constants independent of d.

(b) Let $\delta > 1/2$. Then we can assume that $N/d < c_4 < 2$. We have

$$\sum_{i=0}^{d-2} \binom{N-1}{i} = 2^{N-1} - \sum_{j=0}^{N-d} \binom{N-1}{j}.$$

Since $2(N-d) < N$, Lemma 6.5.1 yields

$$\binom{N-1}{N-d+1}^{-1} \sum_{j=0}^{N-d} \binom{N-1}{j} \leq \frac{\frac{N}{d-1} - 1}{2 - \frac{N}{d}},$$

and hence

$$\binom{N-1}{d-1}^{-1} \sum_{i=0}^{d-2} \binom{N-1}{i} \geq \binom{N-1}{d-1}^{-1} 2^{N-1} - \frac{1}{2 - \frac{N}{d}}.$$

To estimate the last binomial coefficient, we use Stirling's approximation (6.49) together with (6.52). Thus, we get for large d the lower bound

$$\binom{N-1}{d-1}^{-1} \sum_{i=0}^{d-2} \binom{N-1}{i} \geq c_5 H\left(\delta_d, 2 - \delta_d^{-1}\right)^N - c_6.$$

Combining these estimates and starting again from (6.45), we finally obtain

$$\limsup_{d\to\infty} \frac{\mathbb{E} f_k(C_N)}{\binom{N}{k}}$$

$$\leq c_7 \limsup_{d\to\infty} H(\delta_d, \rho_d)^N \frac{1 + c_2}{1 + c_5 H(\delta_d, 2 - \delta_d^{-1})^N - c_6}$$

$$= c_8 \limsup_{d\to\infty} \left(\frac{H(\delta_d, \rho_d)}{H(\delta_d, 2 - \delta_d^{-1})}\right)^N \frac{1}{(1 - c_6)H(\delta_d, 2 - \delta_d^{-1})^{-N} + c_5}$$

$$= 0.$$

Here we have used that

$$\frac{H(\delta, \rho)}{H(\delta, 2 - \delta^{-1})} < 1 \quad \text{for } \rho > 2 - \delta^{-1}$$

by (6.52) and that

$$H(\delta, 2 - \delta^{-1}) > 1 \quad \text{for } \delta > \frac{1}{2}$$

by (6.53). This completes the proof also in the case where $\rho > \rho_W(\delta)$. □

Precisely at the threshold, that is for $\rho = \rho_W(\delta)$, the asymptotic behavior of $\mathbb{E} f_k(C_N)/\binom{N}{k}$ is more complicated. It is similar to that of the random cone D_N, described in Theorem 6.4.2, though with slight differences.

Theorem 6.5.5. *Let $0 < \delta < 1$. Let $k < d < N$ be integers (depending on d) such that*

$$\frac{d}{N} \to \delta, \quad \frac{k}{d} \to \rho = \rho_W(\delta) = \begin{cases} 0 & \text{if } \delta < 1/2, \\ 2 - \frac{1}{\delta} & \text{if } \delta > 1/2, \end{cases} \quad \text{as } d \to \infty.$$

Let $\delta < 1/2$. If $k \to \infty$, then

$$\lim_{d\to\infty} \frac{\mathbb{E} f_k(C_N)}{\binom{N}{k}} = 0. \tag{6.54}$$

If $k \not\to \infty$, then a limit of $\mathbb{E} f_k(C_N)/\binom{N}{k}$ need not exist.
Let $\delta > 1/2$ (and hence $\rho \in (0,1)$).
(a) If

$$N - 2d + k \sim cd^{1/2} \quad with \ c \in \mathbb{R},$$

then

$$\lim_{d \to \infty} \frac{\mathbb{E} f_k(C_N)}{\binom{N}{k}} = \Phi\left(-\frac{c}{\sqrt{2(1-\rho)}}\right) \in (0,1).$$

(b) *If*

$$N - 2d + k \geq cd^{\alpha} \quad with \ \frac{1}{2} < \alpha < 1 \ and \ c > 0,$$

then

$$\lim_{d \to \infty} \frac{\mathbb{E} f_k(C_N)}{\binom{N}{k}} = 0.$$

(c) *If*

$$-cd^{\alpha} \leq N - 2d + k \leq cd^{\alpha} \quad with \ 0 < \alpha < \frac{1}{2} \ and \ c \geq 0,$$

then

$$\lim_{d \to \infty} \frac{\mathbb{E} f_k(C_N)}{\binom{N}{k}} = \frac{1}{2}.$$

(d) *If*

$$N - 2d + k \leq cd^{\alpha} \quad with \ \frac{1}{2} < \alpha < 1 \ and \ c < 0,$$

then

$$\lim_{d \to \infty} \frac{\mathbb{E} f_k(C_N)}{\binom{N}{k}} = 1.$$

Let $\delta = 1/2$ and hence $\rho = 0$. Then a limit of $\mathbb{E} f_k(C_N)/\binom{N}{k}$ need not exist, but by suitable choices of sequences $k(d), N(d)$ any limit in $[0,1]$ can be achieved.

Proof. Let $\delta < 1/2$. If $k \to \infty$, then (6.54) follows from part (b) of the proof of Theorem 6.5.3. If $k \to \infty$ is not required, we can choose a sequence $(k(d))_{d \in \mathbb{N}}$ that attains only two different values, each one infinitely often. Then it follows from Theorem 6.5.3 that the sequence $\left(\mathbb{E} f_k(C_N)/\binom{N}{k}\right)_{d \in \mathbb{N}}$ has two subsequences with different limits.

Let $\delta > 1/2$.

(a) As used in the proof of Theorem 6.5.2, it follows from the Berry–Esseen theorem that

$$\frac{\mathbb{E} f_k(C_N)}{\binom{N}{k}} = \frac{\Phi\left(\frac{2d-N-k-1}{\sqrt{N-k-1}}\right) + O\left(\frac{1}{\sqrt{N-k-1}}\right)}{\Phi\left(\frac{2d-N-1}{\sqrt{N-1}}\right) + O\left(\frac{1}{\sqrt{N-1}}\right)}.$$

We have

$$\frac{N-k-1}{d} \to \frac{1}{\delta} - \rho = 2(1-\rho),$$

$$\frac{2d - N - k - 1}{\sqrt{N - k - 1}} = \frac{2d - N - k}{\sqrt{d}\sqrt{\frac{N-k-1}{d}}} - \frac{\frac{1}{\sqrt{d}}}{\sqrt{\frac{N-k-1}{d}}} \to \frac{-c}{\sqrt{2(1-\rho)}},$$

$$\frac{2d - N - 1}{\sqrt{N - 1}} = \frac{\sqrt{d}\left(2 - \frac{N}{d} - \frac{1}{d}\right)}{\sqrt{\frac{N-1}{d}}} \to \infty,$$

where we have used that $\delta > 1/2$. The assertion follows.

Assertions (b), (c), (d) follow from (a) in combination with the monotonicity provided by Theorem 6.5.1, similarly as in the proof of Theorem 6.3.2.

Finally, we consider the case where $\delta = 1/2$ and hence $\rho = 0$. For arbitrary choices of $a \in \mathbb{R}$ and $c \geq 0$, we consider sequences $k \sim cd^{1/2}$ and $N - 2d \sim ad^{1/2}$, hence $N - 2d + k \sim (a + c)d^{1/2}$. Then the argument for part (a) shows that

$$\frac{\mathbb{E} f_k(C_N)}{\binom{N}{k}} \to \frac{\Phi\left(-\frac{a+c}{\sqrt{2}}\right)}{\Phi\left(-\frac{a}{\sqrt{2}}\right)} \in (0,1] \qquad \text{as } d \to \infty.$$

Thus by suitable choices of a, c, any limit in $(0, 1]$ can be achieved. By choosing $k \sim \log(d)\sqrt{d}$ we see that also the limit 0 can be attained. Choosing sequences with suitable subsequences shows that a limit need not exist. $\qquad\square$

Similarly as for the random cones D_N, we can also treat the difference $\binom{N}{k} - \mathbb{E} f_k(C_N)$. The threshold that appears is the same, namely $\rho_S(\delta)$, defined by (6.25). Before stating the theorem, we prove a lemma.

Lemma 6.5.3. *Writing*

$$A := \frac{1}{2^{N-1} P_{d-k,N-k}} \sum_{j=1}^{k} \binom{k}{j} \sum_{m=0}^{j-1} \binom{N - k - 1}{d - k + m},$$

we have

$$\frac{\mathbb{E} f_k(C_N)}{\binom{N}{k}} = \frac{P_{d-k,N-k}}{P_{d,N}} = \frac{1}{1 + A} \tag{6.55}$$

and

$$\binom{N}{k} - \mathbb{E} f_k(C_N) = \binom{N}{k}\frac{A}{1 + A} \tag{6.56}$$

for $k \in \{1, \ldots, d - 1\}$.

Proof. First we recall (6.35). Next, writing $\binom{N-1}{i} = \binom{N-2}{i-1} + \binom{N-2}{i}$ for $i = 1, \ldots, d - 1$, we have

$$P_{d,N} = \frac{1}{2} P_{d-1,N-1} + \frac{1}{2} P_{d,N-1}.$$

Induction gives

$$P_{d,N} = \frac{1}{2^k} \sum_{j=0}^{k} \binom{k}{j} P_{d-k+j,N-k}$$

for $k \in \{1, \ldots, d-1\}$. For $j \in \{1, \ldots, k\}$ we have

$$P_{d-k+j,N-k} = P_{d-k,N-k} + \frac{1}{2^{N-k-1}} \sum_{m=0}^{j-1} \binom{N-k-1}{d-k+m}.$$

This yields

$$\frac{P_{d,N}}{P_{d-k,N-k}}$$

$$= \frac{1}{2^k} \sum_{j=0}^{k} \binom{k}{j} \frac{P_{d-k+j,N-k}}{P_{d-k,N-k}}$$

$$= \frac{1}{2^k} \left[\binom{k}{0} + \sum_{j=1}^{k} \binom{k}{j} \left(1 + \frac{1}{2^{N-k-1}P_{d-k,N-k}} \sum_{m=0}^{j-1} \binom{N-k-1}{d-k+m} \right) \right]$$

$$= 1 + A$$

and thus the assertion.   □

The function G in the following theorem is defined by (6.24).

Theorem 6.5.6. *Let $1/2 < \delta < 1$ and $0 < \rho < \rho_W(\delta) = 2 - 1/\delta$ be given. Let $k < d < N$ be integers (depending on d) such that*

$$\frac{d}{N} =: \delta_d \to \delta, \qquad \frac{k}{d} =: \rho_d \to \rho \qquad \text{as } d \to \infty.$$

Then

$$\binom{N}{k} - \mathbb{E}\, f_k(C_N) = \frac{1}{N} e^{NG(\delta_d, \rho_d)} \cdot h(d), \tag{6.57}$$

where h remains in a bounded positive interval.

In particular,

$$\lim_{d \to \infty} \left[\binom{N}{k} - \mathbb{E}\, f_k(C_N) \right] = \begin{cases} 0 & \text{if } \rho < \rho_S(\delta), \\ \infty & \text{if } \rho > \rho_S(\delta). \end{cases}$$

If $\rho = \rho_S(\delta)$, then the numbers N and k can be chosen such that

$$\liminf_{d \to \infty} \left[\binom{N}{k} - \mathbb{E}\, f_k(C_N) \right] = 0 \quad \text{and} \quad \limsup_{d \to \infty} \left[\binom{N}{k} - \mathbb{E}\, f_k(C_N) \right] = \infty.$$

Proof. It follows from (6.16) and (6.17) that $\binom{N}{k} - \mathbb{E} f_k(C_N) < \binom{N}{k} -$
$\mathbb{E} f_k(D_N)$, hence Theorem 6.4.3 yields $\binom{N}{k} - \mathbb{E} f_k(C_N) < \frac{1}{N} e^{NG(\delta_d, \rho_d)} \cdot c$ with
a constant c independent of d.

For a corresponding lower bound, we note that Lemma 6.5.3 yields that

$$\binom{N}{k} - \mathbb{E} f_k(C_N) = \binom{N}{k} \frac{A}{1+A}$$

$$= \binom{N}{k} \frac{1}{2^{N-1} P_{d,N}} \sum_{j=1}^{k} \binom{k}{j} \sum_{m=0}^{j-1} \binom{N-k-1}{d-k+m}.$$

Using the trivial estimates $P_{d,N} \leq 1$ and

$$\sum_{m=0}^{j-1} \binom{N-k-1}{d-k+m} \geq \binom{N-k-1}{d-k},$$

we obtain

$$\binom{N}{k} - \mathbb{E} f_k(C_N) \geq 2^{1-N} \binom{N}{k} \sum_{j=1}^{k} \binom{k}{j} \binom{N-k-1}{d-k}$$

$$\geq 2^{k-N} \binom{N}{k} \frac{N-d}{N-k} \binom{N-k}{N-d}.$$

We can now use (6.29), (6.30), (6.31), (6.32) to complete the proof of (6.57).
The last assertion follows as in the proof of Theorem 6.4.3.   □

If $\delta > 1/2$ and $\rho < \rho_S(\delta)$, then one can conclude in the same way as after
(6.34) that

$$\mathbb{P}\left(f_k(C_N) = \binom{N}{k}\right) \geq 1 - \left[\binom{N}{k} - \mathbb{E} f_k(C_N)\right] \to 1$$

as $d \to \infty$.

Notes for Section 6.5

1. In [99], some similar results were obtained for the expectations of the
Grassmann angles. For example the following theorem is proved there.

Theorem. *Suppose that*

$$\frac{d}{N} \to \delta \quad as \ d \to \infty,$$

*with a number $\delta \in [0, 1]$. Let $k \in \mathbb{N}$ be fixed. Then the Cover–Efron cone C_N
satisfies*

$$\lim_{d \to \infty} \mathbb{E}\, 2U_{d-k}(C_N) = \begin{cases} 0 & \text{if } 1/2 < \delta \le 1, \\ 1 - \left(\frac{\delta}{1-\delta}\right)^k & \text{if } 0 \le \delta < 1/2. \end{cases}$$

Also here, the first part of the theorem has a stronger version:

Theorem. *If* $N - 2d$ *is bounded from above, then*

$$\lim_{d \to \infty} \mathbb{E}\, U_{d-k}(C_N) = 0.$$

There is also a counterpart to Theorem 6.5.4 for Grassmann angles.

Theorem. *Let* $0 < \delta, \rho < 1$ *be given. Let* $k < d < N$ *be integers such that*

$$\frac{d}{N} \to \delta, \qquad \frac{k}{d} \to \rho \qquad \text{as } d \to \infty.$$

Then

$$\lim_{d \to \infty} \mathbb{E}\, 2U_{d-k}(C_N) = \begin{cases} 0 \text{ if } \rho < \frac{1}{2}\rho_W(\delta), \\ 1 \text{ if } \rho > \frac{1}{2}\rho_W(\delta). \end{cases}$$

2. The monotonicity property expressed by Theorem 6.5.1 has a counterpart for the function $U_{d-1} = v_d$, the solid angle.

Theorem. *The solid angle of the random Cover–Efron cone satisfies*

$$\mathbb{E}\, v_d(C_{N+1}) > \mathbb{E}\, v_d(C_N)$$

for $N \ge d$.

Proof. By (5.35) and (5.22) we have

$$\mathbb{E}\, v_d(C_N) = \frac{C(N,d) - C(N,d-1)}{2C(N,d)} = \frac{\binom{N-1}{d-1}}{2^N P_{d,N}} =: g(N).$$

This gives

$$g(N+1) - g(N) = \frac{\binom{N}{d-1}}{2^{N+1} P_{d,N+1}} - \frac{\binom{N-1}{d-1}}{2^N P_{d,N}}$$

$$= \frac{1}{2^{N+1} P_{d,N+1} P_{d,N}} \binom{N-1}{d-1} \left[\frac{N}{N-d+1} P_{d,N} - 2P_{d,N+1}\right].$$

Hence, with a positive factor $p(d,N)$,

$$p(d,N)[g(N+1) - g(N)]$$

$$= \frac{N}{N-d+1} \sum_{i=0}^{d-1} \binom{N-1}{i} - \sum_{i=0}^{d-1} \binom{N}{i}$$

$$= \frac{N}{N-d+1} \sum_{i=0}^{d-1} \binom{N-1}{i} - \left[2 \sum_{i=0}^{d-2} \binom{N-1}{i} + \binom{N-1}{d-1} \right]$$

$$= \frac{2d-N-2}{N-d+1} \sum_{i=0}^{d-1} \binom{N-1}{i} + \binom{N-1}{d-1}.$$

This is strictly positive if $2d - N - 2 \geq 0$.

Now suppose that $2d - N - 2 < 0$. Then we can apply Lemma 6.4.1 and obtain

$$\sum_{i=0}^{d-1} \binom{N-1}{i} < \binom{N-1}{d-1} \frac{N-d+1}{N-2d+2}.$$

This again gives $g(N+1) > g(N)$.   □

3. The asymptotic behavior of random Donoho–Tanner and Cover–Efron cones in high dimensions was further thoroughly investigated by Godland, Kabluchko and Thäle [69]. They considered not only face numbers, but also conic intrinsic volumes, Grassmann angles, and the statistical dimension. Their approach is described by the following quotation from the introduction: "In the background of our results are probabilistic parameters of the Cover–Efron and the Donoho–Tanner random cones in terms of probabilities for binomial random variables, ... We can thus build on known limit theorems from probability theory, such as the law of large numbers, the central limit theorem, the local limit theorem, Cramér's theorem for large deviations, mod-ϕ convergence and the Cramér–Petrov theorem for moderate deviations."

4. The investigations of random cones in high dimensions were continued by Godland, Kabluchko and Thäle [70]. They started with independent and identically distributed random vectors $Y_1, \ldots, Y_d$ in $\mathbb{R}^d$ (satisfying some mild assumptions on the distribution) and considered the random cones $\mathrm{pos}\{Y_1 - Y_2, \ldots, Y_{n-1} - Y_n\}$ and $\mathrm{pos}\{Y_1 - Y_2, \ldots, Y_{n-1} - Y_n, Y_n\}$, each under the condition that it is different from $\mathbb{R}^d$. For d, n, and possibly k, tending to infinity in a coordinated way, they found the asymptotic behavior of the number of k-faces, of conic intrinsic volumes and Grassmann angles, including some threshold phenomena. The approach is similar to the one above, but more sophisticated.

6.6 Random cones in halfspaces

In Definition 6.4.1, the Donoho–Tanner cone D_n was defined as the positive hull of independent random vectors $X_1, \ldots, X_n$ in $\mathbb{R}^d$ with a certain distribution ϕ. In contrast, the Cover–Efron cone C_n was defined (see Definition 5.3.1) as the positive hull of the same vectors, under the condition that this positive hull is not the whole space $\mathbb{R}^d$. Thus, the Cover–Efron cone is always contained in a closed halfspace, which may depend on the cone. New phenomena can be observed if we consider random cones in a fixed halfspace. We choose a vector $e \in \mathbb{S}^{d-1}$ and consider the open hemisphere

$$\mathbb{S}_e^+ = \{u \in \mathbb{S}^{d-1} : \langle u, e \rangle > 0\}.$$

On this hemisphere, we consider only the uniform distribution

$$\phi_e := \frac{2}{\omega_d}(\sigma_{d-1} \llcorner \mathbb{S}_e^+).$$

Definition 6.6.1. *The random cone C_n^e is defined by*

$$C_n^e := \mathrm{pos}\{X_1, \dots, X_n\},$$

where $X_1, \dots, X_n$ are i.i.d. random points in $\mathbb{S}_e^+$ with distribution ϕ_e.

The following lemma shows that this random cone is closely related to the random cone S_n^e defined in Definition 5.4.1.

Lemma 6.6.1. *We have*

$$(C_n^e)^\circ \overset{d}{=} -S_n^e.$$

Proof. Using (5.32), for $B \in \mathcal{B}(\mathcal{P}\mathcal{C}^d)$ we get

$$\mathbb{P}\left((C_n^e)^\circ \in B\right) = \mathbb{P}\left(\bigcap_{i=1}^n X_i^- \in B\right)$$

$$= \left(\frac{2}{\omega_d}\right)^n \int_{(\mathbb{S}_e^+)^n} \mathbb{1}_B\left(\bigcap_{i=1}^n x_i^-\right) \sigma_{d-1}^n(\mathrm{d}(x_1, \dots, x_n))$$

$$= \int_{G(d,d-1)} \mathbb{1}_B\left(\bigcap_{i=1}^n H_i^{-e}\right) \nu_{d-1}^n(\mathrm{d}(H_1, \dots, H_n))$$

$$= \mathbb{P}(-S_n^e \in B),$$

as stated.   □

It follows from this Lemma that

$$\mathbb{E}\, f_k(C_n^e) = \mathbb{E}\, f_{d-k}(S_n^e)$$

for $k = 1, \dots, d-1$. Together with (5.57), this gives

$$\mathbb{E}\, f_{d-1}(C_n^e) = 2\binom{n}{d-1}\theta(n-d+1, d). \tag{6.58}$$

We have

$$\lim_{n \to \infty} \mathbb{E}\, f_{d-1}(C_n^e) = 2^{-(d-1)}(d-1)!\kappa_{d-1}^2. \tag{6.59}$$

The finiteness of this limit may seem surprising at first sight. It may seem more plausible if one takes into consideration that the boundary of the halfsphere $\mathbb{S}_e^+$, as a submanifold of the sphere $\mathbb{S}^{d-1}$, has everywhere zero curvature.

Relation (6.59) was first proved, together with a finite limit for $\mathbb{E}\, f_1(C_n^e)$, by Bárány, Hug, Reitzner and Schneider [17]. The limit relation (6.59) follows from (5.51) and the following lemma (with d replaced by $d-1$).

Lemma 6.6.2. *We have*

$$\lim_{n\to\infty} \binom{n}{d} \int_0^\pi \left(1 - \frac{\alpha}{\pi}\right)^{n-d} \sin^{d-1}\alpha \,d\alpha = \frac{\pi^d}{d}$$

and

$$\frac{2\omega_d}{\omega_{d+1}} \frac{\pi^d}{d} = 2^{-d} d!\, \kappa_d^2.$$

Proof. In the integral, we substitute $\alpha = \pi(1 - e^{-s})$ and obtain

$$I := \int_0^\pi \left(1 - \frac{\alpha}{\pi}\right)^{n-d} \sin^{d-1}\alpha \,d\alpha = \pi \int_0^\infty e^{-ns} e^{(d-1)s} \sin^{d-1}(\pi(1 - e^{-s})) \,ds.$$

Since

$$e^{(d-1)s} = 1 + O(s)$$

and

$$\sin^{d-1}(\pi(1 - e^{-s})) = \left(\pi(1 - e^{-s}) + O((1 - e^{-s})^3)\right)^{d-1}$$
$$= \left(\pi\left(s + O(s^2)\right) + O(s^3)\right)^{d-1}$$
$$= \pi^{d-1} s^{d-1}(1 + O(s)),$$

we get

$$I = \pi \int_0^\infty e^{-ns}[1 + O(s)][\pi^{d-1} s^{d-1}(1 + O(s))]\,ds$$
$$= \pi^d \int_0^\infty e^{-ns} s^{d-1}(1 + O(s))\,ds.$$

We substitute $ns = t$ and get

$$I = \left(\frac{\pi}{n}\right)^d \int_0^\infty e^{-t} t^{d-1}(1 + O(t/n))\,dt$$
$$= \left(\frac{\pi}{n}\right)^d \left[(d-1)! + O(1/n)\right].$$

Since

$$\binom{n}{d} = \frac{n^d}{d!}(1 + O(1/n)),$$

we obtain the first statement of the lemma. For the second one, we use the Legendre duplication formula and get

$$\frac{2\omega_d}{\omega_{d+1}}\frac{\pi^d}{d} = 2\frac{\pi^{\frac{d}{2}}}{\pi^{\frac{d+1}{2}}}\frac{\Gamma\left(\frac{d+1}{2}\right)}{\Gamma\left(\frac{d}{2}\right)}\frac{\pi^d}{d} = \frac{1}{\sqrt{\pi}}\frac{\Gamma\left(\frac{d+1}{2}\right)}{\Gamma\left(\frac{d}{2}+1\right)}\pi^d$$

$$= \frac{1}{\sqrt{\pi}}\frac{1}{\Gamma\left(\frac{d}{2}+1\right)}\frac{\sqrt{\pi}\,2^{-d}\Gamma(d+1)}{\Gamma\left(\frac{d}{2}+1\right)}\pi^d = 2^{-d}d!\kappa_d^2,$$

as stated. $\qquad\square$

Since trivially there is a constant $c(d)$, depending only on the dimension d, such that any polytope P in $\mathbb{R}^d$ with $f_{d-1}(P)$ facets has at most $c(d)f_{d-1}(P)$ faces, it follows from (6.59) that

$$\limsup_{n\to\infty} f_k(C_n^e) < \infty \quad \text{for } k = 1,\ldots,d-1.$$

This does not imply that the limits $\lim_{n\to\infty} f_k(C_n^e)$ exist, but this has been shown, with powerful new tools. First we explain a key idea in the work of Kabluchko, Marynych, Temesvari, and Thäle [108].

We use the standard orthonormal basis in $\mathbb{R}^d$ and suppose that e is the first basis vector. We define a map $T_n : \mathbb{R}^d \to \mathbb{R}^d$ by

$$T_n(x_1,\ldots,x_d) := (nx_1, x_2, \ldots, x_d).$$

Let $H_1 := H(e,1)$, the tangent hyperplane to the unit sphere at e. In the following, we identify $e^\perp$ with $\mathbb{R}^{d-1}$.

Let ξ be a Poisson point process in $\mathbb{R}^{d-1} \setminus \{o\}$ whose intensity measure has a density with respect to Lebesgue measure given by

$$x \mapsto \frac{2}{\omega_d}\frac{1}{\|x\|^d}, \quad x \in \mathbb{R}^{d-1} \setminus \{o\}.$$

For any centered ball in $\mathbb{R}^{d-1}$ of radius larger than 1, this Poisson process has almost surely only finitely many points outside the ball. Further, the convex hull $\operatorname{conv}\xi$ is almost surely a polytope. The following is proved in [108].

Theorem. *As $n \to \infty$, the random polytopes $(T_n C_n^e \cap H_1) - e$ converge in distribution to $\operatorname{conv}\xi$.*

From this fact, many strong conclusions are drawn in [108]. Among them is the existence of finite limits $\lim_{n\to\infty} \mathbb{E} f_k(C_n^e)$. For the solid angle α it is proved that

$$n\left(\frac{1}{2} - \alpha(C_n^e)\right) \xrightarrow{d} \frac{1}{\omega_d}\int_{\mathbb{R}^{d-1}\setminus\operatorname{conv}\xi} \frac{1}{\|x\|^d}\,dx,$$

where $\xrightarrow{d}$ means convergence in distribution. Moreover, the authors determine the finite limits

$$\lim_{n \to \infty} n^{d-k} \left(\frac{1}{2} - \mathbb{E} \, U_k(C_n^e) \right)$$

for $k \in \{0, 1, \ldots, d-1\}$ and

$$\lim_{n \to \infty} n^{d-2-k} \mathbb{E} \, v_k(C_n^e), \qquad \lim_{n \to \infty} n^{d-k} \left(1 - \mathbb{E} \, W_{k+1}(C_n^e) \right)$$

for $k \in \{0, 1, \ldots, d-2\}$.

Expressions for $\mathbb{E} \, f_k(C_n^e)$ in terms of certain expected angles were derived by Kabluchko, Thäle and Zaporozhets [112]. These expressions lead to alternative forms of the limits $\lim_{n \to \infty} \mathbb{E} \, f_k(C_n^e)$.

Finally, based on [108] and [112], Kabluchko [106, Thm. 2.2] was able to provide explicit formulas for $\mathbb{E} \, f_k(C_n^e)$. They involve combinatorial quantities which first turned up when Kabluchko succeeded with determining $\mathbb{E} \, f_k(Z_o)$, where Z_o is the zero cell of a stationary and isotropic Poisson hyperplane tessellation in $\mathbb{R}^d$. Kabluchko's [106, Thm. 1.1] result says that

$$\mathbb{E} \, f_k(Z_o) = \frac{\pi^{d-k}}{(d-k)!} A[d, d-k]$$

for $k \in \{0, \ldots, d\}$. For even $d - k$, the number $A[d, d-k]$ is the coefficient of x^{d-k} in the polynomial

$$x \mapsto \prod_{j \in \{1, \ldots, d-1\}, \, d-j \, \text{odd}} (1 + j^2 x^2).$$

For odd $d - k$, the value of $\mathbb{E} \, f_k(Z_o)$, and thus that of $A[d, d-k]$, can be determined from the known cases and the Dehn–Sommerville equations, since Z_o is almost surely a simple polytope (or see Kabluchko [106] for the explicit values).

The formula for $\mathbb{E} \, f_k(C_n^e)$ requires further the numbers

$$B\{n, k\} := \frac{1}{(k-1)!(n-k)!} \int_0^\pi (\sin x)^{k-1} x^{n-k} \, \mathrm{d}x, \quad k \in \{1, \ldots, n\}.$$

Kabluchko's result [106, Thm. 2.2] (slightly reformulated) then says that

$$\mathbb{E} \, f_k(C_n^e) = \frac{n! \pi^{k-n}}{k!} \sum_{s=0}^{\lfloor \frac{d-1-k}{2} \rfloor} B\{n, d-1-2s\}(d-2-2s)^2 A[d-3-2s, k-2].$$

Notes for Section 6.6

1. For fixed $m \in \{0, 1, \ldots\}$, the asymptotic behavior of the expected solid angle of the cone C_{d+m}^e, as $d \to \infty$, was determined by Kabluchko [107, Thm. 1.2]. He found that

$$\mathbb{E} \, v_d(C_{d+m}^e) \sim \sqrt{3} \frac{(d/2)^m}{m!} \pi^{-d} \quad \text{as } d \to \infty.$$

7

Convex hypersurfaces adapted to cones

In this chapter, the viewpoint is distinctly different. We still start with a pointed closed convex cone C with interior points. But our main interest will be in convex hypersurfaces, namely boundaries of closed convex sets, in this cone, whose behavior at infinity is determined by the cone. We may distinguish different types of such determination. Let $K \subset C$ be a closed convex set. We say that K is *C-asymptotic* if for $x \in \operatorname{bd} C$ the distance of x from K tends to zero as $\|x\| \to \infty$. If $C \setminus K$ has finite volume (Lebesgue measure), we say that K is *C-close*. And finally, the set K is called *C-full* if $C \setminus K$ is bounded. Clearly, a C-full set is C-close, and a C-close set is C-asymptotic.

Few but remarkable examples of sets of this kind have appeared in the literature. First, we would like to quote a conjecture of Calabi [42]: "Every complete, n-dimensional affine hypersurface with mean curvature $H < 0$ is asymptotic to the boundary of a convex cone with vertex at the center; conversely, every pointed, nondegenerate convex cone C determines an affine hypersphere of hyperbolic type, asymptotic to the boundary of C, and uniquely determined by the value $H < 0$ of the mean curvature." One can read about the further fate of this conjecture in the book by Li, Simon and Zhao [124].

In the work of Gigena [62], a class of C-asymptotic sets was investigated, whose boundaries were introduced as *constant volume envelopes* of hyperplanes cutting off constant volume from the cone. Interestingly, this construction is analogous to that of convex floating bodies in the theory of convex bodies (for references, see [163, Sect. 10.6]), but appeared about ten years earlier. Gigena proved a number of interesting properties of these C-asymptotic sets.

We wanted to mention these remarkable examples of convex hypersurfaces asymptotic to convex cones, though our starting point is different. Of interest are also the differences $C \setminus K$. Such sets, in the special case where K is C-full, were introduced and studied by Khovanskiĭ and Timorin [115], motivated by some questions from algebraic geometry and singularity theory. Their concept is also of interest from the viewpoint of convex geometry, and we will study an extended version of it under this aspect. In the following, we shall say that

© The Author(s), under exclusive license to Springer Nature Switzerland AG 2022
R. Schneider, *Convex Cones*, Lecture Notes in Mathematics 2319,
https://doi.org/10.1007/978-3-031-15127-9_7

a set A is *coconvex* if it is of the form $A = C \setminus K$, where $K \subset C$ is closed and convex, and A has finite volume. The case $K = C$ is not excluded, so in our terminology, a coconvex set may be empty. In Sections 7.2 to 7.4 we shall consider bounded coconvex sets (the case studied in [115]); later we deal with general coconvex sets.

We assume in this chapter that the reader is familiar with the basic facts of the classical Brunn–Minkowski theory of convex bodies. Our goal is to carry over some elements of this theory to coconvex sets. The present chapter is based on [167, 169].

7.1 Coconvex sets

As already said, $C \in \mathcal{C}^d$ is a fixed pointed closed convex cone with interior points. Since C is pointed, there is a unit vector w such that $\langle w, x \rangle > 0$ for all $x \in C \setminus \{o\}$ (choose $-w \in \operatorname{int} C^\circ$). Then the sets $C \cap H(w, t)$, $C \cap H^-(w, t)$ are bounded for all $t > 0$. We fix the vector w in the following and write

$$C_t := C \cap H^-(w, t) \quad \text{for } t > 0.$$

Now we give formal definitions of the notions already mentioned.

Definition 7.1.1. *A closed convex set $K \subset C$ is called C-asymptotic if*

$$\lim_{\|x\| \to \infty} \operatorname{dist}(x, K) = 0 \quad \text{for } x \in \operatorname{bd} C.$$

If K is C-asymptotic, then

$$x + C \subseteq K \quad \text{for } x \in K. \tag{7.1}$$

In fact, if $x \in K$ and there is a ray $R \subset C$ (with endpoint o) such that $x + R \not\subseteq K$, then we can choose a point $y \in (x + R) \setminus K$. The closed convex set K and the point y can be strongly separated by a hyperplane H, and then the part of $\operatorname{bd} C$ contained in the halfspace bounded by H and not containing K contains points arbitrarily far from K, a contradiction.

Definition 7.1.2. *A set K is called C-close if $K \subseteq C$, K is closed and convex, and $C \setminus K$ has finite volume. If this holds, then $C \setminus K$ is called a C-coconvex set.*

The set K is called C-full if $K \subseteq C$, K is closed and convex, and $C \setminus K$ is bounded.

Thus, if K is a C-full set, then $C \setminus K$ is a bounded C-coconvex set. For each bounded C-coconvex set A there exists a number $t > 0$ such $A \subset C_t$. We note that a C-close set (in contrast to a C-full set) may have empty intersection with the boundary of C. Clearly, a C-close set is C-asymptotic.

Since C will be fixed in this chapter, we say 'coconvex' instead of 'C-coconvex'. In the following, we write a coconvex set often in the form $A = C \setminus A^\bullet$, where $A^\bullet$ always means the convex set that is C-close and determines A.

Let K be a C-close set, and let $x \in \operatorname{bd} K \cap \operatorname{int} C$. Then there is supporting hyperplane of K at x such that o is contained in the open halfspace bounded by this hyperplane that does not contain K (otherwise, $C \setminus K$ would have infinite volume). Hence, the half-open segment $[o, x)$ does not meet K. In other words, the boundary of K inside $\operatorname{int} C$ is 'visible' from o.

The Minkowski addition of convex sets induces an addition of C-asymptotic sets.

Definition 7.1.3. *The co-sum of the C-asymptotic sets $A_i = C \setminus A_i^\bullet$, $i = 1, 2$, is defined by*

$$A_1 \boxplus A_2 := C \setminus (A_1^\bullet + A_2^\bullet).$$

We note that the vector sum $A_1^\bullet + A_2^\bullet$ is contained in C and is closed (as is easy to see), because $A_1^\bullet, A_2^\bullet$ are contained in the cone C. Clearly, $A_1 \boxplus A_2$ is again C-asymptotic, and it is bounded if A_1 and A_2 are bounded. For general coconvex sets, it is not obvious that $A_1 \boxplus A_2$ is again coconvex, that is, that it has finite volume. This follows, however, from Theorem 7.5.1 below.

We point out that $A_1^\bullet = C$ is not excluded, thus $\emptyset$ is a coconvex set (but $\{o\}$ is not a coconvex set). Since $\emptyset^\bullet = C$ and $C + A^\bullet = A^\bullet$ for a C-asymptotic set A, we have $\emptyset \boxplus A = A$. For a C-asymptotic set A, we define

$$\lambda A := \begin{cases} \{\lambda a : a \in A\} & \text{if } \lambda > 0, \\ \emptyset & \text{if } \lambda = 0. \end{cases}$$

Then $\lambda A = C \setminus \lambda A^\bullet$ for $\lambda > 0$. We have $\lambda_1 A \boxplus \lambda_2 A = (\lambda_1 + \lambda_2) A$ for $\lambda_1, \lambda_2 \geq 0$. Together with A, also $\lambda A = \{\lambda a : a \in A\}$ is coconvex for $\lambda > 0$, since $V_d(\lambda A) = \lambda^d V_d(A)$. Here the volume (Lebesgue measure) of a coconvex set A is denoted by $V_d(A)$.

Now we carry basic subjects from the theory of convex bodies, namely support function, area measure, and a volume representation, over to coconvex sets.

We define

$$\Omega_C := \mathbb{S}^{d-1} \cap \operatorname{int} C^\circ,$$

where C° is the polar cone of C. In the following, this set will play for coconvex sets the role that the whole unit sphere $\mathbb{S}^{d-1}$ plays for a convex body.

The support function of a C-asymptotic set $A^\bullet$ is defined by

$$h(A^\bullet, x) := \max\{\langle x, y \rangle : y \in A^\bullet\} \quad \text{for } x \in \operatorname{int} C^\circ.$$

It is easy to see that the maximum exists. Clearly, $h(A^\bullet, \cdot)$ is determined by its values on Ω_C. The support function $h(A^\bullet, \cdot)$ determines $A^\bullet$ uniquely, namely by

$$A^\bullet = C \cap \bigcap_{u \in \Omega_C} H^-(u, h(A^\bullet, u)).$$

For $u \in \Omega_C$, the closed halfspace

$$H^-(A^\bullet, u) = \{x \in \mathbb{R}^d : \langle x, u \rangle \le h(A^\bullet, u)\}$$

is the supporting halfspace of $A^\bullet$ with outer normal vector u. We have

$$o \in A^\bullet \Leftrightarrow A^\bullet = C \Leftrightarrow h(A^\bullet, \cdot) \equiv 0.$$

If $o \notin A^\bullet$, then for each $u \in \Omega_C$ there is a supporting halfspace of $A^\bullet$ with outer normal vector u and not containing o, thus $-\infty < h(A^\bullet, u) < 0$.

Now let $A = C \setminus A^\bullet$ be a coconvex set. We set

$$\overline{h}(A, u) := -h(A^\bullet, u)$$

for $u \in \Omega_C$ and call the function $\overline{h}(A, \cdot) : \Omega_C \to \mathbb{R}$ the *support function* of A.

The σ-algebra of Borel sets in Ω_C is denoted by $\mathcal{B}(\Omega_C)$. The spherical image $\sigma(A^\bullet, \beta)$ of the closed convex set $A^\bullet$ at the set $\beta \in \mathcal{B}(\mathbb{R}^d)$ is the set of all outer unit normal vectors of $A^\bullet$ at points of $A^\bullet \cap \beta$. For the C-asymptotic set $A^\bullet$, we have $\sigma(A^\bullet, \operatorname{int} C) \subseteq \Omega_C$, since a supporting hyperplane of $A^\bullet$ at a point of $\operatorname{int} C \cap \operatorname{bd} A^\bullet$ separates $A^\bullet$ and the origin o. For $\omega \subseteq \Omega_C$, the reverse spherical image $\tau(A^\bullet, \omega)$ is defined as the set of all points in $\operatorname{bd} A^\bullet$ at which there exists an outer unit normal vector belonging to ω. For sets $\omega \in \mathcal{B}(\Omega_C)$ one then defines

$$\overline{S}_{d-1}(A, \omega) := S_{d-1}(A^\bullet, \omega) = \mathcal{H}^{d-1}(\tau(A^\bullet, \omega)),$$

where $\mathcal{H}^{d-1}$ is the $(d-1)$-dimensional Hausdorff measure (so that $S_{d-1}(A^\bullet, \cdot)$ is the usual surface area measure of $A^\bullet$, restricted to Ω_C). Using the theory of surface area measures of convex bodies (see [163, Sect. 4.2]), it is easily seen that this defines a Borel measure on Ω_C, the *surface area measure* $\overline{S}_{d-1}(A, \cdot)$ of A. In contrast to the case of convex bodies, the surface area measure of a C-asymptotic set is only defined on the subset Ω_C of $\mathbb{S}^{d-1}$, and it need not be finite. Note that $\overline{S}_{d-1}(\emptyset, \cdot) = 0$.

The volume of a coconvex set A has an integral representation similar to that in the case of convex bodies, as stated in the following lemma. We point out, however, that here the integration is only over Ω_C.

Lemma 7.1.1. *The volume of a coconvex set A is represented by*

$$V_d(A) = \frac{1}{d} \int_{\Omega_C} \overline{h}(A, u) \overline{S}_{d-1}(A, du). \tag{7.2}$$

Proof. In the following, for sets $M \subseteq C$ we write $M_t := M \cap H^-(w, t)$, in particular, $C_t = C \cap H^-(w, t)$ (as already mentioned) and

$$A_t^\bullet := A^\bullet \cap H^-(w, t), \qquad A_{i,t}^\bullet := A_i^\bullet \cap H^-(w, t).$$

Further, we abbreviate
$$H(w, t) := H_t.$$

Let A be a coconvex set, and let $t > 0$ be such that $A_t^\bullet$ has interior points. Let
$$\omega_t := \sigma(A_t^\bullet, \operatorname{int} C_t);$$
this is the spherical image of the set of boundary points of $A^\bullet$ in the interior of C_t. Further, let
$$\eta_t := \sigma(A_t^\bullet, \operatorname{bd} C) \cap \operatorname{bd} C^\circ.$$

By a standard representation of the volume of convex bodies (formula (5.3) in [163]), we have
$$V_d(A_t^\bullet) = \frac{1}{d} \int_{\mathbb{S}^{d-1}} h(A_t^\bullet, u) \, S_{d-1}(A_t^\bullet, du).$$

Here,
$$\int_{\eta_t} h(A_t^\bullet, u) \, S_{d-1}(A_t^\bullet, du) = 0,$$
since $u \in \eta_t$ implies that $h(A_t^\bullet, u) = 0$. We state that
$$S_{d-1}(A_t^\bullet, \mathbb{S}^{d-1} \setminus (\omega_t \cup \eta_t \cup \{w\})) = 0. \tag{7.3}$$

For the proof, let x be a boundary point of $A_t^\bullet$ where a vector $u \in \mathbb{S}^{d-1} \setminus (\omega_t \cup \eta_t \cup \{w\})$ is attained as outer normal vector. Then $x \notin \operatorname{int} C_t$ and hence $x \in H_t$ or $x \in \operatorname{bd} C$. If $x \in H_t$, then $u \neq w$ implies that x lies in two distinct supporting hyperplanes of $A_t^\bullet$. If $x \in (\operatorname{bd} C) \setminus H_t$, then $u \notin \eta_t$ implies that x lies in two distinct supporting hyperplanes of $A_t^\bullet$. In each case, x is a singular boundary point of $A_t^\bullet$. Now the assertion (7.3) follows from [163, (4.32) and Thm. 2.2.5].

As a result, we have
$$V_d(A_t^\bullet) = \frac{1}{d} \int_{\omega_t \cup \{w\}} h(A_t^\bullet, u) \, S_{d-1}(A_t^\bullet, du).$$

Since
$$h(A_t^\bullet, w) = t, \qquad S_{d-1}(A_t^\bullet, \{w\}) = V_{d-1}(A^\bullet \cap H_t),$$
we obtain
$$V_d(A_t^\bullet) = -\frac{1}{d} \int_{\omega_t} \overline{h}(A, u) \, \overline{S}_{d-1}(A, du) + \frac{1}{d} t V_{d-1}(A^\bullet \cap H_t),$$
by the definition of $\overline{h}(A, \cdot)$ and $\overline{S}_{d-1}(A, \cdot)$. We define
$$B(t) := \operatorname{conv}((A^\bullet \cap H_t) \cup \{o\}) \setminus A_t^\bullet.$$

Recalling that the boundary of $A^\bullet$ inside $\operatorname{int} C$ is visible from o, we have

$$V_d(B(t)) = \frac{1}{d} t V_{d-1}(A^\bullet \cap H_t) - V_d(A_t^\bullet)$$

and thus

$$V_d(B(t)) = \frac{1}{d} \int_{\omega_t} \overline{h}(A, u) \, \overline{S}_{d-1}(A, \mathrm{d}u).$$

On the other hand, writing

$$q(t) := V_{d-1}(C \cap H_t) - V_{d-1}(A^\bullet \cap H_t),$$

we get

$$V_d(A_t) = V_d(B(t)) + \frac{1}{d} t q(t) = \frac{1}{d} \int_{\omega_t} \overline{h}(A, u) \, \overline{S}_{d-1}(A, \mathrm{d}u) + \frac{1}{d} t q(t).$$

Given $\varepsilon > 0$, to each $t_0 > 0$ there exists $t \geq t_0$ with $t q(t) < \varepsilon$. Otherwise, there would exist t_0 with $t q(t) \geq \varepsilon$ for $t \geq t_0$ and hence $\int_{t_0}^{\infty} q(t) \mathrm{d}t = \infty$, which yields $V_d(A) = \infty$, a contradiction. Therefore, we can choose an increasing sequence $(t_i)_{i \in \mathbb{N}}$ with $t_i \to \infty$ for $i \to \infty$ such that $t_i q(t_i) \to 0$. From

$$V_d(A_{t_i}) = \frac{1}{d} \int_{\omega_{t_i}} \overline{h}(A, u) \, \overline{S}_{d-1}(A, \mathrm{d}u) + \frac{1}{d} t_i q(t_i)$$

and $\omega_{t_i} \uparrow \Omega_C$ we then obtain

$$V_d(A) = \frac{1}{d} \int_{\Omega_C} \overline{h}(A, u) \, \overline{S}_{d-1}(A, \mathrm{d}u),$$

as stated.   □

7.2 Mixed volumes involving bounded coconvex sets

In Sections 7.2 to 7.4, we restrict ourselves to bounded coconvex sets.

After the volume, also mixed volumes and their representations can be carried over to bounded coconvex sets. More generally, we consider mixed volumes of at least one bounded coconvex set, where the other arguments are either bounded coconvex sets or convex bodies.

Let $p \in \{1, \ldots, d\}$, let $A_1, \ldots, A_p$ be bounded coconvex sets, and let $K_{p+1}, \ldots, K_d$ be convex bodies in $\mathbb{R}^d$. With $A_i = C \setminus A_i^\bullet$ and $t > 0$, we recall that $A_{i,t}^\bullet := A_i^\bullet \cap H^-(w, t)$, and we choose t so large that $A_i \subset H^-(w, t)$ for $i = 1, \ldots, p$.

We define the *mixed area measure* of $A_2, \ldots, A_p, K_{p+1}, \ldots, K_d$ by

$$\overline{S}(A_2, \ldots, A_p, K_{p+1} \ldots, K_d, \omega) = S(A_{2,t}^\bullet, \ldots, A_{p,t}^\bullet, K_{p+1}, \ldots, K_d, \omega)$$

for Borel sets $\omega \subseteq \Omega_C$, where the right-hand side is the usual mixed area measure of the convex bodies $A_{2,t}^\bullet, \ldots, A_{p,t}^\bullet, K_{p+1}, \ldots, K_d$ (see [163, Sect. 5.1]).

This definition is independent of the choice of t. The mixed area measure of bounded coconvex sets and convex bodies is only defined on Ω_C, and it is finite.

We shall later need the following lemma.

Lemma 7.2.1. *Let* $\omega \subseteq \mathbb{S}^{d-1}$ *be open. Let* $K_1, \ldots, K_{d-1}, L_1, \ldots, L_{d-1} \in \mathcal{K}^d$ *be such that* $h(K_i, u) = h(L_i, u)$ *for* $u \in \omega$ *and* $i = 1, \ldots, d-1$. *Then*

$$S(K_1, \ldots, K_{d-1}, \cdot) = S(L_1, \ldots, L_{d-1}, \cdot) \quad on \ \omega.$$

Proof. First let $K, L \in \mathcal{K}^d$ satisfy $h(K, u) = h(L, u)$ for $u \in \omega$. Since ω is open, it follows from [163, Thm. 1.7.2] that the support sets satisfy $F(K, u) = F(L, u)$ for $u \in \omega$. This implies that $\tau(K, \omega) = \tau(L, \omega)$. Taking the $(d-1)$-dimensional Hausdorff measure of this set, we obtain $S_{d-1}(K, \omega) = S_{d-1}(L, \omega)$. In the same way, we deduce that $S_{d-1}(K, \omega') = S_{d-1}(L, \omega')$ for every open subset $\omega' \subseteq \omega$. It follows that $S_{d-1}(K, \beta) = S_{d-1}(L, \beta)$ for every Borel subset $\beta \subseteq \omega$. Now we observe that the assumptions of the lemma imply that $h(K_{i_1} + \cdots + K_{i_k}, u) = h(L_{i_1} + \cdots + L_{i_k}, u)$ for $u \in \omega$ and $1 \le i_1 < \cdots < i_k \le d-1$, so that the relation [163, (5.21)] gives the assertion. □

Now we define mixed volumes involving bounded coconvex sets.

Definition 7.2.1. *Let* $p \in \{1, \ldots, d\}$, *let* $A_1, \ldots, A_p$ *be bounded coconvex sets, and let* $K_{p+1}, \ldots, K_d \in \mathcal{K}^d$ *be convex bodies. The mixed volume of* $A_1, \ldots, A_p, K_{p+1}, \ldots, K_d$ *is defined by*

$$\overline{V}(A_1, \ldots, A_p, K_{p+1}, \ldots, K_d)$$

$$:= \frac{1}{d} \int_{\Omega_C} \overline{h}(A_1, u) \, \overline{S}(A_2, \ldots, A_p, K_{p+1}, \ldots K_d, \mathrm{d}u). \tag{7.4}$$

Note that the integration in (7.4) is only over Ω_C. We remark that

$$\overline{V}(A_1, \ldots, A_p, K_{p+1}, \ldots, K_d) = 0 \quad \text{if } A_i = \emptyset \text{ for some } i. \tag{7.5}$$

This is clear from $\overline{h}(\emptyset, \cdot) = 0$ if $i = 1$. If, say, $i = 2$, it follows from $\overline{S}(\emptyset, A_3, \ldots, A_p, K_{p+1}, \ldots K_d, \Omega_C) = 0$. The latter follows from $\emptyset^\bullet = C$ and

$$S(C_t, A_{3,t}^\bullet, \ldots, A_{p,t}^\bullet, K_{p+1}, \ldots K_d, \omega) = 0,$$

if ω is an open subset of $\sigma(C_t, \{o\})$, since o is a 0-singular boundary point of C_t. This is seen from [163, (5.22)] in the case of polytopes, and the general case is obtained by using approximation and the weak convergence of the mixed area measures.

The following lemma expresses the new mixed volume as a difference of ordinary mixed volumes.

Lemma 7.2.2. *Choose* $t_i > 0$ *such that* $A_i \subset \operatorname{int} C_{t_i}$ *and set* $A_{i,t_i}^{\bullet} := A_i^{\bullet} \cap H^-(w, t_i)$, *for* $i = 1, \ldots, p$. *Then*

$$\overline{V}(A_1, \ldots, A_p, K_{p+1}, \ldots, K_d)$$
$$= V(C_{t_1}, \ldots, C_{t_p}, K_{p+1}, \ldots, K_d) - V(A_{1,t_1}^{\bullet}, \ldots, A_{p,t_p}^{\bullet}, K_{p+1}, \ldots, K_d).$$

Proof. The ordinary mixed volume of the convex bodies $A_{1,t_1}^{\bullet}, \ldots, A_{p,t_p}^{\bullet}$ and $K_{p+1}, \ldots, K_d$ is given by

$$V(A_{1,t_1}^{\bullet}, \ldots, A_{p,t_p}^{\bullet}, K_{p+1}, \ldots, K_d)$$
$$= \frac{1}{d} \int_{\mathbb{S}^{d-1}} h(A_{1,t_1}^{\bullet}, u)\, S(A_{2,t_2}^{\bullet}, \ldots, A_{p,t_p}^{\bullet}, K_{p+1}, \ldots, K_d, du).$$

Since $h(A_{1,t_1}^{\bullet}, u) = 0$ for $u \in \mathbb{S}^{d-1} \cap \operatorname{bd} C^{\circ}$, we have

$$d\overline{V}(A_1, \ldots, A_p, K_{p+1}, \ldots, K_d)$$
$$= \int_{\operatorname{cl}\Omega_C} \overline{h}(A_1, u)\, \overline{S}(A_2, \ldots, A_p, K_{p+1}, \ldots K_d, du)$$
$$= -\int_{\operatorname{cl}\Omega_C} h(A_{1,t_1}^{\bullet}, u)\, S(A_{2,t_2}^{\bullet}, \ldots, A_{p,t_p}^{\bullet}, K_{p+1}, \ldots K_d, du) =: -\int_{\operatorname{cl}\Omega_C} [\cdot]$$
$$= -\int_{\mathbb{S}^{d-1}} [\cdot] + \int_{\mathbb{S}^{d-1} \setminus \operatorname{cl}\Omega_C} [\cdot].$$

Here,

$$\int_{\mathbb{S}^{d-1}} [\cdot] = dV(A_{1,t_1}^{\bullet}, \ldots, A_{p,t_p}^{\bullet}, K_{p+1}, \ldots, K_d).$$

For $u \in \mathbb{S}^{d-1} \setminus \operatorname{cl}\Omega_C$ we have $h(A_{i,t_i}^{\bullet}, u) = h(C_{t_i}, u)$ for $i = 1, \ldots, p$, hence it follows from Lemma 7.2.1 that

$$\int_{\mathbb{S}^{d-1} \setminus \operatorname{cl}\Omega_C} [\cdot] = \int_{\mathbb{S}^{d-1} \setminus \operatorname{cl}\Omega_C} h(C_{t_1}, u)\, S(C_{t_2}, \ldots, C_{t_p}, K_{p+1}, \ldots, K_d, du)$$
$$= \int_{\mathbb{S}^{d-1}} h(C_{t_1}, u)\, S(C_{t_2}, \ldots, C_{t_p}, K_{p+1}, \ldots, K_d, du)$$
$$= dV(C_{t_1}, \ldots, C_{t_p}, K_{p+1}, \ldots, K_d),$$

where we have used that $h(C_{t_1}, u) = 0$ for $u \in \operatorname{cl}\Omega_C$. This gives the assertion. $\qquad\square$

We show that these mixed volumes have properties similar to those of the classical mixed volumes of convex bodies. First we note that Lemma 7.2.2 gives

$$\overline{V}(A_1, \ldots, A_p, K_{p+1}, \ldots, K_d)$$
$$= V(C_t, \ldots, C_t, K_{p+1}, \ldots, K_d) - V(A_{1,t}^{\bullet}, \ldots, A_{p,t}^{\bullet}, K_{p+1}, \ldots, K_d)$$

for sufficiently large t; consequently:

Corollary 7.2.1. *The mixed volume* $\overline{V}(A_1, \ldots, A_p, K_{p+1}, \ldots, K_d)$ *is symmetric in* $A_1, \ldots, A_p$ *and in* $K_{p+1} \ldots, K_d$.

However, we cannot interchange an A and a K argument.
It is clear from (7.2) that

$$\overline{V}(A, \ldots, A) = V_d(A)$$

for a bounded coconvex set A.

Let $m \in \mathbb{N}$, $p \in \{1, \ldots, d\}$, and $q \in \{1, \ldots, p\}$. Let $A_1, \ldots, A_m$ $(m \in \mathbb{N})$ and $A'_{q+1}, \ldots, A'_p$ be bounded coconvex sets and let $K_{p+1}, \ldots, K_d$ be convex bodies. We abbreviate

$$(A'_{q+1}, \ldots, A'_p, K_{p+1}, \ldots, K_d) =: \mathcal{C}$$

and

$$\overline{V}(\underbrace{A, \ldots, A}_{q}, A'_{q+1}, \ldots, A'_p, K_{p+1}, \ldots, K_d) =: \overline{V}(A[q], \mathcal{C}).$$

We state that for $\lambda_1, \ldots, \lambda_m \geq 0$ we have

$$\overline{V}(\lambda_1 A_1 \boxplus \cdots \boxplus \lambda_m A_m[q], \mathcal{C}) = \sum_{i_1, \ldots, i_q = 1}^{m} \lambda_{i_1} \cdots \lambda_{i_q} \overline{V}(A_{i_1}, \ldots, A_{i_q}, \mathcal{C}). \quad (7.6)$$

For the proof, we first assume that $\lambda_1, \ldots, \lambda_m > 0$. We note that, for $u \in \Omega_{\mathcal{C}}$,

$$\begin{aligned}
\overline{h}(\lambda_1 A_1 \boxplus \cdots \boxplus \lambda_m A_m, u) &= -h((\lambda_1 A_1 \boxplus \cdots \boxplus \lambda_m A_m)^\bullet, u) \\
&= -h(\lambda_1 A_1^\bullet + \cdots + \lambda_m A_m^\bullet, u) \\
&= -[\lambda_1 h(A_1^\bullet, u) + \cdots + \lambda_m h(A_m^\bullet, u)] \\
&= \lambda_1 \overline{h}(A_1, u) + \cdots + \lambda_m \overline{h}(A_m, u).
\end{aligned}$$

For sufficiently large t and for Borel sets $\omega \in \mathcal{B}(\Omega_{\mathcal{C}})$, we have,

$$\begin{aligned}
&\overline{S}(\lambda_1 A_1 \boxplus \cdots \boxplus \lambda_m A_m[q], \mathcal{C}) \\
&= S(\lambda_1 A_{1,t}^\bullet + \cdots + \lambda_m A_{m,t}^\bullet[q], A'^\bullet_{q+1}, \ldots, A'^\bullet_p, K_{p+1}, \ldots, K_d, \omega) \\
&= \sum_{i_1, \ldots, i_q = 1}^{m} \lambda_{i_1} \cdots \lambda_{i_q} S(A_{i_1,t}^\bullet, \ldots, A_{i_q,t}^\bullet, A'^\bullet_{q+1}, \ldots, A'^\bullet_p, K_{p+1}, \ldots, K_d, \omega) \\
&= \sum_{i_1, \ldots, i_q = 1}^{m} \lambda_{i_1} \cdots \lambda_{i_q} \overline{S}(A_{i_1}, \ldots, A_{i_q}, \mathcal{C}),
\end{aligned}$$

by [163, (5.18)]. This yields (7.6) for $\lambda_1, \ldots, \lambda_m > 0$. If some $\lambda_i = 0$, then $\lambda_i A_i = \emptyset$ and $\lambda_i A_i \boxplus A = A$, so that A_i appears on neither side of (7.6). Thus, (7.6) holds for $\lambda_1, \ldots, \lambda_m \geq 0$.

Next we show the following polarization formula.

Lemma 7.2.3.

$$\overline{V}(A_1, \ldots, A_q, \mathcal{C})$$

$$= \frac{1}{q!} \sum_{k=1}^{q} (-1)^{q+k} \sum_{1 \leq i_1 < \cdots < i_k \leq q} \overline{V}(A_{i_1} \boxplus \cdots \boxplus A_{i_k}[q], \mathcal{C}). \qquad (7.7)$$

Proof. We modify the proof of [163, Lem. 5.1.4]. Denote the right-hand side of (7.7) by $f(A_1, \ldots, A_q)$. According to (7.6), the function $(\lambda_1, \ldots, \lambda_q) \mapsto f(\lambda_1 A_1, \ldots, \lambda_q A_q)$ is a homogeneous polynomial for $\lambda_1, \ldots, \lambda_q \geq 0$. We have

$$(-1)^{q+1} q! f(\emptyset, A_2, \ldots, A_q)$$

$$= \sum_{2 \leq i \leq q} \overline{V}(A_i[q], \mathcal{C}) - \left[\sum_{2 \leq j \leq q} \overline{V}(\emptyset \boxplus A_j[q], \mathcal{C}) + \sum_{2 \leq i < j \leq q} \overline{V}(A_i \boxplus A_j[q], \mathcal{C}) \right]$$

$$+ \left[\sum_{2 \leq j < k \leq q} \overline{V}(\emptyset \boxplus A_j \boxplus A_k[q], \mathcal{C}) + \sum_{2 \leq i < j < k \leq q} \overline{V}(A_i \boxplus A_j \boxplus A_k[q], \mathcal{C}) \right]$$

$$- \cdots$$

$$= 0.$$

It follows that $f(\emptyset, \lambda_2 A_2, \ldots, \lambda_p A_q)$ is identically zero for all $\lambda_2, \ldots, \lambda_q \geq 0$. This means that in the polynomial $f(\lambda_1 A_1, \lambda_2 A_2, \ldots, \lambda_p A_q)$ all monomials $\lambda_{i_1} \cdots \lambda_{i_q}$ with $1 \notin \{i_1, \ldots, i_q\}$ have zero coefficients. Here the number 1 can be replaced by any of the numbers $2, \ldots, q$. Therefore, only the monomial $\lambda_1 \cdots \lambda_q$ has a nonzero coefficient, and by (7.6) this is equal to $\overline{V}(A_1, \ldots, A_q, \mathcal{C})$. This completes the proof of (7.7). $\qquad \square$

The special case $q = d$ of (7.7) reads

$$\overline{V}(A_1, \ldots, A_d) = \frac{1}{d!} \sum_{k=1}^{d} (-1)^{d+k} \sum_{1 \leq i_1 < \cdots < i_k \leq d} V_d(A_{i_1} \boxplus \cdots \boxplus A_{i_k}), \qquad (7.8)$$

for bounded coconvex sets $A_1, \ldots, A_d$.

The mixed volumes of bounded coconvex sets (and convex bodies) also satisfy inequalities of Aleksandrov–Fenchel type, but with reversed inequality sign.

Theorem 7.2.1. *Let $p \in \{2, \ldots, d\}$, let $A_1, \ldots, A_p$ be bounded coconvex sets, and let $K_{p+1}, \ldots, K_d$ be convex bodies. Set $\mathcal{C} := (A_3, \ldots, A_p, K_{p+1}, \ldots, K_d)$. Then*

$$\overline{V}(A_1, A_2, \mathcal{C})^2 \leq \overline{V}(A_1, A_1, \mathcal{C}) \overline{V}(A_2, A_2, \mathcal{C}). \qquad (7.9)$$

Proof. We emphasize that the elegant idea of the following proof is due, in a special case, to Khovanskiĭ and Timorin [115], but we modify their presentation in several respects.

We may assume that $K_{p+1}, \ldots, K_d$ have interior points, since the general case then follows by approximation.

Let $s > 0$ be sufficiently large (it will be fixed in the following). For $\lambda_1, \lambda_2 > 0$ and sufficiently large $t > 0$ we define

$$Q_\alpha(\lambda_1, \lambda_2, t)$$
$$:= V(\lambda_1 A_{1,t}^\bullet + \lambda_2 A_{2,t}^\bullet, \lambda_1 A_{1,t}^\bullet + \lambda_2 A_{2,t}^\bullet, A_{3,s}^\bullet, \ldots, A_{p,s}^\bullet, K_{p+1}, \ldots, K_d)$$

and

$$Q_\beta^C(\lambda_1, \lambda_2) := \overline{V}(\lambda_1 A_1 \boxplus \lambda_2 A_2, \lambda_1 A_1 \boxplus \lambda_2 A_2, A_3, \ldots, A_p, K_{p+1}, \ldots, K_d).$$

We have

$$V(C_t, C_t, C_s, \ldots, C_s, K_{p+1}, \ldots, K_d) = ct^2$$

with the positive constant

$$c = V(C_1, C_1, C_s, \ldots, C_s, K_{p+1}, \ldots, K_d).$$

From Lemma 7.2.1 we obtain that

$$Q_\alpha(\lambda_1, \lambda_2, t) = ct^2 - Q_\beta^C(\lambda_1, \lambda_2).$$

By (7.6) this is equal to

$$Q_\alpha(\lambda_1, \lambda_2, t) = ct^2 - \lambda_1^2 \overline{V}(A_1, A_1, C) - 2\lambda_1 \lambda_2 \overline{V}(A_1, A_2, C) - \lambda_2^2 \overline{V}(A_2, A_2, C).$$

We use this to define $Q_\alpha(\lambda_1, \lambda_2, t)$ as a quadratic form for all λ_1, λ_2, t. Let (k, ℓ) (with $k + \ell \le 3$) be the signature of the quadratic form Q_α. Since $c > 0$, we have $k \ge 1$. Suppose that $k \ge 2$. Then the space $\mathbb{R}^3$, with standard coordinates λ_1, λ_2, t, has a two-dimensional subspace E on which the form Q_α is positive definite. Therefore, its corresponding bilinear form B_α satisfies the Cauchy–Schwarz inequality, that is,

$$B_\alpha((\lambda_1, \lambda_2, t), (\lambda_1', \lambda_2', t'))^2 < Q_\alpha(\lambda_1, \lambda_2, t) Q_\alpha(\lambda_1', \lambda_2', t') \tag{7.10}$$

for linearly independent vectors $(\lambda_1, \lambda_2, t), (\lambda_1', \lambda_2', t') \in E$. We can exclude that the subspace E is given by $t = 0$, since that would immediately imply the Cauchy–Schwarz inequality (7.9). Therefore, E contains linearly independent vectors $(\lambda_1, \lambda_2, t)$ and $(\lambda_1', \lambda_2', t')$ with t, t' arbitrarily large. For sufficiently large t, t' we have

$$B_\alpha((\lambda_1, \lambda_2, t), (\lambda_1', \lambda_2', t'))$$
$$= V(\lambda_1 A_{1,t}^\bullet + \lambda_2 A_{2,t}^\bullet, \lambda_1' A_{1,t'}^\bullet + \lambda_2' A_{2,t'}^\bullet, A_{3,s}^\bullet, \ldots, A_{p,s}^\bullet, K_{p+1}, \ldots, K_d).$$

But then (7.10) contradicts the Aleksandrov–Fenchel inequalities. This contradiction proves that $k = 1$, which implies that the quadratic form Q_β^C is positive semi-definite. Its corresponding bilinear form therefore satisfies the Cauchy–Schwarz inequality. This is the assertion of the theorem. $\square$

The obtained inequalities imply the following counterpart to the general Brunn–Minkowski theorem.

Theorem 7.2.2. *Let* $m \in \{2, \ldots, d\}$. *Let* $A_0, A_1, A_{m+1}, \ldots, A_p$ *be bounded coconvex sets and let* $K_{p+1}, \ldots, K_d$ *be convex bodies. Define*

$$A_\lambda := (1 - \lambda)A_0 \boxplus \lambda A_1$$

and

$$f(\lambda) := \overline{V}(A_\lambda[m], A_{m+1}, \ldots, A_p, K_{p+1}, \ldots, K_d)^{1/m}, \quad 0 \leq \lambda \leq 1,$$

(where the argument A_λ *is repeated* m *times). Then* f *is a convex function.*

The proof parallels that in the classical case; see [163, Thm. 7.4.5], where we get a concave function instead of a convex one.

7.3 Wulff shapes in cones

In the theory of convex bodies, Wulff shapes are a useful tool, particularly in connection with a variational lemma going back to Aleksandrov. We develop here a similar concept for the given cone C. In later sections, it will be applied to existence theorems.

We assume in this section that a nonempty compact set $\omega \subset \Omega_C$ is given.

Definition 7.3.1. *A closed convex set* K *is said to be* C-*determined by* ω *if*

$$K = C \cap \bigcap_{u \in \omega} H^-(K, u).$$

By $\mathcal{K}(C, \omega)$ *we denote the set of all closed convex sets that are* C-*determined by* ω.

We show that a set $K \in \mathcal{K}(C, \omega)$ is C-full (that is, $C \setminus K$ is bounded). Since ω is a compact subset of Ω_C, it has a positive distance from the boundary of Ω_C. Hence, there is a number $a_0 > 0$ such that $\langle x, u \rangle \leq -a_0$ for all $x \in C \cap \mathbb{S}^{d-1}$ and all $u \in \omega$. Therefore,

$$|\langle x, u \rangle| \geq a_0 \|x\| \quad \text{for } x \in C \text{ and } u \in \omega. \tag{7.11}$$

Now let $x \in C \setminus K$. Then there exists some $u \in \omega$ with $x \notin H^-(u, h(K, u))$ and thus with $|\langle x, u \rangle| < -h(K, u)$. This gives

$$\|x\| \leq \frac{1}{a_0} \max\{-h(K, v) : v \in \omega\},$$

which proves the boundedness of $C \setminus K$ (since $h(K, \cdot)$ is continuous and ω is compact).

A convex body is determined by its support function. The idea underlying the concept of Wulff shape is to associate a convex body also with a function which is not necessarily a support function. This idea can be carried over to functions on ω and C-full sets determined by ω.

Definition 7.3.2. *Let $f : \omega \to \mathbb{R}$ be a positive and continuous function. The closed convex set*

$$K_f := C \cap \bigcap_{u \in \omega} H^-(u, -f(u))$$

is called the Wulff shape *associated with (C, ω, f).*

The Wulff shape K_f associated with (C, ω, f) is C-determined by ω and is a C-full set. It satisfies

$$-h(K_f, u) \geq f(u) \quad \text{for } u \in \omega. \tag{7.12}$$

The convergence of C-full sets, which is needed in the following, is reduced to the convergence of convex bodies.

Definition 7.3.3. *For C-full sets K_j, $j \in \mathbb{N}_0$, we write $\lim_{j \to \infty} K_j = K_0$ (or $K_j \to K_0$) if there is a number $t > 0$ such that $C \setminus K_j \subset C_t$ for $j \in \mathbb{N}_0$ and $\lim_{j \to \infty}(K_j \cap C_t) = K_0 \cap C_t$.*

The next two lemmas involve this convergence.

Lemma 7.3.1. *If K_j is the Wulff shape associated with (C, ω, f_j), for $j \in \mathbb{N}_0$, and if $(f_j)_{j \in \mathbb{N}}$ converges uniformly (on ω) to f_0, then $K_j \to K_0$.*

Proof. First let $x \in \operatorname{int} K_0$. Then there is some $\varepsilon > 0$ with $\langle x, u \rangle \leq -f_0(u) - \varepsilon$ for all $u \in \omega$. Since $f_j \to f_0$ uniformly, there is some j_0 with $|f_j(u) - f_0(u)| < \varepsilon$ for all $j \geq j_0$ and all $u \in \omega$. Therefore, $\langle x, u \rangle \leq -f_j(u)$ for all $j \geq j_0$ and all $u \in \omega$. This shows that $x \in K_j$ for $j \geq j_0$. If now $x_0 \in K_0$ (but not necessarily $x_0 \in \operatorname{int} K_0$), we can choose a sequence $(x_i)_{i \in \mathbb{N}}$ in $\operatorname{int} K_0$ with $x_i \to x_0$. Using the preceding observation, we easily construct a sequence $(y_j)_{j \in \mathbb{N}}$ with $y_j \in K_j$ and $y_j \to x_0$. Conversely, suppose that $x_{i_j} \in K_{i_j}$ and $x_{i_j} \to x_0$. Then $\langle x_{i_j}, u \rangle \leq -f_{i_j}(u)$ for all $u \in \omega$. It follows that $\langle x_0, u \rangle \leq -f(u)$ for all $u \in \omega$ and thus $x_0 \in K_0$. Now we see from [163, Thm. 1.8.8] that $\lim_{j \to \infty}(K_j \cap C_t) = K_0 \cap C_t$, which means that $K_j \to K_0$. $\square$

Lemma 7.3.2. *If $(K_j)_{j \in \mathbb{N}}$ is a sequence in $\mathcal{K}(C, \omega)$ such that $K_j \to K_0$ for some C-full set K_0, then $K_0 \in \mathcal{K}(C, \omega)$.*

Proof. From $K_j \to K_0$ it follows that $h_{K_j} \to h_{K_0}$ uniformly on ω, hence, by Lemma 7.3.1, $K_j \to K$, where K is the Wulff shape associated with $(C, \omega, -h_{K_0})$. But $K = K_0$, hence $K_0 \in \mathcal{K}(C, \omega)$. $\square$

Lemma 7.3.3. *If K_f is the Wulff shape associated with (C, ω, f) for some positive, continuous function f, then*

$$S_{d-1}(K_f, \Omega_C \setminus \omega) = 0. \tag{7.13}$$

Proof. Let $x \in \operatorname{bd} K_f \cap \operatorname{int} C$. Then there exists a vector $u \in \omega$ such that $x \in H(u, -f(u))$, since otherwise $\langle x, u \rangle < -f(u)$ for all $u \in \omega$, and since $\langle x, \cdot \rangle$ and f are continuous on the closed set ω, there would exist a number $\varepsilon > 0$ with $\langle x, u \rangle \le -f(u) - \varepsilon$ for all $u \in \omega$, hence $x \in \operatorname{int} K_f$, a contradiction. This shows that $x \in H(K_f, u)$ for some $u \in \omega$.

Now let $v \in \Omega_C \setminus \omega$ and $x \in K_f \cap H(K_f, v)$. If $x \in \operatorname{int} C$, then $x \in H(K_f, u)$ for some $u \in \omega$, as just shown. Hence, x is a singular point of K_f. If $x \in \operatorname{bd} C$, then the point x lies also in a supporting hyperplane of C, with a normal vector in $\operatorname{bd} \Omega_C$ and thus different from v, and hence is again a singular point of K_f. The assertion (7.13) now follows from [163, Thm. 2.2.5]. □

Definition 7.3.4. *For a positive, continuous function $f : \omega \to \mathbb{R}$, let*

$$V(f) := V_d(C \setminus K_f),$$

where K_f is the Wulff shape associated with (C, ω, f).

The functional V is continuous in the following sense: if $f_j \to f$ uniformly on ω, then $V(f_j) \to V(f)$. We have

$$V(f) = \frac{1}{d} \int_\omega f(u) \, S_{d-1}(K_f, \mathrm{d}u), \tag{7.14}$$

where K_f is the Wulff shape associated with (C, ω, f). In fact, if $v \in \omega$ is such that $h(K_f, v) \neq -f(v)$, then any point $x \in K_f \cap H(K, v)$ lies in $H(K_f, v)$ and in some distinct supporting hyperplane $H(u, -f(u))$ with $u \in \omega$, hence x is a singular point. This shows that

$$S_{d-1}(K_f, \{v \in \omega : -h(K_f, v) \neq f(v)\}) = 0. \tag{7.15}$$

Now (7.14) follows from (7.2) and (7.13).

We assume now that a continuous function $G : [-\varepsilon, \varepsilon] \times \omega \to \mathbb{R}$, for some $\varepsilon > 0$, with $G(0, \cdot) > 0$ is given and that there is a continuous function $g : \omega \to \mathbb{R}$ such that

$$\lim_{\tau \downarrow 0} \frac{G(\tau, \cdot) - G(0, \cdot)}{\tau} = g \quad \text{uniformly on } \omega. \tag{7.16}$$

For all sufficiently small $|\tau|$, the function $G(\tau, \cdot)$ is positive, hence $V(G(\tau, \cdot))$ is defined.

Theorem 7.3.1. *Let G be as above, and let K_0 be the Wulff shape associated with $(C, \omega, G(0, \cdot))$. Then*

$$\lim_{\tau \downarrow 0} \frac{V(G(\tau, \cdot)) - V(G(0, \cdot))}{\tau} = \int_\omega g(u) \, S_{d-1}(K_0, \mathrm{d}u). \tag{7.17}$$

The same assertion holds if in (7.16) and (7.17) the one-sided limit $\lim_{\tau \downarrow 0}$ is replaced by $\lim_{\tau \uparrow 0}$ or $\lim_{\tau \to 0}$.

Proof. For sufficiently small $\tau \geq 0$, let K_τ be the Wulff shape associated with $(C, \omega, G(\tau, \cdot))$. Let $A_\tau := C \setminus K_\tau$. We abbreviate

$$V_0 := V(G(0, \cdot)) = V_d(A_0),$$
$$V_1(\tau) := \overline{V}(A_\tau, A_0, \ldots, A_0),$$
$$V_{d-1}(\tau) := \overline{V}(A_\tau, \ldots, A_\tau, A_0)$$
$$V_d(\tau) := V_d(A_\tau).$$

By (7.12), we have $-h(K_\tau, u) \geq G(\tau, u)$ for $u \in \omega$. Together with (7.13), this yields

$$\int_{\Omega_C} -h(K_\tau, u)\, S_{d-1}(K_0, du) \geq \int_\omega G(\tau, u)\, S_{d-1}(K_0, du).$$

By (7.4) and (7.14), we get

$$V_1(\tau) - V_0 \geq \frac{1}{d} \int_\omega [G(\tau, u) - G(0, u)]\, S_{d-1}(K_0, du).$$

From (7.16) we deduce that

$$\liminf_{\tau \downarrow 0} \frac{V_1(\tau) - V_0}{\tau} \geq \frac{1}{d} \int_\omega g(u)\, S_{d-1}(K_0, du). \tag{7.18}$$

From (7.14), applied to $G(\tau, \cdot)$, we obtain

$$\frac{1}{d} \int_\omega G(\tau, u)\, S_{d-1}(K_\tau, du) = V_d(\tau).$$

Since $-h(K_0, u) \geq G(0, u)$ for $u \in \omega$, we have, using (7.13) again,

$$\frac{1}{d} \int_\omega G(0, u)\, S_{d-1}(K_\tau, du) \leq \frac{1}{d} \int_{\Omega_C} -h(K_0, u)\, S_{d-1}(K_\tau, du) = V_{d-1}(\tau).$$

Subtraction gives

$$\frac{1}{d} \int_\omega [G(\tau, u) - G(0, u)]\, S_{d-1}(K_\tau, du) \geq V_d(\tau) - V_{d-1}(\tau). \tag{7.19}$$

For $\tau \to 0$ we have $K_\tau \to K_0$, by Lemma 7.3.1. For sufficiently large $t > 0$, this means the convergence $K_\tau \cap C_t \to K_0 \cap C_t$ of convex bodies, and this implies the weak convergence

$$S_{d-1}(K_\tau \cap C_t, \cdot) \xrightarrow{w} S_{d-1}(K_0 \cap C_t, \cdot),$$

equivalently

$$\int_{\mathbb{S}^{d-1}} F\, dS_{d-1}(K_\tau \cap C_t, \cdot) \to \int_{\mathbb{S}^{d-1}} F\, dS_{d-1}(K_0 \cap C_t, \cdot)$$

for every continuous function $F : \mathbb{S}^{d-1} \to \mathbb{R}$. Given a continuous function $h : \omega \to \mathbb{R}$, there is (by Tietze's extension theorem) a continuous function $F : \mathbb{S}^{d-1} \to \mathbb{R}$ with $F = h$ on ω and $F = 0$ on $\mathbb{S}^{d-1} \setminus \Omega_C$. Then

$$\int_{\mathbb{S}^{d-1}} F \, dS_{d-1}(K_\tau \cap C_t, \cdot) = \int_\omega h \, dS_{d-1}(K_\tau, \cdot),$$

because $F = 0$ on $\mathbb{S}^{d-1} \setminus \Omega_C$ and $S_{d-1}(K_\tau, \Omega_C \setminus \omega) = 0$ by (7.13), further $S_{d-1}(K_\tau \cap C_t, \cdot) = S_{d-1}(K_\tau, \cdot)$ on ω. It follows that

$$\int_\omega h \, dS_{d-1}(K_\tau, \cdot) \to \int_\omega h \, dS_{d-1}(K_0, \cdot)$$

as $\tau \to 0$. Now (7.19) (where the convergence is uniform) and (7.16) give

$$\frac{1}{d} \int_\omega g(u) \, S_{d-1}(K_0, du) \geq \limsup_{\tau \downarrow 0} \frac{V_d(\tau) - V_{d-1}(\tau)}{\tau}. \qquad (7.20)$$

The Aleksandrov–Fenchel-type inequalities of Theorem 7.2.1 yield, in a way completely analogous to the classical case, the Minkowski-type inequality

$$\overline{V}(A_0, \ldots, A_0, A_1)^d \leq V_d(A_0)^{d-1} V_d(A_1).$$

It gives

$$[V_1(\tau) - V_0] \sum_{k=0}^{d-1} [V_1(\tau)/V_0]^k = [V_1(\tau)^d - V_0^d] V_0^{1-d}$$

$$\leq [V_0^{d-1} V_d(\tau) - V_0^d] V_0^{1-d} = V_d(\tau) - V_0.$$

For $\tau \to 0$ we have $K_\tau \to K_0$ and hence, as follows from (7.4), $V_1(\tau) \to V_0$. Therefore, we deduce that

$$d \liminf_{\tau \downarrow 0} \frac{V_1(\tau) - V_0}{\tau} \leq \liminf_{\tau \downarrow 0} \frac{V_d(\tau) - V_0}{\tau}. \qquad (7.21)$$

Replacing the pair $(V_1(\tau), V_0)$ by $(V_{d-1}(\tau), V_d(\tau))$, we can argue in a similar way. Here we have to observe that $V_{d-1}(\tau)/V_d(\tau) \to 1$ as $\tau \downarrow 0$, by the weak continuity of S_{d-1} and the uniform convergence $h(K_\tau, \cdot) \to h(K_0, \cdot)$ (on ω). We obtain

$$d \limsup_{\tau \downarrow 0} \frac{V_d(\tau) - V_{d-1}(\tau)}{\tau} \geq \limsup_{\tau \downarrow 0} \frac{V_d(\tau) - V_0}{\tau}. \qquad (7.22)$$

Applying successively (7.20), (7.22), $\limsup \geq \liminf$, (7.21), (7.18), we conclude that in all these inequalities the equality sign is valid, hence

$$\lim_{\tau \downarrow 0} \frac{V_d(\tau) - V_0}{\tau} = \int_\omega g(u) \, S_{d-1}(K_0, du).$$

This yields the main assertion of the theorem. The corresponding assertion for left-sided derivatives follows upon replacing $G(\tau, u)$ by $G(-\tau, u)$, and both results together give the corresponding result for general limits. $\qquad \square$

7.4 A Minkowski-type existence theorem

Minkowski's theorem for convex bodies says that a finite Borel measure φ on the unit sphere $\mathbb{S}^{d-1}$ is the surface area measure of a convex body if and only if it is not concentrated on a great subsphere and satisfies $\int_{\mathbb{S}^{d-1}} u\,\varphi(\mathrm{d}u) = o$. Moreover, this body is uniquely determined up to a translation.

In this section, we prove similar existence and uniqueness theorems for C-full sets. Since the existence proof is based on an extremal problem, we start with a lemma which will be needed to establish its solvability.

Again we assume that a pointed closed convex cone C with interior points and a nonempty compact set $\omega \subset \Omega_C$ are given. We remind the reader that $\mathcal{K}(C,\omega)$ denotes the set of all closed convex sets that are C-determined by ω.

Lemma 7.4.1. *Let $B \subset C$ be a bounded set. There exists a number $t > 0$ such that the relation*

$$H(u,\tau) \cap B \neq \emptyset \quad \text{for some } u \in \omega \text{ and some } \tau$$

implies that $H(u,\tau) \cap C \subset C_t$.

Further, there exists a number $t > 0$ such that the relations

$$K \in \mathcal{K}(C,\omega) \quad \text{and} \quad V_d(C \setminus K) = 1$$

imply that $C \cap H(w,t) \subset K$.

Proof. Since ω is compact, there is by (7.11) a number $a_0 > 0$ with $|\langle x, u\rangle| \geq a_0\|x\|$ for all $x \in C$ and all $u \in \omega$.

Let $B \subset C$ be a bounded set, and suppose that $H(u,\tau) \cap B \neq \emptyset$ with $u \in \omega$ and $\tau \in \mathbb{R}$. Let $x \in H(u,\tau) \cap C$. There is a point $y \in H(u,\tau) \cap B$, and we have

$$|\langle x, u\rangle| = |\langle y, u\rangle| \leq \|y\| \leq \sup\{\|z\| : z \in B\} =: b < \infty,$$

since B is bounded. It follows that

$$\|x\| \leq \frac{|\langle x, u\rangle|}{a_0} \leq \frac{b}{a_0}.$$

Thus, $H(u,\tau) \cap C$ is contained in the ball with center o and radius b/a_0. Therefore, $H(u,\tau) \cap C \subset C_t$ with a number t independent of $H(u,\tau)$. This proves the first part.

For the second part, we choose a number $s > 0$ with $V_d(C_s) > 1$. By the first part, there is a number $t > 0$ such that every hyperplane $H(u,\tau)$ with $H(u,\tau) \cap C_s \neq \emptyset$ and $u \in \omega$ satisfies $H(u,\tau) \cap C \subset C_t$.

Let $K \in \mathcal{K}(C,\omega)$ be a closed convex set with $V_d(C \setminus K) = 1$. Then every supporting hyperplane $H(u,\tau)$ of K with outer normal vector $u \in \omega$ satisfies $H(u,\tau) \cap C_s \neq \emptyset$ (since $V_d(C_s) > V_d(C \setminus K)$) and, therefore, $H(u,\tau) \cap C \subset C_t$. This shows that $C \setminus K \subset C_t$, which implies $C \cap H(w,t) \subset K$. □

Now we formulate an existence theorem.

Theorem 7.4.1. *Let $\omega \subset \Omega_C$ be compact. Every finite Borel measure on Ω_C whose support is contained in ω is the surface area measure of a C-full set.*

We emphasize that the condition of the theorem is sufficient, but not necessary: there are C-full sets for which the support of the area measure contains boundary points of Ω_C and hence is not a compact subset of the open set Ω_C.

Proof of Theorem 7.4.1. Let φ be a finite Borel measure on Ω_C that is concentrated on ω. The case $\varphi \equiv 0$, where $S_{d-1}(C, \cdot) = \varphi$, is trivial, hence we may assume that φ is nonzero. Let $\mathcal{C}(\omega)$ denote the set of positive, continuous functions on ω, equipped with the topology induced by the maximum norm. For $f \in \mathcal{C}(\omega)$, let K_f be the Wulff shape associated with (C, ω, f). We recall that $V(f) = V_d(C \setminus K_f)$. A functional $\Phi : \mathcal{C}(\omega) \to (0, \infty)$ is defined by

$$\Phi(f) := V(f)^{-1/d} \int_\omega f \, d\varphi.$$

Then Φ is continuous. We show that it attains a maximum on the set

$$\mathcal{L}' := \{-h_L : L \in \mathcal{K}(C, \omega), V_d(C \setminus L) = 1\}.$$

Let $L \in \mathcal{K}(C, \omega)$ be such that $-h_L \in \mathcal{L}'$. By Lemma 7.4.1, there is a number $t > 0$, independent of L, such that $C \cap H(w, t) \subset L$. This implies that

$$\Phi(-h_L) \le \int_\omega -h(C \cap H(w, t), u) \, \varphi(du),$$

which is independent of L. It follows that $\sup\{\Phi(f) : f \in \mathcal{L}'\} < \infty$.

Let $(K_i)_{i \in \mathbb{N}}$ be a sequence with $-h_{K_i} \in \mathcal{L}'$ such that

$$\lim_{i \to \infty} \Phi(-h_{K_i}) = \sup\{\Phi(f) : f \in \mathcal{L}'\}.$$

For each i we have $C \cap H(w, t) \subset K_i$, hence $K_i \cap H^-(w, t) \ne \emptyset$. By the Blaschke selection theorem, the bounded sequence $(K_i \cap H^-(w, t))_{i \in \mathbb{N}}$ of convex bodies has a subsequence converging to some convex body. Therefore, a subsequence of the sequence $(K_i)_{i \in \mathbb{N}}$ converges to a C-full set K_0. This set satisfies $V_d(C \setminus K_0) = 1$, hence $-h_{K_0}$ belongs to $\mathcal{L}'$, as follows from Lemma 7.3.3. By continuity, the functional Φ attains its maximum on $\mathcal{L}'$ at $-h_{K_0}$.

Since the functional Φ is homogeneous of degree zero, also its maximum on the larger set $\mathcal{L} := \{-h_L : L \in \mathcal{K}(C, \omega)\}$ is attained at $-h_{K_0}$.

For any $f \in \mathcal{C}(\omega)$ with Wulff shape K_f we have $-h_{K_f}(u) \ge f(u)$ for $u \in \omega$ and $V(-h_{K_f}) = V(f)$, where (7.15) and (7.14) were used. Therefore, $\Phi(f) \le \Phi(-h_{K_f}) \le \Phi(-h_{K_0})$. Thus, also the maximum of Φ on the set $\mathcal{C}(\omega)$ is attained at $-h_{K_0}$.

Let $f \in C(\omega)$. Then $-h_{K_0} + \tau f \in C(\omega)$ for sufficiently small $|\tau|$, hence the function

$$\tau \mapsto \Phi(-h_{K_0} + \tau f)$$
$$= V(-h_{K_0} + \tau f)^{-1/d} \left(\int_\omega -h_{K_0}\, d\varphi + \tau \int_\omega f\, d\varphi \right) \quad (7.23)$$

attains a maximum at $\tau = 0$. Theorem 7.3.1 yields

$$\frac{d}{d\tau} V(-h_{K_0} + \tau f)\Big|_{\tau=0} = \int_\omega f\, dS_{d-1}(K_0, \cdot).$$

Therefore, and since $V(-h_{K_0}) = 1$, the derivative of the function (7.23) at $\tau = 0$ is given by

$$- \int_\omega f\, dS_{d-1}(K_0, \cdot) \cdot \frac{1}{d} \int_\omega -h_{K_0}\, d\varphi + \int_\omega f\, d\varphi,$$

and this is equal to zero. With

$$\lambda := \frac{1}{d} \int_\omega -h_{K_0}\, d\varphi$$

we have $\lambda > 0$ and

$$\int_\omega f\, d\varphi = \lambda \int_\omega f\, dS_{d-1}(K_0, \cdot).$$

Since this holds for all functions $f \in C(\omega)$, it holds for all continuous real functions f on ω. This yields $\varphi = \lambda S_{d-1}(K_0, \cdot) = S_{d-1}(\lambda^{\frac{1}{d-1}} K_0, \cdot)$. Thus, φ is the surface area measure of the C-full set $\lambda^{\frac{1}{d-1}} K_0$. $\qquad\square$

Theorem 7.4.1 provides, to each finite Borel measure φ on Ω_C with $\operatorname{supp} \varphi \subseteq \omega$, a C-full set K with $S_{d-1}(K, \cdot) = \varphi$. For the corresponding uniqueness result, there are several stronger versions. One of these is Theorem 7.7.1 below. For φ with compact support, as considered here, we can show an improved result, namely a corresponding stability theorem (following [169]). We prepare this by two lemmas. We need the Lévy–Prokhorov metric for finite measures and a special norm for real functions on ω.

We assume that two nonzero, finite Borel measures μ, ν on ω are given. For $A \subset \omega$ and for $\varepsilon > 0$, let

$$A_\varepsilon := \{y \in \omega : \|x - y\| < \varepsilon \text{ for some } x \in A\}.$$

The Lévy–Prokhorov distance of μ and ν is defined by

$$\delta_{LP}(\mu, \nu) := \inf\{\varepsilon > 0 : \mu(A) \le \nu(A_\varepsilon) + \varepsilon,\ \nu(A) \le \mu(A_\varepsilon) + \varepsilon\ \forall A \in \mathcal{B}(\omega)\}.$$

For a continuous real function f on ω, we define

$$\|f\|_L := \sup_{u,v \in \omega,\, u \neq v} \frac{|f(u) - f(v)|}{\|u - v\|}, \qquad \|f\|_\infty := \sup_{u \in \omega} |f(u)|,$$

$$\|f\|_{BL} := \|f\|_L + \|f\|_\infty.$$

This depends on ω, which is not expressed by the notation. Note that $\|f\|_L$ need not be finite.

Lemma 7.4.2. *Under the assumptions above, there is a constant c_0, depending only on the total measures $\mu(\omega), \nu(\omega)$ such that*

$$\left| \int_\omega f \, d(\mu - \nu) \right| \leq c_0 \|f\|_{BL} \cdot \delta_{LP}(\mu, \nu).$$

Proof. We modify the argument given in [163, p. 481]. By assumption, $\mu_1 := \mu(\omega)$, $\nu_1 := \nu(\omega)$ are positive. For probability measures, the proof of Corollary 11.6.5 in Dudley [54] gives

$$\left| \int_\omega f \, d\left(\frac{\mu}{\mu_1} - \frac{\nu}{\nu_1} \right) \right| \leq 2\|f\|_{BL} \cdot \delta_{LP}\left(\frac{\mu}{\mu_1}, \frac{\nu}{\nu_1} \right).$$

We have $|\mu_1 - \nu_1| \leq \delta_{LP}(\mu, \nu)$ by definition and, as shown in [163, p. 481],

$$\delta_{LP}\left(\frac{\mu}{\mu_1}, \frac{\nu}{\nu_1} \right) \leq c_1 \delta_{LP}(\mu, \nu)$$

with a constant c_1 depending only on μ_1, ν_1. Writing

$$\mu - \nu = \mu_1 \left(\frac{\mu}{\mu_1} - \frac{\nu}{\nu_1} \right) - \mu_1 \left(\frac{1}{\mu_1} - \frac{1}{\nu_1} \right) \nu,$$

we get

$$\left| \int_\omega f \, d(\mu - \nu) \right| \leq \mu_1 \left\{ \left| \int_\omega f \, d\left(\frac{\mu}{\mu_1} - \frac{\nu}{\nu_1} \right) \right| + \left| \frac{1}{\mu_1} - \frac{1}{\nu_1} \right| \cdot \left| \int_\omega f \, d\nu \right| \right\},$$

from which the assertion follows. □

Now let $K \in \mathcal{K}(C, \omega)$ and recall that $\overline{h}(K, u) := -h(K, u)$ for $u \in \omega$.

Lemma 7.4.3. *There exists a constant c_1, depending only on C, ω and $S_{d-1}(K, \omega)$, such that*

$$\|\overline{h}(K, \cdot)\|_{BL} \leq c_1.$$

Proof. Let B be the largest ball with center o satisfying $B \cap \operatorname{int} K = \emptyset$. Let H be a supporting hyperplane of K, and suppose that $H \cap B = \emptyset$. Since K lies entirely in one of the closed halfspaces bounded by H, the ball B can be increased without intersecting K, a contradiction. Thus, any supporting hyperplane of K intersects B, which means that

$$\overline{h}(K, \cdot) \le r,$$

where r is the radius of B. The mapping

$$f : \text{bd}\, K \cap \text{int}\, C \to \text{bd}\, B, \quad x \mapsto r\frac{x}{\|x\|},$$

has Lipschitz constant 1 and its image is $\text{bd}\, B \cap \text{int}\, C$, hence

$$S_{d-1}(K, \omega) = \mathcal{H}^{d-1}(\text{bd}\, K \cap \text{int}\, C)$$
$$\ge \mathcal{H}^{d-1}(\text{bd}\, B \cap \text{int}\, C) = r^{d-1}\mathcal{H}^{d-1}(\mathbb{S}^{d-1} \cap \text{int}\, C).$$

It follows that

$$\overline{h}(K, \cdot) \le c_2. \tag{7.24}$$

The constant c_2, as c_3 below, depends only on C, ω and $S_{d-1}(K, \omega)$.

We need a Lipschitz constant for the function $\overline{h}(K, \cdot)$. Since ω is a compact subset of Ω_C, there is by (7.12) a positive constant a_0 (depending only on C and ω) such that

$$|\langle x, u \rangle| \ge a_0\|x\| \quad \text{for } x \in C \text{ and } u \in \omega.$$

Let $x \in K$ and $u \in \omega$ be such that $\langle x, u \rangle = h(K, u)$. Then we obtain

$$\|x\| \le \frac{1}{a_0}|\langle x, u \rangle| = \frac{1}{a_0}\overline{h}(K, u) \le \frac{c_2}{a_0} =: c_3. \tag{7.25}$$

Now let $u, v \in \omega$ and choose $x \in K$ with $h(K, u) = \langle x, u \rangle$. We have $\langle x, v \rangle \le h(K, v)$ and hence

$$h(K, u) - h(K, v) \le \langle x, u - v \rangle \le \|x\|\|u - v\| \le c_3\|u - v\|.$$

Therefore,

$$\|\overline{h}(K, \cdot)\|_L \le c_3. \tag{7.26}$$

From (7.24) and (7.26) the assertion follows. $\qquad\square$

To formulate a stability theorem, we need a metric for C-full sets. We can define the Hausdorff distance of C-full sets K, L in the same way as for compact sets, by

$$d_H(K, L) = \max\{\sup_{x \in K} \inf_{y \in L} \|x - y\|, \sup_{x \in L} \inf_{y \in K} \|x - y\|\}.$$

For given K, L and sufficiently large t, we have

$$d_H(K, L) = d_H(K \cap C_t, L \cap C_t),$$

hence d_H defines a metric on the set of C-full sets.

Theorem 7.4.2. *Let $K, L \in \mathcal{K}(C, \omega)$. There is a constant c, depending only on $C, \omega, S_{d-1}(K, \omega), S_{d-1}(L, \omega)$ such that*

$$d_H(K, L) \leq c\delta_{LP}(S_{d-1}(K, \cdot), S_{d-1}(L, \cdot))^{1/d}.$$

Proof. Let $x \in \operatorname{bd} K \cap \operatorname{int} C$. There is an outer normal vector $u \in \omega$ to K at x, hence (7.25) shows that $\|x\| \leq c_3$, where c_3 depends only on $C, \omega, S_{d-1}(K, \omega)$. Thus, $\operatorname{bd} K \cap \operatorname{int} C$, and similarly $\operatorname{bd} L \cap \operatorname{int} C$, is contained in the ball with center o and radius c_3. We choose $t > 0$ so large that

$$\operatorname{bd} K \cap \operatorname{int} C \subset H^-(w, t), \qquad \operatorname{bd} L \cap \operatorname{int} C \subset H^-(w, t)$$

and that

$$K_t := K \cap H^-(w, t), \quad L_t := L \cap H^-(w, t)$$

have inradius at least 1. How large t has to be chosen to achieve this effect, depends only on $C, \omega, S_{d-1}(K, \omega), S_{d-1}(L, \omega)$. Then there is a number R, also depending only on these data, such that K_t, L_t have circumradius at most R.

Now we set $\delta_{LP}(S_{d-1}(K, \cdot), S_{d-1}(L, \cdot)) =: \varepsilon$. By Lemmas 7.4.2 and 7.4.3 we have

$$\left| \int_\omega h(K, \cdot) \, d\left(S_{d-1}(K, \cdot) - S_{d-1}(L, \cdot)\right) \right| \leq c_0 \|h(K, \cdot)\|_{BL} \cdot \varepsilon \leq c_0 c_1 \varepsilon.$$

We have a disjoint decomposition

$$\mathbb{S}^{d-1} = \omega \cup (\mathbb{S}^{d-1} \cap \operatorname{bd} C^\circ) \cup \{w\} \cup N$$

with $N := \mathbb{S}^{d-1} \setminus (\omega \cup \operatorname{bd} C^\circ \cup \{w\})$. Here,

$$h(K_t, u) = h(K, u) \quad \text{for } u \in \omega,$$

$$S_{d-1}(K_t, A) = S_{d-1}(K, A) \quad \text{for } A \in \mathcal{B}(\omega),$$

$$h(K_t, u) = 0 \quad \text{for } u \in \operatorname{bd} C^\circ,$$

and similarly for L, L_t. Further,

$$S_{d-1}(K_t, \{w\}) = V_{d-1}(C \cap H(w, t)) = S_{d-1}(L_t, \{w\}),$$

$$S_{d-1}(K_t, A) = 0 = S_{d-1}(L_t, A) \quad \text{for } A \in \mathcal{B}(N),$$

the latter because a boundary point of K_t or L_t with outer normal vector in N is a singular point. Therefore, for the volume and mixed volumes of the convex bodies K_t and L_t we obtain from [163, (5.19)] that

$$d \left| V_d(K_t) - V(K_t, L_t, \ldots, L_t) \right|$$

$$= \left| \int_{\mathbb{S}^{d-1}} h(K_t, u) \left(S_{d-1}(K_t, du) - S_{d-1}(L_t, du) \right) \right|$$

$$= \left| \int_\omega h(K, u) \left(S_{d-1}(K, du) - S_{d-1}(L, du) \right) \right|.$$

We conclude that

$$|V_d(K_t) - V(K_t, L_t, \dots, L_t)| \le c_4\varepsilon \tag{7.27}$$

with a constant c_4, and since K and L may be interchanged and the constant may be adapted, also

$$|V_d(L_t) - V(L_t, K_t, \dots, K_t)| \le c_4\varepsilon. \tag{7.28}$$

As the proof of Theorem 8.5.1 and Lemma 8.5.2 in [163] shows, the two inequalities (7.27) and (7.28) are sufficient to obtain an inequality of the form

$$d_H(K_t, L_t) \le \gamma(c_4\varepsilon)^{1/d}$$

with a constant γ that depends only on the dimension d, a positive lower bound for the inradii of K_t, L_t and an upper bound for the circumradii of K_t, L_t. The proof requires first a restriction to $\varepsilon \le \varepsilon_0$ with an explicit $\varepsilon_0 > 0$. Later, this restriction can be removed, by adapting the constant. This yields the assertion. $\qquad\square$

Another result that is more general than the uniqueness theorem corresponding to Theorem 7.4.1 is the following analogue of the Aleksandrov–Fenchel–Jessen theorem for C-full sets. The classical Aleksandrov–Fenchel–Jessen theorem refers to the i-th area measure $S_i(K, \cdot)$ of a convex body K. Correspondingly, we now define

$$\overline{S}_i(A, \omega) := \overline{S}(\underbrace{A, \dots, A}_{i}, \underbrace{B^d, \dots, B^d}_{d-1-i}, \omega)$$

for a C-full set A, for $\omega \in \mathcal{B}(\Omega_C)$ and $i = 1, \dots, d-1$, where B^d is the unit ball of $\mathbb{R}^d$. Thus, we have

$$\overline{S}_i(A, \cdot) = S_i(A_t^\bullet, \cdot) \quad \text{on } \Omega_C$$

if t is sufficiently large.

Theorem 7.4.3. *Let* A_1, A_2 *be* C-full *sets with*

$$\overline{S}_i(A_1, \cdot) = \overline{S}_i(A_2, \cdot)$$

for some $i \in \{1, \dots, d-1\}$. *Then* $A_1 = A_2$.

Proof. The proof uses the ordinary mixed volumes of convex bodies. By assumption, we have $\overline{S}_i(A_1, \cdot) = \overline{S}_i(A_2, \cdot)$ on Ω_C, hence, for sufficiently large t it holds that

$$S_i(A_{1,t}^\bullet, \cdot) = S_i(A_{2,t}^\bullet, \cdot) \quad \text{on } \Omega_C.$$

Further,

$$S_i(A_{1,t}^\bullet, \cdot) = S_i(C_t, \cdot) = S_i(A_{2,t}^\bullet, \cdot) \quad \text{on } \mathbb{S}^{d-1} \setminus \text{cl}\, \Omega_C$$

by Lemma 7.2.1, and

$$h(A^{\bullet}_{1,t}, \cdot) = 0 = h(A^{\bullet}_{2,t}, \cdot) \quad \text{on } \operatorname{cl} \Omega_C.$$

Therefore,

$$\int_{\mathbb{S}^{d-1}} h(A^{\bullet}_{1,t}, u) \, S_i(A^{\bullet}_{1,t}, \mathrm{d}u) = \int_{\mathbb{S}^{d-1}} h(A^{\bullet}_{1,t}, u) \, S_i(A^{\bullet}_{2,t}, \mathrm{d}u).$$

By ([163, (5.19)]), this can be written as

$$V(\underbrace{A^{\bullet}_{1,t}, \ldots, A^{\bullet}_{1,t}}_{i+1}, \underbrace{B^d, \ldots, B^d}_{d-1-i}) = V(\underbrace{A^{\bullet}_{1,t}, A^{\bullet}_{2,t}, \ldots, A^{\bullet}_{2,t}}_{i}, \underbrace{B^d, \ldots, B^d}_{d-1-i}).$$

With the abbreviation

$$V_{(r,i)} := V(\underbrace{A^{\bullet}_{1,t}, \ldots, A^{\bullet}_{1,t}}_{r} \underbrace{A^{\bullet}_{2,t}, \ldots, A^{\bullet}_{2,t}}_{i+1-r}, \underbrace{B^d, \ldots, B^d}_{d-1-i})$$

this reads $V_{(i+1,0)} = V_{(1,i)}$. Interchanging A_1 and A_2, we obtain $V_{(0,i+1)} = V_{(i,1)}$. The proof can now be completed as that of Theorems 7.4.6 and 8.1.3 in [163]. □

7.5 A Brunn–Minkowski theorem for coconvex sets

For bounded coconvex sets, the case $m = d$ of Theorem 7.2.2 yields a Brunn–Minkowski type theorem, with reversed inequality sign. It would not be difficult to extend this inequality, by approximation, to general coconvex sets. However, this approach would not yield information about the equality case. Since such information allows further conclusions, we extend below the classical Kneser–Süss approach to the Brunn–Minkowski inequality, though with further details, since we are dealing with unbounded convex sets.

Theorem 7.5.1. *Let A_0, A_1 be coconvex sets, and let $\lambda \in (0,1)$. Then*

$$V_d((1-\lambda)A_0 \boxplus \lambda A_1)^{1/d} \le (1-\lambda)V_d(A_0)^{1/d} + \lambda V_d(A_1)^{1/d}. \tag{7.29}$$

Equality holds if and only if $A_0 = \alpha A_1$ with some $\alpha > 0$.

Proof. Let the coconvex sets A_0, A_1 be given. First we assume that

$$V_d(A_0) = V_d(A_1) = 1. \tag{7.30}$$

Let $0 < \lambda < 1$ and define

$$A^{\bullet}_\lambda := (1-\lambda)A^{\bullet}_0 + \lambda A^{\bullet}_1, \qquad A_\lambda := C \setminus A^{\bullet}_\lambda = (1-\lambda)A_0 \boxplus \lambda A_1.$$

In this proof, we abbreviate

$$H_t := \{x \in \mathbb{R}^d : \langle x, w \rangle = t\}, \qquad H_t^- := \{x \in \mathbb{R}^d : \langle x, w \rangle \le t\}$$

for $t \ge 0$. In the following, $\nu \in \{0, 1\}$. We write

$$v_\nu(\zeta) := V_{d-1}(A_\nu^\bullet \cap H_\zeta), \qquad w_\nu(\zeta) := V_d(A_\nu^\bullet \cap H_\zeta^-)$$

for $\zeta \ge 0$, thus

$$w_\nu(\zeta) = \int_{\alpha_\nu}^\zeta v_\nu(s)\,\mathrm{d}s,$$

where α_ν is the number for which H_{α_ν} supports $A_\nu^\bullet$. On (α_ν, ∞), the function v_ν is continuous, hence w_ν is differentiable and

$$w_\nu'(\zeta) = v_\nu(\zeta) > 0 \quad \text{for } \alpha_\nu < \zeta < \infty.$$

Let z_ν be the inverse function of w_ν, then

$$z_\nu'(\tau) = \frac{1}{v_\nu(z_\nu(\tau))} \quad \text{for } 0 < \tau < \infty.$$

With

$$D_\nu(\tau) := A_\nu^\bullet \cap H_{z_\nu(\tau)}, \quad z_\lambda(\tau) := (1 - \lambda)z_0(\tau) + \lambda z_1(\tau),$$

the inclusion

$$A_\lambda^\bullet \cap H_{z_\lambda(\tau)} \supseteq (1 - \lambda)D_0(\tau) + \lambda D_1(\tau) \tag{7.31}$$

holds (trivially). For $\tau > 0$ we have

$$V_d(A_\nu \cap H_{z_\nu(\tau)}^-) = V_d(C \cap H_{z_\nu(\tau)}^-) - V_d(A_\nu^\bullet \cap H_{z_\nu(\tau)}^-)$$

$$= V_d(C \cap H_{z_\nu(\tau)}^-) - \tau,$$

$$V_d(A_\lambda \cap H_{z_\lambda(\tau)}^-) = V_d(C \cap H_{z_\lambda(\tau)}^-) - V_d(A_\lambda^\bullet \cap H_{z_\lambda(\tau)}^-). \tag{7.32}$$

We write

$$V_d(A_\lambda^\bullet \cap H_{z_\lambda(\tau)}^-) =: f(\tau).$$

Then, with $\alpha_\lambda = (1 - \lambda)\alpha_0 + \lambda\alpha_1$,

$$f(\tau) = \int_{\alpha_\lambda}^{z_\lambda(\tau)} V_{d-1}(A_\lambda^\bullet \cap H_\zeta)\,\mathrm{d}\zeta$$

$$= \int_0^\tau V_{d-1}(A_\lambda^\bullet \cap H_{z_\lambda(t)})z_\lambda'(t)\,\mathrm{d}t$$

$$\ge \int_0^\tau V_{d-1}((1 - \lambda)D_0(t) + \lambda D_1(t))z_\lambda'(t)\,\mathrm{d}t,$$

by (7.31). In the integrand, we use the Brunn–Minkowski inequality in dimension $d - 1$ and obtain

$$f(\tau) \geq \int_0^\tau \left[(1 - \lambda) v_0(z_0(t))^{\frac{1}{d-1}} + \lambda v_1(z_1(t))^{\frac{1}{d-1}} \right]^{d-1}$$

$$\times \left[\frac{1 - \lambda}{v_0(z_0(t))} + \frac{\lambda}{v_1(z_1(t))} \right] dt \geq \tau, \tag{7.33}$$

where the last inequality follows by estimating the integrand according to [163, p. 371].

From (7.32) we have

$$V_d(A_\lambda \cap H^-_{z_\lambda(\tau)}) = V_d(C \cap H^-_{z_\lambda(\tau)}) - f(\tau),$$

and we intend to let $\tau \to \infty$. Since C is a cone, for $\zeta > 0$,

$$C \cap H^-_\zeta = \zeta C_1 \quad \text{with} \quad C_1 := C \cap H^-_1$$

and hence $V_d(C \cap H^-_\zeta) = \zeta^d V_d(C_1)$. Therefore,

$$V_d(C \cap H^-_{z_\lambda(\tau)}) = [(1 - \lambda) z_0(\tau) + \lambda z_1(\tau)]^d V_d(C_1),$$

$$V_d(C \cap H^-_{z_\nu(\tau)}) = z_\nu(\tau)^d V_d(C_1).$$

This gives

$$V_d(A_\lambda \cap H^-_{z_\lambda(\tau)})$$

$$= \left[(1 - \lambda) V_d(C \cap H^-_{z_0(\tau)})^{\frac{1}{d}} + \lambda V_d(C \cap H^-_{z_1(\tau)})^{\frac{1}{d}} \right]^d - f(\tau)$$

$$= \left[(1 - \lambda) [V_d(A_0 \cap H^-_{z_0(\tau)}) + \tau]^{\frac{1}{d}} + \lambda [V_d(A_1 \cap H^-_{z_1(\tau)}) + \tau]^{\frac{1}{d}} \right]^d - f(\tau)$$

$$= \left[(1 - \lambda) [b_0(\tau) + \tau]^{\frac{1}{d}} + \lambda [b_1(\tau) + \tau]^{\frac{1}{d}} \right]^d - f(\tau)$$

with $b_\nu(\tau) = V_d(A_\nu \cap H^-_{z_\nu(\tau)})$ for $\nu = 0, 1$. Note that (7.30) implies

$$\lim_{\tau \to \infty} b_\nu(\tau) = 1.$$

Using the mean value theorem (for each fixed τ), we can write

$$(b_1(\tau) + \tau)^{\frac{1}{d}} - (b_0(\tau) + \tau)^{\frac{1}{d}} = (b_1(\tau) - b_0(\tau)) \frac{1}{d} (b(\tau) + \tau)^{\frac{1}{d} - 1}$$

with $b(\tau)$ between $b_0(\tau)$ and $b_1(\tau)$, and hence tending to 1 as $\tau \to \infty$. With $\frac{1}{d}(b(\tau) + \tau)^{\frac{1}{d} - 1} =: h(\tau) = O\left(\tau^{\frac{1-d}{d}} \right)$ (as $\tau \to \infty$), we get

$$V_d(A_\lambda \cap H^-_{z_\lambda(\tau)})$$

$$= \left[(1 - \lambda)(b_0(\tau) + \tau)^{\frac{1}{d}} + \lambda \left((b_0(\tau) + \tau)^{\frac{1}{d}} + (b_1(\tau) - b_0(\tau)) h(\tau) \right) \right]^d - f(\tau)$$

$$= \left[(b_0(\tau) + \tau)^{\frac{1}{d}} + \lambda(b_1(\tau) - b_0(\tau))h(\tau)\right]^d - f(\tau)$$

$$= b_0(\tau) + \tau - f(\tau) + \sum_{r=1}^{d} \binom{d}{r}(b_0(\tau) + \tau)^{\frac{d-r}{d}} \left[\lambda(b_1(\tau) - b_0(\tau))\right]^r h(\tau)^r.$$

Since $b_0(\tau) \to 1$, $f(\tau) \geq \tau$, $(b_0(\tau)+\tau)^{\frac{d-r}{d}} h(\tau)^r = O(\tau^{1-r})$, and $b_1(\tau) - b_0(\tau) \to 0$ as $\tau \to \infty$, we conclude that

$$V_d(A_\lambda) = \lim_{\tau \to \infty} V_d(A_\lambda \cap H_{z_\lambda(\tau)}^-) \leq 1.$$

This proves that

$$V_d((1 - \lambda)A_0 \boxplus \lambda A_1) \leq 1. \tag{7.34}$$

If there exists a number $\tau_0 > 0$ for which $f(\tau_0) = \tau_0 + \varepsilon$ with $\varepsilon > 0$, then, for $\tau > \tau_0$,

$$f(\tau) = V_d(A_\lambda^\bullet \cap H_{z_\lambda(\tau)}^-)$$

$$= \tau_0 + \varepsilon + \int_{\tau_0}^{\tau} \left[(1 - \lambda)v_0(z_0(t))^{\frac{1}{d-1}} + \lambda v_1(z_1(t))^{\frac{1}{d-1}}\right]^{d-1}$$

$$\times \left[\frac{1 - \lambda}{v_0(z_0(t))} + \frac{\lambda}{v_1(z_1(t))}\right] dt$$

$$\geq \tau_0 + \varepsilon + (\tau - \tau_0) = \tau + \varepsilon,$$

and as above we obtain that $V_d(A_\lambda) \leq 1 - \varepsilon$.

Suppose now that (7.34) holds with equality. Then, as just shown, we have $f(\tau) = \tau$ for all $\tau \geq 0$. Thus, we have equality in (7.33) and hence equality in (7.31), for all $\tau \geq 0$. Explicitly, this means that

$$A_\lambda^\bullet \cap H_{z_\lambda(\tau)} = (1 - \lambda)(A_0^\bullet \cap H_{z_0(\tau)}) + \lambda(A_1^\bullet \cap H_{z_1(\tau)}) \quad \text{for all } \tau \geq 0. \tag{7.35}$$

We claim that this implies

$$A_\lambda^\bullet \cap H_{z_\lambda(\tau)}^- = (1 - \lambda)(A_0^\bullet \cap H_{z_0(\tau)}^-) + \lambda(A_1^\bullet \cap H_{z_1(\tau)}^-) \tag{7.36}$$

for all $\tau \geq 0$. For the proof, let $x \in A_\lambda^\bullet \cap H_{z_\lambda(\tau)}^-$. Then there is a number $\sigma \in [0, \tau]$ such that $x \in A_\lambda^\bullet \cap H_{z_\lambda(\sigma)}$. By (7.35),

$$x \in (1 - \lambda)(A_0^\bullet \cap H_{z_0(\sigma)}) + \lambda(A_1^\bullet \cap H_{z_1(\sigma)})$$

$$\subset (1 - \lambda)(A_0^\bullet \cap H_{z_0(\tau)}^-) + \lambda(A_1^\bullet \cap H_{z_1(\tau)}^-),$$

since $\sigma \leq \tau$ implies $H_{z_\nu(\sigma)} \subset H_{z_\nu(\tau)}^-$. This shows the inclusion $\subseteq$ in (7.36). The inclusion $\supseteq$ is trivial.

To (7.36), we can now apply the Brunn–Minkowski inequality for d-dimensional convex bodies and conclude that

$$V_d(A_\lambda^\bullet \cap H_{z_\lambda(\tau)}^-) \geq \tau.$$

But we know that equality holds here, since equality holds in (7.33), hence the convex bodies $A_0^\bullet \cap H_{z_0(\tau)}^-$ and $A_1^\bullet \cap H_{z_1(\tau)}^-$, which have the same volume, are translates of each other. The translation vector might depend on τ, but in fact, it does not, since for $0 < \sigma < \tau$, the body $A_\nu^\bullet \cap H_{z_\nu(\sigma)}$ is the intersection of $A_\nu^\bullet \cap H_{z_\nu(\tau)}$ with a closed halfspace. We conclude that $A_1^\bullet$ is a translate of $A_0^\bullet$, thus there is a vector v with $A_0^\bullet + v = A_1^\bullet$. We have to show that $v = o$.

Suppose that $v \neq o$. We assert that $v \in C$. If this is wrong, then there is a supporting hyperplane H of C that contains a ray $R \subset C$ and such that v is contained in the open halfspace bounded by H and not containing points of C. For each $x \in R$, we have $x + v \notin C$. There are points in $A_0^\bullet$ arbitrarily close to the ray R (otherwise A_0 would have infinite volume). Hence, there are points $y \in A_0^\bullet$ for which $y + v \notin C$. This is a contradiction to $A_0^\bullet + v = A_1^\bullet \subset C$, which shows that $v \in C$. Each line parallel to v and meeting $A_0^\bullet$ intersects $A_0^\bullet$ in a closed ray (since $v \in C$). Let M be the set of endpoints of all these rays. Then the set $\bigcup_{x \in M}[x, x + v)$ is contained in $C \setminus A_1^\bullet$ and has infinite volume, a contradiction. This shows that $v = o$.

This proves Theorem 7.5.1 under the assumption (7.30). Now let A_0, A_1 be arbitrary coconvex sets. As already mentioned, also the volume of coconvex sets is homogeneous of degree d. Therefore (as in the case of convex bodies, see [163, p. 370]), we define

$$\overline{A}_\nu := V_d(A_\nu)^{-1/d} A_\nu \text{ for } \nu = 0, 1, \quad \overline{\lambda} := \frac{\lambda V_d(A_1)^{1/d}}{(1-\lambda)V_d(A_0)^{1/d} + \lambda V_d(A_1)^{1/d}}.$$

Then $V_d(\overline{A}_\nu) = 1$ for $\nu = 0, 1$, hence $V_d((1-\overline{\lambda})\overline{A}_0 \boxplus \overline{\lambda}\,\overline{A}_1) \leq 1$, as just proved. This gives the assertion. □

Notes for Section 7.5

1. For the special case of Theorem 7.5.1 referring only to bounded coconvex sets, a much shorter proof is possible; see Fillastre [60].

7.6 Mixed volumes of general coconvex sets

In Section 7.2 we have introduced mixed volumes of bounded coconvex sets and convex bodies. Now we extend this notion to not necessarily bounded coconvex sets (but without convex bodies). For this, we use approximation by mixed volumes of bounded coconvex sets.

Let $\omega \subset \Omega_C$ be an open subset whose closure is contained in Ω_C. Let A be a coconvex set, so that $A^\bullet = C \setminus A$ is closed and convex. We define

$$A_{(\omega)}^\bullet := C \cap \bigcap_{u \in \omega} H^-(A^\bullet, u), \qquad A_{(\omega)} := C \setminus A_{(\omega)}^\bullet,$$

where $H^-(A^\bullet, u)$ denotes the supporting halfspace of the closed convex set $A^\bullet$ with outer normal vector u. Since $A^\bullet_{(\omega)}$ is C-determined by the compact set $\mathrm{cl}\,\omega$, the set $A_{(\omega)}$ is a bounded coconvex set.

With A and ω as above, we associate another set, namely

$$A[\omega] := \bigcup_{x \in \tau(A^\bullet, \omega) \cap \mathrm{int}\, C} (o, x),$$

where (o, x) denotes the open line segment with endpoints o and x. We choose an increasing sequence $(\omega_j)_{j \in \mathbb{N}}$ of open subsets of Ω_C with closures in Ω_C and with $\bigcup_{j \in \mathbb{N}} \omega_j = \Omega_C$. Then

$$A[\omega_j] \uparrow \mathrm{int}\, A \qquad \text{as } j \to \infty. \tag{7.37}$$

In fact, that the set sequence is increasing, follows from the definition. Let $y \in \mathrm{int}\, A$. Then there is a boundary point x of $A^\bullet$ with $y \in (o, x)$. Let u be an outer unit normal vector of $A^\bullet$ at x. Then $u \in \Omega_C$, hence $u \in \omega_j$ for some j. For this j, we have $y \in A[\omega_j]$.

Lemma 7.6.1. *If $A_1, \ldots, A_d$ are coconvex sets, then*

$$\lim_{j \to \infty} V_d(A_{1(\omega_j)} \boxplus \cdots \boxplus A_{d(\omega_j)}) = V_d(A_1 \boxplus \cdots \boxplus A_d). \tag{7.38}$$

Proof. We state that

$$(A_1 \boxplus \cdots \boxplus A_d)[\omega_j] \subseteq A_{1(\omega_j)} \boxplus \cdots \boxplus A_{d(\omega_j)} \subseteq A_1 \boxplus \cdots \boxplus A_d. \tag{7.39}$$

For the proof of the first inclusion, let $y \in (A_1 \boxplus \cdots \boxplus A_d)[\omega_j]$. Then there exists a point $x \in \tau(A_1^\bullet + \cdots + A_d^\bullet, \omega_j) \cap \mathrm{int}\, C$ with $y \in (o, x)$. Let $u \in \omega_j$ be an outer unit normal vector of $A_1^\bullet + \cdots + A_d^\bullet$ at x. Denoting by $F(K, u)$ the support set of a closed convex set K with outer normal vector u, we have (by [163, Thm. 1.7.5], here applied to the intersections of the $A_i^\bullet$ with C_t for sufficiently large t)

$$F(A_1^\bullet + \cdots + A_d^\bullet, u) = F(A_1^\bullet, u) + \cdots + F(A_d^\bullet, u),$$

hence there are points $x_i \in F(A_i^\bullet, u)$ $(i = 1, \ldots, d)$ with $x = x_1 + \cdots + x_d$. We have $x_i \in A^\bullet_{i(\omega_j)}$, hence $x \in A_{1(\omega_j)} \boxplus \cdots \boxplus A_{d(\omega_j)}$. This proves the first inclusion of (7.39). The second inclusion follows immediately from the definitions. From (7.39) and (7.37) we obtain

$$A_{1(\omega_j)} \boxplus \cdots \boxplus A_{d(\omega_j)} \uparrow \mathrm{int}\,(A_1 \boxplus \cdots \boxplus A_d) \qquad \text{as } j \to \infty,$$

from which the assertion (7.38) follows. $\qquad\square$

Now let $A_1, \ldots, A_d$ be coconvex sets. From (7.8) and Lemma 7.6.1 it follows that the limit

$$\lim_{j \to \infty} \overline{V}(A_{1(\omega_j)}, \ldots, A_{d(\omega_j)}) =: \overline{V}(A_1, \ldots, A_d)$$

exists. We call $\overline{V}(A_1, \ldots, A_d)$ the *mixed volume* of $A_1, \ldots, A_d$. From (7.6) it follows by approximation that

$$V_d(\lambda_1 A_1 \boxplus \cdots \boxplus \lambda_d A_d) = \sum_{i_1, \ldots, i_d = 1}^{d} \lambda_{i_1} \cdots \lambda_{i_d} \overline{V}(A_{i_1}, \ldots, A_{i_d}) \qquad (7.40)$$

for $\lambda_1, \ldots, \lambda_d > 0$.

For the mixed volume $\overline{V}$, we shall now establish an integral representation. To that end, we note first that the support functions of $A^\bullet$ and $A^\bullet_{(\omega_j)}$ satisfy

$$h(A^\bullet, u) = h(A^\bullet_{(\omega_j)}, u) \quad \text{for } u \in \omega_j. \qquad (7.41)$$

Since ω_j is open, then for $u \in \omega_j$ the support functions of $A^\bullet$ and $A^\bullet_{(\omega_j)}$ coincide in a neighborhood of u. By [163, Thm. 1.7.2], the support sets of $A^\bullet$ and $A^\bullet_{(\omega_j)}$ with outer normal vector u are the same. It follows that $\tau(A^\bullet, \omega_j) = \tau(A^\bullet_{(\omega_j)}, \omega_j)$ and, therefore, that also

$$S_{d-1}(A^\bullet, \cdot) = S_{d-1}(A^\bullet_{(\omega_j)}, \cdot) \quad \text{on } \omega_j. \qquad (7.42)$$

More generally, if $A_1, \ldots, A_{d-1}$ are coconvex sets, we can define their mixed area measure by first defining

$$S(A_1^\bullet, \ldots, A_{d-1}^\bullet, \cdot) = S(A_{1(\omega_j)}^\bullet, \ldots, A_{d-1(\omega_j)}^\bullet, \cdot) \quad \text{on } \omega_j, \qquad (7.43)$$

for $j \in \mathbb{N}$. Since $\omega_j \uparrow \Omega_C$, this yields a Borel measure on all of Ω_C. It need not be finite. Then we define

$$\overline{S}(A_1, \ldots, A_{d-1}, \cdot) := S(A_1^\bullet, \ldots, A_{d-1}^\bullet, \cdot).$$

By Lemma 7.1.1, (7.41) and (7.42) we have

$$V_d(A_{(\omega_j)})$$

$$= \frac{1}{d} \int_{\omega_j} \overline{h}(A_{(\omega_j)}, u) \overline{S}_{d-1}(A_{(\omega_j)}, du) + \frac{1}{d} \int_{\Omega_C \setminus \omega_j} \overline{h}(A_{(\omega_j)}, u) \overline{S}_{d-1}(A_{(\omega_j)}, du)$$

$$= \frac{1}{d} \int_{\omega_j} \overline{h}(A, u) \overline{S}_{d-1}(A, du) + \frac{1}{d} \int_{\Omega_C \setminus \omega_j} \overline{h}(A_{(\omega_j)}, u) \overline{S}_{d-1}(A_{(\omega_j)}, du).$$

From $A_{(\omega_j)} \uparrow A$ we get

$$\lim_{j \to \infty} V_d(A_{(\omega_j)}) = V_d(A),$$

and $\omega_j \uparrow \Omega_C$ gives

$$\lim_{j\to\infty} \frac{1}{d} \int_{\omega_j} \overline{h}(A,u)\,\overline{S}_{d-1}(A,\mathrm{d}u) = \frac{1}{d}\int_{\Omega_C} \overline{h}(A,u)\,\overline{S}_{d-1}(A,\mathrm{d}u) = V_d(A).$$

It follows that

$$\lim_{j\to\infty}\int_{\Omega_C\setminus\omega_j} \overline{h}(A_{(\omega_j)},u)\,\overline{S}_{d-1}(A_{(\omega_j)},\mathrm{d}u)=0. \tag{7.44}$$

From (7.4), and using (7.42) and (7.43), we get

$$\overline{V}(A_{1(\omega_j)},\ldots,A_{d(\omega_j)})$$

$$=\frac{1}{d}\int_{\omega_j} \overline{h}(A_{1(\omega_j)},u)\,\overline{S}(A_{2(\omega_j)},\ldots,A_{d(\omega_j)},\mathrm{d}u)$$

$$+\frac{1}{d}\int_{\Omega_C\setminus\omega_j} \overline{h}(A_{1(\omega_j)},u)\,\overline{S}(A_{2(\omega_j)},\ldots,A_{d(\omega_j)},\mathrm{d}u)$$

$$=\frac{1}{d}\int_{\omega_j} \overline{h}(A_1,u)\,\overline{S}(A_2,\ldots,A_d,\mathrm{d}u)$$

$$+\frac{1}{d}\int_{\Omega_C\setminus\omega_j} \overline{h}(A_{1(\omega_j)},u)\,\overline{S}(A_{2(\omega_j)},\ldots,A_{d(\omega_j)},\mathrm{d}u). \tag{7.45}$$

Writing $A:=A_1\boxplus\cdots\boxplus A_d$, we have the trivial estimates

$$\overline{h}(A_{1(\omega_j)},u)\le \overline{h}(A_{(\omega_j)},u),\quad \overline{S}(A_{2(\omega_j)},\ldots,A_{d(\omega_j)},\cdot)\le \overline{S}_{d-1}(A_{(\omega_j)},\cdot).$$

Hence, the term (7.45) can be estimated from above by

$$\frac{1}{d}\int_{\Omega_C\setminus\omega_j} \overline{h}(A_{1(\omega_j)},u)\,\overline{S}(A_{2(\omega_j)},\ldots,A_{d(\omega_j)},\mathrm{d}u)$$

$$\le \frac{1}{d}\int_{\Omega_C\setminus\omega_j} \overline{h}(A_{(\omega_j)},u)\,\overline{S}_{d-1}(A_{(\omega_j)},\mathrm{d}u),$$

and by (7.44) this tends to zero as $j\to\infty$. We conclude that

$$\overline{V}(A_1,\ldots,A_d)=\frac{1}{d}\int_{\Omega_C} \overline{h}(A_1,u)\,\overline{S}(A_2,\ldots,A_d,\mathrm{d}u). \tag{7.46}$$

7.7 Minkowski's theorem for general coconvex sets

At the beginning of Section 7.4 we have mentioned the classical theorem of Minkowski. This theorem has an existence and a uniqueness part. One can ask the same questions for C-close sets. What are the necessary and sufficient conditions on a Borel measure on Ω_C to be the surface area measure of some C-close set? And is such a set uniquely determined?

Theorem 7.4.1 provides a sufficient condition: every finite Borel measure with compact support in Ω_C is the surface area measure of a C-full set K. This set is uniquely determined, as follows from the stronger stability theorem 7.4.2. However, a characterization of surface area measures of C-close sets (or, equivalently, of not necessarily bounded coconvex sets), remains an open problem. The uniqueness question, however, can be answered.

Theorem 7.7.1. *If K_0, K_1 are C-close sets with $S_{d-1}(K_0, \cdot) = S_{d-1}(K_1, \cdot)$, then $K_0 = K_1$.*

Proof. As said, we can equivalently consider coconvex sets A_0, A_1 with $\overline{S}_{d-1}(A_0, \cdot) = \overline{S}_{d-1}(A_1, \cdot)$. By (7.46) we then have

$$\overline{V}(A_0, \ldots, A_0, A_1) = \frac{1}{d} \int_{\Omega_C} \overline{h}(A_1, u) \, \overline{S}_{d-1}(A_0, \mathrm{d}u).$$

Therefore, the assumption gives $\overline{V}(A_0, \ldots, A_0, A_1) = V_d(A_1)$. Similarly, we obtain $\overline{V}(A_1, \ldots, A_1, A_0) = V_d(A_0)$. Multiplication yields

$$\overline{V}(A_0, \ldots, A_0, A_1)\overline{V}(A_1, \ldots, A_1, A_0) = V_d(A_0)V_d(A_1). \qquad (7.47)$$

Now we set $A_\lambda := (1 - \lambda)A_0 \boxplus \lambda A_1$ for $0 < \lambda < 1$ and recall that a special case of (7.40) says that

$$V_d(A_\lambda) = \sum_{i=0}^{d} \binom{d}{i}(1 - \lambda)^{d-i}\lambda^i\overline{V}(\underbrace{A_0, \ldots, A_0}_{d-i}, \underbrace{A_1, \ldots, A_1}_{i}).$$

The function f defined by

$$f(\lambda) = V_d(A_\lambda)^{1/d} - (1 - \lambda)V_d(A_0)^{1/d} - \lambda V_d(A_1)^{1/d} \quad \text{for } 0 \leq \lambda \leq 1$$

is convex, as follows from Theorem 7.5.1 and a similar argument as in the case of convex bodies (see [163, pp. 369–370]). Also as in the convex body case (see [163, p. 382]), one obtains the counterpart to Minkowski's first inequality, namely

$$\overline{V}(A_0, \ldots, A_0, A_1)^d \leq V_d(A_0)^{d-1}V_d(A_1), \qquad (7.48)$$

with equality if and only if $A_0 = \alpha A_1$ with some $\alpha > 0$. Here we can interchange A_0 and A_1 and then multiply both inequalities, to obtain

$$\overline{V}(A_0, \ldots, A_0, A_1)\overline{V}(A_1, \ldots, A_1, A_0) \leq V_d(A_0)V_d(A_1).$$

According to (7.47), equality holds here. Therefore, equality holds in (7.48), which implies that $A_0 = \alpha A_1$ with $\alpha > 0$. Since $\overline{S}_{d-1}(A_0, \cdot) = \overline{S}_{d-1}(A_1, \cdot)$, we must have $\alpha = 1$. This proves Theorem 7.7.1. $\qquad \Box$

Concerning the existence part, we can prove that the finiteness of the measure is a sufficient condition. Here we follow [169].

Theorem 7.7.2. *Every finite Borel measure on Ω_C is the surface area mea-sure of a C-close set.*

The condition of finiteness is, of course, not necessary. One should also note that from the finiteness of the surface area measure of the C-close set K one cannot conclude that $C \setminus K$ is bounded. It is not difficult to construct unbounded coconvex sets with finite surface area measure.

Proof of Theorem 7.7.2. Let φ be a finite Borel measure on Ω_C. The idea is to use Theorem 7.4.1, where the given measure has compact support. Therefore, we choose a sequence $(\omega_j)_{j\in\mathbb{N}}$ of nonempty open subsets of Ω_C such that $\operatorname{cl}\omega_j \subset \omega_{j+1}$ for all $j \in \mathbb{N}$ and $\bigcup_{j\in\mathbb{N}} \omega_j = \Omega_C$. For each $j \in \mathbb{N}$, the measure φ_j defined by $\varphi_j(A) := \varphi(A \cap \omega_j)$ for $A \in \mathcal{B}(\Omega_C)$ is defined on Ω_C and has compact support. Therefore, Theorem 7.4.1 can be applied to it. It yields a C-full set L_j (uniquely determined according to Theorem 7.7.1) such that $S_{d-1}(L_j, \cdot) = \varphi_j$.

We need a number t_1 such that (7.49) below holds. We can argue similarly as in the proof of Lemma 7.4.3. Let $j \in \mathbb{N}$ be given. Let B_r be the largest ball with center o such that $B_r \cap \operatorname{int} L_j = \emptyset$; let r be its radius (it may depend on j). Then there is a point $z_j \in B_r \cap L_j$. The mapping from $\operatorname{bd} L_j \cap \operatorname{int} C$ to $\operatorname{bd} B_r$ that is defined by $x \mapsto rx/\|x\|$ does not increase distances. Therefore, $\mathcal{H}^{d-1}(\operatorname{bd} B_r \cap \operatorname{int} C) \le \mathcal{H}^{d-1}(\operatorname{bd} L_j \cap \operatorname{int} C)$, which gives

$$r^{d-1}\mathcal{H}(\mathbb{S}^{d-1} \cap \operatorname{int} C) \le S_{d-1}(L_j, \Omega_C) = \varphi_j(\Omega_C) \le \varphi(\Omega_C).$$

It follows that $r \le t_0$ for a constant t_0 that depends only on C and $\varphi(\Omega_C)$ (and not on j). If we choose $t_1 > t_0$, then $z_j \in H^-(w, t_1)$. This shows that

$$L_j \cap H^-(w, t_1) \ne \emptyset \quad \text{for all } j \in \mathbb{N}. \tag{7.49}$$

Let $(t_k)_{k\in\mathbb{N}}$ be an increasing sequence with $t_k \uparrow \infty$ as $k \to \infty$. We have $L_j \cap C_{t_1} \ne \emptyset$ for $j \in \mathbb{N}$, hence the bounded sequence $(L_j \cap C_{t_1})_{j\in\mathbb{N}}$ of convex bodies has a convergent subsequence. Thus, for a subsequence $(j_i^{(1)})_{i\in\mathbb{N}}$ of $\mathbb{N}$, there is a convex body K_1 satisfying

$$L_{j_i^{(1)}} \cap C_{t_1} \to K_1 \quad \text{as } i \to \infty.$$

For the same reason, there are a subsequence $(j_i^{(2)})_{i\in\mathbb{N}}$ of $(j_i^{(1)})_{i\in\mathbb{N}}$ and a convex body K_2 such that

$$L_{j_i^{(2)}} \cap C_{t_2} \to K_2 \quad \text{as } i \to \infty.$$

By induction, we obtain for each $k \in \mathbb{N}$ a subsequence $(j_i^{(k)})_{i\in\mathbb{N}}$ of $(j_i^{(k-1)})_{i\in\mathbb{N}}$ and a convex body K_k such that

$$L_{j_i^{(k)}} \cap C_{t_k} \to K_k \quad \text{as } i \to \infty.$$

The diagonal sequence $(\ell_i)_{i \in \mathbb{N}} := (j_i^{(i)})_{i \in \mathbb{N}}$ then satisfies

$$L_{\ell_i} \cap C_{t_k} \to K_k \quad \text{as } i \to \infty, \text{ for each } k \in \mathbb{N}.$$

For $1 \le m < k$ we have

$$L_{\ell_i} \cap C_{t_m} \to K_m, \quad L_{\ell_i} \cap C_{t_k} \to K_k \quad \text{as } i \to \infty.$$

Using [163, Thm. 1.8.10], we obtain

$$K_m = \lim_{i \to \infty} (L_{\ell_i} \cap C_{t_m}) = \lim_{i \to \infty} [(L_{\ell_i} \cap C_{t_k}) \cap C_{t_m}]$$
$$= \left[\lim_{i \to \infty} (L_{\ell_i} \cap C_{t_k}) \right] \cap C_{t_m} = L_k \cap C_{t_m}.$$

Therefore, if we define

$$K := \bigcup_{k \in \mathbb{N}} K_k,$$

then

$$K \cap C_{t_k} = K_k \quad \text{for } k \in \mathbb{N}.$$

This implies, in particular, that $K \subset C$ is a closed convex set.

We have to show that

$$S_{d-1}(K, \cdot) = \varphi. \tag{7.50}$$

Let $j \in \mathbb{N}$, and let $\omega \subset \omega_{\ell_j}$ be an open set. The set $\tau(K, \omega)$ of all boundary points of K with an outer normal vector in ω is bounded, as follows from the compactness of $\mathrm{cl}\,\omega$ and (7.11). Hence, there is a number $k \in \mathbb{N}$ with $\tau(K, \omega) = \tau(K_k, \omega)$. From $\lim_{i \to \infty} (L_{\ell_i} \cap C_{t_k}) = K_k$ and the weak continuity of the surface area measure, it follows that

$$S_{d-1}(K_k, \omega) \le \liminf_{i \to \infty} S_{d-1}(L_{\ell_i} \cap C_{t_k}, \omega).$$

By the definition of $\overline{L}_{\ell_i}$ we have $S_{d-1}(L_{\ell_i} \cap C_{t_k}, \omega) = \varphi(\omega)$ for sufficiently large i, thus

$$S_{d-1}(K_k, \omega) \le \varphi(\omega). \tag{7.51}$$

If $\beta \subset \omega_j$ is a closed set, then a similar argument yields

$$S_{d-1}(K_k, \beta) \ge \limsup_{i \to \infty} S_{d-1}(L_{\ell_i} \cap C_{t_k}, \beta) = \varphi(\beta). \tag{7.52}$$

Let $\beta \subset \omega_{\ell_j}$ be closed. We choose a sequence $(\eta_r)_{r \in \mathbb{N}}$ of open neighborhoods of β with $\eta_r \subset \omega_{\ell_{j+1}}$ and $\eta_r \downarrow \beta$ as $r \to \infty$. By (7.51), we have $S_{d-1}(K_k, \eta_r) \le \varphi(\eta_r)$. Since $\eta_r \downarrow \beta$, this gives $S_{d-1}(K_k, \beta) \le \varphi(\beta)$, and from (7.52) we then conclude that $S_{d-1}(K_k, \beta) = \varphi(\beta)$.

For a closed set $\beta \subset \Omega_C$ with $\beta \subset \omega_{\ell_j}$ for some $j \in \mathbb{N}$, we have $\tau(K, \omega) = \tau(K_k, \omega)$ for suitable $k \in \mathbb{N}$, hence $S_{d-1}(K, \beta) = S_{d-1}(K_k, \beta) = \varphi(\beta)$. Since $\omega_{\ell_j} \uparrow \Omega_C$ as $j \to \infty$, the equality $S_{d-1}(K, \beta) = \varphi(\beta)$ holds for every closed set $\beta \in \mathcal{B}(\Omega_C)$ and thus for every Borel set in $\mathcal{B}(\Omega_C)$. We have proved that $S_{d-1}(K, \cdot) = \varphi$.

It remains to show that K is C-close. For this, we note that by (7.2) we have, for each $j \in \mathbb{N}$,

$$V_d(C \setminus L_j) = -\frac{1}{d} \int_{\Omega_C} h(L_j, u) \, S_{d-1}(L_j, du)$$

$$= -\frac{1}{d} \int_{\omega_j} h(L_j, u) \, \varphi(du).$$

It follows from (7.49) that there is a constant c, depending only on C and t_1, such that $|h(L_j, \cdot)| \le c$. This yields

$$V_d(C \setminus L_j) \le c\varphi(\Omega_C).$$

For $i, k \in \mathbb{N}$ we get

$$V_d(C_{t_k} \setminus L_{\ell_i}) \le V_d(C \setminus L_{\ell_i}) \le c\varphi(\Omega_C).$$

Since $L_{\ell_i} \cap C_{t_k} \to K_k$ as $i \to \infty$, this gives

$$V_d(C_{t_k} \setminus K) = V_d(C_{t_k} \setminus K_k) \le c\varphi(\Omega_C).$$

This holds for all $k \in \mathbb{N}$, hence we deduce that

$$V_d(C \setminus K) \le c\varphi(\Omega_C) < \infty.$$

Thus, K is a C-close set. $\qquad\square$

If the Borel measure φ on Ω_C is the surface area measure of a C-asymptotic set, then it is finite on compact subsets of Ω_C. This condition, however, is not sufficient. The measure has also to satisfy a growth restriction on compact sets ω approaching Ω_C. For a precise formulation, let $\Delta(\omega)$ denote the spherical distance of the nonempty compact set $\omega \subset \Omega_C$ from the boundary of Ω_C. The following was shown in [169]. This theorem is not restricted to C-asymptotic sets, but holds for arbitrary nonempty closed convex sets $K \subseteq C$. But it should be observed that it only says something about the surface area measure of K in compact subsets of Ω_C, which is an open subset of a hemisphere.

Theorem 7.7.3. *Let φ be a Borel measure on Ω_C. If φ is the surface area measure of some nonempty closed convex set $K \subset C$, then there is a constant c, depending only on C and K, such that*

$$\Delta(\omega)^{d-1} \varphi(\omega) \le c \quad \text{for all nonempty compact sets } \omega \subset \Omega_C.$$

Proof. As always in this chapter, we assume that $C \in \mathcal{C}^d$ is a pointed closed convex cone with nonempty interior. Let $K \subseteq C$ be a nonempty closed convex set with $S_{d-1}(K, \cdot) \llcorner \Omega_C = \varphi$. We exclude the trivial case where $o \in K$.

The unit vector w was chosen such that $w \in -\operatorname{int} C^o$. We choose it now such that $w \in \operatorname{int} C \cap (-\operatorname{int} C^\circ)$. This is possible, since $\operatorname{int} C \cap (-\operatorname{int} C^\circ) = \emptyset$ would imply that C and $-C^\circ$ could be separated by a hyperplane (by [163, Thm. 1.3.8]), so that C and C° would lie in the same closed halfspace, a contradiction. Since now also $w \in \operatorname{int} C$, there is a constant $c_1 > 0$ such that

$$|\langle w, u \rangle| \geq c_1 \quad \text{for all } u \in \Omega_C. \tag{7.53}$$

Here and in the following, $c_1, \dots, c_4$ denote positive constants that may depend on C and K (and on w, but this may be chosen such that it depends only on K and C), but not on the considered compact subsets $\omega \subset \Omega_C$.

Claim. If r is the radius of the largest ball with center o that does not meet $\operatorname{int} K$, then

$$\tau(K, \omega) \subset H^- \left(w, \frac{r}{\sin \Delta(\omega)} \right)$$

for every nonempty compact set $\omega \subset \Omega_C$.

For the proof, let $\omega \subset \Omega_C$ be a nonempty compact set. Let B denote the largest ball with center o that does not meet $\operatorname{int} K$, then r is its radius. Let $x \in \tau(K, \omega)$ and let $H(u, s)$ be a supporting hyperplane of K at x with outer normal vector $u \in \omega$. Then $H(u, s) \cap B \neq \emptyset$, by the maximality of B, hence

$$s \leq r. \tag{7.54}$$

Case 1: $u = -w$.

Since $H(u, s) \cap B \neq \emptyset$, we have $\langle x, w \rangle \leq r$, so that

$$\langle x, w \rangle \leq \frac{r}{\sin \Delta(\omega)} \tag{7.55}$$

holds trivially.

Case 2: $u \neq -w$.

Let $y \in H(u, s) \cap \operatorname{bd} C$ be a point for which $\langle w, y \rangle$ is maximal, and set $t := \langle w, y \rangle$. Then $y \in H(w, t)$ and $C \cap H(u, s) \subset H^-(w, t)$. Define

$$E := H(u, s) \cap H(w, t).$$

The $(d - 2)$-plane E is a supporting hyperplane of $C \cap H(w, t)$ in $H(w, t)$. Therefore, there is a supporting hyperplane $H(v, 0)$ of the cone C, with outer normal vector v, such that $H(v, 0) \cap H(w, t) = E$. Let α be the angle between u and v. Since $u \in \omega$ and $v \in \operatorname{bd} \Omega_C$, we have

$$\alpha \geq \Delta(\omega). \tag{7.56}$$

Let z be the image of o under orthogonal projection to $H(u, s)$ and let p be the image of z under orthogonal projection to E. Let γ be the angle of the triangle with vertices o, z, p at p. We state that

$$\gamma \geq \alpha. \tag{7.57}$$

For $d = 2$ this is clear, since then $p = y$ and $\gamma = \alpha$. Let $d \geq 3$. The two-dimensional plane through p orthogonal to E contains z. We choose $q \in H(v, 0) \setminus \{p\}$ such that p is the orthogonal projection of q to E. Then the unit vectors

$$a_1 := p + \frac{z - p}{\|z - p\|}, \quad a_2 := p - \frac{p}{\|p\|}, \quad a_3 := p + \frac{q - p}{\|q - p\|}$$

are the vertices of a spherical triangle on a two-dimensional unit sphere with center p. This spherical triangle has side lengths γ (opposite to a_3) and α (opposite to a_2; note that α is the angle between the hyperplanes $H(u, s)$ and $H(v, t)$, and hence the angle between the vectors $z - p$ and $q - p$). The side length opposite to a_1 we denote by β. At a_3, the spherical triangle has a right angle, since the planes spanned by $p, q, z - p$ and by p, q, o are orthogonal. It follows that $\cos \gamma = \cos \alpha \cos \beta \leq \cos \alpha$, which proves (7.57).

From (7.56) and (7.57) (together with the fact that $\gamma < \pi/2$) we deduce that

$$\sin \Delta(\omega) \leq \sin \alpha \leq \sin \gamma = \frac{s}{\|p\|}$$

and hence, by (7.54), that

$$\|p\| \sin \Delta(\omega) \leq r.$$

This gives

$$\langle w, x \rangle \leq \langle w, y \rangle = \langle w, p \rangle \leq \|p\| \leq \frac{r}{\sin \Delta(\omega)}. \tag{7.58}$$

For arbitrary $x \in \tau(K, \omega)$ we have proved that (7.56) respectively (7.58) holds, hence the Claim is proved.

Now we consider the orthogonal projection of $\tau(K, \omega)$ to the hyperplane $H(w, t)$ with $t := r/\sin \Delta(\omega)$. It is contained in C, by the Claim and since $w \in \operatorname{int} C$. Its $(d - 1)$-dimensional measure is given by

$$\int_{\tau(K,\omega)} |\langle w, n(K, x) \rangle| \, \mathcal{H}^{d-1}(\mathrm{d}x),$$

where $\mathcal{H}^{d-1}$ denotes the $(d-1)$-dimensional Hausdorff measure and $n(K, x)$ is the ($\mathcal{H}^{d-1}$-everywhere unique) outer unit normal vector of K at x. It follows that

$$\int_{\omega} |\langle w, u \rangle| \, S_{d-1}(K, \mathrm{d}u) \leq V_{d-1}(C \cap H(w, t))$$

$$= t^{d-1} V_{d-1}(C \cap H(w, 1)) \leq \frac{c_3}{\sin^{d-1} \Delta(w)},$$

with a constant c_3 that depends only on C, K, w, but not on ω. From (7.53) it follows that the measure $\varphi = S_{d-1}(K, \cdot)$ satisfies

$$\varphi \leq \frac{c_4}{\sin^{d-1} \Delta(\omega)},$$

with a constant c_4 independent of ω. Since $\Delta(\omega) \leq \pi/2$ and hence $\sin \Delta(\omega) \geq (2/\pi)\Delta(\omega)$, the assertion follows. $\qquad\square$

Notes for Section 7.7

1. As mentioned, the finiteness condition in Theorem 7.7.2 is sufficient, but not necessary. It is also not possible to conclude from the finiteness of $S_{d-1}(K, \cdot)$ for a C-close set K that K would be C-full. A counterexample is given in [169].

We do not know of a sufficient condition for an infinite Borel measure on Ω_C to be the surface area measure of a C-asymptotic or C-coconvex set.

7.8 The cone-volume measure

When the classical Brunn–Minkowski theory of convex bodies was generalized to the L_p Brunn–Minkowski theory, also the notion of surface area measure was appropriately extended. The case $p = 0$ leads to the cone-volume measure. Also for C-close convex sets K we can define such a cone-volume measure, where now this refers not to an interior point of K but to the origin o, which is outside K.

Again, $C \subset \mathbb{R}^d$ is a pointed closed convex cone with nonempty interior.

Definition 7.8.1. *Let K be a C-close convex set, and let $\omega \in \mathcal{B}(\Omega_C)$. Let*

$$M(K, \omega) := \bigcup_{x \in \tau(K, \omega)} [o, x]$$

and

$$V_K(\omega) := \mathcal{H}^d(M(K, \omega)).$$

This defines the cone-volume measure V_K *of K.*

A few remarks are necessary to justify this definition. First, if $x \in \tau(K, \omega)$, then the half-open segment $[o, x)$ belongs to $C \setminus K$. Let $\mathcal{M}$ denote the set of all sets $\omega \subset \Omega_C$ for which $M(K, \omega)$ is Lebesgue measurable. If ω is closed, then $M(K, \omega)$ is closed and hence ω belongs to $\mathcal{M}$. If $\omega_1 \cap \omega_2 = \emptyset$, then $\tau(K, \omega_1) \cap \tau(K, \omega_2)$ contains only singular points of K and hence has $\mathcal{H}^{d-1}$ measure zero. Then $M(K, \omega_1) \cap M(K, \omega_2)$ has $\mathcal{H}^d$ measure zero. (For any $\varepsilon > 0$, $\tau(K, \omega_1) \cap \tau(K, \omega_2)$ can be covered by countably many balls of radii $r_1, r_2, \ldots$ with $\sum r_i^{d-1} < \varepsilon$. Then $M(K, \omega_1) \cap M(K, \omega_2)$ can be covered by countably many balls of radii $r_1, r_2, \ldots$ with $\sum r_i^d < \varepsilon$.) Now it follows that $\mathcal{M}$ is a σ-algebra containing the closed sets and hence the Borel sets and that V_K is a measure on $\mathcal{M}$.

The next lemma establishes an integral representation of the cone-volume measure.

Lemma 7.8.1. *The cone-volume measure of a C-close set K can be represented by*

$$V_K(\omega) = \frac{1}{d} \int_\omega -h(K, u) \, S_{d-1}(K, \mathrm{d}u) \tag{7.59}$$

for $\omega \in \mathcal{B}(\Omega_C)$.

Proof. Let K be C-close set. We define a measure ψ_K on Ω_C by

$$\psi_K(\omega) := \frac{1}{d} \int_\omega -h(K, u) \, S_{d-1}(K, \mathrm{d}u) \quad \text{for } \omega \in \mathcal{B}(\Omega_C).$$

First we assume that $C \setminus K$ is bounded, thus K is a C-full set. Then $C \setminus K \subset C_t$ for suitable $t > 0$. We choose a dense sequence $(u_i)_{i \in \mathbb{N}}$ in Ω_C and define

$$K_j := C \cap \bigcap_{i=1}^{j} H^-(K, u_j),$$

recalling that $H^-(K, u)$ denotes the supporting halfspace of K with outer normal vector u. Clearly, K_j is a K-full set, and we have

$$K_j \to K \quad \text{as } j \to \infty,$$

since the sequence $(u_i)_{i \in \mathbb{N}}$ is dense in Ω_C. Therefore,

$$V_d(C \setminus K_j) \to V_d(C \setminus K) \quad \text{as } j \to \infty.$$

Since K_j is the intersection of C with a polyhedron, we see that

$$V_{K_j} = \psi_{K_j}. \tag{7.60}$$

The idea is now to show the weak convergences

$$V_{K_j} \xrightarrow{w} V_K \quad \text{as } j \to \infty \tag{7.61}$$

and

$$\psi_{K_j} \xrightarrow{w} \psi_K \quad \text{as } j \to \infty. \tag{7.62}$$

For the proof of (7.61), we define the radial function $\rho(K, \cdot)$ of K by

$$\rho(K, u) := \min\{r \geq 0 : ru \in K\} \quad \text{for } u \in C \cap \mathbb{S}^{d-1}.$$

Let $\nu(x) := x/\|x\|$ for $x \in \mathbb{R}^d \setminus \{o\}$. By using polar coordinates, we obtain for $\omega \in \mathcal{B}(\Omega_C)$ that

$$V_K(\omega) = \mathcal{H}^d \left(\bigcup_{x \in \tau(K, \omega)} [o, x] \right) = \int_{\nu(\tau(K, \omega))} \int_0^{\rho(K, u)} s^{d-1} \, \mathrm{d}s \, \mathcal{H}^{d-1}(\mathrm{d}u)$$

$$= \frac{1}{d} \int_{\nu(\tau(K, \omega))} \rho(K, u)^d \, \mathcal{H}^{d-1}(\mathrm{d}u).$$

Similarly,

$$V_{K_j}(\omega) = \frac{1}{d} \int_{\nu(\tau(K_j,\omega))} \rho(K_j,u)^d \, \mathscr{H}^{d-1}(\mathrm{d}u).$$

For $\mathscr{H}^{d-1}$-almost all $u \in C \cap \mathbb{S}^{d-1}$, the outer unit normal vector $n(K_j,u)$ of K_j at the boundary point $\rho(K_j,u)u$ is uniquely determined for all $j \in \mathbb{N}_0$. (This follows from [163, Thm. 2.2.5] and holds for general convex K_j, not using that here they are intersections of C with polyhedra.). Since $K_j \to K$, for almost all $u \in C \cap \mathbb{S}^{d-1}$, we have $n(K_j,u) \to n(K,u)$ for $j \to \infty$. For an open set $\omega \subset \Omega_C$, this implies that for almost all $u \in C \cap \mathbb{S}^{d-1}$, the inequality

$$\mathbb{1}_{\nu(\tau(K,\omega))}(u) \le \liminf_{j\to\infty} \mathbb{1}_{\nu(\tau(K_j,\omega))}(u)$$

holds. Fatou's lemma and the continuous dependence of $\rho(K,\cdot)$ on $K \cap C_t$ give

$$V_K(\omega) \le \liminf_{j\to\infty} V_{K_j}(\omega).$$

Since

$$V_K(\Omega_C) = V_d(C \setminus K) = \lim_{j\to\infty} V_d(C \setminus K_j) = \lim_{j\to\infty} V_{K_j}(\Omega_C),$$

this completes the proof of the weak convergence (7.61).

For the proof of (7.62), let $\omega \in \mathcal{B}(\Omega_C)$. We have

$$\psi_K(\omega) = \frac{1}{d} \int_\omega -h(K,u) \, S_{d-1}(K,\mathrm{d}u)$$

$$= \frac{1}{d} \int_\omega |h(K \cap C_t, u)| \, S_{d-1}(K \cap C_t, \mathrm{d}u)$$

and similarly

$$\psi_{K_j}(\omega) = \frac{1}{d} \int_\omega |h(K_j \cap C_t, u)| \, S_{d-1}(K_j \cap C_t, \mathrm{d}u).$$

For $j \in \mathbb{N}_0$, let η_j be the measure defined by

$$\eta_j(\omega) := \frac{1}{d} \int_\omega |h(K \cap C_t, u)| \, S_{d-1}(K_j \cap C_t, \mathrm{d}u), \quad \omega \in \mathcal{B}(\Omega_C), \, j \in \mathbb{N},$$

$$\eta_0(\omega) := \frac{1}{d} \int_\omega |h(K \cap C_t, u)| \, S_{d-1}(K \cap C_t, \mathrm{d}u), \quad \omega \in \mathcal{B}(\Omega_C).$$

Since $h(K \cap C_t, \cdot)$ is continuous and the area measure S_{d-1} is weakly continuous, we have $\eta_j \overset{w}{\to} \eta_0$ as $j \to \infty$. By [163, Lemma 1.8.14], the sequence $(h(K_j \cap C_t, \cdot))_{j\in\mathbb{N}}$ converges uniformly on $\mathbb{S}^{d-1}$ to $h(K \cap C_t, \cdot)$. Hence, for each $\varepsilon > 0$ we have $|h(K \cap C_t, \cdot)| \le |h(K_j \cap C_t, \cdot)| + \varepsilon$ for all $u \in \mathbb{S}^{d-1}$ and hence $\eta_j(\omega) \le \psi_{K_j}(\omega) + c\varepsilon$, if j is sufficiently large; here c is a constant independent of j. Since this holds for all $\varepsilon > 0$ and since $\eta_j \overset{w}{\to} \eta_0$, we deduce that for each open set $\omega \subset \Omega_C$ we get

$$\psi_K(\omega) = \eta_0(\omega) \le \liminf_{j\to\infty} \eta_j(\omega) \le \liminf_{j\to\infty} \psi_{K_j}(\omega).$$

Using Lemma 7.1.1,

$$\psi_K(\Omega_C) = V_d(C \setminus K) = \lim_{j\to\infty} V_d(C \setminus K_j) = \lim_{j\to\infty} \psi_{K_j}(\Omega_C).$$

This completes the proof of the weak convergence (7.62).

From (7.60), (7.61), (7.62) it now follows that $V_K = \psi_K$. Up to now, K was a C-full set.

Now let K be a C-close set for which $C \setminus K$ is unbounded. Consider an open set $\omega \subset \Omega_C$ with $\mathrm{cl}\,\omega \subset \Omega_C$. Then the set $\tau(K, \omega)$ is bounded. The set

$$M := C \cap \bigcap_{u \in \mathrm{cl}\,\omega} H^-(K, u)$$

is C-full and satisfies $\tau(K, \omega) = \tau(M, \omega)$ and hence $V_K(\omega) = V_M(\omega)$, moreover, $\psi_K(\omega) = \psi_M(\omega)$. Since $C \setminus M$ is bounded, we have $V_M(\omega) = \psi_M(\omega)$. Therefore, $V_K(\omega) = \psi_K(\omega)$. Since V_K and ψ_K are both measures on Ω_C, the equality $V_K(\omega) = \psi_K(\omega)$ extends to arbitrary Borel subsets $\omega \in \mathcal{B}(\Omega_C)$. This completes the proof of Lemma 7.8.1. □

Now we can prove an existence theorem.

Theorem 7.8.1. *Every finite Borel measure on Ω_C that is concentrated on a compact subset of Ω_C is the cone-volume measure of a C-full set.*

Every finite Borel measure on Ω_C is the cone-volume measure of a C-close set.

Proof. To prove the first part, we follow the procedure in the proof of Theorem 7.4.1, modified in the way the logarithmic Minkowski problem was treated by Böröczky, Lutwak, Yang and Zhang in [30].

Let $\omega \subset \Omega_C$ be a nonempty, compact set, and let $\mathcal{C}(\omega)$ denote the space of positive, continuous functions on ω. Let φ be a finite Borel measure on Ω_C that is concentrated on ω.

Without loss of generality, we assume that $\varphi(\omega) = 1$ (the case $\varphi \equiv 0$ is trivial). Define a function $\Phi : \mathcal{C}(\omega) \to (0, \infty)$ by

$$\Phi(f) := V(f)^{-1/d} \exp \int_\omega \log f \, d\varphi, \quad f \in \mathcal{C}(\omega).$$

The following assertions are verified as in the proof of Theorem 7.4.1. The functional Φ is continuous and attains a maximum on the set $\mathcal{L}' := \{-h_L : L \in \mathcal{K}(C, \omega), V_d(C \setminus L) = 1)\}$, say at $-h_{K_0}$. Since Φ is homogeneous of degree zero (here it is used that $\varphi(\omega) = 1$) this is also the maximum of Φ on the set $\mathcal{L} := \{-h_L : L \in \mathcal{K}(C, \omega)\}$. Let $f \in \mathcal{C}(\omega)$, and let K_f be the Wulff shape associated with (C, ω, f). Then $-h_{K_f} \geq f$ and $V(f) = V(-h_{K_f})$, hence $\Phi(f) \leq \Phi(-h_{K_f}) \leq \Phi(-h_{K_0})$, since $-h_{K_f} \in \mathcal{L}$. Thus, Φ attains its maximum on $\mathcal{C}(\omega)$ at $-h_{K_0}$.

Now let $f \in \mathcal{C}(\omega)$ and define

$$G(\tau, \cdot) := -h_{K_0} e^{\tau f}.$$

Then $G(\tau, \cdot) \in \mathcal{C}(\omega)$, hence the function

$$\tau \mapsto \Phi(G(\tau, \cdot)) = V(G(\tau, \cdot))^{-1/d} \exp \int_\omega \log G(\tau, \cdot) \, d\varphi \qquad (7.63)$$

attains its maximum at $\tau = 0$. Since

$$\frac{G(\tau, \cdot) - G(0, \cdot)}{\tau} \to -f h_{K_0} \quad \text{uniformly on } \omega$$

as $\tau \to 0$, we can conclude from Theorem 7.3.1 that

$$\frac{\mathrm{d}}{\mathrm{d}\tau} V(G(\tau, \cdot))\Big|_{\tau=0} = \int_\omega -f h_{K_0} \, dS_{d-1}(K_0, \cdot).$$

Therefore, the derivative of the function (7.63) at $\tau = 0$ is given by

$$\left[-\frac{1}{d} \int_\omega -f h_{K_0} \, dS_{d-1}(K_0, \cdot) + \int_\omega f \, d\varphi \right] \exp \int_\omega \log(-h_{K_0}) \, d\varphi$$

(note that $V(G(0, \cdot)) = V(-h_{K_0}) = V_d(C \setminus K_0) = 1$). Since this is equal to zero, we obtain, in view of Lemma 7.8.1, that

$$\int_\omega f \, dV_{K_0} = \int_\omega f \, d\varphi.$$

This holds for all $f \in \mathcal{C}(\omega)$, hence we conclude that $V_{K_0} = \varphi$.

Given the first part of Theorem 7.8.1, the proof of the second part proceeds essentially as that of Theorem 7.7.2. We start with a sequence $(\omega_j)_{j \in \mathbb{N}}$ of nonempty open subsets of Ω_C with $\mathrm{cl}\,\omega_j \subset \omega_{j+1}$ and $\bigcup_{j \in \mathbb{N}} \omega_j = \Omega_C$, define $\varphi_j(A) := \varphi(A \cap \omega_j)$ for $A \in \mathcal{B}(\Omega_C)$, and apply the first part of the theorem. This yields a C-full set L_j satisfying $V_{L_j} = \varphi_j$.

Now we choose $t > 0$ with $V_d(C_t) > \varphi(\Omega_C)$. If $L_j \cap C_t = \emptyset$ for some j, then

$$\varphi(\omega_j) = V_{L_j}(\omega_j) = V_{L_j}(\Omega_C) = V_d(C \setminus L_j) \geq V_d(C_t) > \varphi(\Omega_C) \geq \varphi(\omega_j),$$

a contradiction. This shows that $L_j \cap C_t \neq \emptyset$ for all $j \in \mathbb{N}$. Now we can proceed precisely as in the proof of Theorem 7.7.2. This need not be carried out here. The required weak continuity of the cone-volume measure was proved in (7.61), where it was remarked that the proof holds for general convex K_j. Finally, that the solution set is C-close, follows from the finiteness of φ. $\qquad\square$

8

Appendix: Open questions

We have occasionally mentioned open questions, and in this Appendix we want to repeat them and present them as a brief collection, for the reader's convenience.

1. In Note 1 for Section 2.4, we have defined a polyhedral vertex curvature. It is an open question to determine all translation invariant polyhedral vertex curvatures.

2. Is the outer angle Γ, as defined at the beginning of Section 2.4, the only polyhedral vertex curvature that is invariant under rigid motions?

3. It has often been asked and hence has no traces of originality left, but is still very interesting: Is there a 'spherical Hadwiger theorem'? In other words, is every real valuation on spherically convex bodies in $\mathbb{S}^{d-1}$ that is continuous (with respect to the spherical Hausdorff metric) and invariant under rotations necessarily a linear combination (with constant coefficients) of the spherical intrinsic volumes?

4. The same as before, with 'continuous' replaced by 'increasing', 'constant coefficients' replaced by 'nonnegative constant coefficients' and 'spherical intrinsic volumes' replaced by 'spherical quermassintegrals' (since these are increasing).

5. Is every real valuation on spherically convex polytopes in $\mathbb{S}^{d-1}$ that is continuous, invariant under rotations and simple necessarily a constant multiple of the spherical volume?

6. For inequality (3.25), the 'spherical isoperimetric inequality', a proof has been given in Section 3.4 that uses the method of 'two-point symmetrization'. Can this proof be modified to settle the equality case? Since this is not obvious, and no proof has been published, we consider this as an open question.

© The Author(s), under exclusive license to Springer Nature Switzerland AG 2022
R. Schneider, *Convex Cones*, Lecture Notes in Mathematics 2319,
https://doi.org/10.1007/978-3-031-15127-9_8

7. In Note 1 for Section 4.4, we have asked the following question. The Wills functional of a convex body $K \subset \mathbb{R}^d$ is defined by

$$W(K) := \sum_{k=0}^{d} V_k(K),$$

where $V_0, \ldots, V_d$ are the intrinsic volumes. It was proved by P. McMullen that

$$W(K) \leq e^{V_1(K)}.$$

Is there a similar inequality for conic intrinsic volumes?

8. The following question was posed in Section 4.5

Which of the functionals v_j, U_j, W_j on $\mathcal{C}^d$ have the property that they attain an extremum on the set of cones $C \in \mathcal{C}^d$ with given $v_d(C) > 0$ precisely at spherical cones?

9. The following question was posed by Amelunxen [5, Conjecture 4.4.16]: Is the sequence of conic intrinsic volumes log-concave?

10. From Note 1 for Section 5.5, we repeat the following question, which was posed (in an equivalent form) by Kabluchko and Thäle [110]. For the notions of the typical k-face and the weighted typical k-face, we refer to Definitions 5.5.1, 5.5.3.

Is it true, in the isotropic case, that

$$\mathbb{E} f_\ell \left(W_{n,d}^{(k)} \right) > \mathbb{E} f_\ell \left(S_{n,d}^{(k)} \right),$$

whenever $n \geq d + 1$ and $\ell \in \{1, \ldots, k-1\}$?

11. What is a necessary and sufficient condition on a Borel measure μ on Ω_C to be the surface area measure of a C-full set? Theorem 7.4.1 gives only a sufficient condition, namely that the support of μ be contained in Ω_C.

12. Can the Aleksandrov–Fenchel–Jessen-type theorem 7.4.3 be extended from C-full sets to C-close sets?

13. What is a necessary and sufficient condition on a Borel measure on Ω_C to be the surface area measure of a C-close set? Theorem 7.7.2 shows only that finiteness of the measure is sufficient.

14. What is a necessary and sufficient condition on a Borel measure on Ω_C to be the surface area measure of some closed convex subset of C? Theorem 7.7.3 gives a necessary condition.

15. In Note 1 for Section 7.7 it was asked for a sufficient condition for an infinite Borel measure on Ω_C to be the surface area measure of a C-asymptotic or C-coconvex set.

16. In Theorem 7.8.1 it was proved that every finite Borel measure on Ω_C is the cone-volume measure of a C-close set. Is this set uniquely determined?

References

1. Adiprasito, K.A., Sanyal, R., An Alexander-type duality for valuations. *Proc. Amer. Math. Soc.* **143** (2015), 833–843.
2. Affentranger, F., Schneider, R., Random projections of regular simplices. *Discrete Comput. Geom.* **7** (1992), 219–226.
3. Aliprantis, C.D., Tourky, R., *Cones and Duality*. Amer. Math. Soc., Providence, RI, 2007.
4. Alonso–Gutiérrez, D., Hernández Cifre, M.A., Yepes Nicolás, J., Further inequalities for the (generalized) Wills functional. *Commun. Contemp. Math.* **23** (2021), no. 3, 2050011, 35 pp.
5. Amelunxen, D., Geometric analysis of the condition of the convex feasibility problem. PhD thesis, Univ. Paderborn, 2011.
6. Amelunxen, D., Measures on polyhedral cones: characterizations and kinematic formulas. arXiv:1412.1569v2 (2015).
7. Amelunxen, D., Bürgisser, P., Intrinsic volumes of symmetric cones. (Expanded version of [9]) arXiv:1205.1863.
8. Amelunxen, D., Bürgisser, P., Probabilistic analysis of the Grassmann condition number. *Found. Comput. Math.* **15** (2015), 3–51.
9. Amelunxen, D., Bürgisser, P., Intrinsic volumes of symmetric cones and applications in convex programming. *Math. Programm., Ser. A* **149** (2015), 105–130. (Abridged version of [7])
10. Amelunxen, D., Lotz, M., Intrinsic volumes of polyhedral cones: a combinatorial perspective. *Discrete Comput. Geom.* **58** (2017), 371–409.
11. Amelunxen, D., Lotz, M., McCoy, M.B., Tropp, J.A., Living on the edge: a geometric theory of phase transitions in convex optimization. *Information and Inference* **3** (2014), 224–294.
12. Artstein–Avidan, S., Giannopoulos, A., Milman, V., *Asymptotic Geometric Analysis, Part I*. Amer. Math. Soc., Providence, RI, 2015.
13. Artstein–Avidan, S., Slomka, B., Order isomorphisms in cones and a characterization of duality for ellipsoids. *Selecta Math.* **18** (2012), 391–415.
14. Ash, R.B., *Measure, Integration, and Functional Analysis*. Academic Press, New York, 1972.
15. Backman, S., Manecke, S., Sanyal, R., Cone valuations, Gram's relation, and flag-angles. arXiv:1809.00956v1

© The Author(s), under exclusive license to Springer Nature Switzerland AG 2022 329
R. Schneider, *Convex Cones*, Lecture Notes in Mathematics 2319,
https://doi.org/10.1007/978-3-031-15127-9

16. Banchoff, T., Critical points and curvature for embedded polyhedra. *J. Differential Geom.* **1** (1967), 245–256.

17. Bárány, I., Hug, D., Reitzner, M., Schneider, R., Random points in halfspheres. *Random Structures Algorithms* **50** (2017), 3–22.

18. Barvinok, A., *A Course in Convexity*. Amer. Math. Soc., Providence, RI, 2002.

19. Barvinok, A., *Integer Points in Polyhedra*. Eur. Math. Soc., Zürich, 2008.

20. Barvinok, A., Pommersheim, J.E., An algorithmic theory of lattice points in polyhedra. *New Perspectioves in Algebraic Combinatorics* (Berkeley, CA, 1996–97), Math. Sci. Res. Inst. Publ., **38**, Cambridge Univ. Press, Cambridge, 1999, pp. 91–147.

21. Bauer, H., *Maß- und Integrationstheorie*. Walter de Gruyter, Berlin, 1990.

22. Beck, M., Haase, C., Sottile, F., Formulas of Brion, Lawrence, and Varchenko on rational generating functions for cones. *Math. Intelligencer* **31** (2009), 9–17.

23. Beck, M., Robins, S., *Computing the Continuous Discretely: Integer-Point Enumeration in Polyhedra*. 2nd ed., Springer, New York, 2015.

24. Beck, M., Sanyal, R., *Combinatorial Reciprocity Theorems: An Invitation To Enumerative Geometric Combinatorics*. Graduate Studies in Mathematics, vol. 195. Amer. Math. Soc., Providence, RI, 2018.

25. Becker, R., *Convex Cones in Analysis*. Hermann, Paris, 2006 (French original: 1999).

26. Benyamini, Y., Two point symmetrization, the isoperimetric inequality on the sphere and some applications. *Longhorn Notes, Univ. of Texas, Texas Funct. Anal. Seminar* (1983–1984), 53–76.

27. Birkhoff, G., *Lattice Theory*. Third edition. Amer. Math. Soc. Colloquium Publications, Vol. XXV, Amer. Math. Soc., Providence, RI, 1967.

28. Bloch, E.D., The angle defect for arbitrary polyhedra. *Beitr. Algebra Geom.* **39** (1998), 379–393.

29. Bogachev, V.I., *Gaussian Measures*. Mathematical Surveys and Monographs, vol. 62, Amer. Math. Soc., Providence, RI, 1998.

30. Böröczky, K.J., Lutwak, E., Yang, D., Zhang, G., The logarithmic Minkowski problem. *J. Amer. Math. Soc.* **26** (2013), 831–852.

31. Böröczky, K.J., Sagmeister, Á., The isodiametric problem on the sphere and in the hyperbolic space. *Acta Math. Hungar.* **160** (2020), 13–32.

32. Böröczky, K.J., Schneider, R., A characterization of the duality mapping for convex bodies. *Geom. Funct. Anal.* **18** (2008), 657–667.

33. Borwein, J.M., Moors, W.B., Stability of closedness of convex cones under linear mappings. *J. Convex Anal.* **16** (2009), 699–705.

34. Borwein, J.M., Moors, W.B., Stability of closedness of convex cones under linear mappings II. *J. Nonlinear Anal. Optim.* **1** (2010), 1–7.

35. Boucheron, S., Lugosi, G., Massart, P., *Concentration Inequalities. A Nonasymptotic Theory of Independence*. Oxford University Press, Oxford, 2013.

36. Brianchon, C.J., Théorème nouveau sur les polyèdres. *J. Ècole (Royale) Polytechnique* **15** (1837), 317–319.

37. Brin, I.A., Gauss–Bonnet theorems for polyhedra. (Russian) *Uspekhi Mat. Nauk* **3** (1948), 226–227.

38. Brion, M., Points entiers dans les polyèdres convexes. *Ann. Sci. École Norm. Sup.* (4) **21** (1988), 653–663.

39. Brøndsted, A., *An Introduction to Convex Polytopes*. Springer, New York, 1983.

40. Budach, L., Lipschitz–Killing curvatures of angular partially ordered sets. *Adv. Math.* **78** (1989), 140–167.

41. Burago, Yu.D., Zalgaller, V.A., *Geometric Inequalities.* Springer, Berlin, 1988.
42. Calabi, E., Complete affine hypersurfaces. I. *Symposia Mathematica, Vol. X (Convegno di Geometria Differenziale, INDAM, Rome, 1971)*, pp. 19–38. Academic Press, London, 1972.
43. Callahan, K., Hann, K., An Euler-type volume identity. *Bull. Austral. Math. Soc.* **59** (1999), 495–508.
44. Cheeger, J., Spectral geometry of singular Riemannian spaces. *J. Differential Geom.* **18** (1983), 575–657.
45. Cheeger, J., Müller, W., Schrader, R., On the curvature of piecewise flat spaces. *Commun. Math. Phys.* **92** (1984), 405–454.
46. Cheeger, J., Müller, W., Schrader, R., Kinematic and tube fomulas for piecewise linear spaces. *Indiana University Math. J.* **35** (1986), 737–754.
47. Chen, B., The Gram–Sommerville and Gauss–Bonnet theorems and combinatorial geometric measures for noncompact polyhedra. *Adv. Math.* **91** (1992), 269–291.
48. Chen, B., On the Euler characteristic of finite unions of convex sets. *Discrete Comput. Geom.* **10** (1993), 79–93.
49. Chen, B., The incidence algebra of polyhedra over the Minkowski algebra. *Adv. Math.* **118** (1996), 337–365.
50. Cover, T.M., Efron, B., Geometrical probability and random points on a hypersphere. *Ann. Math. Stat.* **38** (1967), 213–220.
51. Diestel, J., Spalsbury, A., *The Joys of Haar Measure.* Amer. Math. Soc., Providence, RI, 2014.
52. Donoho, D.L., Tanner, J., Counting the faces of randomly-projected hypercubes and orthants, with applications. *Discrete Comput. Geom.* **43** (2010), 522–541.
53. Drton, M., Klivans, C.J., A geometric interpretation of the characteristic polynomial of reflection arrangements. *Proc. Amer. Math. Soc.* **138** (2010), 2873–2887.
54. Dudley, R.M., *Real Analysis and Probability.* Cambridge University Press, New York, 2002.
55. Elstrodt, J., *Maß- und Integrationstheorie.* Springer, Berlin, 1996.
56. Faraut, J., Koranyi, A., *Analysis on Symmetric Cones.* Oxford University Press, London and New York, 1994.
57. Faure, C.-A., An elementary proof of the fundamental theorem of projective geometry. *Geom. Dedicata* **90** (2002), 145–151.
58. Federer, H., Curvature measures. *Trans. Amer. Math. Soc.* **93** (1959), 418–491.
59. Figiel, T., Lindenstrauss, J., Milman, V.D., The dimension of almost spherical sections of convex bodies. *Acta Math.* **139** (1977), 53–94.
60. Fillastre, F., A short elementary proof of reversed Brunn–Minkowski inequality for coconvex bodies. *Séminaire de théorie spectrale et géométrie, Grenoble* **34** (2016–2017), 93–96.
61. Gao, F., Hug, D., Schneider, R., Intrinsic volumes and polar sets in spherical space. *Math. Notae* **41** (2001/2002), 159–176.
62. Gigena, S., Integral invariants of convex cones. *J. Differential Geom.* **13** (1978), 191–222.
63. Glasauer, S., Integralgeometrie konvexer Körper im sphärischen Raum. PhD Thesis, University of Freiburg, 1995.
64. Glasauer, S., Integral geometry of spherically convex bodies. *Diss. Summ. Math.* **1** (1996), 219–226.

65. Glasauer, S., An Euler-type version of the local Steiner formula for convex bodies. *Bull. London Math. Soc.* **30** (1998), 618–622.

66. Godland, T., Kabluchko, Z., Conical tessellations associated with Weyl chambers. *Trans. Amer. Math. Soc.* **374** (2021), 7161–7186.

67. Godland, T., Kabluchko, Z., Conical intrinsic volumes of Weyl chambers. arXiv:2005.06205.

68. Godland, T., Kabluchko, Z., Positive hulls of random walks and bridges. *Stochastic Process. Appl.* **147** (2022), 327–362.

69. Godland, T., Kabluchko, Z., Thäle, C., Random cones in high-dimensions I: Donoho–Tanner and Cover–Efron cones. arXiv:2012.06189v1.

70. Godland, T., Kabluchko, Z., Thäle, C., Random cones in high-dimensions II: Weyl cones. *Mathematika* **68** (2022), 710–737.

71. Goldstein, L., Nourdin, I., Peccati, G., Gaussian phase transitions and conic intrinsic volumes: Steining the Steiner formula. *Ann. Appl. Prob.* **27** (2017), 1–47.

72. Götze, F., Kabluchko, Z., Zaporozhets, D., Grassmann angles and absorption probabilities of Gaussian convex hulls. *Zap. Nauchn. Sem. POMI* **501** (2021), 126–148.

73. Gourion, D., Seeger, A., Critical angles in random polyhedral cones. *J. Math. Anal. Appl.* **374** (2011), 8–21.

74. Gourion, D., Seeger, A., Solidity indices for convex cones. *Positivity* **16** (2012), 685–705.

75. Gram, J.P., Om Rumvinklerne i et Polyeder. *Tidsskr. Math. (Copenhagen)* (3) **4** (1874), 161–163.

76. Gray, A., *Tubes.* Addison–Wesley Publishing Company, Redwood City, CA, 1990.

77. Gröbner, W., Hofreiter, N., *Integraltafel. Zweiter Teil, Bestimmte Integrale.* Springer, Wien, 1950.

78. Groemer, H., On the extension of additive functionals on classes of convex sets. *Pacific J. Math.* **75** (1978), 397–410.

79. Groemer, H., *Geometric Applications of Fourier Series and Spherical Harmonics.* Encyclopedia of Mathematics and Its Applications **61**, Cambridge University Press, Cambridge, 1996.

80. Gruber, P.M., Schneider, R., Problems in geometric convexity. In *Contributions to Geometry, Siegen 1978* (J. Tölke, J. M. Wills, eds.), pp. 255–278, Birkhäuser, Basel, 1979.

81. Grünbaum, B., Grassmann angles of convex polytopes. *Acta Math.* **121** (1968), 293–302.

82. Grünbaum, B., *Convex Polytopes.* Second Edition. Springer, New York, 2003.

83. Haase, C., Polar decomposition and Brion's theorem. In *Integer points in polyhedra—geometry, number theory, algebra, optimization* (A. Barvinok et al., eds.), pp. 91–99, *Contemp. Math.* **374**, Amer. Math. Soc., Providence, RI, 2005.

84. Hadwiger, H., Eulers Charakteristik und kombinatorische Geometrie. *J. reine angew. Math.* **194** (1955), 101–110.

85. Hadwiger, H., *Vorlesungen über Inhalt, Oberfläche und Isoperimetrie.* Springer, Berlin, 1957.

86. Hadwiger, H., Eckenkrümmung beliebiger kompakter euklidischer Polyeder und Charakteristik von Euler–Poincaré. *L'Enseignement math.* **15** (1969), 147–151.

87. Hadwiger, H., Das Wills'sche Funktional. *Monatsh. Math.* **79** (1975), 213–221.
88. Hann, K., The average number of normals through a point in a convex body and a related Euler-type identity. *Geom. Dedicata* **48** (1993), 27–55.
89. Henrion, R., Seeger, A., On properties of different notions of centers for convex cones. *Set-Valued Var. Anal.* **18** (2010), 205–231.
90. Henrion, R., Seeger, A., Inradius and circumradius of various convex cones arising in applications. *Set-Valued Var. Anal.* **18** (2010), 483–511.
91. Henrion, R., Seeger, A., Condition number and eccentricity of a closed convex cone. *Math. Scand.* **109** (2011), 285–308.
92. Hewitt, E., Ross, K.A., *Abstract Harmonic Analysis, Vol. 1.* Springer, Berlin, 1963.
93. Hug, D., On the mean number of normals through a point in the interior of a convex body. *Geom. Dedicata* **55** (1995), 319–340.
94. Hug, D., Kabluchko, Z., An inclusion-exclusion identity for normal cones of polyhedral sets. *Mathematika* **64** (2018), 124–136.
95. Hug, D., Reichenbacher, A., Geometric inequalities, stability results and Kendall's problem in spherical space. arXiv:1709.06522v1.
96. Hug, D., Schneider, R., Stability results involving surface area measures of convex bodies. *Rend. Circ. Mat. Palermo* (2) *Suppl.* **70** (2002), vol. II, 21 – 51.
97. Hug, D., Schneider, R., Hölder continuity for support measures of convex bodies. *Arch. Math.* **104** (2015), 83–92.
98. Hug, D., Schneider, R., Random conical tessellations. *Discrete Comput. Geom.* **56** (2016), 395–426.
99. Hug, D., Schneider, R., Threshold phenomena for random cones. *Discrete Comput. Geom.* **67** (2022), 564–594.
100. Hug, D., Schneider, R., Another look at threshold phenomena for random cones. *Studia. Sci. Math. Hungar.* **58** (2021), 489–504.
101. Isac, G., Németh, A.B., Monotonicity of metric projections onto positive cones of ordered Euclidean spaces. *Arch. Math.* **46** (1986), 568–576.
102. Isac, G., Németh, A.B., Isotone projection cones in Euclidean spaces. *Ann. Sci. Math. Québec* **16** (1992), 35–52.
103. Iusem, A.N., Seeger, A., Axiomatization of the index of pointedness for closed convex cones. *Comput. Appl. Math.* **24** (2005), 245–283.
104. Iusem, A.N., Seeger, A., Measuring the degree of pointedness of a closed convex cone: a metric approach. *Math. Nachr.* **279** (2006), 599–618.
105. Iusem, A.N., Seeger, A., Distances between closed convex cones: old and new results. *J. Convex Anal.* **17** (2020), 1033–1055.
106. Kabluchko, Z., Expected *f*-vector of the Poisson zero polytope and random convex hulls in the half-sphere. *Mathematika* **66** (2020), 1028–1053.
107. Kabluchko, Z., Face numbers of high-dimensional Poisson zero cells. (Preprint) arXiv:2110.08201v1.
108. Kabluchko, Z., Marynych, A., Temesvari, D., Thäle, Ch., Cones generated by random points on half-spheres and convex hulls of Poisson point processes. *Probab. Theory Related Fields* **175** (2019), 1021–1061.
109. Kabluchko, Z., Seidel, H., Convex cones spanned by regular polytopes. *Adv. Geom.* **22** (2022), 245–267.
110. Kabluchko, Z., Thäle, Ch., Faces in random great hypersphere tessellations. *Electron. J. Probab.* **26** (2021), Paper No. 3, 35pp.

111. Kabluchko, Z., Temesvari, D., Thäle, Ch., A new approach to weak convergence of random cones and polytopes. *Canad. J. Math.* **73** (2021), 1627–1647.

112. Kabluchko, Z., Thäle, Ch., Zaporozhets, D., Beta polytopes and Poisson polyhedra: f-vectors and angles. *Adv. Math.* **374** (2020), 107333, 63 pp.

113. Kabluchko, Z., Vysotsky, V., Zaporozhets, D., Convex hulls of random walks, hyperplane arrangements, and Weyl chambers. *Geom. Funct. Anal.* **27** (2017), 880–918.

114. Kabluchko, Z., Vysotsky, V., Zaporozhets, D., Convex hulls of random walks: Expected number of faces and face probabilities. *Adv. Math.* **320** (2017), 595–629.

115. Khovanskiĭ, A., Timorin, V., On the theory of coconvex bodies. *Discrete Comput. Geom.* **52** (2014), 806–823.

116. Klain, D.A., Rota, G.–C., *Introduction to Geometric Probability.* Cambridge University Press, Cambridge, 1997.

117. Klee, V., Some characterizations of convex polyhedra. *Acta Math.* **102** (1959), 79–107.

118. Klivans, C.E., Swartz, E., Projection volumes of hyperplane arrangements. *Discrete Comput. Geom.* **46** (2011), 417–426.

119. Lawrence, J., Valuations and polarity. *Discrete Comput. Geom.* **3** (1988), 307–324.

120. Lawrence, J., Polytope volume computation. *Math. Comp.* **57**, no. 195 (1991), 259–271.

121. Lawrence, J., A short proof of Euler's relation for convex polytopes. *Canad. Math. Bull.* **40** (1997), 471–474.

122. Lenz, H., Mengenalgebra und Eulersche Charakteristik. *Abh. Math. Sem. Univ. Hamburg* **34** (1970), 135–147.

123. Lévy, P., *Problèmes concrets d'analyse fonctionnelle. Avec un complément sur les fonctionnelles analytiques par F. Pellegrino.*, 2d ed., Gauthier–Villars, Paris, 1951.

124. Li, A.M., Simon, U., Zhao, G.S., *Global Affine Differential Geometry of Hypersurfaces.* Walter de Gruyter, Berlin, 1993.

125. Lieb, E.L., Loss, M., *Analysis.* 2nd ed., Graduate Studies in Mathematics, vol. 14, Amer. Math. Soc., Providence, RI, 2001.

126. Matheron, G., *Random Sets and Integral Geometry.* Wiley, New York, 1975.

127. McCoy, M.B., A geometric analysis of convex demixing. PhD thesis. California Institute of Technology, 2013. http://thesis.library.caltech.edu/7726/

128. McCoy, M.B., Tropp, J.A., Sharp recovery bounds for convex demixing, with applications. *Found. Comput. Math.* **14** (2014), 503–567.

129. McCoy, M.B., Tropp, J.A., From Steiner formulas for cones to concentration of intrinsic volumes. *Discrete Comput. Geom.* **51** (2014), 926–963.

130. McMullen, P., Non-linear angle-sum relations for polyhedral cones and polytopes. *Math. Proc. Camb. Phil. Soc.* **78** (1975), 247–261.

131. McMullen, P., Angle-sum relations for polyhedral sets. *Mathematika* **33** (1986), 173–188.

132. McMullen, P., Inequalities between mixed volumes. *Monatsh. Math.* **111** (1991), 47–53.

133. McMullen, P., Valuations and dissections. In *Handbook of Convex Geometry* (P.M. Gruber, J.M. Wills, eds.), vol. B, pp. 933–988, North-Holland, Amsterdam, 1993.

134. McMullen, P., *Convex Polytopes and Polyhedra*. (Provisional title – in preparation).

135. McMullen, P., Schneider, R., Valuations on convex bodies. In *Convexity and Its Applications* (P.M. Gruber, J.M. Wills, eds.), pp. 170–247, Birkhäuser, Basel, 1983.

136. Miles, R.E., Random polytopes: the generalisation to n dimensions of the intervals of a Poisson process. Ph.D. Thesis, Cambridge Univ., 1961.

137. Mirkil, H., New characterizations of polyhedral cones. *Canad. J. Math.* **9** (1957), 1–4.

138. Moreau, J.J., Décomposition orthogonale d'un espace hilbertien selon deux cones mutuellement polaires. *C. R. Acad. Sci.* **255** (1962), 238–240.

139. Morvan, J.-M., *Generalized Curvatures*. Springer, Berlin, 2008.

140. Nachbin, L., *The Haar Integral*. Van Nostrand, Princeton, 1965.

141. Nef, W., Zur Einführung der Eulerschen Charakteristik. *Monatsh. Math.* **92** (1981), 41–46.

142. Németh, A.B., Németh, S.Z., A duality between the metric projection onto a convex cone and the metric projection onto its dual. *J. Math. Anal. Appl.* **392** (2012), 171–178.

143. Okamoto, M., Some inequalities relating to the partial sum of binomial probabilities. *Ann. Inst. Statist. Math.* **10** (1958), 29–35.

144. Pataki, G., On the closedness of the linear image of a closed convex cone. *Math. Oper. Res.* **32** (2007), no. 2, 395–412.

145. Perles, M.A., Shephard, G.C., Angle sums of convex polytopes. *Math. Scand.* **21** (1967), 199–218.

146. Przesławski, K., Linear algebra of convex sets and the Euler characteristic. *Linear and Multilinear Algebra* **31** (1992), 153–191.

147. Reiter, H., *Classical Harmonic Analysis and Locally Compact Groups*. Clarendon Press, Oxford, 1968.

148. Rockafellar, R.T., *Convex Analysis*. Princeton Univ. Press, Princeton, NJ, 1970.

149. Rota, G.-C., On the foundations of combinatorial theory I. Theory of Möbius functions. *Z. Wahrscheinlichkeitstheorie verw. Geb.* **2** (1964), 340–368.

150. Sallee, G.T., Polytopes, valuations, and the Euler relation. *Canad. J. Math.* **20** (1968), 1412–1424.

151. Sandgren, L., On convex cones. *Math. Scand.* **2** (1954), 19–28.

152. Santaló, L.A., Sobre la formula de Gauss–Bonnet para poliedros en espacios de curvatura constante. *Rev. Un. Mat. Argentina* **20** (1962), 79–91.

153. Santaló, L.A., Sobre la formula fundamental cinematica de la geometria integral en espacios de curvatura constante. *Math. Notae* **18** (1962), 79–94.

154. Santaló, L.A., *Integral Geometry and Geometric Probability*. Encyclopedia of Mathematics and Its Applications, vol. **1**, Addison–Wesley, Reading, MA (1976).

155. Schechtman, G., Concentration results and applications. In *Handbook of the Geometry of Banach Spaces*, (W.B. Johnson, J. Lindenstrauss, eds.), vol. 2, 1603–1634, North-Holland, Amsterdam, 2003.

156. Schläfli, L., *Gesammelte Mathematische Abhandlungen*, I. Birkhäuser, Basel, 1950.

157. Schmidt, E., Beweis der isoperimetrischen Eigenschaft der Kugel im hyperbolischen und sphärischen Raum jeder Dimensionszahl. *Math. Z.* **49** (1943/44), 1–109.

158. Schmidt, E., Die Brunn–Minkowskische Ungleichung und ihr Spiegelbild sowie die isoperimetrische Eigenschaft der Kugel in der euklidischen und nichteuklidischen Geometrie. I. *Math. Nachr.* **1** (1948), 81–157.

159. Schneider, R., Gleitkörper in konvexen Polytopen. *J. reine angew. Math.* **248** (1971), 193–220.

160. Schneider, R., Curvature measures of convex bodies. *Ann. Mat. Pura Appl.* **116** (1978), 101–134.

161. Schneider, R., Parallelmengen mit Vielfachheit und Steiner-Formeln. *Geom. Dedicata* **9** (1980), 111-127.

162. Schneider, R., The endomorphisms of the lattice of closed convex cones. *Beitr. Algebra Geom.* **49** (2008), 541–547.

163. Schneider, R., *Convex Bodies: The Brunn-Minkowski Theory.* 2nd edn, Encyclopedia of Mathematics and Its Applications **151**, Cambridge University Press, Cambridge, 2014.

164. Schneider, R., Combinatorial identities for polyhedral cones. *Algebra i Analiz* **29** (2017), 279–295. *St. Petersburg Math. J.* **92** (2018), 209–221.

165. Schneider, R., Intersection probabilities and kinematic formulas for polyhedral cones. *Acta Math. Hungar.* **155** (2018), 3–24.

166. Schneider, R., Polyhedral Gauss–Bonnet theorems and valuations. *Beitr. Algebra Geom.* **59** (2018), 199–210.

167. Schneider, R., A Brunn–Minkowski theory for coconvex sets of finite volume. *Adv. Math.* **332** (2018), 199–234.

168. Schneider, R., Conic support measures. *J. Math. Anal. Appl.* **471** (2019), 812–835.

169. Schneider, R., Minkowski type theorems for convex sets in cones. *Acta Math. Hungar.* **164** (2021), 282–295.

170. Schneider, R., Weil, W., *Stochastic and Integral Geometry.* Springer, Berlin, 2008.

171. Schneider, R., Wieacker, J.A., Integral geometry. In *Handbook of Convex Geometry* (P.M. Gruber, J.M. Wills, eds.), vol. B, pp. 1349–1390, North-Holland, Amsterdam, 1993.

172. Seeger, A., On measures of size for convex cones. *J. Convex Anal.* **23** (2016), 947–980.

173. Seeger, A., Torki, M., On highly eccentric cones. *Beitr. Algebra Geom.* **55** (2014), 521–544.

174. Seeger, A., Torki, M., Centers and partial volumes of convex cones I. Basic theory. *Beitr. Algebra Geom.* **56** (2015), 227–248.

175. Seeger, A., Torki, M., Centers and partial volumes of convex cones II. Advanced topics. *Beitr. Algebra Geom.* **56** (2015), 491–514.

176. Seeger, A., Torki, M., Conic versions of Loewner–John ellipsoid theorem. *Math. Programm.* **155** (2016), 403–433.

177. Seeger, A., Torki, M., Inscribed and circumscribed ellipsoidal cones: volume ratio analysis. *Beitr. Algebra Geom.* **59** (2018), 717–737.

178. Seeger, A., Torki, M., Measuring axial symmetry in convex cones. *J. Convex Anal.* **25** (2018), 983–1011.

179. Seeger, A., Torki, M., Axial symmetry indices for convex cones: axiomatic formalism and applications. *J. Convex Anal.* **26** (2019), 217–244.

180. Shephard, G.C., An elementary proof of Gram's theorem for polytopes. *Canad. J. Math.* **19** (1967), 1214–1217.

181. Shiryaev, A.N., *Probability*, vol. 1., 3rd edn., *Graduate Texts in Mathematics* **95**, Springer, New York, (2006).

182. Sommerville, D.M.Y., The relations connecting the angle-sum and volume of a polytope in space of n dimensions. *Proc. Roy. Soc. London*, A, **115** (1927), 103–119.

183. Stanley, R.P., An introduction to hyperplane arrangements. *Geometric combinatorics*, 389–496, *IAS/Park City Ser.*, 13, Amer. Math. Soc., Providence, RI, 2007.

184. Stanley, R.P., *Enumerative Combinatorics, Volume I*. 2nd edn., Cambridge University Press, Cambridge, 2012.

185. Steiner, J., Einige Gesetze über die Theilung der Ebene und des Raumes. *J. reine angew. Math.* **1** (1826), 349–364.

186. Varchenko, A.N., Combinatorics and topology of the arrangement of affine hyperplanes in the real space. *Funktional. Anal. i Prizholen.* **21** (1987), 11–22. English translation: *Functional Anal. Appl.* **21** (1987), 9–19.

187. Vershik, A.M., Sporyshev, P.V., Asymptotic behavior of the number of faces of random polyhedra and the neighborliness problem. *Selecta Math. Soviet.* **11**, vol. 2 (1992), 181–201.

188. Waksman, Z., Epelman, M., On point classification in convex sets. *Math. Scand.* **38** (1976), 83–96.

189. Welzl, E., Gram's equation—a probabilistic proof. *Results and trends in theoretical computer science (Graz, 1994)*, pp. 422–424, *Lecture Notes in Comput. Sci.* **812**, Springer, Berlin, 1994.

190. Wendel, J.G., A problem in geometric probability. *Math. Scand.* **11** (1962), 109–111.

191. Zähle, M., Approximation and characterization of generalized Lipschitz–Killing curvatures. *Ann. Global Anal. Geom.* **8** (1990), 249–260.

192. Zaslavsky, T., Facing up to arrangements: face-count formulas for partitions of space by hyperplanes. *Mem. Amer. Math. Soc.* **1** (1975), issue 1, no. 154.

193. Zheng, H., Zydor, M., A new valuation on polyhedral cones. *Algebra i Analiz* **32** (2020), 1–11.

194. Ziegler, G.M., *Lectures on Polytopes*. Springer, New York, 1995.

Notation Index

© The Author(s), under exclusive license to Springer Nature Switzerland AG 2022
R. Schneider, *Convex Cones*, Lecture Notes in Mathematics 2319,
https://doi.org/10.1007/978-3-031-15127-9

Author Index

© The Author(s), under exclusive license to Springer Nature Switzerland AG 2022
R. Schneider, *Convex Cones*, Lecture Notes in Mathematics 2319,
https://doi.org/10.1007/978-3-031-15127-9

Subject Index

Printed in the United States
by Baker & Taylor Publisher Services